T0190506

Space Physics

Advanced Texts in Physics

This program of advanced texts covers a broad spectrum of topics which are of current and emerging interest in physics. Each book provides a comprehensive and yet accessible introduction to a field at the forefront of modern research. As such, these texts are intended for senior undergraduate and graduate students at the MS and PhD level; however, research scientists seeking an introduction to particular areas of physics will also benefit from the titles in this collection.

Springer-Verlag Berlin Heidelberg GmbH

Physics and Astronomy | ONLINE LIBRARY

springeronline.com

May-Britt Kallenrode

Space Physics

An Introduction
to Plasmas and Particles
in the Heliosphere
and Magnetospheres

Third, Enlarged Edition

With 211 Figures, 12 Tables,
Numerous Exercises and Problems

 Springer

Professor Dr. May-Britt Kallenrode
Universität Osnabrück
Fachbereich Physik
Barbarastraße 7
49069 Osnabrück
Germany
E-mail: mkallenr@uos.de

ISSN 1439-2674
ISBN 978-3-642-05829-5

Library of Congress Cataloging-in-Publication Data
Kallenrode, May-Britt, 1962-
Space physics : an introduction to plasmas and particles in the heliosphere and magnetospheres /
May-Britt Kallenrode.– 3rd, enl. ed.
p. cm. – (Advanced texts in physics, ISSN 1439-2674)
ISBN 978-3-642-05829-5   ISBN 978-3-662-09959-9 (eBook)
DOI 10.1007/978-3-662-09959-9
1. Plasma (Ionized gases) 2. Space plasmas. 3. Heliosphere (Astrophysics) 4. Magnetosphere.
5. Cosmic physics. I. Title. II. Series.
QC718.K28 2004   523.01–dc22   2003064770

This work is subject to copyright. All rights are reserved, whether the whole or part of the material
is concerned, specifically the rights of translation, reprinting, reuse of illustrations, recitation, broad-
casting, reproduction on microfilm or in any other way, and storage in data banks. Duplication of
this publication or parts thereof is permitted only under the provisions of the German Copyright Law
of September 9, 1965, in its current version, and permission for use must always be obtained from
Springer-Verlag Berlin Heidelberg GmbH.
Violations are liable for prosecution under the German Copyright Law.
springeronline.com

© Springer-Verlag Berlin Heidelberg 1998, 2001, 2004
Originally published by Springer-Verlag Berlin Heidelberg New York in 2004
Softcover reprint of the hardcover 3rd edition 2004

The use of general descriptive names, registered names, trademarks, etc. in this publication does not
imply, even in the absence of a specific statement, that such names are exempt from the relevant pro-
tective laws and regulations and therefore free for general use.

Typesetting: Data prepared by the author using a Springer TEX macro package
Final processing: Frank Herweg, Leutershausen
Cover design: design & production GmbH, Heidelberg

Printed on acid-free paper   SPIN 10973561   57/3141/di   5 4 3 2 1 0

Preface to the Third Edition

This book is a revised and expanded version of the second edition of *Space Physics*. The first part introduces basic concepts and formalisms which are used in almost all branches of space physics. The second part is concerned with the application of these concepts to plasmas in space and in the heliosphere. More specialized concepts, such as collisionless shocks and particle acceleration, are also introduced. The third part deals with methodological considerations. It consists of an expanded chapter on space measurement methods and a new chapter on general methodological problems. This last chapter is relevant in that it points out the differences between laboratory physics and physics in a complex natural environment, in particular the problems of limited knowledge – or as Pollack [415] puts it, "Uncertain Science ... Uncertain World". In Part II, in most chapters a section "What I Did Not Tell You" has been added – it should help the reader to understand some crucial assumptions underlying the basic ideas introduced in the text and might help you to appreciate the limitations of our knowledge and our models. These sections also give illustrative examples that help to understand the last chapter.

This edition has also been expanded by numerous examples, in particular in Part I. They illustrate basic concepts and aid the reader in the application of these concepts to real problems. In addition, new results from recent space missions, such as ACE, TRACE, and Wind, have been added. In the appendix, a list of Internet resources has been added. This list can also be found (in a "clickable" version) at www.physik.uni-osnabrueck. de/sotere/spacebook/intro.html. On that page, supplementary material to this course can be found, too.

The idea of the book is an introduction to many aspects of space plasmas. Obviously, this approach has the disadvantage that a specialist in any of the subfields will be disappointed that his or her field is dealt with in only a brief and very elementary way. That is, without doubt, true. My idea, however, is to introduce the basic concepts to the novice not already specialized in any field and to help the specialist to easily grasp some ideas in other fields. Therefore the focus is on concepts rather than on detailed mathematical analysis. References should help both the novice who is looking for a deeper formal treatment and the specialist who wants to find reviews giving more details.

There are also a few good books on plasma physics and/or space physics which can be recommended to the reader. A very accessible, unmatched in its style and consciencious approach, book on plasma physics is *Plasma Physics and Controlled Fusion* by F.F. Chen [97]; a well-written and up-to-date account of the phenomena in space plasmas is given in *Introduction to Space Physics*, edited by M.G. Kivelson and C.T. Russell [290]. These two books cannot be matched by the present one and can serve as valuable supplements. More formal introductions to plasma physics are *Plasma Dynamics* by R.O. Dendy [128] and *Plasma Physics* by R.J. Goldston and P.H. Rutherford [192]. Very good introductions to plasma physics of the kind required by a space scientist are given in *Physics of Space Plasmas* by G.K. Parks [397], *Physics of Solar System Plasmas* by T.E. Cravens [113], and *Basic Space Plasma Physics* by W. Baumjohann and R.A. Treumann [36] and its sequel *Advanced Space Plasma Physics* [520]. A useful collection of plasma formulas can be found at wwwppd.nrl.navy.mil/nrlformulary/nrlformulary.html.

As in the earlier editions, symbols in the margin help to guide you through the text. The symbols are

- This section contains an example from space plasmas to illustrate a physical concept. Such a section might be skipped by a reader who is interested primarily in the concepts and less in space science.

- This section is more formal, but is not vital for an understanding of basic observations and ideas. It might be skipped by a reader who is mainly interested in an introduction to space physics.

- An apparently confused reader, "Whatnow", marks supplementary sections: although the ideas presented here are important in space physics, the theoretical background is complicated and only briefly sketched. In particular, the beginner in space physics should feel free to skip these sections on first reading and return to them later after becoming more acquainted with the topic.

- This text points to hotly debated topics and fundamental open problems.

I am grateful to the following persons, who all contributed to the development of this book: Andre Balogh, R.A. Cairns, Stanley H. Cowley, Ulrich Fischer, Roman Hatzky, Bernd Heber, Eberhard Moebius, Reinhold Müller-Mellin, Constantinos Paizes, Wilfried Schröder, Günter Virkus, C.L. Waters, Gerd Wibberenz, and even an anonymous reader who sent hints about errors without being traceable. I am grateful to the helpful team at Springer, in particular Claus Ascheron, Adelheid Duhm, Gertrud Dimler, Ian Mulvany, and Frank Holzwarth. And – last but not least – a big thank-you to Klaus Betzler.

Osnabrück, December 2003   *May-Britt Kallenrode*

Contents

Part I

Plasmas: The Basics

1 Introduction

We shall not cease from exploration.
And the end of all our exploring
will be to arrive where we started
and know the place for the first time.
T.S. Eliot, *Little Gidding*

1.1 Neutral Gases and Plasmas

Matter, in our daily experience, can be divided into solids, liquids, and gases. Manipulation of matter has shaped our scientific world view as well as our intuitive understanding of its different states and their behavior. But moving upwards from the surface of the Earth, our environment changes and no longer fits into this picture: starting at a height of about 80 km, the atmosphere contains an ionized particle component, the ionosphere. With increasing height, the relative importance of the neutral component decreases and ionized matter becomes dominant. Farther out in the magnetosphere and in interplanetary space almost all gas is ionized: the hard electromagnetic radiation from the Sun immediately ionizes almost all matter. Space therefore is dominated by a plasma, the "fourth state of matter".

A plasma differs from a neutral gas in so far as it (also) contains charged particles. The number of charged particles is large enough to allow for electromagnetic interactions. In addition, the number of positive and negative charges is nearly equal, a property which is called quasi-neutrality: viewed from the outside the plasma appears to be electrically neutral. The reason for this quasi-neutrality can be understood from the electrostatic forces between charged particles. For instance, in a gas discharge a typical length scale is $L = 0.01$ m and a typical number density number density of the electron gas is $n_{\mathrm{e}} = 10^{20}$ m^{-3}. The electric field on the surface of a sphere with $r = L$ containing only the electron gas but no ions is then $E \approx 10^{10}$ V/m. Such a strong field will immediately cause a rearrangement of charges and quasi-neutrality will be restored. In the rarefied plasmas in space, number densities are smaller by many orders of magnitude (see Fig. 1.1); however, since the spatial scales are measured in kilometers or even thousands of kilometers, the same argument can be applied: on the relevant spatial scales the plasma is quasi-neutral even in the rarefied plasmas in space, although this is not necessarily the case on the centimeter scale.

Because a plasma (partly) consists of free charges, it is a conductor. Moving electric charges are currents. These currents induce magnetic fields which in turn influence the motion of the very particles forming the field-generating currents. Thus the particle motion in a plasma is not only controlled by external electric and magnetic fields, but also creates fields which add to the external ones and modify the motion of the particles: a plasma can interact with itself. Consequently, dynamics in a plasma are more complex than in a neutral gas. This is most obvious in the large number of different types of plasma waves (Chap. 4).

In apparently simple situations, a plasma can behave counter-intuitively. Pouring milk into our coffee, we expect the milk to heat up and mix with the coffee. A sunspot is a sharply bordered volume of cool gas embedded in the hot solar photosphere; but it stays stable for several months prevented by strong magnetic fields from warming or mixing with its environment. A cold and dense volume of gas or liquid in a hot environment sinks. A solar filament is cold and dense compared with the ambient corona but it is held in position against gravity by strong magnetic fields. Such discrepancies between our daily experience and the behavior of ionized gases clearly show that plasmas do not form a significant part of our environment. Why then do we study such exotic phenomena? Are there applications for plasmas?

First, plasmas are not exotic but quite common. The interplanetary and interstellar medium and the stars are made of ionized gases. Thus about 99% of matter in the universe is plasma. Nearest regions dominated by plasmas are the magnetosphere with its radiation belts, the ionosphere, lightning bolts in the atmosphere, and, in a wider sense, the Earth's core; thus even in the system Earth plasmas are not uncommon. Plasma physics, therefore, contributes to the understanding of our environment. In turn, the natural plasma laboratories, i.e. the ionosphere, the magnetosphere, and interplanetary space, help to test the concepts of plasma physics on spatial scales and at densities unattainable in a laboratory.

Even some everyday materials can be described as plasmas because they show similarities to the free-electron plasma described above: the conduction electrons in metals and electron–hole pairs in semiconductors are charges which can move quasi-freely and lead to a behavior of the matter which can be described in the same way as for a plasma. The free-electron gas in metals is therefore also included as example of a plasma in Fig. 1.1.

Second, plasmas can be used for quite worldly applications. One of the most ambitious projects is nuclear fusion: to merge hydrogen atoms to helium, imitating the processes inside the Sun (Sect. 6.1) and the stars, in order to create a clean and long-lasting power source. The main aspects of this project are the production of a plasma with suitable properties (density, temperature, losses) and its confinement inside a magnetic field.

There are also less spectacular applications of plasma physics. Chemistry utilizes the different chemical reactions in plasmas and neutral gases: for instance, cyan gas can be synthesized by burning coal dust in a nitrogen electric arc plasma. Plasma beams are used for ion implantation in microchip production. Plasma burners and pistols are used to cut, weld, or clean metals. Other technical applications of plasmas are as diverse as lasers, capacitors, oscillators, and particle accelerators [323].

1.2 Characterization of a Plasma

The main characteristics of a plasma are its electron temperature T_e and the electron number density n_e. The first gives a measure of the thermal energy or more correctly the average kinetic energy of the particles (see Sect. 5.1): $E_{th} = mv_{th}/2 = 3k_B T_e/2$, where k_B is Boltzmann's constant. The temperature is often given in the units of particle energy, electronvolts (eV), where $T_e[eV] = 3E_{th}/3[eV]$. The temperature and number density are given for the electron component: whereas ions, owing to their larger mass, are rather immobile and on many occasions can be regarded as a fixed background of positive charges, the electrons are the mobile part of a plasma.

The second parameter, the electron number density n_e, is an indicator of the particle motion. In a low density plasma, particle motion is determined by the electric and magnetic fields only, while for high densities the interactions between the particles dominate. Thus the two parameters T_e and n_e combined classify a plasma with respect to (a) interactions between plasma constituents, (b) the relative importance of electromagnetic fields for the particle motion, and (c) the range over which particles can propagate freely.

Figure 1.1 shows such an n_e/T_e diagram. Some typical plasmas are indicated. Note that both parameters extend over many orders of magnitude. Astrophysical plasmas can be found anywhere in this diagram: the rarefied ionospheric plasma has a rather low temperature, while the rarefied plasmas in the magnetosphere and in the solar wind have much higher temperatures. All three have densities far below those in terrestrial plasmas such as gas discharges and lightning. Other astrophysical objects, such as the interior of the Sun, have higher densities and temperatures. The only terrestrial plasmas with comparable or higher temperatures are fusion plasmas.

The dotted and dashed lines in Fig. 1.1 indicate two additional plasma parameters: the Debye length λ_D and the number N_D of particles inside a sphere of radius λ_D. The Debye length gives the spatial scale over which particles in a plasma exert electrostatic forces on each other (Sect. 3.7). λ_D increases with decreasing density because in a dense plasma charges of opposite sign screen each other, and it increases with increasing temperature because, ow-

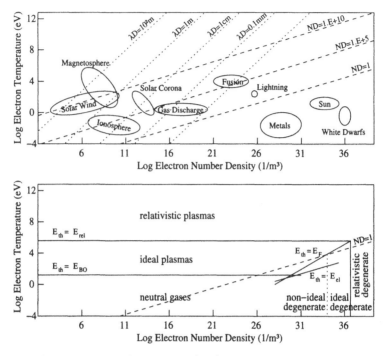

Fig. 1.1. Characteristics of a plasma. (*Top*) Electron temperature T_e, electron number density n_e, Debye length λ_D, and number N_D of particles inside a sphere of radius λ_D, for different plasmas. (*Bottom*) Definition of different plasmas using the characteristic energies

ing to the increased thermal motion of the particles, quasi-neutrality can be violated on a larger spatial scale.

An n_e/T_e diagram can also be used for a general classification of plasmas as shown in the lower panel of Fig. 1.1. Five characteristic energies provide the reference frame for classification: the thermal energy E_{th}, the non-relativistic Fermi energy E_F, the electrostatic energy E_{el}, the energy of the ground state E_{B0}, and the relativistic electron energy E_{rel}. Since these energies depend on the particle species under study, we shall not discuss the characteristic energies in general but shall do so only for the example of hydrogen ($Z = 1$). This is also the most common element in space plasmas.

The first characteristic energy is given by a thermal energy equal to that of the ground state, that is, the ionization energy: $E_{th} = E_{B0}$. This characteristic energy is marked by the lower horizontal line in the lower panel of Fig. 1.1: above this line the plasma is (fully) ionized, below it is (almost) neutral. This characteristic energy therefore divides neutral gases from ideal plasmas. Note that this is a very simple description for two reasons. (a) Certainly, there will be no sharp boundary between neutral and fully ionized, and even for a given temperature there will be stochastic variations in the degree of ionization. (b) The ionization also depends on the density, as described by the Saha equation:

$$\frac{\chi^2}{\chi - 1} = \frac{(2\pi m_e)^{3/2}}{h^3} \frac{(k_B T)^{5/2}}{p_{gas}} \exp\left\{-\frac{E_{B0}}{k_B T}\right\}, \tag{1.1}$$

where χ is the degree of ionization, m_e is the electron mass, p_{gas} is the total gas pressure, and h is Planck's constant. A high degree of ionization might be obtained even if the temperature was only $1/10$ of the ionization temperature. Nonetheless, since we are mainly concerned with rarefied plasmas at rather high temperatures, the above distinction is sufficient for the purpose of this book.

If we increase the temperature further, we obtain a characteristic energy where the thermal energy equals the electron's relativistic energy: $E_{th} = E_{rel}$. This is indicated by the upper horizontal line in Fig. 1.1. The plasma above this line is said to be relativistic. Here the scattering of photons from electrons has to be described as Compton scattering, and the equilibrium radiation field has enough energy to allow pair production.

With increasing density, the plasma degenerates: the energy distribution is no longer Maxwellian but is described by a distribution in which all phase space cells up to the Fermi energy E_F are filled, while at higher energies the population density decreases rapidly. This characteristic energy is indicated by a solid inclined line and separates degenerate and non-degenerate plasmas.

The last characteristic energy relates the thermal energy to the electrostatic energy: $E_{th} = E_{el}$. Plasmas to the left of the corresponding characteristic line are ideal: here the kinetic energy of a particle is larger than its potential energy. In the non-ideal plasmas to the right of the characteristic line the electrostatic interaction is predominant.

1.3 Plasmas in Space

This book focuses on natural plasmas in space. Depending on their location, these plasmas exhibit different properties as characterized by the plasma parameters (see Fig. 1.1); all space plasmas, except for stellar interiors, can be characterized as ideal plasmas. Stellar interiors consist of hot and extremely dense plasmas with the highest densities inside white dwarfs; these plasmas are ideal but degenerate. Plasma density and temperature decrease in the stellar coronae which are still hot enough to be blown away as stellar winds. The combined action of the stellar wind and the magnetic field slows down the star's rotation and winds up the magnetic field lines. This "starsphere" is a void in space, filled by plasma and magnetic flux from the star. The interstellar medium, which fills the space between the starspheres, is most likely an even more attenuated plasma than stellar winds are.

The spatially closest example for such a starsphere is the heliosphere (Chap. 6), see Fig. 1.2: a void in the interstellar medium structured by the

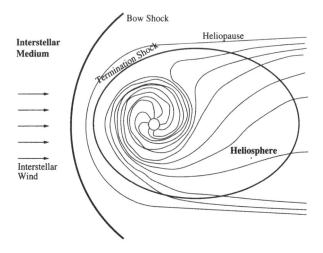

Interstellar Medium

Bow Shock

Heliopause

Termination Shock

Interstellar Wind

Heliosphere

Fig. 1.2. Structure of the heliosphere

solar wind and the frozen-in solar magnetic field. It is separated from the interstellar medium by the heliopause. The heliosphere has an extent of at least about 100 AU[1]. Voids in the heliosphere exist too: the interaction between the solar wind and a planetary magnetic field forms a magnetosphere: a cavity in the solar wind, dominated by the planet's magnetic field.

The topology of the magnetosphere (Chap. 8), is even more complex; however, the physical processes, although on smaller spatial scales, are the same. A magnetosphere is defined as a spatial region where the motion of particles is governed by the planet's magnetic field.[2] This brief definition contains a lot of information. We learn the obvious: the very existence of the magnetosphere requires a magnetic field. Particles in the magnetosphere are at least partly charged: the motion of neutrals would not be influenced by the magnetic field. Their density is low: in a dense medium, collisions between the particles would determine their motion, and the influence of the magnetic field would be negligible. In addition, the energy density of the charged particles is small compared with the energy density of the field: otherwise the particle motion would distort the field instead of the field guiding the particles.

The inner boundary of the magnetosphere is determined by the density: getting closer to the planet, the density increases. Brownian motion becomes dominant and the magnetic field no longer guides the particles. In the Earth's magnetosphere this transition happens at a height of a few hundred kilometers. The upper ionosphere, extending to a height of about 1000 km, lies well inside the magnetosphere. The outer boundary of the magnetosphere is the magnetopause. It separates the planetary and the interplanetary mag-

---

[1] AU is short for astronomical unit which is the average distance between the Sun and the Earth: 1 AU = $149.6 \cdot 10^6$ km.

[2] With substituting "planet's" by "Sun's" or "star's" we could use this statement as a definition for the heliosphere or a starsphere, too.

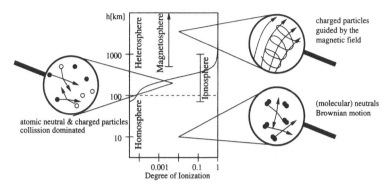

Fig. 1.3. Height scales in the near-Earth environment

netic fields and prevents most of the solar wind plasma from entering the Earth's magnetosphere and atmosphere. Moving towards the Sun, we would detect this boundary at a distance of about 10 Earth's radii. In the opposite direction, the magnetosphere extends far beyond the orbit of the moon.

Three different height regimes are summarized in Fig. 1.3. Regimes can be distinguished according to charge or "mixedness". Below about 80 km, the atmosphere is completely neutral. In the ionosphere, the relative number of ionized particles increases with height. Below 100 km, in the homosphere, particles collide frequently. Thus the different atmospheric constituents are mixed thoroughly. In the heterosphere, above 100 km, particle motion is still dominated by collisions but different constituents start to separate, the degree of ionization increases, and neutrals are atomic rather than molecular. Above 500 km, in the magnetosphere, collisions are infrequent, particles are charged, and particle motion is determined by the magnetic field.

The main components of a magnetosphere are summarized in Fig. 1.4. The magnetosphere is formed by the interaction between the solar wind and the geomagnetic field. A boundary sheet, the magnetopause, forms where the pressure of the solar wind equals the magnetic field pressure. The solar wind streams around the magnetopause but it does not penetrate into the magnetosphere. Where the supersonic solar wind is slowed down to subsonic speed, the bow shock develops. At high latitudes, polar cusps form, separating closed magnetic field lines in front of the magnetosphere from open field lines pulled away to the magnetotail by the solar wind. At these cusps, particles and plasmas can penetrate into the magnetosphere and subsequently precipitate down into the atmosphere.

The basic ingredient of the magnetosphere, the geomagnetic field, originates in a magnetohydrodynamic (MHD) dynamo in the ionized fluid inside the Earth's core. A similar process can be found inside other planets; solar and stellar magnetic fields originate in dynamo processes too. In the solar system all planets except for our two neighbors, Mars and Venus, house a sufficiently strong MHD dynamo to build a planetary magnetic field and to

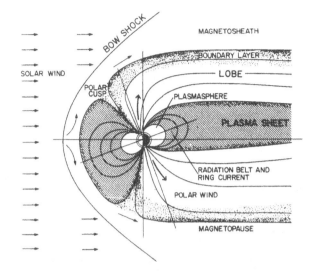

Fig. 1.4. Structure of the Earth's magnetosphere. Reprinted from H. Rosenbauer et al., [447], *J. Geophys. Res.* **80**, Copyright 1975, American Geophysical Union

form a magnetosphere. But even planets without magnetospheres and comets are shielded against the solar wind: when the supersonic solar wind hits their atmospheres, it is slowed down and deflected around the obstacle, forming a bow shock in front of it.

Magnetospheres are not stationary but change as the basic ingredients, the planetary magnetic field and the solar wind, vary in time. Disturbances in the solar wind, caused either by temporal and spatial variations or by transient phenomena, shake the magnetosphere and lead to geomagnetic storms and aurorae. When the terrestrial magnetic field varies in strength, the spatial extent of the magnetosphere changes too. And when the dipole axis drifts, the structure of the magnetosphere is modified. All these different modes of magnetospheres can be observed in the solar system (Chap. 9).

Space physics is not only concerned with plasmas and fields but also with energetic particles (Chap. 7). Their energy by far exceeds the kinetic energy of plasma particles although, owing to its larger density, the energy density of the plasma exceeds that of the energetic particles. An understanding of the acceleration and propagation of these particles also is an important topic. For instance, fluctuations in plasma motions on the Sun, interactions of different plasma streams, and plasma clouds ejected from the Sun excite different kinds of waves. Some of these waves steepen during their propagation, forming shock waves. Energetic particles of solar and galactic origin interact with these waves: this results in spatial scattering as well as scattering in momentum space. These wave–particle interactions are of foremost interest in understanding space plasmas; however, they are also formally difficult because they delve deeply into non-linear processes.

Plasmas and particles also influence the terrestrial environment. For instance, geomagnetic disturbances related to the arrival of large plasma clouds ejected from the Sun (coronal mass ejections) disrupt power and computer lines; energetic particles from a solar flare can ionize the atmosphere and reduce the ozone concentration. Such questions are addressed in solar–terrestrial relationships (Chap. 10), also called "space weather" for short.

1.4 A Brief History of Space Research

In situ observations of plasmas and particles in the magnetosphere and in interplanetary space became possible with the advance of satellite technology in the late 1950s and early 1960s. Many of these observations are discussed within the broad topic of solar–terrestrial relationships. Solar–terrestrial relationships is an old field of science; it dates back to the first correlations between sunspots and aurora in ancient China a few thousand years ago. Aurorae are the prime example of solar–terrestrial relationships: they can be detected easily, even with the naked eye; they are closely correlated with solar activity; and they also have an aesthetic and even mythological quality. Nonetheless, the big steps in understanding solar–terrestrial relationships required observations made on board rockets and satellites: measurements of plasmas and particles in the ionosphere, the magnetosphere, and interplanetary space, and the measurement of the Sun's electromagnetic radiation in frequency ranges not observable from the ground.

The solar wind and the magnetosphere act as coupling devices in solar–terrestrial relationships. Both have been studied, though only indirectly, long before the space age. Magnetism was detected more than 2000 years ago when the ancient Greeks found stones that attracted iron. The first reliable description of a compass dates back to the eleventh century when Shon-Kau (1030–1093) wrote in a Chinese encyclopedia: "fortune-tellers rub the point of a needle with the stone of the magnet in order to make it properly indicate the south". The first written account of a compass in Europe dates back to Alexander Neekam (1157–1217), a monk at St. Albans. He describes the compass and its application in navigation as common. Neekam's compass is a second generation instrument and quite similar to the ones used today: while in the first compass a small piece of magnetic stone floated in water on a piece of wood or cork, in Neekam's compass a needle is placed on a pivot, allowing it to rotate freely and align itself along the north–south direction. In the fourteenth century the compass was common on ships. The declination, the difference between magnetic and geographic north, was well known by the early fifteenth century [94]; its temporal variation was reported in 1634 by Henry Gellibrand (1597–1636). Magnetic inclination was discovered in the second half of the sixteenth century independently by Georg Hartmann (1489–1564) and Robert Norman.

The reason for the north-south-directivity of the compass needle was less well understood. Philosophers of the early thirteenth century suggested some connection by virtue between the loadstone used to rub the compass needle and the polar star – the latter being a special star since, unlike other stars, it is fixed. Later that century, the idea of polar loadstone mountains was proposed. Again, the polar star was believed to give his virtue to the loadstone mountains which in turn imparted it to the compass needle leading it to point towards the polar star. This idea was questioned by Petrus Peregrinus (Peter the Wayfarer, ca. 1240–?): loadstone deposits can be found in many parts of the world. Why should the polar ones be the only ones that attract a compass needle? Peregrinus attacked this question in a manner which can be termed scientific by present day standards: he performed experiments with loadstone, reported in his *Epistola de Magnete*. In particular, he introduced the concept of polarity, discovered magnetic meridians, and described different methods to determine the positions of the poles of a spherical loadstone.

But only in 1600, the basic ingredient of the magnetosphere, the geomagnetic field, was detected: in his treatise *De Magnete* [185], William Gilbert (1544–1604) suggested that the north–south alignment of the compass results from the magnetic field of the Earth. In the middle of the nineteenth century, scientists began to understand the terrestrial magnetic field. A global network of observatories started continuous registrations of the magnetic field and its fluctuations. From these data, Carl Friedrich Gauss (1777–1855) proposed that the Earth's magnetic field consists of two components, one from its interior and a second one generated in the atmosphere [182]. To first order, the internal geomagnetic field can be described as that of a homogeneously magnetized sphere. Hans Christian Ørstæd (1777–1851) and Andre Ampére (1775–1836) suggested ring currents inside the Earth as source of the internal field. This dipole-field approach survived up to the early 1960s when satellite and rocket observations in the upper atmosphere gave a more detailed picture of the true field and suggested modifications to the model.

Gauss and Wilhelm Eduard Weber (1804–1891) initiated very precise measurements of the Earth's magnetic field with relative accuracies of at least 10^{-5} [553,554]. Small fluctuations could be detected, showing systematic variations in time and location as well as superimposed stochastic changes. These latter strongly indicated that the Earth is not an isolated object in space but that strong forces from outside act on spaceship Earth.

In its heyday in the second half of the nineteenth century, solar–terrestrial relationships were an illustrative example of the development of science from correlations and apparently uncorrelated, sometimes even seemingly contradictory observations into a consistent picture of a complex environment. The converging developments included, for example, the discovery of the 11-year sunspot cycle by Heinrich Samuel Schwabe (1789–1875) in 1844 [470, 471] and the correlation between sunspot numbers and the frequency of geomagnetic disturbances by Edward Sabine (1788–1883) in 1852. In the same year,

Rudolf Wolf (1816–1893) found a correlation between sunspots and geomagnetic disturbances. A relationship between individual aurorae and accompanying geomagnetic disturbances had already been noticed by Anders Celsius (1701–1744) and Olof Peter Hiorter (1696–1750) in 1747 [88, 223] and by Alexander von Humboldt (1769–1859) in 1806 [243]. The spatial distribution of aurorae suggested an involvement of the geomagnetic field, too. Aurorae always were known as a high-latitude phenomenon. But in the 1840s the ill-fated Arctic explorer John Franklin (1786–1847) noticed that the frequency of aurorae does not increase all the way towards the pole [169]. In 1860, Elias Loomis (1811–1889) showed that the highest incidence of aurora is seen inside an oval of 20°–25° around the magnetic pole [328]. In 1881 Hermann Fritz (1830–1883) [173] published similar results with his famous map of isochasms (see Fig. 8.43).

During some ten years, out of these observations and statistical correlations a closed picture of solar–terrestrial relationships emerged with the Sun as a source of geomagnetic activity as well as aurorae. In the late 1870s, Henri Becquerel (1852–1908) offered the first physical explanation: the sunspots are assumed to be a source of fast protons [45]. On hitting the Earth's magnetic field, these particles are guided towards the auroral oval by the magnetic field. Despite its simplicity, the model contains a revolutionary aspect: the Sun is not only a source of electromagnetic radiation but its blemishes, the sunspots, are also a source of energetic particles which could affect the terrestrial environment. In the early twentieth century, a similar idea led Kristian Birkeland (1867–1917) to build the terrella, a model of the Earth which allows simulations of the aurora in the laboratory: a cathode-ray tube substitutes for the Sun as a source of energetic particles and a magnetic dipole inside a sphere covered by a fluorescent material simulates the Earth's magnetic field surrounded by its atmosphere. With these ingredients, Birkeland showed that the geomagnetic field was responsible for the formation of the aurora ovals. From the correlation between aurorae and the number of active regions on the Sun, Birkeland suggested that sunspots might be the source of a continuous stream of particles [50]. This idea was an early introduction of the concept of a plasma flow from the Sun, which later evolved into the concept of the solar wind.

In the 1930s, Sydney Chapman and Victor Ferraro developed an idea about solar–terrestrial relationships which comes closer to our current understanding: sunspots are indicators of solar activity [95]. Solar activity manifests itself in violent eruptions, called solar flares. Chapman and Ferraro suggested that flares not only emit electromagnetic radiation but also fling out clouds of ionized matter which in size dwarf the Earth. After a travel time of 1 to 3 days, such a cloud might hit the magnetosphere and compress it, leading to geomagnetic disturbances. Some of the cloud matter might penetrate the magnetosphere close to the poles, causing the aurora. The existence of these proposed clouds, today called coronal mass ejections (CMEs), was first confirmed in the early 1970s by Skylab.

While their basic idea is accepted even today, Chapman and Ferraro's ansatz still assumes that the geomagnetic field is a dipole. Only the first in situ measurements in the magnetosphere and in interplanetary space revealed the complexity of the magnetic field surrounding the Earth. But again, indirect evidence had been found long before. In the 1920s, the existence of a region of charged particles in the atmosphere, the ionosphere, had been discovered because of its effect on radio waves: they propagate far beyond the horizon due to reflection off the conducting ionosphere. Motions and variations in charge density also can be used to explain the atmospheric contribution to the Earth's magnetic field which had been proposed 70 years earlier by Gauss. Thus, early researchers in magnetospheric physics knew of a conductive layer at a height of some tens of kilometers. In the early 1950s, in Arctica and Antarctica, van Allen and colleagues launched rockets to a height of about 110 km. Their instruments confirmed the existence of energetic electrons in this region, either directly or by observing the electron bremsstrahlung: the existence of the ionosphere had been confirmed by in situ measurements. In 1958, a Geiger counter on board the first US satellite, Explorer 1, detected the Earth's radiation belts, later named van Allen belts to honor their discoverer. In the same year, the Soviet lunar probe made the first measurements in interplanetary space, confirming the existence of the long-proposed solar wind. The first detailed studies of the solar wind were made by Mariner 2 in 1962. The boundary between interplanetary space and the Earth system, the magnetopause, was first studied by Explorer 10 in 1961; the bow shock in front of it was detected by Explorer 12 in 1962 and studied in detail by OGO (Orbiting Geophysical Observatory) in 1964. These and subsequent observations led to the identification of the main components of the magnetosphere, as discussed above.

Exercises and Problems

1.1. Define a plasma. Discuss the importance of the density. Is there a limit for the relative or the absolute size of the neutral component?

1.2. What parameters are used to characterize a plasma? Briefly discuss their physical meaning.

1.3. What energies/temperatures can be used to classify a plasma?

1.4. What do you need to explain a magnetosphere? What determines its spatial extent?

1.5. Describe the basic features of a magnetosphere. How do these properties change if the axis of the magnetic field changes with respect to the solar wind direction; if the terrestrial magnetic field decreases; if solar wind pressure and speed increase?

1.6. Why is space dominated by plasmas?

1.7. Where in the near-Earth environment do plasmas exist? Would we miss them if they were neutral matter instead?

2 Charged Particles in Electromagnetic Fields

I've gotten a rock, I've gotten a reel,
I've gotten a wee bit spinning-wheel;
An' by the whirling rim I've found
how the weary, weary warl goes round.
S. Blamire, *I've gotten a rock*

In space physics the motion of charged particles in electric and magnetic fields often is described by a test particle approach: the particles are guided by the field but their motion does not affect the field. This approach is valid if the energy density of the magnetic field exceeds that of the particles. In the test particle approach the motion can be separated into two parts: the motion of a guiding center of the particle orbit and a gyration around it. The guiding center motion can be interpreted as the effective motion of the particle, averaged over many gyrations. This concept is applied to drifts in stationary electromagnetic fields. The adiabatic invariants allow simple estimates of the particle motion in slowly varying fields; they are applied to the motion of particles in the Earth's radiation belts. This chapter starts with a brief recapitulation of the basics of electromagnetic field theory.

2.1 Electromagnetic Fields

Particle densities in interplanetary space and in the magnetosphere are low. Thus a description of the electromagnetic field in a vacuum is sufficient. Then the permeability μ and the permittivity ε both equal 1: the medium can be neither magnetized nor polarized. In principle, we can assign a permittivity to a plasma, but this is just another description of the existence of charged particles. A plasma can also be magnetized: for instance, an axial-symmetric ring current of charged particles in a dipole field leads to a reduction of the magnetic moment, which is a diamagnetic effect. An equivalent description is the magnetic induction B in a vacuum, consisting of both the dipole field and the field disturbance, and the current arising from the particle motion.

2.1.1 Maxwell's Equations in Vacuum

In 1873 Maxwell (1831–1879) introduced a unified theory of the electromagnetic field giving for the first time the set of four partial differential equations that today bear his name [338]. The sources of the electric and magnetic fields are charges, magnetized bodies, and currents, which can be either discrete or continuous and either stationary or time-dependent.

The electric field $\boldsymbol{E}$ generated by a charge density ϱ_c is described by Poisson's equation:[1]

$$\nabla \cdot \boldsymbol{E} = \varrho_c / \varepsilon_0 . \tag{2.1}$$

Since the electric field is non-rotational, it can be expressed by the gradient of the scalar Coulomb potential φ: $\boldsymbol{E} = -\nabla\varphi$. Integrating (2.1) over a volume V and using Gauss's theorem (A.33), Poisson's equation can be rewritten as Gauss's law for the electric field:

$$\oint_{O(V)} \boldsymbol{E} \cdot \mathrm{d}\boldsymbol{S} = \int_V \frac{\varrho_c}{\varepsilon_0} \mathrm{d}^3\boldsymbol{r} . \tag{2.2}$$

Gauss's law states: the electric flux through a surface $\boldsymbol{S}$ enclosing a volume V is determined by the total charge inside V. If there are no net charges enclosed in V, the flux through $\boldsymbol{S}$ is zero. But V is not necessarily field-free, e.g. if V is placed inside a dipole field either enclosing none of the charges or both of them. Using a spherical test volume V with radius r, Coulomb's law can be derived from Gauss's law.

Gauss's law for a magnetic field is formally analogous, $\nabla \cdot \boldsymbol{B} = 0$, or

$$\oint_{O(V)} \boldsymbol{B} \cdot \mathrm{d}\boldsymbol{S} = 0 . \tag{2.3}$$

It states that there are no magnetic monopoles.

Faraday's law describes the electric field (or electro-motoric force EMF) generated by a changing magnetic field:

$$\nabla \times \boldsymbol{E} = -\frac{\partial \boldsymbol{B}}{\partial t} . \tag{2.4}$$

The magnetic flux Φ through a surface $\boldsymbol{S}$ in the magnetic field is defined as

---

[1] Equations are given in SI units throughout. However, since the cgs system still is widely used in geophysics and space physics, equations which are frequently used to determine parameters also are given in cgs units in App. A.2. The cgs system is advantageous in so far as the absolute permittivity and the absolute permeability both equal 1 compared with $\varepsilon_0 = (4\pi \cdot 9 \cdot 10^9)^{-1}$ F/m $= 8.854 \cdot 10^{-12}$ F/m and $\mu_0 = 4\pi \cdot 10^{-7}$ H/m $= 1.256 \cdot 10^{-6}$ H/m in the SI system. In the cgs system, on the other hand, occasionally a factor $c = 1/\sqrt{\varepsilon_0\mu_0}$ appears. If quantities in SI units are inserted into equations given in the cgs system, or vice versa, the analysis of units automatically leads to the correct consideration of ε_0 and μ_0.

$$\Phi = \int \boldsymbol{B} \cdot \mathrm{d}\boldsymbol{S} \ . \tag{2.5}$$

Graphically, the magnetic flux can be interpreted as the number of field lines going through $\boldsymbol{S}$. Using this definition and Stokes's theorem (A.39), Faraday's law can be rewritten as

$$\oint_C \boldsymbol{E} \cdot \mathrm{d}\boldsymbol{l} = -\frac{\partial}{\partial t} \int_O \boldsymbol{B} \cdot \mathrm{d}\boldsymbol{S} = -\frac{\partial \Phi}{\partial t} \ . \tag{2.6}$$

It states that a change in the magnetic flux through a surface creates an EMF in its circumference. The minus sign indicates that a current generated by the EMF causes a magnetic field anti-parallel to the original one (Lenz's rule). Faraday's law has two consequences: a stationary magnetic field does not produce an electric field. And if the electric field is zero, the magnetic field is stationary.

Ampére's law describes magnetic field generated by a time-dependent electric field $\boldsymbol{E}$ and a current density $\boldsymbol{j}$:

$$\nabla \times \boldsymbol{B} = \mu_0 \boldsymbol{j} + \varepsilon_0 \mu_0 \frac{\partial \boldsymbol{E}}{\partial t} \ . \tag{2.7}$$

The last term on the right-hand side is the displacement current. It is related to the equation of continuity: by taking the divergence of (2.7) we get

$$\nabla \cdot (\nabla \times \boldsymbol{B}) = \mu_0 \nabla \cdot \boldsymbol{j} + \mu_0 \varepsilon_0 \nabla \cdot \frac{\partial \boldsymbol{E}}{\partial t} = \mu_0 \nabla \cdot (\varrho_c \boldsymbol{v}) + \mu_0 \frac{\partial \varrho_c}{\partial t} \ . \tag{2.8}$$

Since the divergence of a rotational field (left-hand side) vanishes (A.25), this gives the equation of continuity for charges (see Sect. 3.1.3):

$$\frac{\partial \varrho_c}{\partial t} + \nabla \cdot (\varrho_c \boldsymbol{v}) = 0 \quad \text{or} \quad \frac{\mathrm{d}\varrho_c}{\mathrm{d}t} + \varrho_c \nabla \cdot \boldsymbol{v} = 0 \ . \tag{2.9}$$

Using Stokes' theorem, (2.7) can be written as

$$\oint_C \boldsymbol{B} \cdot \mathrm{d}\boldsymbol{l} = \mu_0 \int \boldsymbol{j} \cdot \mathrm{d}\boldsymbol{S} + \mu_0 \varepsilon_0 \int \frac{\partial \boldsymbol{E}}{\partial t} \cdot \mathrm{d}\boldsymbol{S} \ . \tag{2.10}$$

Ampére's law states that a changing electric field and/or a current creates a rotational magnetic field.

Maxwell's equations for the electric and magnetic field are symmetric, except for the fact that there are neither magnetic charges (there are no magnetic monopoles) nor magnetic currents.

2.1.2 Transformation of Field Equations

In space physics, fields and plasmas move with respect to the observer: the solar wind is swept across a spacecraft in interplanetary space, another space-craft crosses through the radiation belts of a planet. The fields $\boldsymbol{E}, \boldsymbol{B}$ and

E', B' in two reference frames C and C', with C' moving with velocity v with respect to C, are related by the relativistic transformations

$$E' = \frac{1}{\gamma} \left[E + \frac{v}{v^2} (v \cdot E)(1 - \gamma) + v \times B \right] \quad \text{and}$$

$$B' = \frac{1}{\gamma} \left[B + \frac{v}{v^2} (v \cdot B)(1 - \gamma) - \frac{1}{c^2} v \times E \right] , \tag{2.11}$$

with $\gamma = \sqrt{1 - v^2/c^2}$. In the non-relativistic case, all terms in v^2/c^2 can be ignored ($\gamma \to 1$), reducing the equations to

$$E' = E + v \times B \quad \text{and} \quad B' = B - \frac{1}{c^2} v \times E . \tag{2.12}$$

Equation (2.12) implies that field components parallel to the direction of motion remain unchanged.

Within the framework of this book, the second set of transformations will be sufficient, considering effects of order v/c (e.g. the Doppler effect) but ignoring effects of order v^2/c^2 (e.g. the Lorentz contraction).

We shall frequently encounter one consequence of the field transformations for space plasmas, namely the $v \times B$ electric induction field (the second term on the right-hand side of the first equation in (2.12)): the convection of a magnetic field B with a plasma moving at speed v leads to an electric induction field $v \times B$. Applications include the electric field in the front of non-parallel shocks (Sect. 7.5.1) and the electric field in interplanetary space leading to the corotation of energetic particles (Sect. 6.3).

2.1.3 Generalized Ohm's Law

Ohm's law connects the current density j and the electric field E by a constant, the conductivity σ:

$$j = \sigma E . \tag{2.13}$$

Note that often, e.g. in the ionosphere, the conductivity is anisotropic and should be described by a tensor rather than a scalar (Sect. 8.3.2).

Equation (2.13) is valid in the plasma rest frame only. If an observer is moving with velocity v with respect to the plasma frame, a generalized form of Ohm's law is required. It can be obtained by applying (2.12). The plasma has a high conductivity, that is all electric fields except induction fields vanish immediately, and thus the current on the left-hand side of (2.13) transforms under consideration of (2.7) as $j = j'$. The generalized form of Ohm's law then reads

$$j = \sigma (E + v \times B) . \tag{2.14}$$

The second term on the right-hand side describes the electric induction field which gives rise to the Hall current.

2.1.4 Energy Equation of the Electromagnetic Field

From Faraday's law we can derive an energy equation for the electromagnetic field. Multiplication of Faraday's law (2.4) by $\boldsymbol{B}$ and integration over a volume V gives

$$\int_V \boldsymbol{B} \cdot \frac{\partial \boldsymbol{B}}{\partial t}\, \mathrm{d}^3 r = -\int_V \boldsymbol{B} \cdot (\nabla \times \boldsymbol{E})\, \mathrm{d}^3 r \;. \tag{2.15}$$

The divergence of a vector product can be written as (see (A.29))

$$\nabla \cdot (\boldsymbol{E} \times \boldsymbol{B}) = \boldsymbol{B} \cdot (\nabla \times \boldsymbol{E}) - \boldsymbol{E} \cdot (\nabla \times \boldsymbol{B}) \;. \tag{2.16}$$

The second term on the right-hand side can be rewritten using Ampére's law (2.7). Solving for the first term on the right-hand side we get

$$\boldsymbol{B} \cdot (\nabla \times \boldsymbol{E}) = \nabla \cdot (\boldsymbol{E} \times \boldsymbol{B}) + \mu_0\, \boldsymbol{E} \cdot \boldsymbol{j} + \frac{1}{c^2}\, \boldsymbol{E} \cdot \frac{\partial \boldsymbol{E}}{\partial t} \;. \tag{2.17}$$

Inserting into (2.15) gives the the energy equation

$$\int_V \boldsymbol{B} \cdot \frac{\partial \boldsymbol{B}}{\partial t}\, \mathrm{d}^3 r = -\int_V \nabla \cdot (\boldsymbol{E} \times \boldsymbol{B})\, \mathrm{d}^3 r - \int_V \mu_0 \boldsymbol{E} \cdot \boldsymbol{j}\, \mathrm{d}^3 r$$
$$-\frac{1}{c^2} \int_V \boldsymbol{E} \cdot \frac{\partial \boldsymbol{E}}{\partial t}\, \mathrm{d}^3 r \;. \tag{2.18}$$

With Gauss's theorem (A.33), the volume integral in the first term on the right-hand side can be changed into a surface integral. The first term on the left-hand side and the last terms on the right-hand side contain the product of a vector and its temporal derivative. This is equal to half the temporal derivative of the squared vector as can be seen by differentiating the middle term in

$$\boldsymbol{a} \cdot \frac{\partial \boldsymbol{a}}{\partial t} = \frac{1}{2} \frac{\partial (\boldsymbol{a} \cdot \boldsymbol{a})}{\partial t} = \frac{1}{2} \frac{\partial a^2}{\partial t} \;. \tag{2.19}$$

The energy equation then reads

$$\frac{\partial}{\partial t} \int_V \left(\frac{B^2}{2\mu_0} + \frac{\varepsilon_0 E^2}{2} \right) \mathrm{d}^3 r = -\oint_{O(V)} \frac{(\boldsymbol{E} \times \boldsymbol{B})}{\mu_0} \cdot \mathrm{d}\boldsymbol{S} - \int_V \boldsymbol{E} \cdot \boldsymbol{j}\, \mathrm{d}^3 r \;. \tag{2.20}$$

Here $B^2/2\mu_0$ and $\varepsilon_0 E^2/2$ are the energy densities of the magnetic and the electric fields. Equation (2.20) can be interpreted as an equation for the continuity of the electromagnetic field: the change in the energy density of the electromagnetic field inside a volume V is given by the energy flux (Poynting vector $\boldsymbol{S}_\mathrm{P}$),

$$\boldsymbol{S}_\mathrm{P} = \frac{\boldsymbol{E} \times \boldsymbol{B}}{\mu_0} \;, \tag{2.21}$$

through the surface of the volume and ohmic losses $\boldsymbol{E} \cdot \boldsymbol{j}$ inside V; positive values of $\boldsymbol{E} \cdot \boldsymbol{j}$ indicate losses, while negative ones describe a generator.

If we consider electromagnetic fields in matter, the field exerts a force on the particles, described by the $\boldsymbol{j} \times \boldsymbol{B}$ term in the equation of motion (3.28). Thus, the energy equation has to be supplemented by a term describing the work done by the field:

$$
\frac{\partial}{\partial t} \int_V \left(\frac{B^2}{2\mu_0} + \frac{\epsilon_0 E^2}{2} \right) \mathrm{d}^3 r = - \oint_{O(V)} \frac{\boldsymbol{E} \times \boldsymbol{B}}{\mu_0} \mathrm{d}\boldsymbol{S} - \int_V \boldsymbol{E} \cdot \boldsymbol{j} \, \mathrm{d}^3 r
$$
$$
- \int_V \boldsymbol{u} \cdot (\boldsymbol{j} \times \boldsymbol{B}) \, \mathrm{d}^3 r . \tag{2.22}
$$

A comparison of the energy densities of the field and the plasma shows whether the particle motion will be governed by the electromagnetic fields or by the gas laws: if the plasma's energy density is high and the conductivity is infinite, the field is frozen-in (Sect. 3.4.1) and carried away by the plasma (e.g. the interplanetary magnetic field frozen-into the solar wind). If the energy density of the field is high, the field determines the motion of the particles (e.g. energetic particles in interplanetary space or in the radiation belts). This latter case is discussed in this chapter.

2.2 Particle Motion in Electromagnetic Fields

Let us now turn to the motion of individual charged particles in a prescribed electromagnetic field. Particle densities are assumed to be small: there are no collisions between particles, the particle motion is determined by the fields only. The energy density of the particles is small too; thus their motion does not modify the external field.

Although these limitations are strong, the resulting motion and their formal description are basic and instructive. Table 2.1 places them into the general framework of particle motion in electromagnetic fields. Fields can either be smooth or turbulent. For smooth fields, two cases of particle motion can be distinguished: (a) The field varies only weakly on the temporal and spatial scales of the gyration and the particle motion can be described by the concepts of guiding center motion and adiabatic invariants (Sect. 2.4). These concepts can be applied to particles in the radiation belts (Sect. 8.7.1). (b) The field changes significantly during one gyration of the particle. The above concepts are no longer valid and the equation of motion has to be integrated. One example are the Størmer orbits of galactic cosmic rays in the magnetosphere (Sect. 8.7.2). Turbulent fields require an entirely different approach. The particle motion is determined not only by the average magnetic field but also by scattering at field fluctuations, a stochastic process. Formally, we have to consider particle ensembles instead of single particles and transport

Table 2.1. Particle motion in different types of magnetic fields. The characteristic length scale L for changes in the magnetic field is defined as $1/L = (1/|B|)\,\partial B/\partial x$

Field			
	smooth field analytical description possible		turbulent, irregular fast fluctuations
Scales	small variations in space and time $r_{\rm L} \ll L$	large variations $B = <B> + b,$ $r_{\rm L} \geq L$	weak turbulence $b \ll |<B>|$ strong turbulence: $b \approx |<B>|$
Formalism	adiabatic invariants; guiding center motion; drifts	integration of the equation of motion	transport equations pitch-angle scattering resonance interaction
Occurrence	periodic motion in radiation belts, magnetic mirrors and bottles	Størmer orbits	particle propagation in interplanetary space

equations instead of equations of motion. Propagation then can be understood as a diffusive process (Chap. 7.3); applications are the interplanetary transport (Sect. 7.4) or particle scattering into the loss cone in the radiation belts (Sect. 8.7.1).

2.2.1 Lorentz Force and Gyration

The general equation of motion is Newton's second law $\boldsymbol{F} = \mathrm{d}\boldsymbol{p}/\mathrm{d}t$. The net force on a particle can consist of different components, e.g. gravitation, electromagnetic forces, and a pressure gradient. In a pure electromagnetic field only the Lorentz force acts on a particle

$$\boldsymbol{F}_{\rm L} = m\frac{\mathrm{d}\boldsymbol{v}}{\mathrm{d}t} = q\,(\boldsymbol{E} + \boldsymbol{v} \times \boldsymbol{B})\;. \qquad (2.23)$$

First Integral of Motion. In a pure magnetic field the electric field is zero. According to Faraday's law (2.4), the field then is stationary. The equation of motion reduces to

$$m\frac{\mathrm{d}\boldsymbol{v}}{\mathrm{d}t} = q\boldsymbol{v} \times \boldsymbol{B}\;. \qquad (2.24)$$

Multiplication by the particle speed v and consideration of (2.19) gives the first integral of motion:

$$\frac{1}{2}m\frac{\mathrm{d}v^2}{\mathrm{d}t} = \frac{\mathrm{d}W_{\mathrm{kin}}}{\mathrm{d}t} = qv \cdot (v \times B) = 0 \qquad (2.25)$$

because the cross-product $v \times B$ is perpendicular to v and thus its scalar product with v vanishes. The first integral of motion (2.25) states that in a pure magnetic field the kinetic energy W_{kin} of a particle is constant. W_{kin} is given in electronvolts (eV) where $1\ \mathrm{eV} = 1.602 \cdot 10^{-19}$ J is the energy gained by an electron after traversing a potential difference of 1 V.

Gyration. Let us now assume a homogeneous magnetic field along the z axis: $B = Be_z$. The equation of motion for the components then reads

$$m\dot{v}_x = qBv_y , \qquad m\dot{v}_y = -qBv_x , \qquad \text{and} \qquad m\dot{v}_z = 0 . \qquad (2.26)$$

Integration of the last equation gives $v_\parallel = v_z = \mathrm{const}$: the particle moves parallel to the field line with constant speed v_z. The other two equations are coupled ordinary differential equations (ODEs) of first order. They can be combined into two separate ODEs of second order by first differentiating them with respect to t and then inserting the other ODE:

$$\ddot{v}_x = \frac{qB}{m}\dot{v}_y = -\left(\frac{qB}{m}\right)^2 v_x \quad \text{and} \quad \ddot{v}_y = -\frac{qB}{m}\dot{v}_x = -\left(\frac{qB}{m}\right)^2 v_y . \qquad (2.27)$$

These second order ODEs can be solved with an ansatz $v_i = v_{0,i}\,\mathrm{e}^{\mathrm{i}\omega t}$. They describe a harmonic oscillator with a cyclotron frequency ω_{c}

$$\omega_{\mathrm{c}} = \frac{|q|B}{m} . \qquad (2.28)$$

The solution of the equation of motion is a circular orbit around the magnetic field lines in the xy plane. The components of the trajectory are

$$x(t) = r_{\mathrm{L}}\sin\omega_{\mathrm{c}}t \quad \text{and} \quad y(t) = r_{\mathrm{L}}\cos\omega_{\mathrm{c}}t , \qquad (2.29)$$

and the components of the particle velocity are

$$v_x(t) = r_{\mathrm{L}}\omega_{\mathrm{c}}\cos\omega_{\mathrm{c}}t \quad \text{and} \quad v_y(t) = -r_{\mathrm{L}}\omega_{\mathrm{c}}\sin\omega_{\mathrm{c}}t . \qquad (2.30)$$

The particle speed $v_\perp$ perpendicular to the magnetic field then is

$$v_\perp^2 = v_x^2 + v_y^2 = r_{\mathrm{L}}^2\omega_{\mathrm{c}}^2 , \qquad (2.31)$$

and the radius of the particle orbit, the Larmor radius r_{L}, is

$$r_{\mathrm{L}} = \frac{v_\perp}{\omega_{\mathrm{c}}} = \frac{mv_\perp}{|q|B} . \qquad (2.32)$$

The direction of motion depends on the particle's charge and has to obey Lenz's rule: the ring current associated with the particle motion creates a magnetic field opposite to the external one. An electron obeys the right-hand rule (see Table 2.2): if the thumb is directed along the magnetic field line, the tips of the curved fingers give the direction of the electron motion. If the initial velocity has a component parallel to the magnetic field, the particle follows a helical path around the line of force.

Example 1. A hot plasma with $T = 1$ keV is confined in a homogeneous magnetic field $B = 1$ T. The thermal speeds are given by $v = \sqrt{2W_{\text{kin}}/m}$. Thus we obtain $v_e = 18.7 \times 10^6$ km/s for the electrons and $v_p = 4.37 \times 10^5$ m/s for the protons. The Gyro-radii then are determined from (2.32) to be $r_e = 0.1$ mm and $r_p = 4.6$ mm. From (2.28), we get obtain $\omega_{c,e} = 1.8 \times 10^{11}$ s^{-1} and $\omega_{c,p} = 10^8$ s^{-1} for the cyclotron frequencies. Note that both ω_c and r_L scale with B; thus in a much weaker field, such as the interplanetary medium, which is of the order of a few nT, the gyro radii would increase by many orders of magnitude while the cyclotron frequencies would decrease by orders of magnitude.   □

2.2.2 Useful Definitions

The magnetic rigidity P [V] describes the resistance of a particle to change its direction of motion under the influence of a magnetic field. It is defined as the ratio of the momentum $p_\perp$ perpendicular to $\boldsymbol{B}$ and the charge q:

$$P = \frac{p_\perp}{q} . \tag{2.33}$$

The Larmor radius (2.32) then is the ratio between the magnetic rigidity and the magnetic field strength $r_L = P/B$.

The gyration of the particle is determined by its speed $v_\perp$ perpendicular to the magnetic field while the particle itself is characterized by its energy or velocity. The relative sizes of the velocity components parallel and perpendicular to the magnetic field can be described by the pitch angle α with

$$\tan \alpha = \frac{v_\perp}{v_\parallel} . \tag{2.34}$$

The velocity components perpendicular and parallel to the field therefore are $v_\perp = v \sin \alpha$ and $v_\parallel = v \cos \alpha$.

A particle with charge q moving in a circular orbit with radius r_L and gyration time T_c gives rise to a ring current $I = q/T_c = q\omega_c/(2\pi)$. The magnetic moment μ is defined as the product of the ring current and the enclosed area $A = \pi r_L^2$:

$$\mu = |I| A = \frac{mv_\perp^2}{2B} = \frac{W_{\text{kin}\perp}}{B} . \tag{2.35}$$

The orientation of the magnetic moment is determined by the direction of the ring current:

$$\boldsymbol{\mu} = \frac{1}{2} q \boldsymbol{r}_{\mathrm{L}} \times \boldsymbol{v}_{\perp} = -\frac{m v_{\perp}^2}{2} \frac{\boldsymbol{B}}{B^2} = -W_{\mathrm{kin}\perp} \frac{\boldsymbol{B}}{B^2} = -\frac{W_{\mathrm{kin}\perp}}{B} \boldsymbol{t}_{\mathrm{B}} \tag{2.36}$$

where $\boldsymbol{t}_{\mathrm{B}} = \boldsymbol{B}/B$ is the tangential unit vector to the magnetic field. The magnetic moment does not depend on the charge of the particle; its direction is opposite to that of the external field. A plasma, therefore, is diamagnetic.

In a highly conductive plasma, no electric field exists and the particle energy is conserved (see (2.25)). Then $v_{\parallel}$ is constant. Therefore $v_{\perp}$ is constant, too. If the magnetic field is constant, the magnetic moment will also be constant. Even in the case of a slowly (adiabatically) varying field μ is constant. It is therefore called an adiabatic invariant (Sect. 2.4.1).

Excursion 1. *Relativistic Quantities.* The highly energetic particles accelerated in solar flares or coming from the galactic cosmic radiation have speeds close to the speed of light. Thus these particles have to be described in terms of relativistic quantities. The mass of a particle with rest mass m_0 increases with increasing speed v, as described by the relativistic mass equation

$$m(v) = \frac{m_0}{\sqrt{1 - \frac{v^2}{c^2}}} = \gamma m_0 . \tag{2.37}$$

The relativistic energy is then

$$E = mc^2 , \tag{2.38}$$

the relativistic momentum is

$$\boldsymbol{p} = m(v) \, \boldsymbol{v} = \gamma m_0 \boldsymbol{v} , \quad \text{and} \quad p^2 = \frac{W_{\mathrm{kin}}^2}{c^2} - m_0 c^2 . \tag{2.39}$$

Consequently, the relativistic force can be written as

$$\boldsymbol{F} = \frac{\mathrm{d}}{\mathrm{d}t} \left(\gamma m_0 \boldsymbol{v} \right) . \tag{2.40}$$

The cyclotron frequency of a relativistic particle then is simply

$$\omega_{\mathrm{c,rel}} = \frac{|q| \, B}{m_{\mathrm{rel}}} = \frac{|q| \, B}{\gamma m_0} , \tag{2.41}$$

and the Larmor radius is

$$r_{\mathrm{L,rel}} = \frac{v_{\perp}}{\omega_{\mathrm{c,rel}}} = \frac{p_{\perp}}{|q| \, B} . \tag{2.42}$$

□

Example 2. So far, the highest proton energies observed in the galactic cosmic radiation are 10^{20} eV. These protons gyrate in an interstellar magnetic field of about $B = 3 \times 10^{-10}$ T. Since for such high energies the kinetic energy W_{kin} is approximately equal to the total energy E_{total}, the second term on the right-hand side of the second equation in (2.39) can be ignored and the particle momentum is $p = W_{\mathrm{kin}}/c$. With (2.42), we obtain a maximum Larmor radius of

$$r_{\mathrm{L}} = \frac{p}{eB} = \frac{W_{\mathrm{kin}}}{ceB} = 10^{21} \text{ m} . \tag{2.43}$$

This is approximately the size of the Milky Way.   □

Excursion 2. *Local Gyration Radius.* The example above points to a general problem, already mentioned in connection with Table 2.2: for a particle with pitch angle $\alpha = 90°$, a closed gyro-orbit with a constant r_{L} results only if the magnetic field is homogeneous across the particle orbit. This is certainly not true for the Milky Way, and it is also not true for the path of cosmic rays in the magnetosphere. In these cases, a different approach is required. Again, the particle motion is decomposed into motions parallel and perpendicular to the field, and the equation of motion is separated into two parts,

$$\frac{\mathrm{d}v_{\parallel}}{\mathrm{d}t} = 0 \quad \text{and} \quad \frac{\mathrm{d}\boldsymbol{v}_{\perp}}{\mathrm{d}t} = \frac{q}{m} \left(\boldsymbol{v}_{\perp} \times \boldsymbol{B} \right) . \tag{2.44}$$

We assume a stationary magnetic field, $\mathrm{d}\boldsymbol{B}/\mathrm{d}t = 0$; the direction parallel to the field lines is given by the tangential vector $\boldsymbol{t}_{\mathrm{B}} = \boldsymbol{B}/B$. Instead of a gyration radius for the entire orbit, we now can determine a "local gyration radius" which gives the local curvature of the particle path; see Fig. 2.1.

The equation of motion gives us a closed gyro-orbit for $\boldsymbol{B} = \mathrm{const}$ over the entire orbit. For $\boldsymbol{B} = \boldsymbol{B}(\boldsymbol{r}, t)$ it gives us the local Lorentz-force from which we obtain the local gyro-radius. Locally, its magnitude is the same as for a constant field with the local value of $\boldsymbol{B}$. We can derive this also graphically: During a time interval δt the radius of curvature r_{c} changes in accordance

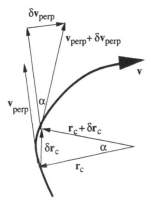

Fig. 2.1. Particle orbit in a stationary but non-homogeneous magnetic field

with $\delta r_c = v_\perp \, \delta t$, while the perpendicular speed changes in accordance with $\delta v_\perp = (q/m) \, (v_\perp \times B) \, \delta t$. Since the triangles in Fig. 2.1 are similar, we obtain $\delta r_c/r_c = \delta v_\perp/v_\perp$. Solving for r_c gives the curvature radius or local Larmor radius

$$r_{\mathrm{lL}} = \frac{mv_\perp}{|q|B} \tag{2.45}$$

or, in vector form, where $t_v = v/v$ is the tangent vector to the velocity,

$$r_{\mathrm{lL}} = -\frac{mv_\perp}{|q|B} \, (t_v \times t_B) \ . \tag{2.46}$$

The equation gives essentially the same result as (2.32). The only difference is that now r_{lL} depends on position: $r_{\mathrm{lL}} = r_{\mathrm{lL}}(r)$ because $B = B(r)$. Thus r_{lL} and r_{lL} both vary along the particle orbit, while in (2.32) r_{L} is constant along the particle path.   □

2.3 Drifts of Particles in Electromagnetic Fields

Particle drifts in electromagnetic fields result from changes in the Larmor radius during one gyration, either due to changes in particle speed (as in $E \times B$ and in the gravitational drift) or in the magnetic field (as in the gradient drift). For particles with pitch angle $0°$ the drifts vanish, except for the curvature drift, because the latter arises from the field-parallel motion.

2.3.1 The Concept of the Guiding Center

The concept of the guiding center, introduced in the 1940s by H. Alfvén, separates the motion v of a particle into motions $v_\|$ parallel and $v_\perp$ perpendicular to the field. The latter can consist of a drift v_D and a gyration ω:

$$v = v_\| + v_\perp = v_\| + v_D + \omega = v_{\mathrm{gc}} + \omega_c \tag{2.47}$$

with v_{gc} being the motion of the guiding center. Thus the motions v_{gc} of the guiding center and the gyration ω_c around the magnetic field are decoupled. If we follow the particle for a long time, the gyration is of minor importance. In some sense it is averaged out, and the particle motion is described by the motion v_{gc} of the guiding center, consisting of a field-parallel motion and a drift. The particle always is within a gyro-radius of this position.

2.3.2 Crossed Magnetic and Electric Fields: $E \times B$ Drift

Assume an electric field E perpendicular to the magnetic field B, both homogeneous and constant in time. The magnetic field forces the particle into a gyration around the field line. During half of its orbit, the particle motion

Table 2.2. Drift in various types of fields

$\vec{B}$ upward through the paper	Charge positive	negative	eq.
B homogeneous			(2.32)
E homogeneous	$\vec{E}$ ... $\vec{v}_D$	$\vec{E}$ ... $\vec{v}_D$	(2.55)
Gravitation	$\vec{g}$ ... $\vec{v}_D$	$\vec{g}$ ... $\vec{v}_D$	(2.57)
Inhomogeneous B $\uparrow$grad $|\vec{B}|$	$\uparrow$ ... $\vec{v}_D$	$\uparrow$ ... $\vec{v}_D$	(2.58)

has a component parallel to the electric field, during the other half it is anti-parallel. Therefore the particle alternately is accelerated and decelerated.

Let us now look at this motion in detail. The second row of Table 2.2 gives the motion of particles in a magnetic field pointing upwards through the paper and an electric field pointing downwards. An electron (right column) starts in the upper right corner and is forced into a downward motion by the magnetic field. Since this motion is parallel to the electric field, the electron decelerates and its gyro-radius decreases. At the lowest point of its orbit the electron therefore is shifted towards the left with respect to its starting point. In the upward part of its gyro-orbit the electron moves anti-parallel to the field. Thus it is accelerated and its Larmor radius increases, shifting the electron farther to the left. This $\boldsymbol{E} \times \boldsymbol{B}$ drift results from a continuous change of the particle's gyro-radius and is perpendicular to both the electric and magnetic fields. Therefore, in spite of the presence of an electric field the particle does not gain energy. For a particle with positive charge, the direction of the drift is the same: its gyro-motion is opposite to that of an electron, as are the parts of its orbit with the smallest and largest gyro-radii. Since electrons and protons drift into the same direction, no current results.

To derive a quantitative description of the $\boldsymbol{E} \times \boldsymbol{B}$-drift we substitute

$$\boldsymbol{v} = \boldsymbol{w} + \frac{\boldsymbol{E} \times \boldsymbol{B}}{B^2} \ . \tag{2.48}$$

This corresponds to the transformation into a frame of reference moving with velocity $\boldsymbol{E} \times \boldsymbol{B}/B^2$. It is $\dot{\boldsymbol{v}} = \dot{\boldsymbol{w}}$ because both the electric and magnetic fields are constant. The equation of motion therefore reads

$$m\dot{\boldsymbol{w}} = m\dot{\boldsymbol{v}} = q\boldsymbol{E} + q\boldsymbol{w} \times \boldsymbol{B} + \frac{q}{B^2}(\boldsymbol{E} \times \boldsymbol{B}) \times \boldsymbol{B} \ . \tag{2.49}$$

Because $\boldsymbol{E}$ is perpendicular to $\boldsymbol{B}$, the double cross-product in the last term can be written as $-\boldsymbol{E}B^2$ (A.20). Therefore, the first and last terms on the right-hand side cancel and the equation of motion reduces to

$$m\dot{\boldsymbol{w}} = q\boldsymbol{w} \times \boldsymbol{B} \ . \tag{2.50}$$

Now $\boldsymbol{w}$ fulfills (2.24). In the new reference system the particle motion therefore is a gyration and the motion of this system gives the drift velocity:

$$v_{\boldsymbol{E} \times \boldsymbol{B}} = \frac{\boldsymbol{E} \times \boldsymbol{B}}{B^2} \ . \tag{2.51}$$

The direction and size of the drift depend on the fields alone: particle properties such as mass, charge or velocity do not enter into the drift.

We can also derive a general equation for the drift velocity in the presence of a general force $\boldsymbol{F}$ perpendicular to the magnetic field by substituting the electric field $\boldsymbol{E}$ by the general force $\boldsymbol{F}/q$ in (2.48) and (2.49):

$$v_{\mathrm{F}} = \frac{\boldsymbol{F} \times \boldsymbol{B}}{qB^2} = \frac{1}{\omega_{\mathrm{c}}}\left(\frac{\boldsymbol{F}}{m} \times \frac{\boldsymbol{B}}{B}\right) = \frac{\boldsymbol{F}}{\omega_{\mathrm{c}}m} \times \boldsymbol{t}_{\mathrm{B}} \ . \tag{2.52}$$

In the above example $\boldsymbol{F} = q\boldsymbol{E}$ has been used.

Example 3. Wien filter. A magnetic field $B = 5 \times 10^{-4}$ T is perpendicular to an electric field $E = 1000$ V/m. This configuration provides a simple example of $\boldsymbol{E} \times \boldsymbol{B}$ drift. Since the two fields are perpendicular, we obtain from (2.51) a drift velocity $v_{\boldsymbol{E} \times \boldsymbol{B}} = (EB)/B^2 = E/B = 2 \times 10^6$ m/s. The drift velocity is perpendicular to both the electric and the magnetic field. An electron approaches perpendicular to both fields. On hitting the field combination, the electron will start to gyrate around the magnetic field and drift along its original direction of motion. The size of the gyro-orbit depends on the electron speed v_{e}. For $v_{\mathrm{e}} = v_{\boldsymbol{E} \times \boldsymbol{B}}$ the electron moves along a straight line. This can be understood as follows: if the electron moves along a straight line at constant speed, no force acts on the electron. Thus the forces exerted by the electric and the magnetic field must be equal: $eE = ev_{\mathrm{e}}B$ or $v_{\mathrm{e}} = E/B$, which is the drift velocity derived above. $\square$

2.3.3 Magnetic and Gravitational Fields

Now consider a gravitational field perpendicular to a homogeneous magnetic field (third row in Table 2.2). As in the previous case, the Larmor radius changes continuously. But here the external force does not depend on the charge sign, and the points of the largest and smallest gyro-radii are the same for both positive and negative particles. Because particles with different charge signs have opposite directions of gyration, they drift into opposite directions. The direction of drift therefore depends on the charge q, and its size depends on the particle mass m, because the external force $\boldsymbol{F} = m\boldsymbol{g}$ depends on the particle mass. Equation (2.52) yields for the drift velocity

$$\boldsymbol{v}_g = \frac{m}{q}\frac{\boldsymbol{g} \times \boldsymbol{B}}{B^2} \, . \tag{2.53}$$

A gravitational field perpendicular to a magnetic field therefore allows the separation of particles with positive and negative charges. This drift leads to a current. In addition, the magnetic field prevents the particles from "falling down". There is no net acceleration along $\boldsymbol{g}$ and the potential and kinetic energies averaged over a gyration both are constant.

Example 4. A proton with a kinetic energy of 1 keV (and also a 10 keV electron) gyrates in the equatorial plane of the terrestrial magnetic field at a radial distance of five Earth radii from the center of the Earth. All its kinetic energy is in the gyration. The equatorial magnetic field at the surface of the Earth is 3.11×10^{-5} T; it falls off with radial distance as r^{-3}. Thus the local field at the proton orbit is $B = 2.5 \times 10^{-7}$ T. The proton speed is $v_{\mathrm{p}} = \sqrt{2W_{\mathrm{kin}}/m} = 4.4 \times 10^5$ m/s (and the electron speed is $v_{\mathrm{e}} = 5.9 \times 10^7$ m/s); its gyro-radius according to (2.32) is $r_{\mathrm{L,p}} = 1.8 \times 10^4$ m ($r_{\mathrm{L,e}} = 1.4 \times 10^3$ m/s), which is still small compared with the scales of the system; for instance, the drift path around the Earth has a length of $l_{\mathrm{drift}} = 2\pi r = 50\pi r_{\mathrm{E}} = 2 \times 10^8$ m. The gravitational acceleration scales with r^2, and thus at the particles position it is only $g/25$, g being the gravitational acceleration at the surface of Earth. Since at and above the equator the magnetic field is parallel to the surface, the gravitational field is perpendicular to the magnetic field. The drift speed from (2.53) is then $v_g = 1.6$ cm/s perpendicular to both fields: the particle drifts along a circle in the equatorial plane. Since the geomagnetic field has its south pole close to the geographic north, the field in the equatorial plane is directed northward. Thus a proton drifts to the west while an electron drifts eastward. The $\boldsymbol{g} \times \boldsymbol{B}$ drift thus would give the particle such a small speed that it would take 1.25×10^{10} s or about 400 years to drift around the entire Earth. Note that the drift speed depends only on the particle mass. Thus all protons drift with the same speed, independent of their energy. To be more precise, almost all protons: if the energy becomes too large, the gyration radius of the particle may become so large that it either hits the atmosphere and is absorbed during an interaction or suddenly finds itself in interplanetary space and escapes. The same argument holds also for all other particle species; only their drift speeds must be scaled by the ratio of their mass to the proton mass. The drift speed of an electron is therefore smaller by a factor of 1836 than that of the proton. □

2.3.4 Inhomogeneous Magnetic Fields

In an inhomogeneous magnetic field (bottom row in Table 2.2) particles with positive and negative charges also drift in opposite directions. In contrast to the previous example, here the particle speed stays constant during the gyration but the gyro-radius increases/decreases with decreasing/increasing field strength. This change in gyro-radius is independent of the charge sign, leading to drifts of electrons and protons in opposite directions. The resulting gradient drift is

$$\boldsymbol{v}_{\nabla B} = \frac{\mu}{qB^2} \, \boldsymbol{B} \times \nabla B = \pm \frac{v_\perp r_L}{2B^2} \, \boldsymbol{B} \times \nabla B \; ; \tag{2.54}$$

the $\pm$ sign reflects the charge dependence of the drift direction. In contrast to the gravitational drift discussed above, here the drift speed depends on the particles kinetic energy and thus on both speed and mass.

A very efficient drift develops in the configuration of two opposing magnetic fields. Imagine the guiding center of the particle orbit on the neutral line between the fields. The particle starts its gyration in the upper half of the field, but after crossing the neutral line, the direction of gyration is reversed. Therefore, instead of a closed circular orbit two semicircles result, leading to displacement by 4 Larmor radii during one gyration (see Fig. 2.2). In interplanetary space such a drift takes place along the heliospheric current sheet and contributes to the modulation of galactic cosmic rays (Sect. 7.7).

In the magnetosphere this drift leads to the ring current and the motion of particles trapped in the radiation belts (Sect. 8.7.1).

 Example 5. Let us briefly return to example 4, where we saw that the $\boldsymbol{g} \times \boldsymbol{B}$ drift is extremely slow. Since the magnetic field decreases as r^{-3}, during its gyration the particle scans regions of different magnetic field strength, and thus a gradient drift results. The gradient of the magnetic field has the same direction as the gravitational acceleration, and thus the gradient drift is in the same direction as the $\boldsymbol{g} \times \boldsymbol{B}$ drift. Although the magnetic field is roughly a dipole field, in this special case it can be treated as spherically symmetric since the particle gyrates in the equatorial plane and therefore is not influenced by the latitudinal variation of B. From (A.43) we obtain the gradient of the magnetic field $B(r) = B_0 \, (r_0/r)^3$ as $\nabla B = (-B/r, 0, 0)$. The cross product in (2.54) then gives $|\boldsymbol{B} \times \nabla B| = B^2/r$ and the drift speed

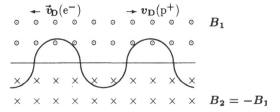

Fig. 2.2. An efficient gradient drift occurs in adjacent fields of opposite polarity, for instance along the heliospheric current sheet

becomes $v_D = r_L v_\perp/(2r) = 220$ km/s for the proton. For the 10 keV electron we get a drift speed of 259 km/s. Thus drift frequencies are of the order of mHz (see also Fig. 8.47). □

2.3.5 Curvature Drift

Imagine a homogeneous magnetic field with the lines of force are curved with a radius r_c. In a vacuum such a field would obey Maxwell's equations only if combined with a magnetic field gradient. The net drift therefore would be the sum of the curvature drift and the gradient drift. To derive the curvature drift alone, let us start from a simplified configuration without a magnetic field gradient. The drift arises from the centrifugal force $\boldsymbol{F}_{cf}$ and thus is determined by $v_\parallel$ and not by $v_\perp$ as in the previous examples. Inserting the centrifugal force $\boldsymbol{F}_{cf} = mv_\parallel^2 \boldsymbol{e}_r/r_c = mv_\parallel^2 \boldsymbol{r}_c/r_c^2$ into (2.52) yields

$$v_R = \frac{mv_\parallel^2}{qB^2} \frac{\boldsymbol{r}_c \times \boldsymbol{B}}{r_c^2} . \tag{2.55}$$

The curvature drift depends on the charge, mass, and speed of the particle.

In a real field, a field gradient exists in addition to the curvature. Ampére's law (2.7) in a vacuum without electric current gives $\nabla \times \boldsymbol{B} = 0$. In cylindrical coordinates (see Sect. A.3.2 and Fig. 2.3), $\boldsymbol{B}$ has only one component in θ, ∇B only one in r, and $\nabla \times \boldsymbol{B}$ only one in z, given as

$$(\nabla \times \boldsymbol{B})_z = \frac{1}{r} \frac{\partial(rB_\theta)}{\partial r} = 0 . \tag{2.56}$$

This is zero because B_θ is proportional to $1/r$. Then $|\boldsymbol{B}|$ is proportional to $1/r_c$ and it is $\nabla|\boldsymbol{B}|/|\boldsymbol{B}| = -\boldsymbol{r}_c/r_c^2$. The gradient drift (2.54) then reads

$$\boldsymbol{v}_{\nabla B} = \frac{m}{2q} v_\perp^2 \frac{\boldsymbol{r}_c \times \boldsymbol{B}}{r_c^2 B^2} . \tag{2.57}$$

Combined with (2.55), the drift in a curved magnetic field is

$$\boldsymbol{v}_R + \boldsymbol{v}_{\nabla B} = \frac{m}{q} \frac{\boldsymbol{r}_c \times \boldsymbol{B}}{r_c^2 B^2} \left(v_\parallel^2 + \frac{1}{2}v_\perp^2\right) = \frac{v_\parallel^2 + \frac{1}{2}v_\perp^2}{\omega_c} \boldsymbol{e}_B \times \frac{\nabla B}{|\boldsymbol{B}|} . \tag{2.58}$$

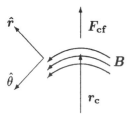

Fig. 2.3. Curved magnetic field and coordinate system

This equation has an important implication: if particles are trapped inside a magnetic field torus, the best and finest adjustments in temperature and field cannot prevent the particles from drifting across the field lines and out of the torus sooner or later. This is one of the fundamental problems in fusion research. Instead of a simple torus, extremely complex field configurations are required, as for instance in the stellarator Wendelstein 7-X (see e.g. www.ipp-garching.mpg.de/eng/pr/publikationen/broschuere. engl.pdf or www.ipp-garching.mpg.de/eng/pr/exptypen/stellarator/ pr_exp_ste.html).

Example 6. A proton plasma with a temperature of 10 MeV (corresponding to a speed of $v = \sqrt{2W_{\text{kin}}/m} = 4.3 \times 10^7$ m/s) is confined by a homogeneous magnetic field inside a torus of diameter 1 m. The dimensions of the torus also give the diameter of the gyro-orbit. Thus a magnetic field $B = mv_\perp/(qr_{\text{L}}) = 0.9$ T is required to keep the proton inside the torus. According to (2.55), the curvature drift is then $v_{\text{r}} = mv^2 r_{\text{L}}/(qB) = 0.24$ m/s tangential to the particle path.   □

2.3.6 Drifts Combined with Changes in Particle Energy

The drifts discussed so far have been associated with acceleration in the sense of a change in the direction of motion but not in average speed. Certain combinations of fields, however, can lead to changes in average speed and therefore also in particle energy. Figure 2.4 shows the drift of a particle in a crossed electric (pointing upward in the drawing plane) and magnetic (pointing upward out of the drawing plane) field with a gradient perpendicular to both the magnetic and electric fields. The motion under the influence of two combined fields has been discussed above and is indicated by thin lines: for a particle with positive charge the $\boldsymbol{E} \times \boldsymbol{B}$ drift leads to a horizontal motion towards the right while the grad$B$ drift leads to an upward motion. A combination of both motions gives the $\boldsymbol{E}$grad$B$ drift oblique to the fields and to the magnetic field gradient. Owing to this drift, the particle moves from one potential to another, gaining energy. Note that the use of a gravitational field instead of the electric field would yield similar results.

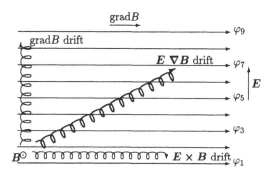

Fig. 2.4. One example of a configuration of fields in which the drift of a charged particle leads to changes in particle energy

2.3.7 Drift Currents in Plasmas

Some of the drifts discussed above lead to the separation of positive and negative charges, giving rise to a drift current. In a plasma, instead of single particles we have to consider k different particle species with number densities n_k and masses m_k. With the general equation for the drift velocity (2.52) we can derive a drift current:

$$j_D = \frac{\sum_k n_k F_k}{B^2} \times B = \sum_k n_k q\, v_{Dk} \, . \tag{2.59}$$

The drift current due to the gradient drift, for instance, is given as

$$j_{D\nabla B} = \sum_k \frac{n_k v_{k\perp}^2}{2B^3} \left(B \times \nabla B\right) \, . \tag{2.60}$$

This current results from the inhomogeneity of the field and leads to a charge separation. An example of a natural drift current is the ring current in the magnetosphere (Sect. 8.3.3).

Example 7. In example 5, we calculated the drift speeds of 1 keV protons and 10 keV electrons. In the radiation belts, both have a number density $n = 10^7$ m^{-3}. With the drift speeds from example 5, we can determine the ring current densities by using (2.60) and $j = 7.7 \times 10^{-7}$ A/m^2. □

2.4 Adiabatic Invariants

So far we have only considered static and homogeneous fields. In weakly and slowly varying fields, the concept of adiabatic invariants provides a powerful tool to describe the periodic motions of particles. A simple analogy is the mathematical pendulum: if its length increases only weakly during one swing, the ratio of the pendulum's energy and frequency is a constant of the motion, called an adiabatic invariant. For charged particles in a magnetic field, three types of motion can be identified:

1. The field changes only slowly during one gyration:

$$\frac{1}{B}\frac{\partial B}{\partial t} \ll \frac{\omega_c}{2\pi} \, . \tag{2.61}$$

2. The field varies only weakly on a scale comparable with the distance traveled along the field by the particle during one gyration (bounce motion, longitudinal oscillation):

$$\frac{\nabla B_{\|}}{B} \ll \frac{\omega_c}{2\pi v_{\|}} \, . \tag{2.62}$$

3. The field varies only weakly in the area encircled by the particle during the gyration or drift motion:

$$\frac{\nabla B_\perp}{B} \ll \frac{\omega_c}{2\pi v_\perp} \quad \text{or} \quad \frac{\nabla B_\perp}{B} \ll \frac{\omega_c}{2\pi v_D} . \tag{2.63}$$

If the particle motion is described by a pair of variables (p_i, q_i) which are generalized momentums and coordinates, for each periodic coordinate q_i the action integral

$$J_i = \oint p_i dq_i , \tag{2.64}$$

integrated over a complete cycle of motion, is approximately an invariant or constant of the motion, provided changes in the variables occur slowly compared with the relevant periods of the system and the rate of change is almost constant (for a proof of this statement, see, for example, [192, 303, 307, 382, 508]). Thus a system can change from one state of motion into another and still have the same action integral.

2.4.1 First Adiabatic Invariant: The Magnetic Moment

The first adiabatic invariant states that in a slowly varying magnetic field the magnetic moment $\mu = W_\perp/B$ is almost constant. It finds applications in magnetic mirrors and bottles. We can derive it by inserting the generalized momentum $p = mv_\perp$ and the generalized coordinate $q = r_L\psi$ (with ψ being the azimuthal angle along the gyro-orbit) into the action integral (2.64)

$$J_1 = \oint p_1 \, dq_1 = \oint mv_\perp r_L \, d\psi = 2\pi m r_L v_\perp . \tag{2.65}$$

With (2.32) and (2.35) this yields

$$J_1 = 4\pi \frac{m}{|q|} \mu = \text{const} \tag{2.66}$$

under the tacit assumption that $\omega/\omega_c \ll 1$ with ω characterizing the change in $\boldsymbol{B}$. Thus (2.66) states: as long as $m/|q|$ is constant, the magnetic moment is constant, too.

A less formal but more illustrative proof, without using the action integral, starts from the motion of a particle in a magnetic field that varies slowly in space or time. Let us choose the latter attempt, i.e. $\partial \boldsymbol{B}/\partial t \neq 0$. Then, according to Faraday's law (2.4), a rotational electric field arises with a component parallel to the orbit of a gyrating particle (or completely parallel to the motion of a gyrating particle if the latter has pitch angle zero). If we further assume $v_\parallel$ to be zero, the particle speed is $v = v_\perp = d\boldsymbol{l}/dt$ with $d\boldsymbol{l}$ being a small element of the particle path. Multiplying the equation of motion (2.23) by the particle speed $\boldsymbol{v}_\perp$, we derive the temporal change in kinetic energy:

$$\frac{d}{dt}\left(\frac{mv^2}{2}\right) = q\boldsymbol{E}\cdot\boldsymbol{v} = q\boldsymbol{E}\cdot\frac{d\boldsymbol{l}}{dt} \ . \tag{2.67}$$

Integration over a gyro-orbit gives the energy change during a gyration

$$\delta W_{\text{kin}} = \int\limits_0^{2\pi/\omega} q\boldsymbol{E}\,\frac{d\boldsymbol{l}}{dt}\,dt \ . \tag{2.68}$$

According to our assumptions, the field changes only weakly during one gyration. Therefore, the particle orbit still is nearly circular and the integration in time can be substituted by a line integral along the particle path:

$$\delta W_{\text{kin}} = \oint q\boldsymbol{E}\cdot d\boldsymbol{l} \ . \tag{2.69}$$

With Stokes' theorem (A.39) and Faraday's law (2.4) we obtain

$$\delta W_{\text{kin}} = q\int\limits_S (\nabla\times\boldsymbol{E})\cdot d\boldsymbol{S} = -q\int\limits_S \frac{\partial\boldsymbol{B}}{\partial t}\cdot d\boldsymbol{S} \ , \tag{2.70}$$

with $\boldsymbol{S}$ being the surface enclosed by the Larmor orbit, its direction given by the right-hand rule. The integral is positive for negatively charged particles and negative for positively charged ones. It then can be written as

$$\delta W_{\text{kin}} = \pm q\frac{\partial B}{\partial t}\,\pi r_{\text{L}}^2 \ . \tag{2.71}$$

With the Larmor radius expressed in terms of the cyclotron frequency this is

$$\delta W_{\text{kin}} = \pm\pi q\frac{\partial B}{\partial t}\frac{v_{\perp}^2}{\omega_{\text{c}}}\frac{m}{\pm qB} = \frac{W_{\text{kin}}}{B}\cdot\frac{2\pi\partial B/\partial t}{\omega_{\text{c}}} \ . \tag{2.72}$$

The first part of the right-hand side gives the magnetic moment, the second part the variation $\delta\boldsymbol{B}$ of the magnetic field during one gyration: $\delta W_{\text{kin}} = \mu\delta B$. Inserting the definition of the magnetic moment (2.35) into the left-hand side, we get $\delta(\mu B) = \mu\delta B + B\delta\mu = \mu\delta B$, which implies

$$B\delta\mu = 0 \ . \tag{2.73}$$

Because $B \neq 0$, this equation gives the first adiabatic invariant: in a slowly varying magnetic field the magnetic moment is constant.

Changes in the magnetic field result in variations in the Larmor radius. Thus the question arises: does the magnetic flux through a Larmor orbit change in a slowly varying field? With the definition of the magnetic flux (2.5) and the area $\boldsymbol{S}$ enclosed by the gyro-orbit we can write

$$\Phi = \int \boldsymbol{B}\cdot d\boldsymbol{o} = \pi B\frac{v_{\perp}^2}{\omega_{\text{c}}^2} \ . \tag{2.74}$$

Inserting (2.28) yields

$$\Phi = \frac{2\pi m}{q^2} \mu .$$ (2.75)

Thus as long as μ is an invariant of the motion, the magnetic flux Φ through a gyro-orbit is constant.

2.4.2 Magnetic Mirrors and Bottles

Magnetic mirrors and bottles are applications of the first adiabatic invariant. In its top left panel Fig. 2.5 shows the configuration of a magnetic mirror together with the path of a particle. The panel below shows the dependence of $v_\parallel$, $v_\perp$, and α on the particle's location. The panel on the right illustrates the restoring force at the mirror point.

Let us start with a particle on the left-hand border of the graph with initial speeds $v_{\parallel,1}$ and $v_{\perp,1}$ parallel and perpendicular to the magnetic field and initial pitch angle α_1. The particle gyrates around the field line, and its guiding center moves to the right. Since the electric field is zero, the particle's kinetic energy is constant:

$$\tfrac{1}{2}mv^2 = \tfrac{1}{2}m(v_\parallel^2 + v_\perp^2) = \text{const} .$$ (2.76)

The kinetic energy perpendicular to the magnetic field can be expressed by the magnetic moment μ (2.35):

$$\tfrac{1}{2}mv^2 = \tfrac{1}{2}mv_\parallel^2 + \mu B .$$ (2.77)

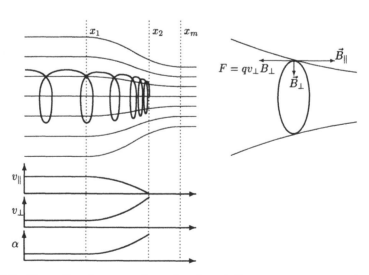

Fig. 2.5. Magnetic mirror: configuration (*top left*), variations in particle speed parallel and perpendicular to the magnetic field and pitch angle α with position (*bottom left*), and restoring force at the mirror point (*right*)

The first term on the right-hand side can be interpreted as the drift energy or the kinetic energy of the guiding center, while the second term μB is the gyration energy. Since μ is invariant, an increase in B has to be compensated for by a decrease in the drift energy until $v_\parallel$ becomes zero. At this mirror point the energy conservation yields

$$\mu B_{\mathrm{mp}} = \tfrac{1}{2}mv^2 = \tfrac{1}{2}mv_{\perp\mathrm{mp}}^2 . \tag{2.78}$$

Thus, at the mirror point the drift energy entirely is transformed into gyration energy, which therefore is the particle's total kinetic energy. The guiding center has come to a standstill and eventually will be reflected back towards the diverging field. The right panel in Fig. 2.5 illustrates the origin of the restoring force: at the mirror point the magnetic field is inhomogeneous. Thus, the plane of gyration is not perpendicular to the field and $\boldsymbol{B}$ has a component $\boldsymbol{B}_\perp$ in this plane, leading to the restoring force

$$\boldsymbol{F}_{\mathrm{r}} = q\boldsymbol{v} \times \boldsymbol{B} = qv_\perp B_\perp \boldsymbol{e}_x . \tag{2.79}$$

The particle is pushed back into regions of decreasing field strength and its pitch angle decreases as gyration energy is transferred back into drift energy. Note that (2.79) also can be written as $\boldsymbol{F} = \boldsymbol{\mu} \cdot \nabla \boldsymbol{B}$ which describes the force an inhomogeneous magnetic field exerts on a dipole magnetic moment.

The location of the mirror point B_{mp} depends on the initial pitch angle α_1 of the particle. If α_1 is zero, the magnetic moment is zero too, and the particle's total kinetic energy is drift energy. An increase in the magnetic field strength does not transform drift energy into gyration energy and the particle traverses the magnetic mirror. If, on the other hand, α_1 is 90°, all the particle's energy is contained in the gyration and the guiding center already is at a standstill. For values of α_1 in between these two extremes either reflection occurs at a point $x_{\mathrm{mp}}(\alpha_1)$ or the particle is transmitted if the increase in the magnetic field strength is not sufficient to convert all the drift energy into gyration energy. Let us now determine whether a particle will be reflected or transmitted. The constancy of the magnetic moment implies that the ratio of the energy of gyration and the magnetic field strength is constant. Thus for any two points in the magnetic field we have

$$\mu = \frac{mv_{\perp 1}^2}{2B_1} = \frac{mv_{\perp 2}^2}{2B_2} \tag{2.80}$$

or, taking the pitch angle into consideration,

$$\frac{v_1^2 \sin^2 \alpha_1}{B_1} = \frac{v_2^2 \sin^2 \alpha_2}{B_2} . \tag{2.81}$$

The kinetic energy is constant; thus it is $v_1 = v_2$ and the quantity

$$\frac{2\mu}{mv^2} = \frac{\sin^2 \alpha_1}{B_1} = \frac{\sin^2 \alpha_2}{B_2} \tag{2.82}$$

becomes an invariant of the motion. At the mirror point, the particle's pitch angle α_{mp} is $90°$. Thus reflection at the position B_{mp} requires an initial pitch angle at B_1 of

$$\sin^2 \alpha_1^r = \frac{B_1}{B_{mp}} = \frac{1}{R_{mp}} \quad \text{or} \quad \alpha_1^r = \arcsin \sqrt{\frac{B_1}{B_{mp}}} = \arcsin \sqrt{\frac{1}{R_{mp}}}. \quad (2.83)$$

Here R_{mp} is the mirror ratio. Particles with an initial pitch angle α_1^r are reflected exactly at B_{mp}; particles with larger α_1 are reflected earlier and particles with smaller α_1 pass through the mirror point. Thus (2.83) defines the boundary of a region in velocity space in the shape of a cone, called the loss cone (see Fig. 2.6): particles inside this cone are not confined by the magnetic mirror. This loss cone is an important concept to describe the dynamics of radiation belt particle populations. Note that the loss cone depends on pitch angle only, and does not depend on other particle parameters, such as mass, speed or charge.

Example 8. Assume an isotropic distribution (that is, the pitch angles are distributed uniformly) of 10 keV electrons injected on a field line at the equator, and $L_0 = 5$ Earth radii from the Earth's center. The magnetic field varies as

$$B(L_0, \Phi) = \frac{B_E}{L_0^3} \frac{\sqrt{1 + 3 \sin^2 \Phi}}{\cos^6 \Phi}, \quad (2.84)$$

where Φ is the geomagnetic latitude and $B_E = 3.11 \times 10^{-5}$ T is the equatorial magnetic field strength at the surface. The equation of the magnetic field line is $L = L_0 \cos^2 \Phi$. From the conservation of the magnetic moment, we can determine the number of particles reflected at geomagnetic latitudes of $30°$ and $60°$. We need only the ratio between the magnetic field strengths at the injection site and at the reflection point. From (2.84) we obtain $B_{refl}/B_{eq} = \sqrt{1 + \sin^2 \Phi_{refl}}/\cos^6 \Phi_{refl}$, which gives a magnetic field ratio of 2.65 for $\Phi = 30°$ and 84.7 for $\Phi = 60°$. From (2.83) we then find that all particles with a pitch angle larger than $38°$ and $6.3°$ are deflected at geomagnetic latitudes below $30°$ and $60°$, respectively (that is, 48% and 83%, respectively, of the initial population). From the equation of the field line, we find that it intersects the surface of the Earth ($L = 1$) at a geomagnetic latitude of $63°$.   □

Example 9. Magnetic pumping is a process in which an adiabatic invariant is used to accelerate particles. To illustrate the idea, let us start from an

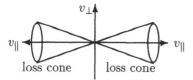

Fig. 2.6. Definition of the loss cone

isotropic plasma gyrating in a homogeneous magnetic field. The plasma temperatures and thus the particle energies parallel and perpendicular to the field are the same: $W_{\mathrm{kin},\|,\mathrm{o}} = W_{\mathrm{kin},\perp,\mathrm{o}} = W_0$. Let us now increase the magnetic field slowly by a factor of two, such that the concept of adiabatic invariants can be applied but fast enough to prevent an exchange between parallel and perpendicular energy. The energy parallel to the field remains unchanged, but since the magnetic moment is conserved, the perpendicular kinetic energy increases as B increases: $W_{\mathrm{kin},\perp} \sim B$. If we wait for a sufficiently long time to allow temperature exchange between the parallel and perpendicular motions, we again have an isotropic plasma, now with $W_{\mathrm{kin},\|,1} = W_{\mathrm{kin},\perp,1} = 1.5\,W_0$ and thus $W_{\mathrm{kin},\|,1} = 1.5W_{\mathrm{kin},\|,0}$ and $W_{\mathrm{kin},\perp,1} = 1.5W_{\mathrm{kin},\perp,0}$. We can now allow the magnetic field to relax to its original value with the same speed as before. Again, the parallel kinetic energy remains constant while that of the motion perpendicular to the field is reduced by a factor of two: $W_{\mathrm{kin},\|,1} = 0.75W_{\mathrm{kin},\|,0}$ and $W_{\mathrm{kin},\perp,1} = 1.5W_{\mathrm{kin},\perp,0}$. The total energy is then $W = 1.25W_0$, that is the plasma has gained energy during this process, which can be repeated for further energy gain. □

2.4.3 Second Adiabatic Invariant: Longitudinal Invariant

Two magnetic mirrors combined give a magnetic bottle (see Fig. 2.7): particles are confined due to repeated reflection between the mirrors. This is a simple configuration in so far as the field is stationary, rotation-free and has a rotational symmetry around the field line. The second adiabatic invariant is related to the drift motion inside the bottle and the distance between the mirrors. The longitudinal invariant can be derived from (2.64) with the momentum $p_2 = mv_\|$ and the distance s along the field as the spatial coordinate:

$$J_2 = \int_{s_1}^{s_2} mv_\| \, \mathrm{d}s = \text{const}. \tag{2.85}$$

With (2.77) this can be written as

$$J_2 = \int_{s_1}^{s_2} m\sqrt{v^2 - \frac{2\mu B}{m}}\,\mathrm{d}s. \tag{2.86}$$

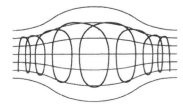

Fig. 2.7. Magnetic bottle

The second adiabatic invariant leads to an efficient mechanism for particle acceleration: for the acceleration of galactic cosmic rays, Fermi proposed a configuration of two converging magnetic mirrors (first-order Fermi effect [150, 151]). In terms of energy, the reflection of a particle at a fixed magnetic mirror is equivalent to a ball bouncing off a wall. Thus in a magnetic bottle with two fixed mirrors the particle oscillates between these mirrors without changes in total energy. The interaction with a moving magnetic mirror, however, is equivalent to the reflection off a moving wall. Depending on the relative speeds between particle and mirror, an energy gain or loss results: head-on collisions lead to an energy gain; if the particle and mirror move in the same direction, an energy loss results. According to (2.85) in each pair of collisions the total energy gain is determined by the shortening of the distance between the mirrors, independent of whether the particle meets both mirrors head-on or only one of them. A familiar equivalent is the warming of a gas during compression, e.g. inside a tire pump.

The application of the second adiabatic invariant in the Earth's magnetosphere is not concerned with acceleration but with the asymmetry of the field. The dipole field in the inner magnetosphere forms magnetic bottles: particles moving from the equatorial regions towards the poles "see" a converging magnetic field, they move into a magnetic mirror (see Sect. 8.7.1). If their pitch angle is sufficiently large to prevent them from entering the loss cone, they eventually are reflected back, cross the equator and travel towards the other pole. This bouncing motion, however, is not exactly the same as in Fig. 2.7: the field lines are curved to follow the Earth's dipole field and the particle therefore faces a stronger magnetic field during that part of the Larmor orbit closest to Earth than in the other half of the orbit. The resulting gradient drift leads to a guiding center motion around the Earth: electrons drift from west to east, protons from east to west, forming a ring current.

In a symmetric magnetic field, the particles should return to a certain field line after each drift period. But, as we shall see in Chap. 8, the Earth's magnetosphere is neither symmetric nor constant in time. Why then should a particle return to a certain field line after one drift period? The particle's energy is conserved during the motion. The first adiabatic invariant then requires that at the mirror point $|\boldsymbol{B}|$ is constant. If a particle has drifted back to a certain longitude, in principle it can be on a field line in an altitude different from its initial one. This is ruled out by the second adiabatic invariant, which determines the length of the field line between the mirror points. Combination of both invariants requires finding field lines with the same $|\boldsymbol{B}|$ at the reflection point and the same length. But at a fixed longitude this is fulfilled by one field line only. Thus even in an asymmetric field the particle returns to its original field line after each drift period.

2.4.4 Third Adiabatic Invariant: Flux Invariant

The third adiabatic invariant, the flux invariant, is also related to the guiding center drift. It states that the magnetic flux enclosed by the drift orbit is constant. The particle moves on a surface which adjusts itself to variations in B so that the flux enclosed by this surface stays constant. Formally, the third adiabatic invariant can be derived from the action integral (2.64) by using the drift motion to define the generalized momentum $p = mv_D$ and ψ as the azimuthal angle of the particles orbit with radius r:

$$J_3 = \oint mv_D r \mathrm{d}\psi \ . \tag{2.87}$$

Similar to (2.65) we then obtain

$$J_3 = \frac{4\pi m}{|q|} M = \text{const} \ , \tag{2.88}$$

with M being the magnetic moment of the axisymmetric field.

The third adiabatic invariant is also used in the Tokamak geometry of a fusion reactor [262, 558, 559]

2.5 Summary

This chapter has introduced the motion of individual charged particles in electromagnetic fields. The particles are considered as test particles: they do not interact with other particles and their motion does not influence the fields. The main results are: (i) The elementary motion of a charged particle in a homogeneous magnetic field is the gyration around the field line, characterized by the Larmor radius (2.32) and the cyclotron frequency (2.28). If the particle has a velocity component parallel to the field, a helical orbit results. (ii) The pitch angle α describes the relation between the particle's motion parallel and perpendicular to the field. (iii) The motion of a charged particle can be separated into the motion of its guiding center and the gyration around the guiding center. The guiding center motion can consist of a field parallel motion and drifts. (iv) In slowly and weakly varying fields, three adiabatic invariants can be defined, each associated with a typical mode of motion: the first adiabatic invariant (constancy of the magnetic moment) is associated with the gyration, the second (longitudinal invariant) with the field parallel motion of the guiding center, and the third (flux invariant) with the drift of the guiding center. The adiabatic invariants can be applied to the motion of particles in the radiation belts.

Exercises and Problems

2.1. Describe the concept of the guiding center. What is the reason for drifts?

2.2. What is an adiabatic invariant? Describe some examples.

2.3. Derive (2.52) and discuss the conditions under which it can be applied.

2.4. Derive an expression for the gyro-radius and frequency of a relativistic particle. The relativistic momentum is $p = \gamma m_0 v$, where $\gamma = \sqrt{1 - v^2/c^2}$ and m_0 is the rest mass. What is the expression for the magnetic moment of a relativistic particle? Show that the relativistic magnetic moment is conserved.

2.5. How is the magnetic moment defined? Give examples of magnetic moments. Why is the magnetic moment an important physical quantity?

2.6. Show that $\boldsymbol{j}' = \boldsymbol{j}$ in the derivation of Ohm's generalized law (2.14).

2.7. Derive Coulomb's law from (2.2).

2.8. Why does (2.43) give a maximum Larmor radius?

2.9. Solve the equation of motion (2.27).

2.10. Develop a simple (numerical) model for the depletion of the radiation belt. Start with the information from examples 5 and 8, and assume losses to occur at a height of 1.05 Earth radii from the center of the Earth (about 300 km height in the atmosphere) and that during each bounce period an amount n, with n being 50% of the number of particles lost, of the remaining particles are scattered into the loss cone.

2.11. Determine the gyro-radii and frequencies for electrons and protons moving with thermal speeds (see Sect. 5.1.2) in the following fields: (a) the Earth's magnetosphere, with $n_e = n_p = 10^4$ cm^{-3}, $T_e = T_p = 10^3$ K, $B = 10^{-2}$ G; (b) the core of the Sun with $n_e = n_p = 10^{26}$ cm^{-3}, $T_e = T_p = 10^{7.2}$ K, $B = 10^6$ G; (c) the solar corona with $n_e = n_p = 10^8$ cm^{-3}, $T_e = T_p = 10^6$ K, $B = 1$ G; (d) the solar wind with $n_e = n_p = 10$ cm^{-3}, $T_e = T_p = 10^5$ K, $B = 10^{-5}$ G.

2.12. A particle gyrates in a homogeneous magnetic field. (a) Determine the size of a volume V which contains an amount of magnetic energy equal to the particle's kinetic energy. (b) Determine the height of a cylinder with this volume and a base given by the Larmor orbit. (c) Discuss this result.

2.13. In the equatorial plane, the Earth's magnetic field can be described as $B = B_0(R_E/r)^3$ with $B_0 = 0.3$ G, R_E being the Earth's radius, and r being the geocentric distance. Determine the time a particle with pitch angle 90° needs to drift around the Earth in the equatorial plane. What is the meaning of this time? Determine the period for electrons and protons with an energy of 1 keV drifting in a height of $5r_E$ above the center of the Earth. Compare with the drift due to the gravitational field and the period of an uncharged particle (e.g. a satellite) in the same orbit.

2.14. A proton of cosmic radiation is trapped between two magnetic mirrors with $R_m = 5$. Initially, it has an energy of 1 keV and $v_\perp = v_\parallel$ in the meridional plane between the two mirrors. Each mirror moves with $v_m = 10$ km/s towards the other. Draw a sketch of the configuration. Determine the acceleration of the proton. (a) Does the acceleration continue until the mirrors are in contact with each other or does the particle escape? Determine the maximum energy acquired by the particle. Determine the maximum energy for other pitch angles, too. (b) How long does the particle need to acquire maximum energy? (Hint: assume the mirrors to be planes moving with speed v_m and show that the energy gain in each interaction is $2v_m$. How many interactions are required for the particle to acquire maximum speed?)

2.15. The magnetic field of a magnetic mirror varies as $B_z = B_0(1 + \alpha z^2)$ along the axis. (a) At $z = 0$ an electron has a speed of $v^2 = 3v_\parallel^2 = 1.5v_\perp^2$. Where does reflection occur? (b) Determine the motion of the guiding center. (c) Show that the motion is sinusoidal. Determine the frequency. (d) Determine the longitudinal invariant belonging to this motion.

2.16. A particle of mass m and charge q is at rest in a uniform magnetic field $\boldsymbol{B}$. At time $t = 0$, a uniform electric field perpendicular to $\boldsymbol{B}$ is switched on. Show that the maximum energy gain is $2m(E/B)^2$.

2.17. A solar proton with energy 1 MeV starts with an initial pitch angle of $85°$ at 2.5 solar radii. The interplanetary magnetic field decreases as r^{-2}. Determine the proton's pitch angle at the Earth's orbit (213 solar radii) from the conservation of the magnetic moment.

2.18. A 10 keV α-particle is trapped inside the radiation belt at a height of 10^6 km above the surface of the Earth in a magnetic field of about 10^{-6} T. Determine the drift speeds for the curvature and the gradient drift. Compare these speeds with the drift caused by the gravitational field.

2.19. One model for particle acceleration in solar flares uses the second adiabatic invariant. A shock propagates outward through a magnetic field loop of sinusoidal form. Particles gyrate on this loop and bounce back and forth from the shock front. Develop a model for particle acceleration.

3 Magnetohydrodynamics

> 'Glorious stirring sight!' murmured Toad, never offering
> to move. 'The poetry of motion! The *real* way to travel!
> The *only* way to travel!'
> Here today-in next week tomorrow! Villages skipped,
> towns and cities jumped – always somebody else's
> horizon! O bliss! O poop-poop! Oh my! Oh my!
> K. Grahame, *The Wind in the Willows*

In the previous chapter we have discussed the motion of individual charged particles in prescribed E- and B-fields. Magnetohydrodynamics is different for two reasons: (a) it considers an ensemble of particles instead of just a single particle and (b) the E- and B-fields are not prescribed but determined by the positions and motions of these particles. Thus the field equations and the equation of motion have to be solved simultaneously and self-consistently: we are looking for a set of particle trajectories and field patterns such that the particles generate the field patterns as they move along their orbits and the field patterns force the particles to move in exactly these orbits. And all this has to be done in a time-varying situation.

While magnetohydrodynamics describes many useful and important concepts, it is only a simplistic approach to plasma physics: it describes the plasma as a fluid with all particles having the same speed, the bulk speed. The thermal motion of particles is neglected. Kinetic theory (Chap. 5) also considers the velocity distribution of the particles.

This chapter consists of four parts. It starts with a brief recapitulation of hydrodynamics and an introduction to the basic equations. Subsequently, magnetohydrostatics will be concerned with the energetics of the field and the particles without allowing for the collective motion of a plasma. Concepts such as magnetic pressure and magnetic tension are introduced. In magnetohydrokinematics we shall discuss the reaction of the field to a fluid with a given velocity field. Basic concepts such as frozen-in fields and the dissipation of fields are introduced. An application of these concepts is the merging of magnetic field lines, also called reconnection. In magnetohydrodynamics fields and particles can interact freely. In this chapter, the magnetohydrodynamic dynamo will be discussed; magnetohydrodynamic waves will be treated in the next chapter together with other types of plasma waves.

3.1 From Hydrodynamics to Magnetohydrodynamics

In a gas or fluid, the motion of each individual particle is described by an equation of motion. If only electromagnetic forces act on a single particle, the equation of motion is given by (2.23). In a plasma, the equation of motion might be even more complex because the interaction between the particles has to be considered: there are not only external forces acting on the particle ensemble but also internal ones. In a plasma we would have to solve the equations of motion simultaneously for all particles, which might be billions inside a volume as small as 1 mm^3. Such a task is impossible to complete. Instead, we can treat the plasma as a fluid: we are no longer interested in the motion of individual particles but only in the motion of a fluid element or the fluid as a whole. So, to understand magnetohydrodynamics, a sound knowledge of hydrodynamics is helpful.

This section recapitulates the basics of hydrodynamics: partial and convective derivatives, the pressure-gradient force, and the momentum balance in different forms, such as Euler's equation or the Navier–Stokes equation. The equations of continuity and state are also recapitulated.

3.1.1 Partial and Convective Derivatives

The equation of motion for a particle, $\boldsymbol{F} = \mathrm{d}\boldsymbol{p}/\mathrm{d}t$, contains a total derivative of the momentum and the external forces acting on the particle. In a fluid, in principle, we can use the same approach: single out a volume element, follow its path, and calculate the local forces acting on the moving volume. This corresponds to Lagrange's description of particle motion. Here we simply would have to multiply the transport equation by the number density n of the particles and obtain

$$nm\frac{\mathrm{d}\boldsymbol{u}}{\mathrm{d}t} = nq\left(\boldsymbol{E} + \boldsymbol{u} \times \boldsymbol{B}\right) , \tag{3.1}$$

with $\boldsymbol{u} = \langle\boldsymbol{v}\rangle$ being the bulk velocity or average velocity of the particles.[1]

---

[1] The bulk velocity gives the velocity with which the fluid element moves. If the thermal motion is ignored, all particles move with the bulk velocity. If the thermal motion is considered, each individual particle moves with the sum of its thermal velocity $\boldsymbol{v}_{\mathrm{th}}$ and the bulk velocity $\boldsymbol{u}$: $\boldsymbol{v}_{\mathrm{p}} = \boldsymbol{v}_{\mathrm{th}} + \boldsymbol{u}$. Averaged over all particles in the fluid element, the individual particle velocities give the bulk velocity $\langle\boldsymbol{v}_{\mathrm{p}}\rangle = \boldsymbol{u}$ because $\langle\boldsymbol{v}_{\mathrm{th}}\rangle = 0$.

For practical purposes, we normally consider a volume fixed in space and measure the properties of the fluid streaming through the volume. A property ε of the fluid then is given as $\varepsilon = \varepsilon(x, y, z, t)$ with the spatial coordinates and the time being independent variables. This corresponds to Euler's description of a fluid. In contrast, in Lagrange's description, the spatial coordinate depends on time too, and a property ε of the fluid is given as $\varepsilon = \varepsilon(x(t), y(t), z(t), t)$. In the atmosphere, Euler's description could be applied to a stationary thermometer while Lagrange's description could be applied to a thermometer on a radio-sonde carried by the prevailing winds.

If we are interested in changes in ε, we have to calculate its derivative. In Euler's description, the total derivative $d\varepsilon/dt$ and the partial derivative $\partial\varepsilon/\partial t$ are equal because all temporal derivatives of the spatial coordinates vanish. In Lagrange's description, the chain rule has to be applied:

$$\frac{d\varepsilon}{dt} = \frac{dx}{dt}\frac{\partial\varepsilon}{\partial x} + \frac{dy}{dt}\frac{\partial\varepsilon}{\partial y} + \frac{dz}{dt}\frac{\partial\varepsilon}{\partial z} + \frac{\partial\varepsilon}{\partial t} = (\boldsymbol{u} \cdot \nabla)\varepsilon + \frac{\partial\varepsilon}{\partial t} . \tag{3.2}$$

The change of a property ε in a moving fluid element therefore consists of two parts: (a) a change in ε at a fixed position in space (second term on the right-hand side); and (b) the relative motion between the observer and the medium (first term). Or, more formally: the total temporal derivative consists of a local temporal derivative and advection; it is also called the convective derivative. Note that the product $(\boldsymbol{u} \cdot \nabla)$ is a scalar differential operator. Occasionally, the total derivative is written as D/Dt instead of d/dt.

To understand the difference between a convective and a local temporal derivative let us take a look at the property of water as, for example, salinity or temperature. Let us first consider a closed volume, e.g. a fish-pond. The temperature then might change due to absorbed solar radiation, and the salinity might change due to evaporation. These changes are local temporal changes. Now think of this volume of water as a segment of a river. The local changes are still the same but there are also changes due to the advection of water from other sites: warmer water might be advected into the volume from a power station upstream or the salinity might increase as the incoming tide carries water with higher salinity into the volume.

3.1.2 Equation of Motion or Momentum Balance

The motion of a fluid element in hydrodynamics can be described by Euler's equation or the Navier–Stokes equation. All these different equations of motion have one basic ingredient, the pressure-gradient force. In single-body motions, only external forces act on the body. In a fluid, on the other hand, regions of different pressure, for instance related to temperature differences, can exist, exerting forces on fluid elements. Thus, before inserting the external forces into the equation of motion, let us have a look at this internal force, the pressure-gradient force.

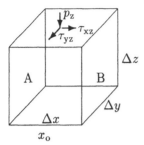

Fig. 3.1. Normal forces p and shear stresses τ acting on a cubic volume

Pressure-Gradient Force. Regions of different pressure in a gas exert forces: particles move from the high pressure towards the low one. This force is proportional to the pressure gradient $-\nabla p$ and is called the pressure-gradient force. Here we give its derivation, closely following Chen [97].

Pressure is related to the thermal motion of particles. The pressure-gradient force leads to a transport in momentum resulting from the motion of particles in and out of a fluid element $V|_{x_0} = \Delta x \Delta y \Delta z$ at position x_0 (see Fig. 3.1). If the random thermal motion is limited to the x-axis, particles enter and leave the volume through surfaces A and B only. The fluid particles are characterized by their mass m, their speed v and their number density n. During a time interval

$$\Delta n = \Delta n_v\, v_x \Delta y \Delta z \tag{3.3}$$

particles with speed v_x pass through surface A with area $A = \Delta y \Delta z$ into the positive x-direction. Here

$$\Delta n_v = \Delta v_x \int\int f(v_x, v_y, v_z)\, \mathrm{d}v_y\, \mathrm{d}v_z \tag{3.4}$$

is the number density of particles with speed v_x, with f being the distribution function (Chap. 5).

Each particle carries a momentum mv_x. The total momentum P_A^+ transported through A into the positive x-direction then is

$$P_\mathrm{A}^+ = \sum \Delta n_v m v_x^2 \Delta y \Delta z = \Delta y \Delta z \left[\tfrac{1}{2} m \langle v_x^2 \rangle n\right]_{x_0 - \Delta x} . \tag{3.5}$$

Here the sum over Δn_v is expressed by the average $\langle v_x^2 \rangle$ of the distribution times the particle number density. The factor $1/2$ indicates that only half of the particles in the adjacent volume element $V|_{x_0 - \Delta x}$ at $x_0 - \Delta x$ have a speed in the positive x-direction and transport momentum through A into $V|_{x_0}$. But particles inside $V|_{x_0}$ also have a momentum in the positive x-direction which is carried out of the volume through the surface B. Their number is given as

$$P_\mathrm{B}^+ = \Delta y \Delta z \left[\tfrac{1}{2} m \langle v_x^2 \rangle n\right]_{x_0} . \tag{3.6}$$

Therefore, the net gain of the positive x-momentum in $V|_{x_0}$ is

$$P_{\rm A}^+ - P_{\rm B}^+ = \Delta y \Delta z \tfrac{1}{2} m (-\Delta x) \frac{\partial n \langle v_x^2 \rangle}{\partial x} . \tag{3.7}$$

Particles moving into the negative x-direction double the momentum gain in (3.7) because the negative x-momentum is transported into the negative x-direction:

$$\frac{\partial}{\partial t}(nmv_x)\Delta x \Delta y \Delta z = -m \frac{\partial}{\partial x}(n\langle v_x^2 \rangle)\Delta x \Delta y \Delta z . \tag{3.8}$$

The particle speed $v_x = u_x + v_{x_{\rm th}}$ consists of two parts, the bulk speed u_x of the fluid element with $u_x = \langle v_x \rangle$ and the superimposed thermal speed $v_{x_{\rm th}}$ with $\langle v_{x_{\rm th}} \rangle = 0$. The latter is described by a one-dimensional Maxwell distribution (Sect. 5.1.2). The relationship between average thermal speed and temperature is:

$$\tfrac{1}{2} m \langle v_{x_{\rm th}}^2 \rangle = \tfrac{1}{2} k_{\rm B} T . \tag{3.9}$$

With (3.8) we obtain

$$\frac{\partial}{\partial t}(nmu_x) = -m \frac{\partial}{\partial x} \left[n(\langle u_x^2 \rangle + 2\langle u_x v_{x_{\rm th}} \rangle + \langle v_{x_{\rm th}}^2 \rangle) \right] . \tag{3.10}$$

The last term on the right-hand side can be substituted by (3.9). The term in the middle is zero because u_x is constant and thus $\langle u_x v_{x_{\rm th}} \rangle = u_x \langle v_{x_{\rm th}} \rangle = 0$ (see Sect. 4.1.3):

$$\frac{\partial}{\partial t}(nmu_x) = -m \frac{\partial}{\partial x} \left(nu_x^2 + \frac{nk_{\rm B}T}{m} \right) . \tag{3.11}$$

The partial differentiation on the right-hand side with $nu_x^2 = nu_x\, u_x$ gives

$$mn \frac{\partial u_x}{\partial t} + mu_x \frac{\partial n}{\partial t} = -mu_x \frac{\partial (nu_x)}{\partial x} - mnu_x \frac{\partial u_x}{\partial x} - \frac{\partial (nkT)}{\partial x} . \tag{3.12}$$

The second term on the left-hand side and the first term on the right-hand side cancel (see the equation of continuity (3.34)). With the pressure p defined as $p = nk_{\rm B}T$ rearrangement leads to

$$mn \left(\frac{\partial u_x}{\partial t} + u_x \frac{\partial u_x}{\partial x} \right) = mn \frac{{\rm d}u_x}{{\rm d}t} = -\frac{\partial p}{\partial x} . \tag{3.13}$$

Generalization to three dimensions gives the pressure-gradient force density

$$mn \left(\frac{\partial \boldsymbol{u}}{\partial t} + \boldsymbol{u}(\nabla \cdot \boldsymbol{u}) \right) = mn \frac{{\rm d}\boldsymbol{u}}{{\rm d}t} = -\nabla p . \tag{3.14}$$

Since n is a number density (unit m^{-3}), the product nm gives the density ϱ and we can write alternatively for the acceleration due to the pressure gradient force

$$\frac{{\rm d}\boldsymbol{u}}{{\rm d}t} = -\frac{1}{\varrho} \nabla p . \tag{3.15}$$

Equation of Motion: Euler and Navier–Stokes. The simplest equation of motion for a fluid considers the acceleration due to the pressure-gradient force and gravitation

$$\frac{d\boldsymbol{u}}{dt} = -\frac{1}{\varrho}\nabla p + \boldsymbol{g} \ . \tag{3.16}$$

This equation is known as Euler's equation and often is used for simple estimates in atmospheric or oceanic motion. Euler's equation can be applied to ideal fluids only. In a real fluid, viscous forces have to be considered too. Here the Navier–Stokes equation is useful:

$$\frac{d\boldsymbol{u}}{dt} = -\frac{1}{\varrho}\nabla p + \nu\nabla^2\boldsymbol{u} \tag{3.17}$$

with ν being the kinematic viscosity. Often, other forces, depending on the situation under study, are added to this equation. Some of these forces will be discussed below where we also shall have a closer look at the viscous forces.

Stress Tensor and Viscosity. In the generalization of (3.13) we tacitly assumed that x_i-momentum is transported in x_i-direction only and that the fluid is isotropic. This is true in an ideal gas or fluid but not in a viscous one, where momentum can be transported in directions perpendicular to the particle motion, and momentum transport is not necessarily isotropic. Then the scalar property p has to be replaced by a tensor P, and the pressure-gradient force ∇p has to be replaced by $\nabla\mathsf{P}$. P not only considers the pressure, which is orthogonal to the surface of a volume element, but also shear stresses, which are forces parallel to the element's surface (see Fig. 3.1). The stress tensor P has the dimensions of a pressure or an energy density. It is symmetric with six independent components P_{ij} for each point: $P_{ij} = mnv_iv_j$; i being the direction of the momentum transport and j the component of the momentum involved. A more compact method to write the stress tensor is

$$\mathsf{P} = mn\langle\boldsymbol{v}_{\mathrm{th}}\boldsymbol{v}_{\mathrm{th}}\rangle \ . \tag{3.18}$$

Here $\boldsymbol{v}_{\mathrm{th}}\boldsymbol{v}_{\mathrm{th}}$ is not a shorthand for a scalar product but the tensor product (dyad) of two vectors: such tensor products $\boldsymbol{ab}$ of two vectors are tensors T, where

$$\mathsf{T} = \boldsymbol{ab} = \begin{pmatrix} a_x \\ a_y \\ a_z \end{pmatrix}\begin{pmatrix} b_x \\ b_y \\ b_z \end{pmatrix} = \begin{pmatrix} a_xb_x & a_xb_y & a_xb_z \\ a_yb_x & a_yb_y & a_yb_z \\ a_zb_x & a_zb_y & a_zb_z \end{pmatrix} \ . \tag{3.19}$$

In the simplest case, the particle distribution is an isotropic Maxwellian and the stress tensor P can be written as

$$\mathsf{P} = \begin{pmatrix} p & 0 & 0 \\ 0 & p & 0 \\ 0 & 0 & p \end{pmatrix} = p\mathsf{E} \ , \tag{3.20}$$

where E is the unit tensor. Here ∇P equals ∇p. In the presence of a magnetic field, a plasma can have two different temperatures $T_\parallel$ and $T_\perp$ parallel and perpendicular to the magnetic field, leading to different pressures $p_\parallel = nk_B T_\parallel$ and $p_\perp = nk_B T_\perp$. In a coordinate system oriented with its z-axis parallel to B, the stress tensor can be written as

$$\mathsf{P} = \begin{pmatrix} p_\perp & 0 & 0 \\ 0 & p_\perp & 0 \\ 0 & 0 & p_\parallel \end{pmatrix} . \tag{3.21}$$

This tensor is diagonal and it is isotropic in a plane perpendicular to $\boldsymbol{B}$.

The off-diagonal elements of the stress tensor in an ordinary fluid are associated with viscosity. Viscosity results from collisions between particles and tends to make the flow more uniform. Quantitatively, the effect of viscosity is described by a kinematic viscosity coefficient $\nu = v_{\text{th}} \lambda$ where v_{th} is the thermal speed and λ the mean free path between collisions. Alternatively, a viscosity coefficient $\eta = \nu \varrho$ can be used. In a fluid, friction is described by

$$\boldsymbol{f}_{\text{frict}} = \eta \nabla^2 \boldsymbol{u} + \frac{1}{3} \eta \nabla (\nabla \times \boldsymbol{u}) . \tag{3.22}$$

In an incompressible fluid, the second term on the right-hand side vanishes:

$$\boldsymbol{f}_{\text{frict}} = \eta \nabla^2 \boldsymbol{u} = \nu \varrho \nabla^2 \boldsymbol{u} . \tag{3.23}$$

This can be interpreted as the collisional part of ∇P $-$ ∇p. Note that the inclusion of viscosity into the momentum balance has two consequences: (a) in agreement with the irreversible character of the transport process, the transport equation is no longer time-reversible: if $\boldsymbol{u}(\boldsymbol{r}, t)$ is a solution of the transport equation, then $\boldsymbol{u}(\boldsymbol{r}, -t)$ is not. (b) Viscosity increases the order of the partial differential equation. Therefore, to determine solutions we need more boundary conditions than in the case of a non-viscous fluid.

In a plasma, off-diagonal elements can arise without collisions: gyration brings particles into different parts of the plasma, a process which tends to equalize the fluid speeds. The scale of this "collisionless viscosity" is given by the Larmor radius rather than by the particle mean free path.

Fictitious Forces in Rotating Systems. The forces discussed so far are sufficient to give the equation of motion for a plasma in the laboratory setting. In large-scale natural plasmas, such as the ionosphere or stellar atmospheres, additional forces act: the Coriolis force and the centrifugal force.

Consider two frames of reference C and C', with C rotating with an angular velocity $\boldsymbol{\Omega}$ with respect to C'. A vector $\boldsymbol{r}$ fixed in C, in C' moves with a speed $\boldsymbol{\Omega} \times \boldsymbol{r}$. The temporal derivative of $\boldsymbol{r}$ in C' s

$$\left(\frac{\mathrm{d}\boldsymbol{r}}{\mathrm{d}t} \right)_{C'} = \left(\frac{\mathrm{d}\boldsymbol{r}}{\mathrm{d}t} \right)_C + \boldsymbol{\Omega} \times \boldsymbol{r} \quad \text{or} \quad \boldsymbol{v}' = \boldsymbol{v} + \boldsymbol{\Omega} \times \boldsymbol{r} . \tag{3.24}$$

The temporal derivative gives the acceleration in the rotating frame:

$$a' = \left(\frac{dv'}{dt}\right)_{C'} = \frac{d'v'}{dt} = \frac{dv'}{dt} + \boldsymbol{\Omega} \times v' = \frac{dv}{dt} + 2\boldsymbol{\Omega} \times v + \boldsymbol{\Omega} \times (\boldsymbol{\Omega} \times r) . \quad (3.25)$$

Thus the density of the fictitious forces in a rotating frame of reference is

$$\boldsymbol{f}_{\text{rot}} = -\varrho 2\,\boldsymbol{\Omega} \times v - \varrho\,\boldsymbol{\Omega} \times (\boldsymbol{\Omega} \times r) \quad (3.26)$$

with the first term on the right-hand side describing the Coriolis force and the second term the centrifugal force.

In the near-Earth environment the Coriolis force has to be considered in the atmospheric motion and in the ionospheric and magnetospheric current systems; it is of vital importance in the dynamo process inside the Sun and the planets. The influence of the Coriolis force can be illustrated by its effect on the atmospheric motion. In the northern hemisphere, wind is deflected towards the right. On a global scale, this deflection leads to the break-up of the Hadley cell driven by the temperature gradient between the equator and the pole into three separate cells, which determine the global atmospheric circulation and govern the energy transport from equator to pole. The Coriolis force, and therefore the size of the deflection, depends on the wind speed: with increasing speed, the distance travelled by a volume of air during a time interval increases. A longer trajectory also means a larger displacement. The Coriolis force becomes effective only if the scales of the system are large enough. Contrary to popular belief, the eddy at the outflow of a bath-tub is not due to the Coriolis force: its direction depends on residual motions in the water or the motion induced by pulling the plug.

Electromagnetic Forces. A charged particle in an electromagnetic field experiences the Lorentz force (2.23). With n being the number density, the force on a volume element can then be written as

$$mn\frac{d\boldsymbol{u}}{dt} = mn\left[\frac{\partial \boldsymbol{u}}{\partial t} + (\boldsymbol{u} \cdot \nabla)\boldsymbol{u}\right] = qn\left(\boldsymbol{E} + \boldsymbol{u} \times \boldsymbol{B}\right) . \quad (3.27)$$

The dimension of n is m^{-3}, thus (3.27) can also be written as a force density

$$\boldsymbol{f}_{\text{elmag}} = \varrho\frac{d\boldsymbol{u}}{dt} = \varrho\left[\frac{\partial \boldsymbol{u}}{\partial t} + (\boldsymbol{u} \cdot \nabla)\boldsymbol{u}\right] = \varrho_{\text{c}}\boldsymbol{E} + \boldsymbol{j} \times \boldsymbol{B} , \quad (3.28)$$

with $\varrho = mn$ being the density, $\varrho_{\text{c}} = qn$ the charge density, and $\boldsymbol{j} = nq\boldsymbol{u}$ the current density. Equation (3.28) gives the force density of the electromagnetic field. For infinite conductivity, the charges immediately rearrange and cancel out the electric field. The force density then reduces to

$$\boldsymbol{f}_{\text{elmag}} = \boldsymbol{j} \times \boldsymbol{B} . \quad (3.29)$$

Putting it all Together. Adding these forces gives the equation of motion or momentum balance:

$$\varrho \frac{d\boldsymbol{u}}{dt} = \varrho \left(\frac{\partial \boldsymbol{u}}{\partial t} + (\boldsymbol{u} \cdot \nabla)\boldsymbol{u} \right)$$
$$= -\nabla \mathsf{P} + \varrho \boldsymbol{E} + \boldsymbol{j} \times \boldsymbol{B} + \varrho \boldsymbol{g} - 2\varrho \boldsymbol{\Omega} \times \boldsymbol{u} - \varrho \boldsymbol{\Omega} \times (\boldsymbol{\Omega} \times \boldsymbol{r}) \,. \quad (3.30)$$

If we neglect the electric field and the fictitious forces and split the stress tensor into the pressure-gradient force and friction, (3.30) can be written as

$$\varrho \frac{d\boldsymbol{u}}{dt} = \varrho \left[\frac{\partial \boldsymbol{u}}{\partial t} + (\boldsymbol{u} \cdot \nabla)\boldsymbol{u} \right] = -\nabla p + \varrho \nu \nabla^2 \boldsymbol{u} + \boldsymbol{j} \times \boldsymbol{B} + \varrho \boldsymbol{g} \,. \quad (3.31)$$

This equation is the Navier–Stokes equation used in hydrodynamics complemented by the forces exerted by the electromagnetic field.

The momentum balance (3.30) still is relatively simple: (a) it does not consider sources and sinks, e.g. due to ionization or recombination, which might involve a net gain or loss of momentum; (b) it does not consider momentum transport due to Coulomb collisions between charged particles; and (c) it does not consider momentum transport arising from the forces exerted by a particle component of opposite charge inside the plasma. The latter will be discussed briefly in the two-fluid description of a plasma (see Sect. 3.2.1).

3.1.3 Equation of Continuity

An equation of continuity is concerned with the conservation of a property ε, such as mass or charge. A change in ε inside a volume V can result from the convergence of a flux $C(\varepsilon)$ into or out of the volume or sources and sinks $S(\varepsilon)$ inside the volume. The general form of an equation of continuity therefore is

$$\frac{\partial \varepsilon}{\partial t} + \nabla C(\varepsilon) = S(\varepsilon) \,. \quad (3.32)$$

The most common application is the conservation of mass:

$$\frac{\partial \varrho}{\partial t} = -\nabla(\varrho \boldsymbol{u}) = -\nabla \boldsymbol{j} \,, \quad (3.33)$$

where ϱ is the density and $\boldsymbol{j} = \varrho \boldsymbol{u}$ is the mass current density. Using (3.2) the conservation of mass can be rewritten as

$$\frac{d\varrho}{dt} = \frac{\partial \varrho}{\partial t} + \boldsymbol{u} \, \nabla \varrho = -\varrho \nabla \boldsymbol{u} \,. \quad (3.34)$$

It states that a change of mass inside a volume is a consequence of the flow of matter into or out of the volume. Local sources and sinks are not considered because, except for elementary particle physics, there are none. With Gauss's theorem (A.33) the integral form of the equation of continuity is

$$\frac{\partial}{\partial t} \int_V \varrho \, \mathrm{d}V = - \oint_{O(V)} \boldsymbol{j} \cdot \mathrm{d}\boldsymbol{o} \, . \tag{3.35}$$

The equation of continuity for the electric charge is formally analogous, with ϱ_c replacing ϱ in (3.33):

$$\frac{\partial \varrho_c}{\partial t} + \nabla(\boldsymbol{u}\varrho_c) = 0 \, . \tag{3.36}$$

3.1.4 Equation of State

Finally, we need a relationship connecting the scalar pressure p and the density ϱ: $p = p(\varrho, T)$. The equation of state describes how the temperature changes during the motion or compression of a gas. In case of an isothermal ideal gas, the equation of state can be written as

$$p = C(T)\,\varrho \, , \tag{3.37}$$

where C is a constant proportional to temperature. If the compression is slow compared with thermal conduction (isothermal compression), the pressure increase results from the density increase but not from the temperature increase. In a plasma particles can freely flow along $\boldsymbol{B}$. Thus conduction parallel to $\boldsymbol{B}$ provides the possibility for a plasma to remain isothermal, especially if the compression is periodic or wave-like along $\boldsymbol{B}$.

A fast moving gas might not be able to exchange energy with its environment. The equation of state for such an adiabatic compression is

$$p = C\,\varrho^{\gamma_a} \, , \tag{3.38}$$

where $\gamma_a = c_p/c_V$ is the specific heat ratio or adiabatic exponent. For an ideal gas, γ_a equals $(N+2)/N$, with N being the number of degrees of freedom. For a three-dimensional ideal gas consisting of atoms, γ_a is $5/3$. Both cases, isothermal as well as adiabatic compression, are of importance in different types of plasma waves.

A third important case arises if adiabatic compression is fast compared with heat conduction and also is anisotropic. Now the degrees of freedom parallel and perpendicular to the field are separated, and $T_\parallel$ ($N = 1, \gamma_a = 3$) can be heated more efficiently than $T_\perp$ ($N = 2, \gamma_a = 2$). The adiabatic invariants can be used to derive generalizations of this relationship.

The perpendicular pressure $p_\perp$ can be expressed by the average magnetic moment μ: $p_\perp = \frac{1}{2}\varrho\langle v_\perp^2 \rangle = n\langle \mu \rangle B$. If the compression is fast compared with the heat conduction but slow compared with the gyration period, the magnetic moment is conserved, leading to the adiabatic relation

$$\frac{\mathrm{d}}{\mathrm{d}t}\left(\frac{p_\perp}{nB}\right) = \text{const} \, . \tag{3.39}$$

Pure perpendicular compression in general is equivalent to an increase in the magnetic field strength. For each area A, the conservation of particles implies $nA = $ const while conservation of the magnetic flux yields $BA = $ const. Thus n/B is constant, too, and (3.39) reduces to $p = C\varrho^{\gamma_a}$ with γ_a being 2 as expected for two-dimensional adiabatic compression.

The pressure $p_\| = \varrho \langle v_\|^2 \rangle$ parallel to the magnetic field is related to the second adiabatic invariant $J_2 \sim v_\| L = $ const, with L being the scale length along the magnetic field. If the compression is slow compared with the particle's oscillation along the field line, then J_2 is conserved. Now length L, area A and volume $V = AL$ change. The conservation of particles and the magnetic flux yield $nV = $ const and $BA = $ const. Thus L can be expressed as $L = V/A \sim B/n$ and we finally get

$$\frac{\mathrm{d}}{\mathrm{d}t} \left(\frac{p_\| B^2}{n^3} \right) = 0 \ . \tag{3.40}$$

Pure parallel compression with $B = $ const then leads to $p = C\varrho^{\gamma_a}$ with $\gamma_a = 3$ as expected for one-dimensional compression.

The two adiabatic relations (3.39) and (3.40) are called the "double adiabatic" equations of state.

3.2 Basic Equations of MHD

We shall start with the one-fluid description of a plasma, i.e. the fluid consists of one particle species only. This is entirely sufficient to introduce the basic concepts (see Sects. 3.3–4.2). In a real plasma, quasi-neutrality suggests the existence of two fluids with positive and negative charges, respectively. For certain phenomena, such as ion waves, a description in the framework of a two-fluid theory will be required, and this is briefly sketched in Sect. 3.2.1.

In magnetohydrodynamics some assumptions about the properties of the system are made: (a) The medium can be neither polarized nor magnetized: $\varepsilon = \mu = 0$. (b) Flow speeds and speeds of changes in field properties are small compared with the speed of light: $u/c \ll 1$ and $v_{\mathrm{ph}}/c \ll 1$. As a consequence, electromagnetic waves cannot be treated in the framework of MHD theory. (c) Conductivity is high, thus strong electric fields are immediately cancelled out: $E/B \ll 1$. As a consequence, the displacement current $\partial E/\partial t$ can be ignored compared with the induction current. MHD is a theory linear in u/c, v_{ph}/c, and E/B and ignores all terms of higher order in these quantities. MHD considers the conservation laws of fluid mechanics which are concerned with mass, momentum, energy, and magnetic flux. The formal description is then based on the following set of equations:

• Maxwell's equations (Sect. 2.1.1):

$$\nabla \cdot \boldsymbol{E} = \varrho_{\mathrm{c}}/\varepsilon_0 \ , \tag{3.41}$$

$$\nabla \cdot \boldsymbol{B} = 0 \, , \tag{3.42}$$

$$\nabla \times \boldsymbol{E} = -\frac{\partial B}{\partial t} \, , \tag{3.43}$$

$$\nabla \times \boldsymbol{B} = \mu_0 \boldsymbol{j} \, ; \tag{3.44}$$

- Ohm's law (Sect. 2.1.3)

$$\boldsymbol{j} = \sigma \left(\boldsymbol{E} + \boldsymbol{u} \times \boldsymbol{B} \right) \, ; \tag{3.45}$$

- equation of continuity (Sect. 3.1.3)

$$\frac{\partial \varrho_c}{\partial t} + \nabla (\boldsymbol{u}\varrho_c) = 0 \, ; \tag{3.46}$$

- equation of motion (momentum balance, Sect. 3.1.2)

$$\varrho \frac{\partial \boldsymbol{u}}{\partial t} + \varrho(\boldsymbol{u} \cdot \nabla)\boldsymbol{u} = -\nabla p + \boldsymbol{j} \times \boldsymbol{B} + \varrho \boldsymbol{g} + \varrho \nu \nabla^2 \boldsymbol{u} \, ; \tag{3.47}$$

- equation of state (Sect. 3.1.4)

$$\frac{\mathrm{d}}{\mathrm{d}t} \left(\frac{p}{\varrho^{\gamma_a}} \right) = 0 \, . \tag{3.48}$$

This set of partial non-linear differential equations can be solved for given boundary conditions. For certain applications only a part of the equations is required, or some equations can be used in a simplified form: in magnetohydrostatics (Sect. 3.3) the left-hand side of the momentum balance vanishes while in magnetohydrokinematics (Sect. 3.4) an external velocity field is prescribed and therefore the momentum balance can be ignored completely.

The momentum balance gives us hints on the kind of motion: in certain slow motions the inertial term $\varrho \dot{\boldsymbol{u}}$ can be ignored while in weak magnetic fields the Lorentz force can be ignored. The relative strength of these two forces is determined by the ratio

$$S = \frac{B^2/2\mu_0}{\varrho u^2/2} = \frac{\text{magnetic field energy density}}{\text{kinetic energy density}} \, . \tag{3.49}$$

For $S \gg 1$ the magnetic field determines the motion of the particles and the single-particle approach can be used. For $S \ll 1$, the magnetic field is swept away by the plasma motion, in accordance with the concept of the frozen-in field described in Sect. 3.4.1. S is another expression for the plasma-β, giving the ratio between the gas dynamic pressure and the magnetic pressure: $\beta = 2\mu_0 p / B^2$.

It should be noted that these two definitions are useful only for an isotropic plasma. If the plasma is anisotropic, frequently a parallel and a perpendicular plasma-β are defined as

$$\beta_{\parallel} = \frac{2\mu_0 p_{\parallel}}{B^2} \quad \text{and} \quad \beta_{\perp} = \frac{2\mu_0 p_{\perp}}{B^2} \, . \tag{3.50}$$

In a low-β plasma ($\beta \ll 1$), the energy density in the thermal motion is much larger that in the magnetic field, while in a high-β plasma ($\beta \gg 1$) the opposite is true.

3.2.1 Two-Fluid Description

So far, we have treated the plasma as a fluid consisting of one kind of charged particles only. A real plasma, however, contains electrons, ions, and possibly also neutral particles. Each particle component has its own speed, temperature, and partial pressure.

Since a plasma is expected to be quasi-neutral, the number of positive and negative charges has to be equal. The charge density is $\varrho_c = n_i q_i + n_e q_e = \varrho_i + \varrho_e$ with n_i and n_e being the number densities of ions and electrons with charges q_i and q_e. The current density is $\boldsymbol{j} = n_i q_i \boldsymbol{u}_i + n_e q_e \boldsymbol{u}_e = \boldsymbol{j}_i + \boldsymbol{j}_e$. If we limit ourselves to a two-fluid plasma, we have to deal with an electron and an ion component; the neutral component is ignored. In addition to the assumptions made in the one-fluid description we assume: (a) the fluid is in thermal equilibrium ($T_i = T_e$), and (b) the plasma is quasi-neutral ($\varrho_i = \varrho_e$). The basic equations in two-fluid MHD are

- Maxwell's equations

$$\nabla \cdot \boldsymbol{E} = (\varrho_i + \varrho_e)/\varepsilon_0 \, , \tag{3.51}$$

$$\nabla \cdot \boldsymbol{B} = 0 \, , \tag{3.52}$$

$$\nabla \times \boldsymbol{E} = -\frac{\partial \boldsymbol{B}}{\partial t} \, , \tag{3.53}$$

$$\nabla \times \boldsymbol{B} = \mu_0 (\boldsymbol{j}_i + \boldsymbol{j}_e) + \varepsilon_0 \mu_0 \frac{\partial \boldsymbol{E}}{\partial t} \, ; \tag{3.54}$$

- Ohm's law

$$\frac{m_e}{e^2 n} \frac{\partial \boldsymbol{j}}{\partial t} = \boldsymbol{E} + \boldsymbol{u} \times \boldsymbol{B} - \frac{\boldsymbol{j} \times \boldsymbol{B}}{en} + \frac{\nabla p_e}{en} - \frac{\boldsymbol{j}}{\sigma} \, ; \tag{3.55}$$

- equation of continuity

$$\frac{\partial n_j}{\partial t} + \nabla \cdot (n_j \boldsymbol{u}_j) = 0 \, , \quad j = \text{i}, \text{e} \, ; \tag{3.56}$$

- momentum balance (equation of motion)

$$m_j n_j \frac{\mathrm{d}\boldsymbol{u}_j}{\mathrm{d}t} = q_j n_j \left(\boldsymbol{E} + \boldsymbol{u}_j \times \boldsymbol{B}\right) - \nabla p_j \pm \beta(\boldsymbol{u}_i - \boldsymbol{u}_e) \, , \, j = \text{i}, \text{e} \, ; \tag{3.57}$$

- equation of state

$$p_j = p_j(\varrho_j, T_j) \, , \quad j = \text{i}, \text{e} \, . \tag{3.58}$$

Compared with the equations in one-fluid MHD we find the following differences: (a) The equations of state, motion and continuity are given for each component separately. (b) The equation of motion contains an additional term coupling the two components to consider momentum transfer arising from Coulomb collisions. The force between the two components depends on their relative speed, therefore $f_i = -f_e = \beta(u_i - u_e)$. (c) Gauss's law for the electric field contains both charge densities as Ampére's law contains both current densities. (d) Ohm's law has become unrecognizable. A derivation of Ohm's law from the equation of motion can be found in [285]; here only the terms will be explained. The left-hand side gives the current acceleration. The first, second and last terms on the right-hand side are expressions already known from Ohm's law in one-fluid MHD. The $j \times B$ term is called the Hall term and describes the Hall effect: in a magnetic field the current created by the moving charges is deflected by the Lorentz force, resulting in an additional electric field perpendicular to both j and B. The fourth term on the right-hand side gives the pressure diffusion: in the presence of a pressure gradient, both particle species diffuse with respect to each other, creating a current along ∇p.

3.3 Magnetohydrostatics

Magnetohydrostatics deals with the energetics of particles and fields. It does not require the entire set of MHD equations; instead, the field equations and the equation of motion (with vanishing inertial term) are sufficient. Important concepts are magnetic pressure and magnetic tension.

3.3.1 Magnetic Pressure

Let us now take a closer look at a magnetic field such as shown in Fig. 3.2. The lines of force are parallel to the z-axis with the field strength varying along the x-axis: $B = (0, 0, B(x))$. The force density exerted by the field is $f = j \times B$. With j expressed by Ampére's law (3.44) we obtain:

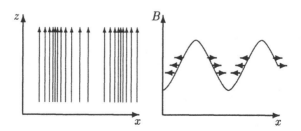

Fig. 3.2. Magnetic pressure: field gradient perpendicular to the field (*left*) and the resulting spatial distribution of the field strength (*right*), with *arrows* indicating the direction of the magnetic pressure

$$f = j \times B = (\nabla \times B) \times B/\mu_0 . \tag{3.59}$$

For a general derivation of the magnetic pressure and tension, we can use (A.20) and (2.19) and obtain

$$j \times B = -\frac{1}{\mu_0} B \times (\nabla \times B) = -\frac{1}{\mu_0} \nabla(BB) + \frac{1}{\mu_0} B(\nabla \cdot B)$$
$$= -\frac{1}{\mu_0} \nabla(BB) + \frac{1}{2\mu_0} \nabla B^2 . \tag{3.60}$$

The first term on the right-hand side gives the force density arising from the magnetic stress tensor BB which describes magnetic tension and torsion. The second term is formally equivalent to the pressure-gradient force, but instead of gas pressure p a magnetic pressure $B^2/(2\mu_0)$ is used. Both the magnetic pressure and the magnetic tension can also be derived more graphically from (3.59) if we use simplified geometries.

For the field defined above, (3.59) yields

$$f = \frac{1}{\mu_0} \left(-B\frac{\partial B}{\partial x}, 0, 0 \right) . \tag{3.61}$$

Thus, the force density only has a component along the x-axis (or, more generally, perpendicular to B and parallel to the field gradient):

$$f_x = -\frac{1}{\mu_0} B\frac{\partial B}{\partial x} = -\frac{\partial}{\partial x}\frac{B^2}{2\mu_0} . \tag{3.62}$$

Therefore, an inhomogeneity in the magnetic field gives rise to a force density pushing field lines back from regions of high density into low density areas. Such behavior is well known from an isothermal gas where a restoring force $f \sim \nabla p$ tries to cancel out pressure gradients. Therefore, (3.62) can be interpreted as the magnetic pressure:

$$p_M = \frac{B^2}{2\mu_0} . \tag{3.63}$$

Graphically, this magnetic pressure can be described as the tendency of neighboring field lines to repulse each other. Note that, in contrast to the gas-dynamic pressure, the magnetic pressure is not isotropic but is always perpendicular to the field.

The analogy with gas-dynamic pressure can be pushed even further if we invoke the concept of frozen-in magnetic fields (Sect. 3.4.1). Imagine a magnetic field frozen-into a plasma: each plasma parcel contains a certain amount of magnetic flux which is tied to this plasma element and follows its path as the plasma parcels are shuffled around. Thus a field gradient always has to be combined with a gradient in gas-dynamic pressure. As the plasma attempts to reduce the pressure gradient, the field will be homogenized, too.

Formally, the magnetic pressure also could be inferred from Maxwell's stress tensor, as is shown in [36].

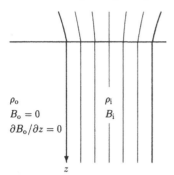

ρ_o
$B_o = 0$
$\partial B_o / \partial z = 0$

ρ_i
B_i

z

Fig. 3.3. Model of a sunspot: the gas-dynamic pressure from the outside is balanced by the magnetic pressure inside the sunspot

Example 10. A homogeneous magnetic field of 5 T, according to (3.63), exerts a magnetic pressure $p = B^2/2\mu_0 = 9.95 \times 10^6$ N/m^2 = 995 hPa $\times$ 100, which is a hundred times the atmospheric pressure at sea level.   □

Example 11. Sunspots are a prime example of the apparently paradoxical behavior of plasmas as well as a good illustration of the concept of magnetic pressure. Sunspots (Sect. 6.6, Fig. 6.26) are cool and dark patches on the visible solar disk. Temperatures in sunspots are 1000 K to 2000 K below the temperature of the ambient photosphere (5700 K). Despite this temperature gradient, the sunspot does not mix with the ambient plasma. Instead, it is very stable and can survive for many solar rotations. This longevity results from the magnetic pressure: inside the sunspot the magnetic field is about 3000 gauss compared to a few gauss at the outside. The boundaries of a sunspot are sharp in both magnetic field strength and temperature.

Figure 3.3 shows a simple model of a sunspot. The indices "i" and "o" refer to plasma and field properties inside and outside of the sunspot, respectively. To prevent hot photospheric material from streaming into the sunspot, the gas-dynamic pressure p_o of the photospheric plasma has to be balanced by the combined magnetic and gas-dynamic pressure inside the spot: $p_i + B_i^2/(2\mu_0) = p_o$. Here we have assumed that the magnetic field pressure outside the sunspot is negligible and that $\partial B/\partial z$ is zero, as is suggested by observations. The variation of pressure with height is described by the hydrostatic equation: $\partial p/\partial z = \varrho g_{Sun}$. Since B is independent of height, this yields $\partial p_i/\partial z = \partial p_o/\partial z$ and $\varrho_i = \varrho_o$. On the other hand, the universal gas law yields for the pressure $p = \varrho k_B T/m$. Because the densities inside and outside the sunspot are equal, the gas law requires T_i to be smaller than T_o to fulfill $p_i < p_o$. Thus, in agreement with the observations, we find: for longevity, the higher magnetic field inside the sunspot has to be combined with a lower temperature and, consequently, less electromagnetic emission.   □

3.3.2 Magnetic Tension

Let us now have a closer look at a simple interpretation of the first term in (3.60) which is concerned with magnetic tension. The upper panel of Fig. 3.4 shows a homogeneous magnetic field parallel to the x-axis: $\boldsymbol{B}_0 = (B_0, 0, 0,)$. The field is assumed to be frozen-into a plasma. The plasma motion $\boldsymbol{u} = (0, 0, u_z(x))$ (middle panel) leads to the deformation of the field shown in the lower panel. The distorted field can be described as the superposition of the undisturbed field $\boldsymbol{B}_0$ and a disturbance $\delta\boldsymbol{B}$ (see (3.94) in Sect. 3.4.2):

$$\boldsymbol{B} = \boldsymbol{B}_0 + \frac{\partial\boldsymbol{B}}{\partial t}dt = \boldsymbol{B}_0 + \nabla \times (\boldsymbol{u} \times \boldsymbol{B})dt . \tag{3.64}$$

With (3.59) and the field described above we find a force density parallel to the disturbing velocity field

$$f_z = \frac{1}{\mu_0}\frac{\partial^2 u_z}{\partial x^2}B_0^2 dt . \tag{3.65}$$

The force, called the magnetic tension, always is a restoring force: if the field lines have a convex curvature into the upward direction, $\partial^2 u_z/\partial x^2$ is less than zero, leading to a force directed downwards. If the curvature is opposite, the force also is in the opposite direction. The magnetic tension can also be interpreted graphically: magnetic field lines have a tendency to shorten.

Example 12. Again, consider a 5 T magnetic field. It is disturbed by a sinusoidal velocity field $v = v_0 \sin kx$, where $v_0 = 1$ m/s and $k = 5$ m^{-1}, acting for $\delta t = 1$ μs. To determine the force density, we need the second derivative of the velocity: $v'' = -v_0 k^2 \sin kx$. With (3.65), we then obtain $f = 20$ N/m^2 $\sin((5/\text{m})\,x)$ and thus for $x = 0$ m, $f = 0$ m, because here the magnetic field is not displaced from its original position and no restoring force acts on it; for $x = 2.5$ m we obtain 14 N/m^2, and for $x = 5$ m (maximum displacement), $f = 20$ N/m^2. ☐

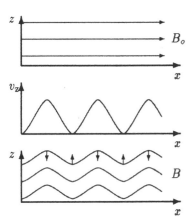

Fig. 3.4. Magnetic tension: the lines of force in a homogeneous magnetic field (*upper panel*) are distorted by a velocity field (*middle panel*), giving rise to a restoring force indicated by the arrows (*lower panel*)

Example 13. Solar Filaments: Another example of the unusual behavior of a plasma are solar filaments or protuberances which are cold and dense matter suspended from magnetic arcades high above the photosphere. Such a structure is called a filament as long as it is seen as a dark (because cold) stripe in front of the photosphere. As the Sun rotates, this structure becomes visible as a bright (because dense) arc extending high above the photosphere: a protuberance. The spatial structure clearly becomes visible: the filament extends vertically above the photosphere, covering angular distances of some 10 degrees. Filaments are stable and can last for a few solar rotations. Under certain conditions they become unstable and are blown out violently as coronal mass ejections (Sect. 6.6). Typical temperatures are about 7000 K (ambient corona: 10^6 K). The density is about 100 times larger than the ambient density; the typical vertical extension is up to 30 Mm, that is about 100 times the scale height in the corona.

The first theoretical description of a filament goes back to Kippenhahn and Schlüter [286]. Here we shall limit ourselves to a much shorter, more general discussion. Filaments are roughly aligned along the neutral line between regions of opposing magnetic fields (Sect. 6.7). Thus the magnetic field seems to play an important role in the existence as well as stability of the filament. Figure 3.5 sketches the situation: the filament (thick vertical line) is supported by magnetic arcades connecting opposite polarities in the photosphere. The magnetic field lines do not form perfect arcades, instead they are ditched-in at the position of the filament: gravity pulls down the filament which in turn pulls down the frozen-in magnetic field. The deformation of the magnetic field causes magnetic tension in the opposite direction. Thus the filament is held at a certain height by an equilibrium between gravity and magnetic tension.

 We can obtain a more quantitative statement from the basic MHD equations. Here we need the equation of motion in the stationary case,

$$\nabla p = \frac{1}{\mu_0}(\nabla \times \boldsymbol{B}) \times \boldsymbol{B} + \varrho \boldsymbol{g} \,, \tag{3.66}$$

and the equation of state $p = nk_{\mathrm{B}}T$, with T being spatially constant.

Following [285] let us define a coordinate system with the xy-plane tangential to the surface of the photosphere and the y-direction extending along the filament into the drawing plane. The z-direction points upward; thus Fig. 3.5 is a cut through the filament in the xz-plane. All quantities are assumed to be independent of y, thus $\partial/\partial y$ is zero. The magnetic field is given as $\boldsymbol{B} = (B_x, 0, B_y)$ and $\boldsymbol{g}$ is $(0,0,-g)$. The double cross product in the equation of motion then can be written as

$$(\nabla \times \boldsymbol{B}) \times \boldsymbol{B} = \left(\frac{\partial B_x}{\partial z} - \frac{\partial B_z}{\partial x} \right) (B_z, 0, B_x) \,. \tag{3.67}$$

Combination of the equation of motion and the equation of state gives

$$k_B T \frac{\partial n}{\partial x} = \frac{1}{\mu_0} B_z \left(\frac{\partial B_x}{\partial z} - \frac{\partial B_z}{\partial x} \right) \tag{3.68}$$

and

$$k_B T \frac{\partial n}{\partial z} = -\frac{1}{\mu_0} B_x \left(\frac{\partial B_x}{\partial z} - \frac{\partial B_z}{\partial x} \right) - mng . \tag{3.69}$$

Differentiation of (3.68) to z and of (3.69) to x and subtraction of these equations yields

$$0 = \left(\frac{\partial B_z}{\partial z} + \frac{\partial B_x}{\partial x} \right) \left(\frac{\partial B_x}{\partial z} - \frac{\partial B_z}{\partial x} \right)$$

$$+ B_z \left(\frac{\partial^2 B_x}{\partial z \partial x} - \frac{\partial^2 B_z}{\partial x^2} \right) + \mu_0 \frac{\partial p}{\partial x} g . \tag{3.70}$$

Gauss's law for a magnetic field in the two-dimensional case gives

$$\frac{\partial^2 B_z}{\partial x \partial z} = -\frac{\partial^2 B_x}{\partial x^2} \quad \text{and} \quad \frac{\partial^2 B_x}{\partial x \partial z} = -\frac{\partial^2 B_z}{\partial z^2} . \tag{3.71}$$

With $H_p = kT/mg$ being the scale height in the barometric height formula $p = p_0 \exp\{- \int dz / H_p\}$ and substitution of $\partial n / \partial x$ according to (3.68), (3.70) can be written as

$$B_z \nabla^2 B_x - B_x \nabla^2 B_z + \frac{1}{H_p} B_z \left(\frac{\partial B_x}{\partial z} - \frac{\partial B_z}{\partial x} \right) = 0 . \tag{3.72}$$

From this equation the details of the magnetic field as well as the density inside the filament can be determined and compared with observations. These results confirm the sharp bend in the magnetic field: inside the filament the lines of force are bent according to tanh. Thus there is still a steady field and not a discontinuity. In addition, the model predicts a decrease in density with increasing height, as is evident from the observations. This decrease leads to a flattening of the ditch in the field lines with increasing height, which also is indicated in Fig. 3.5.

Formally, the fine structure inside the filament, which is also a crucial factor for its stability, can be derived by simplifying (3.72). Since we are

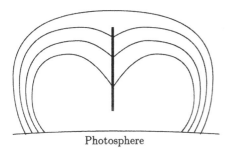

Photosphere

Fig. 3.5. Model of a solar filament (*thick vertical line*) suspended from magnetic arcades

interested only in the details inside the filament, we are concerned with a height range small compared with the total extension of the structure. In this case we can assume $\partial/\partial z = 0$. From Gauss's law, $\nabla \cdot \boldsymbol{B} = 0$ for the magnetic field, we obtain $B_x = \text{const}$ and $B_z = B_z(x)$. In this case, (3.72) gives

$$-B_x \frac{\partial^2 B_z}{\partial x^2} - \frac{B_z}{H_p} \frac{\partial B_z}{\partial x} = 0 \tag{3.73}$$

or, with $\alpha = 1/(H_p B_x) = \text{const}$,

$$\frac{\partial^2 B_z}{\partial x^2} + \alpha B_z \frac{\partial B_z}{\partial x} = 0 \ . \tag{3.74}$$

Integration gives

$$\frac{\partial B_z}{\partial x} + \frac{\alpha}{2} B_z^2 = \text{const} \ . \tag{3.75}$$

A solution of this differential equation is

$$B_z = B_z^\infty \tanh \xi \ , \quad \text{where} \quad \xi = \frac{B_z^\infty}{2 H_p B_x} x \tag{3.76}$$

and $B_z^\infty = \text{const}$ is the value of B_z for $z \to \infty$. Inside the filament, the field lines therefore do not exhibit a sharp kink but the smooth evolution of a tanh function.   □

3.4 Magnetohydrokinematics

Magnetohydrokinematics deals with the reaction of the electromagnetic field to a prescribed velocity field such that the electromagnetic field does not influence the velocity field. Thus, we do not have to solve the equation of motion. Such a situation corresponds to a large plasma-β or a small value of S in (3.49). The basic equations to derive concepts such as frozen-in fields and the dissipation of fields are Maxwell's equations and Ohm's law.

3.4.1 Frozen-in Magnetic Fields

What happens to an electromagnetic field embedded in a moving medium with high conductivity? Let us assume a magnetic field $\boldsymbol{B}(\boldsymbol{r}, t_0)$ at a time t_0 and a prescribed velocity field $\boldsymbol{u}(\boldsymbol{r}, t)$. The magnetic flux through a surface S enclosed by a curve C then is $\Phi = \int \boldsymbol{B} \, d\boldsymbol{S}$. Let us follow the motion of C (see Fig. 3.6): as C moves, the magnetic flux through S changes because (a) the magnetic field varies in time and (b) the field lines move into or out of S. As C moves, it creates a cylinder with a mantle surface M. All changes in flux through S due to the field lines entering or leaving C is associated with a flux of the very same magnetic field lines through M. Thus the total

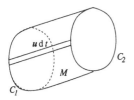

Fig. 3.6. Moving fluid line C to derive the concept of the frozen-in magnetic field

change in magnetic flux through S can be described by

$$\mathrm{d}\Phi = \Phi_2 - \Phi_1 = \mathrm{d}t \int\limits_{S_1} \frac{\partial \boldsymbol{B}}{\partial t}\,\mathrm{d}\boldsymbol{S}_1 + \int\limits_{M} \boldsymbol{B}\,\mathrm{d}\boldsymbol{S}_M\ . \tag{3.77}$$

A surface element S_M on M is given as $\mathrm{d}\boldsymbol{S}_M = \boldsymbol{u} \times \mathrm{d}\boldsymbol{l}\,\mathrm{d}t$ with $\mathrm{d}\boldsymbol{l}$ being the path along C and $\boldsymbol{u}\,\mathrm{d}t$ being the path along the direction of motion of C. Equation (3.77) therefore yields

$$\frac{\mathrm{d}\Phi}{\mathrm{d}t} = \int\limits_{S_1} \frac{\partial \boldsymbol{B}}{\partial t}\,\mathrm{d}\boldsymbol{S}_1 + \int\limits_{C_1} \boldsymbol{B} \cdot \boldsymbol{u} \times \mathrm{d}\boldsymbol{l}_1\ . \tag{3.78}$$

With Stokes' theorem the last term can be written as

$$\int\limits_{C_1} \boldsymbol{B} \cdot \boldsymbol{u} \times \mathrm{d}\boldsymbol{l}_1 = -\int\limits_{C_1} \boldsymbol{u} \times \boldsymbol{B} \cdot \mathrm{d}\boldsymbol{l}_1 = -\int\limits_{S_1} \nabla \times (\boldsymbol{u} \times \boldsymbol{B})\,\mathrm{d}\boldsymbol{S}_1\ . \tag{3.79}$$

Inserting into (3.78) gives

$$\frac{\mathrm{d}\Phi}{\mathrm{d}t} = \int\limits_{S_1} \left[\frac{\partial \boldsymbol{B}}{\partial t} - \nabla \times (\boldsymbol{u} \times \boldsymbol{B}) \right] \mathrm{d}\boldsymbol{S}_1\ . \tag{3.80}$$

The $\boldsymbol{u} \times \boldsymbol{B}$ term can be expressed by Ohm's law (3.45) while the $\partial \boldsymbol{B}/\partial t$ term can be expressed by Faraday's law (3.43), and we obtain

$$\frac{\mathrm{d}\Phi}{\mathrm{d}t} = -\int\limits_{S_1} \nabla \times \boldsymbol{j}\frac{1}{\sigma}\mathrm{d}\boldsymbol{S}_1 = -\int\limits_{C_1} \frac{1}{\sigma}\boldsymbol{j} \cdot \mathrm{d}\boldsymbol{l}_1\ . \tag{3.81}$$

The change in magnetic flux through a moving surface therefore is proportional to $1/\sigma$. If σ converges towards infinity, $\mathrm{d}\Phi/\mathrm{d}t$ converges towards zero: in a medium with infinite conductivity σ, the magnetic field is frozen-into the plasma and carried away by the matter as if glued to it (left side in Fig. 3.7). A prime example of the application of this concept is the interplanetary magnetic field frozen-into the solar wind (Sect. 6.3). A reversal of the concept, the frozen-out field, exists too. In the right panel of Fig. 3.7 a field-free plasma bubble moves towards a region filled with a magnetic field and pushes the field away until its kinetic energy is transferred to additional field energy as evidenced by an increase in magnetic pressure as well as magnetic tension. The field cannot enter into the bubble because then the magnetic flux inside the bubble would change. An example is the solar wind frozen-out of the Earth's magnetic field (Sect. 8.2).

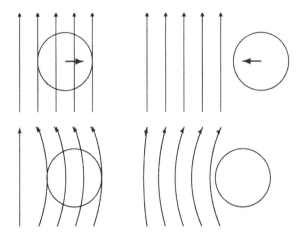

Fig. 3.7. Frozen-in (*left*) and frozen-out (*right*) magnetic fields

3.4.2 Deformation and Dissipation of Fields

Frozen-in magnetic fields always are connected with an infinite conductivity. But what happens to a magnetic field embedded in a flow with finite conductivity? For simplicity let us assume σ to be spatially and temporally constant. The combination of Faraday's law (3.43) and Ohm's law (3.45) yields

$$\frac{\partial \boldsymbol{B}}{\partial t} - \nabla \times (\boldsymbol{u} \times \boldsymbol{B}) = -\frac{1}{\sigma} \nabla \times \boldsymbol{j} \ . \tag{3.82}$$

The current density can be expressed by Ampére's law (3.44), leading to

$$\frac{\partial \boldsymbol{B}}{\partial t} - \nabla \times (\boldsymbol{u} \times \boldsymbol{B}) = -\frac{1}{\sigma} \nabla \times \frac{\nabla \times \boldsymbol{B}}{\mu_0} = -\frac{1}{\mu_0 \sigma} \nabla \times (\nabla \times \boldsymbol{B}) \ . \tag{3.83}$$

The double cross product can be simplified with (A.26):

$$\frac{\partial \boldsymbol{B}}{\partial t} - \nabla \times (\boldsymbol{u} \times \boldsymbol{B}) = \frac{1}{\mu_0 \sigma} \nabla^2 \boldsymbol{B}. \tag{3.84}$$

This equation allows us to determine how a given velocity field $\boldsymbol{u}$ deforms a magnetic field $\boldsymbol{B}$.

Deformation of the Field in a Plasma Flow. If we assume both magnetic field and plasma flow to be independent of time, a stationary solution exists with

$$-\nabla \times (\boldsymbol{u} \times \boldsymbol{B}) = \frac{1}{\mu_0 \sigma} \nabla^2 \boldsymbol{B} \ . \tag{3.85}$$

With a characteristic time scale τ and a characteristic length scale L, the flow speed $u_\perp$ perpendicular to the field can be estimated:

$$u_\perp \approx \frac{L}{\tau} \approx \frac{1}{\mu_0 \sigma L} \ . \tag{3.86}$$

The physical interpretation is simple: if a plasma flows perpendicular to the magnetic field, it deforms the lines of force until their characteristic scale length is small enough to fulfill (3.86). Then the plasma starts to flow across the lines of force.

Excursion 3. *Dimensionless Variables and Dimensional Stability.* To deter- mine the deformation of a line of force quantitatively, we shall use the technique of dimensionless variables. This technique is quite common in fluid dynamics (see e.g. [149]). It is helpful to determine not only one solution of the differential equation but an entire manifold of solutions which can be scaled to the situation under study. This is particularly helpful in hydrodynamics when the solution for a certain size of syringe or nozzle is known and we are looking for a dynamically similar flow on a different scale.

The idea is quite simple: all equations representing scientific laws can be expressed such that both sides are dimensionless. In its simplest case, just divide one side of the equation by the other: the result, one, is dimensionless. To take advantage of dimensionless variables, first identify the physical variables relevant to the problem and combine them into dimensionless groups $A, B, C....$ These groups have to be independent of one another. If the groups are dimensionless, combinations of groups such as AB or A/B^2 are dimensionless, too. But they are not independent of either A or B, though any one of them might be included instead of A or B if this seems advantageous. If the groups are chosen in such a way that the quantity of interest occurs in only one of them, it can be expressed by the function $A = f(B, C, ...)$. The nature of this unknown function can be determined analytically (as demonstrated below) or by computational methods. In an analytical solution, the advantage of the use of dimensionless variables is small; it only shows which parameters are important in scaling. If the solution has to be obtained by numerical simulations, the advantage of this method is more obvious: the procedure to determine a solution for one particular set of parameters can be quite time consuming. Each other set of parameters would require a new run. If dimensionless variables are used instead, the nature of the solution becomes obvious and it can be scaled to suit different sets of parameters. □

Let us now follow this principle and introduce new variables as suggested in [285]:

$$\boldsymbol{B} = b\tilde{\boldsymbol{B}} , \quad \boldsymbol{r} = r\tilde{\boldsymbol{r}} , \quad t = \tau\tilde{t} , \quad \boldsymbol{u} = U\tilde{\boldsymbol{u}} , \quad U = \frac{L}{\tau} . \tag{3.87}$$

The quantities with a tilde are dimensionless. With the abbreviation

$$\eta = \frac{1}{\mu_0\sigma} , \tag{3.88}$$

which can be interpreted as a magnetic viscosity, (3.84) yields

$$\frac{b}{\tau}\frac{\partial\tilde{\boldsymbol{B}}}{\partial\tilde{t}} - \frac{Ub}{L}\tilde{\nabla}\times(\tilde{\boldsymbol{u}}\times\tilde{\boldsymbol{B}}) = \frac{\eta b}{L^2}\tilde{\nabla}^2\tilde{\boldsymbol{B}} . \tag{3.89}$$

Here a tilde above the differential operator indicates that the operator refers to a dimensionless variable.

In ordinary hydrodynamics, the Reynolds number is a measure of the ratio between inertial and viscous forces. If the Reynolds number exceeds a critical value, the flow becomes turbulent. In its definition, the Reynolds number contains typical scales which have to be adjusted to the problem under study. Here we shall use a magnetic Reynolds number:

$$R_{\mathrm{M}} = \frac{UL}{\eta} = \mu_0 \sigma UL . \tag{3.90}$$

It differs from the ordinary Reynolds number in so far as the viscous forces described by η do not depend on forces between particles but on the conductivity of the medium (see (3.88)). The magnetic Reynolds number can also be interpreted as the ratio of the time scale of ohmic diffusion

$$\tau_{\mathrm{diff}} = \frac{4\pi L^2}{c^2 \eta} \tag{3.91}$$

to the advective time scale

$$\tau_{\mathrm{adv}} = \frac{L}{v} \tag{3.92}$$

equal to

$$R_{\mathrm{M}} = \frac{\tau_{\mathrm{diff}}}{\tau_{\mathrm{adv}}} = \frac{4\pi v L}{c^2 \eta} \sim \frac{4\pi}{c^2 \eta} \frac{\nabla \times (\boldsymbol{v} \times \boldsymbol{B})}{\nabla \times (\nabla \times \boldsymbol{B})} . \tag{3.93}$$

The latter is the ratio of the induction term to the dissipation term of the induction equation (3.84).

Now we can rewrite (3.89) as

$$\frac{\partial \tilde{\boldsymbol{B}}}{\partial \tilde{t}} - \tilde{\nabla} \times (\tilde{\boldsymbol{u}} \times \tilde{\boldsymbol{B}}) = \frac{1}{R_{\mathrm{M}}} \tilde{\nabla}^2 \tilde{\boldsymbol{B}} . \tag{3.94}$$

This dimensionless form has an advantage: it shows directly that a three-fold set of solutions exits. What does this mean? Assume we know a solution $\tilde{\boldsymbol{B}}(\tilde{\boldsymbol{r}}, \tilde{t})$ in dimensionless variables for a fixed Reynolds number R_{M} and a velocity field $\tilde{\boldsymbol{u}}$. In this case, $b\tilde{\boldsymbol{B}}(L\tilde{\boldsymbol{r}}, \tau\tilde{t})$ also is a solution of the same Reynolds number and the velocity field $U\tilde{\boldsymbol{u}}$ as long as the conditions $UL/\eta = R_{\mathrm{M}}$ and $L/\tau = U$ are fulfilled. For instance, a free choice of U and L for a given R_{M} determines the values τ and η of the solution. But we still have a free choice for b. Thus one solution of (3.94) contains a three-fold infinite manifold of solutions, characterized, for instance, by U, η, and b.

Let us now determine the solution of (3.94) for a stationary parallel flow perpendicular to a homogeneous magnetic field. Since the flow is stationary, (3.94) reduces to

$$-\tilde{\nabla} \times (\tilde{\boldsymbol{u}} \times \tilde{\boldsymbol{B}}) = \frac{1}{R_{\mathrm{M}}} \tilde{\nabla}^2 \tilde{\boldsymbol{B}} . \tag{3.95}$$

Let us orientate the x-axis of a Cartesian coordinate system along the flow: $\tilde{\boldsymbol{u}} = (\tilde{u}_x(z), 0, 0)$. The magnetic field $\tilde{\boldsymbol{B}} = (\tilde{B}_x(z), 0, \tilde{B}_z(z))$ has one component parallel and another perpendicular to the flow. Note that the flow varies along the perpendicular component. From $\nabla \times \boldsymbol{B} = 0$ we find that $\tilde{B}_z$ is constant. Equation (3.95) is a second-order linear inhomogeneous partial differential equation for $\tilde{B}_x$ as a function of z:

$$\frac{\partial^2 \tilde{B}_x}{\partial z^2} = -\tilde{B}_z R_{\mathrm{M}} \frac{\partial \tilde{u}_x}{\partial z} \ . \tag{3.96}$$

Integrating twice we get

$$\tilde{B}_x = -\tilde{B}_z R_{\mathrm{M}} \left(\int \tilde{u}_x \, \mathrm{d}z + C\tilde{z} + D \right) , \tag{3.97}$$

with C and D to be determined to fulfill the boundary conditions. Let us now assume the flow to have a cosine profile around $z = 0$, that is $\tilde{u}_x = \cos z$ for $|\tilde{z}| \leq \pi/2$ and $\tilde{u}_{\mathrm{x}} = 0$ for $|\tilde{z}| \geq \pi/2$. Equation (3.97) then reads

$$\tilde{B}_x = -\tilde{B}_z R_{\mathrm{M}} (\sin \tilde{z} + C\tilde{z} + D) \ . \tag{3.98}$$

The tangential component of $\tilde{B}$ should be steady at the boundary of the flow to avoid currents; thus one boundary condition is $\tilde{B}_x(\tilde{z} = \pi/2) = 0$. In addition, the flow is assumed to be symmetric around $\tilde{z} = 0$; thus the second boundary condition is $\tilde{B}_x(0) = 0$. The integration constants therefore are $D = 0$ and $C = -\pi/2$ and (3.98) can be written as

$$\tilde{B}_x = -\tilde{B}_z R_{\mathrm{M}} \left(\sin \tilde{z} - \frac{1}{\pi} \tilde{z} \right) \ . \tag{3.99}$$

We can now define a magnetic stream function $\tilde{\psi}$

$$\frac{\partial \tilde{\psi}}{\partial \tilde{x}} = -\tilde{B}_z \quad \text{and} \quad \frac{\partial \tilde{\psi}}{\partial \tilde{z}} = \tilde{B}_x \ . \tag{3.100}$$

Then $\nabla \tilde{\psi} \times \nabla \tilde{B}$ is zero and lines with $\tilde{\psi} = \mathrm{const}$ are the field lines. Integration of the second part of (3.100) combined with (3.99) gives the line of force as

$$\tilde{\psi} = \tilde{B}_z R_{\mathrm{M}} \left(\cos \tilde{z} + \frac{\tilde{z}^2}{\pi} \right) - \tilde{B}_z \tilde{x} \ . \tag{3.101}$$

Let us now determine the maximum displacement of a line of force (see Fig. 3.8). Because $\tilde{\psi}$ is constant along a field line, it is $\tilde{\psi}(\Delta \tilde{x}, 0) = \tilde{\psi}(0, \pi/2)$. Therefore the maximal displacement is given as

$$\Delta \tilde{x} = \left(1 - \frac{\pi}{4} \right) R_{\mathrm{M}} \ . \tag{3.102}$$

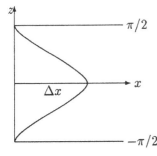

Fig. 3.8. Deformation of a magnetic field line by a plasma flow. The maximal displacement Δx is determined by the Reynolds number R_M

Thus the deformation of the magnetic field line increases with increasing Reynolds number. This is not surprising because the Reynolds number depends linearly on the conductivity (see (3.90)): if the conductivity and therefore the Reynolds number is infinite, the magnetic field is frozen into the fluid and deformation of the field lines becomes maximal.

Note that, in contrast to the frozen-in case, for finite conductivity matter starts to flow across the field after it is curved according to (3.101). The flow across the field is largest for small dimensions because on small scales the condition for frozen-in fields can be violated more easily. Thus we can confirm the suggestion made in connection with (3.86): at each point the flow curves the field such that the radius of curvature becomes small enough to allow for a flow of matter across the field. The physical reason is a reduction of the dissipation time with decreasing spatial scales as will be described below.

Dissipation of Fields. Let us now have a look at a vanishing external velocity field. Then the second term on the left-hand side of (3.84) vanishes and we get

$$\frac{\partial \boldsymbol{B}}{\partial t} = \frac{1}{\mu_0 \sigma} \nabla^2 \boldsymbol{B} \ . \tag{3.103}$$

Formally, this equation is equivalent to the heat conduction equation

$$\frac{\partial T}{\partial t} = \chi \nabla^2 T \ , \tag{3.104}$$

where χ is the thermal conductivity, and to the vorticity equation

$$\frac{\partial \boldsymbol{\omega}}{\partial t} = \nu \nabla^2 \boldsymbol{\omega} \ , \tag{3.105}$$

where $\boldsymbol{\omega} = \nabla \times \boldsymbol{u}$ is vorticity that describes the rotational state of the fluid. Note that the coefficients $(\mu_0 \sigma)^{-1}$, κ, and ν all have the same dimensions $(\mathrm{m}^2/\mathrm{s})$.

While Parks [397] gives a formal description of the consequences of (3.103), we shall use a more graphical approach. Let us align the x-axis of our coordinate system parallel to the magnetic field direction: $B = (B_x(y,t), 0, 0)$. Equation (3.103) then can be simplified to

$$\frac{\partial B_x}{\partial t} = \frac{1}{\mu_0 \sigma} \frac{\partial^2 B_x}{\partial y^2} \,. \tag{3.106}$$

Formally, this is equivalent to the one-dimensional heat conduction equation

$$\frac{\partial T}{\partial t} = \chi \frac{\partial^2 T}{\partial y^2} \,, \tag{3.107}$$

where $T(y)$ is the one-dimensional distribution of temperature and χ is the thermal conductivity. Equation (3.107) gives the temporal change in temperature as the heat is transported away by conduction. Therefore, (3.106) gives the temporal change in magnetic field strength as the magnetic field is transported by a process which depends on conductivity: the field dissolves. Assume that B is particularly strong at a certain position, say $y = 0$. This is analogous to a very hot spot on a metal rod; here we would expect the hot spot to cool down while the other parts of the rod warm up as heat is transported towards them. The same thing happens with the magnetic field: it dissolves to larger values of $|y|$. Note that, while the magnetic flux inside the yz-plane stays constant during this process, the magnetic energy decreases because the field-generating currents are associated with ohmic losses.

If τ is a characteristic time scale for magnetic field changes (e.g. the dissipation time during which the field strength decreases to $1/e$) and L is the characteristic scale length, the change in B can be estimated from (3.106):

$$\frac{B}{\tau} \approx \frac{1}{\mu_0 \sigma} \frac{B}{L^2} \,. \tag{3.108}$$

Thus the dissipation time is

$$\tau \approx \mu_0 \sigma L^2 = L^2 / D_\mathrm{M} \,, \tag{3.109}$$

where $D_\mathrm{M} = 1/\mu_0 \sigma = \eta$ can be interpreted as a magnetic diffusion coefficient. τ depends on the square of the characteristic scale length of the field: smaller fields dissipate faster than larger ones. That is the reason why with reduced spatial scales plasma starts to flow across the field. In addition, the dissipation time increases with increasing conductivity: for infinite conductivity, the dissipation time becomes infinite too, leading to the frozen-in field.

The Sun, for instance, has a linear dimension of about 7×10^8 m and an average conductivity of 2.6 A/Vm. This gives a dissipation time of about 1.2×10^{10} years, nearly three times the age of the Sun. Thus if during its creation the Sun had received a magnetic field, this still would be present today as first suggested in [302]. On the other hand, the solar magnetic field is highly variable on time scales of months to years (Sect. 6.6), making the presence of a fossil field very unlikely. Instead, a MHD dynamo (Sect. 3.6) seems to be responsible for the solar magnetic field.

Fig. 3.9. The dissipation time of the magnetic field decreases as the spatial scales decrease

According to (3.109) the dissipation time depends on the scale of the field. Thus, if the field on the left-hand side of Fig. 3.9 is divided into smaller patches of length L/n instead of L, it dissipates n^2 times faster than the original field. Such a redistribution of field lines smearing out the boundaries between regions of opposite polarity and leading to structures on smaller scales can result from turbulent plasma motions. For instance, the stochastic motions in the photospheric and chromospheric network on the Sun might contribute to the dissipation of magnetic fields, in particular in the declining phase of the solar cycle.

 Example 14. A sunspot with a radius of about 20 000 km has, from (3.109), with the conductivity given above, a lifetime of about 1000 years. If we look more closely at the sunspot, in particular the granules around the spot, we find a spatial scale of about 1000 km, that is, $1/20$ of the scale of the sunspot. Since the dissipation time depends on the square of the length scale, in the granules it is only $1/400$ of the value for the whole sunspot, that is 2.5 years – which comes closer to the observed lifetime of a sunspot.   □

A vortex in the plasma flow might even create a field-free region inside an otherwise relatively undisturbed field [556].

3.5 Reconnection

 The dissipation of magnetic field lines is important, e.g. in reconnection, which is assumed to take place in many locations in the solar system, such as solar flares, the tails of magnetospheres, and in the exchange of solar wind and magnetospheric plasma at the day-side magnetopause (flux-transfer events). Reconnection not only plays an important role in the rearrangement of magnetic fields but also in the formation of shock waves and the acceleration of energetic particles.

The concept of reconnection goes back to Petschek [405]. It is widely used in magnetospheric and solar physics, although the physics behind the process still is under debate; sometimes it is even questioned whether reconnection really exists. The basics of reconnection are outlined in Fig. 3.10. Reconnection requires a topology where two magnetic flux tubes of opposite polarity meet (a). According to Ampére's law, in the neutral line between these flux tubes a current flows perpendicular to the drawing plane with a current density

$$j = \frac{1}{\mu_0} \frac{\Delta B}{d} . \tag{3.110}$$

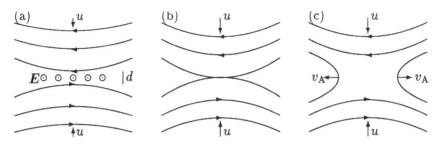

Fig. 3.10. Reconnection: merging of magnetic field lines leads to a rearrangement of fields. In addition, magnetic field energy is released, heating the plasma, creating a shock wave, and accelerating particles

The flux tubes are frozen into a plasma with infinite conductivity. As a plasma flow u pushes the flux tubes towards the neutral line, an X-point configuration arises where anti-parallel field lines meet (b). As the distance between the flux tubes decreases, the current density (3.110) increases and may surpass the current density the plasma can carry. Then the current becomes unstable, leading to a finite conductivity. Now the frozen-in approximation breaks down and magnetic field diffusion starts. At the X-point, magnetic field lines merge. Magnetic tension leads to a shortening of the merged field lines, pulling them away from the former X-point (c). The energy of the terminated neutral line current is converted to high-speed tangential flows, indicated by v_A. The speed of this plasma flow might exceed the local Alfvén speed, forming two shock waves propagating away from the reconnection site. The shocks, in turn, might lead to particle acceleration (Sect. 7.5).

The properties of the current sheet are determined by (3.84). For an infinitesimally thin current sheet and a uniform resistivity, a self-similar solution for the magnetic field component $B_\perp$ perpendicular to the current sheet and parallel to the flow can be determined [14, 110, 113]:

$$B_\perp = B_0 \, \mathrm{erf}(\xi) \, , \quad \text{with} \quad \xi = \sqrt{\frac{\mu_0 \sigma}{t}} \, l_\perp \, , \tag{3.111}$$

where $l_\perp$ is the spatial coordinate along $B_\perp$ and erf the error function

$$\mathrm{erf} = \frac{2}{\sqrt{\pi}} \int_0^\xi e^{\xi^2} \, \mathrm{d}\xi \, . \tag{3.112}$$

Since the parameter ξ depends also on the time t, there is a temporal variation in the width d of the current sheet. The latter can be determined by setting $\xi = 1$:

$$d = \sqrt{\frac{4t}{\mu_0 \sigma}} \, . \tag{3.113}$$

The current density associated with this magnetic field profile is a Gaussian centered around the middle of the current sheet and spreading with time t.

According to [113], the magnetic field energy available for a slice of the current sheet is $W_B = \int w_B \, dl_\perp = \int B^2/(2\mu_0) \, dl_\perp$; the rate of energy conversion can be determined from (3.84) and Ampére's law (3.44) as

$$\frac{\partial W_B}{\partial t} = -\int \boldsymbol{E} \cdot \boldsymbol{j} \, dl_\perp \; . \tag{3.114}$$

If the onset of reconnection does not modify the general field and plasma configuration, stationary reconnection results, as an equilibrium between inflowing mass and magnetic flux, magnetic diffusion, and out-flowing mass and magnetic flux. This is also called steady-state reconnection. In Sweet–Parker reconnection [392, 513], a diffusion region of width d and length L is assumed with $L \gg d$, similar to the configuration in Fig. 3.10. The rate of reconnection and the properties of the outflow can be determined from the conservation of mass, momentum, energy, and magnetic flux.

Solving Ohm's law (3.45) for the electric field and expressing the current density $\boldsymbol{j}$ by Ampére's law (3.44), we obtain the following for the electric field sheet:

$$\boldsymbol{E} = \frac{\boldsymbol{j}}{\sigma} - \boldsymbol{u} \times \boldsymbol{B} = \frac{\nabla \times \boldsymbol{B}}{\mu_0 \sigma} - \boldsymbol{u} \times \boldsymbol{B} = \boldsymbol{u}_0 \times \boldsymbol{B}_0 = \text{const} \; . \tag{3.115}$$

Since we assume steady-state conditions, Faraday's law (3.43) gives $\boldsymbol{E} = \text{const} = \boldsymbol{E}_0$. Outside the current sheet the conductivity is high and the magnetic field is frozen into the plasma flow. Here all electric fields vanish immediately, except for the electric induction field $\boldsymbol{u} \times \boldsymbol{B}$. Therefore we have $\boldsymbol{E}_0 = \boldsymbol{u} \times \boldsymbol{B} = \boldsymbol{u}_0 \times \boldsymbol{B}_0$ outside the current sheet. Inside the current sheet the situation is different: here the conductivity is finite, the frozen-in condition breaks down, and the plasma speed vanishes. Thus the induction field vanishes too, and according to (3.115) the electric field is

$$\boldsymbol{E} = \frac{\nabla \times \boldsymbol{B}}{\mu_0 \sigma} \; . \tag{3.116}$$

The diffusion region is characterized by $R_M < 1$. Taking the width (3.113) of the diffusion layer as the characteristic length scale and assuming $R_M \to 1$, we obtain the width of the layer as

$$d \approx \frac{1}{\mu_0 \sigma u_0} \; . \tag{3.117}$$

For a current sheet of infinite length L, as indicated in the left-hand panel in Fig. 3.11, the converging plasma streams would lead to a pileup of plasma density and magnetic field inside the current sheet. This is incompatible with the assumption of a steady state. Instead, an outflow of mass and magnetic field out of the diffusion region is required, as sketched earlier in the right-hand panel in Fig. 3.10. This outflow is possible only for a finite extent of

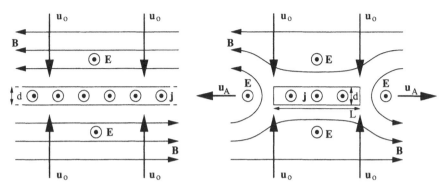

Fig. 3.11. Fields and currents in reconnection for a current sheet of infinitesimal length (*left*), and a current sheet of finite length L larger than the width d (*right*)

the diffusion region, as already mentioned above. The right-hand panel of Fig. 3.11 allows a closer look into the diffusion region. To reach a steady state, the outflow of plasma and magnetic field must equal its inflow, i.e.

$$\boldsymbol{u} \times \boldsymbol{B} = \boldsymbol{u}_A \times \boldsymbol{B}_A = \boldsymbol{u}_0 \times \boldsymbol{B}_0 \, , \tag{3.118}$$

or in scalar form (because all velocities are perpendicular to the fields),

$$uB = u_A B_A = u_0 B_0 \, , \tag{3.119}$$

and from the equation of continuity,

$$u_A d = u_0 L \, . \tag{3.120}$$

With (3.117), we therefore obtain

$$u_0 = \frac{u_A d}{L} = u_A^2 \frac{L}{\sigma \mu_{0} u_0} = \frac{u_A}{\sqrt{R_M^{\text{out}}}} \, , \tag{3.121}$$

where $R_M^{\text{out}} = \sigma \mu_0 u_A L$ is the magnetic Reynold's number in the outflow region. The magnetic field in the outflow region can be determined from (3.119) as

$$B_A = B_0 \frac{u_0}{u_A} = B \frac{d}{L} \, . \tag{3.122}$$

Since we start from $d \ll L$, we also obtain $B_A \ll B_0$ and $u_0 \ll u_A$. The outflow speed can be determined from the energy balance: the inflow of kinetic and magnetic energy must be balanced by its outflow or, formally,

$$2Lu_0 \left(\frac{1}{2} \varrho u_0^2 + \frac{B_0}{2\mu_0} \right) = 2du_A \left(\frac{1}{2} \varrho u_A^2 + \frac{B_A}{2\mu_0} \right) \, . \tag{3.123}$$

With (3.120), this gives

$$\varrho u_0^2 + \frac{B_0}{\mu_0} = \varrho u_A^2 + \frac{B_A}{\mu_0} \ . \tag{3.124}$$

Solving for u_A gives

$$u_A^2 = u_0^2 + v_A^2 \left(1 - \frac{u_0^2}{u_A^2} \right) \ , \tag{3.125}$$

where $v_A = B_0/\sqrt{\mu_0 \varrho}$ is the Alfvén speed (4.38) of the incoming flow. The equation has two solutions, $u_A = u_0$ (for $d = L$) and $u_A = v_A$. The outflow speed of the plasma is equal to the Alfvén speed $v_{A,in}$ in the incoming plasma flow, and the rate of reconnection R_{SP} equals the Mach number of the incident flow:

$$R_{SP} = \sqrt{\frac{1}{L\sigma v_{A,in}\mu_0}} \ . \tag{3.126}$$

Thus the reconnection process depends on the conductivity. For space plasmas, where the conductivity is high, a low rate of reconnection results. The Sweet–Parker reconnection therefore is a slow process in which about half of the incoming magnetic energy is converted into kinetic energy of the out-flowing plasma. This acceleration leads to the two high-speed plasma flows indicated in panel (c) in Fig. 3.10.

Petschek reconnection occurs in more localized regions; the process is faster because the length scale L is smaller; or more correctly: the length scale of the diffusion region equals the length scale of the system. In Petschek reconnection about three-fifth of the inflowing magnetic energy are converted into kinetic energy behind the shock waves, the remaining two-fifths is used to heat the plasma. The reconnection rate R_P is given as

$$R_P = \frac{\pi}{8} \ln \left(\sqrt{\frac{1}{L\sigma v_{A,in}\mu_0}} \right) \ . \tag{3.127}$$

Petschek reconnection varies less with conductivity and therefore is much more efficient in mixing plasmas and fields. And, with a more efficient reconnection, the resulting acceleration becomes more violent, too.

Sweet–Parker reconnection appears to play an important role at the magnetopause where the high-speed flows streaming away from the reconnection side can be detected in situ. Petschek reconnection probably does not play a role in magnetospheric plasmas but might be important in solar flares. Note that the geometries sketched in Figs. 3.10 and 3.11 probably best are realized in the current sheet of the magnetotail. Geometries in flares and on the day side of the magnetosphere are less symmetric.

Magnetic reconnection is not only a theoretical concept applied to various space plasmas; see e.g. [463]. There exist also some laboratory experiments, as summarized in [61].

3.6 The Magnetohydrodynamic Dynamo

Magnetic fields can be found almost everywhere in space. The magnetosphere could not exist without the magnetic field of the Earth, interplanetary space is structured by the solar magnetic field frozen-into the solar wind, and the Sun itself would be a boring star were it not for the magnetic field. But these fields are not permanent: the Sun reverses polarity in an 11-year cycle and polarity reversals of the Earth's magnetic field are known too. Thus these fields cannot be remnants of a fossil field left from the time of the big bang. Instead, a mechanism is required that generates these fields and also allows for the (quasi-cyclic) variations. In such a magnetohydrodynamic (MHD) dynamo a residual seed field is amplified. The energy required for this process is drawn from the rotational energy of the star or the planet. Thus the motion of the plasma drives a dynamo, which amplifies a seed field and preserves it against losses. If we use the solar radius as the scale length and a conductivity of 2.6 A/V m in (3.86), a velocity of the order of 10^{-9} m/s results: thus very small flow speeds are sufficient to compensate for the dissipation of magnetic energy. Our current understanding of MHD dynamos is summarized in [425].

3.6.1 The Idea

In principle, a dynamo consists of a permanent magnet and a rotating circuit loop in which the current is induced. In the hot interior of the Sun and the planets, permanent magnets cannot exist. Thus the static magnetic field must be created by a current, too. Part of the current induced into the circuit loop than is fed back into the system to support the static field. Without such a feedback, the MHD dynamo would not work.

In the core of the Sun or the planets such well-defined parts as coils or rotating wires do not exist. Instead, we find a homogeneous and highly conductive fluid, rotating with the star or planet. Thus the dynamo also is called a homogeneous dynamo. Since the matter inside the core is liquid, the question of how to create a magnetic field can be reduced to a simpler form: What is the nature of the plasma flow that allows to support the required currents?

Since we want to apply the dynamo to planets and stars, the model has to explain the most important features of their magnetic fields, such as: (a) the magnetic flux density increases with increasing rotation speed, (b) to first-order, the field is dipole like, (c) the dipole axis and the axis of rotation are nearly parallel, (d) the dynamo should allow for fluctuations in the magnetic field direction and flux density, and (e) polarity reversals with quasi-periodic but nonetheless stochastic character should be allowed. This latter point means that the reversal period can be identified (for instance 11 years for the Sun and about 500 000 years for the Earth), but that the individual cycle lengths are distributed stochastically around this average.

Since the fields are axial-symmetric, a configuration as the uni-polar inductor in tempting. There a metal cylinder rotates parallel to a homogeneous magnetic field, leading to a potential difference between the center and the mantle of the cylinder. But in the uni-polar inductor the field cannot be amplified. For astrophysical plasmas this is expressed by Cowling's theorem [111], dating back to 1934: there is no finite velocity field that can maintain a stationary axial-symmetric magnetic field. The proof of this theorem is based on the induction equation (3.84) which, under the conditions cited in Cowling's theorem, would allow for decaying magnetic fields only.

3.6.2 The Statistical Dynamo

The situation is different in a statistical magnetic field: on the Sun, for instance, the turbulent motion in the convection zone modifies the field. The average field $\boldsymbol{B}_0 = \langle \boldsymbol{B} \rangle$ still is axial-symmetric but it is modified by fluctuations $\boldsymbol{B}_1$ with $\langle \boldsymbol{B}_1 \rangle = 0$. Thus the magnetic field is $\boldsymbol{B} = \boldsymbol{B}_0 + \boldsymbol{B}_1$ and the velocity field is $\boldsymbol{u} = \boldsymbol{u}_0 + \boldsymbol{u}_1$.[2] The cross product of the speed and the magnetic field reads

$$\langle \boldsymbol{u} \times \boldsymbol{B} \rangle = \boldsymbol{u}_0 \times \boldsymbol{B}_0 + \langle \boldsymbol{u}_1 \times \boldsymbol{B}_1 \rangle . \tag{3.128}$$

The products $\langle \boldsymbol{u}_1 \times \boldsymbol{B}_0 \rangle$ and $\langle \boldsymbol{u}_0 \times \boldsymbol{B}_1 \rangle$ vanish because the quantities with index 'o' are constant and the average of the other quantity equals zero. The product $\langle \boldsymbol{u}_1 \times \boldsymbol{B}_1 \rangle$, which is the correlation function, does not vanish because the fluctuations are not independent: because the matter has a high conductivity, the magnetic field is frozen-into the plasma, and a change in the velocity field leads to a corresponding change in the magnetic field. To first order, the correlation function can be approximated as

$$\langle \boldsymbol{u}_1 \times \boldsymbol{B}_1 \rangle = \alpha \boldsymbol{B}_0 - \beta \nabla \times \boldsymbol{B}_0 . \tag{3.129}$$

 Excursion 4. As suggested by Parker [390], (3.129) can be derived as follows. The magnetic field equations (3.41)–(3.44) are linear in $\boldsymbol{E}$, $\boldsymbol{B}$, ϱ_c, and $\boldsymbol{j}$. The quantities can be split into average and fluctuating quantities and we have two formally identical sets of equations, one for the average field and one for the fluctuating field (see Sect. 4.1.4). Ohm's law (3.45) has to be handled differently because it contains a product of fluctuating quantities $\boldsymbol{u} \times \boldsymbol{B}$. Splitting Ohm's law into an average current $\boldsymbol{j}_0$ and a fluctuating current $\boldsymbol{j}_1$ yields

$$\begin{aligned} \boldsymbol{j} &= \boldsymbol{j}_0 + \boldsymbol{j}_1 \\ &= \sigma \left(\boldsymbol{E}_0 + \boldsymbol{E}_1 + \boldsymbol{u}_0 \times \boldsymbol{B}_0 + \boldsymbol{u}_0 \times \boldsymbol{B}_1 + \boldsymbol{u}_1 \times \boldsymbol{B}_0 + \boldsymbol{u}_1 \times \boldsymbol{B}_1 \right) \end{aligned} \tag{3.130}$$

---

[2] A brief introduction to the mathematical basics of instantaneous quantities, averages, and fluctuations is given in Sect. 4.1.3.

Taking the average gives

$$\boldsymbol{j}_0 = \sigma(\boldsymbol{E}_0 + \boldsymbol{u}_0 \times \boldsymbol{B}_0 + \boldsymbol{u}_1 \times \boldsymbol{B}_1) \,. \qquad (3.131)$$

Thus Ohm's law for the average quantities contains an additional term, the correlation function between the fluctuating velocity field and the fluctuating magnetic field. The expression (3.129) for this term is derived under the assumption that the average velocity $\boldsymbol{u}_0$ vanishes and that the fluctuating velocity field is homogeneous and isotropic: neither are there points in space with extremely high or low levels of fluctuations nor are the fluctuations preferentially in one direction.

The induction equation (3.84) for the instantaneous quantities can be written

$$\frac{1}{\mu_0\sigma}\nabla^2\boldsymbol{B}_0 + \nabla \times (\boldsymbol{u}_1 \times \boldsymbol{B}_0) - \frac{\partial \boldsymbol{B}_0}{\partial t}$$
$$= -\frac{1}{\mu_0\sigma}\nabla^2\boldsymbol{B}_1 - \nabla \times (\boldsymbol{u}_1 \times \boldsymbol{B}_1) + \frac{\partial \boldsymbol{B}_1}{\partial t} \,. \qquad (3.132)$$

This equation still holds if $\boldsymbol{B}_0$ and $\boldsymbol{B}_1$ are multiplied by the same factor: the fluctuating part $\boldsymbol{B}_1$ thus depends linearly and homogeneously on the average field $\boldsymbol{B}_0$. This is also true for $\langle \boldsymbol{u}_1 \times \boldsymbol{B}_1 \rangle$, since averaging does not change the dependence:

$$\langle \boldsymbol{u}_1 \times \boldsymbol{B}_1 \rangle \sim \alpha\boldsymbol{B}_0 \,. \qquad (3.133)$$

Let us now assume that, to first order, $\boldsymbol{B}_1$ and thus also $\langle \boldsymbol{u}_1 \times \boldsymbol{B}_1 \rangle$ at a certain position P depend only on $\boldsymbol{B}_0$ and $\boldsymbol{u}_1$ in a small neighborhood. Then $\langle \boldsymbol{u}_1 \times \boldsymbol{B}_1 \rangle|_P$ depends only on $\boldsymbol{B}_0|_P$ and $(\partial \boldsymbol{B}_0/\partial x_i)|_P$. Thus $\langle \boldsymbol{u}_1 \times \boldsymbol{B}_1 \rangle$ must be proportional to $\nabla \times \boldsymbol{B}_0$:

$$\langle \boldsymbol{u}_1 \times \boldsymbol{B}_1 \rangle \sim \beta\nabla \times \boldsymbol{B}_0 \,. \qquad (3.134)$$

Thus, in sum, we obtain (3.129).   □

Both α and β are determined by the properties of the turbulent velocity field. The β-term describes the increase in magnetic diffusion due to the turbulent motion, leading to a faster dissipation of the field. For a mirror-symmetric velocity field, α would vanish, but not in a rotating system, where the velocity field is not symmetric. Taking the average of the induction equation (3.132) and considering the magnetic viscosity (3.88), we get

$$\frac{\partial \boldsymbol{B}_0}{\partial t} - \nabla \times (\boldsymbol{u}_0 \times \boldsymbol{B}_0 + \alpha\boldsymbol{B}_0) = -(\eta + \beta)\nabla \times (\nabla \times \boldsymbol{B}_0) \,. \qquad (3.135)$$

While β modifies the viscosity, the α-term contains the basic difference compared with (3.84): it allows for an electro-motoric force parallel to the average magnetic field; Cowling's theorem does not apply to this equation.

3.6.3 The $\alpha\Omega$ Dynamo

The basic idea of the MHD dynamo can be applied to different geometries and to stationary as well as periodically varying magnetic fields. Because we are interested in axially symmetric fields, it is reasonable to describe the magnetic field as consisting of a toroidal and a poloidal part:

$$\langle \boldsymbol{B} \rangle = \langle \boldsymbol{B}_{\text{tor}} \rangle + \langle \boldsymbol{B}_{\text{pol}} \rangle = B\,\boldsymbol{e}_\Phi + \nabla \times A\,\boldsymbol{e}_\Phi\,, \qquad (3.136)$$

where $\boldsymbol{e}_\Phi$ is the unit vector in the toroidal direction. Thus two scalar quantities, A and B, determine the three field components. With this ansatz, the induction equation gives two equations: one describing the ohmic dissipation of B and the generation of B out of A due to the α-effect and the differential rotation $\nabla\Omega$, the other describing the ohmic dissipation of A combined with the generation of A out of B.

Differential rotation can occur for various reasons. The Sun, for instance, has a higher angular speed at the equator than at higher latitudes, and thus the rotation depends on latitude. The differential rotation inside the Earth is due to the differences in angular speed between the faster inner and the slower outer core. In both cases, because the field is frozen into the plasma, a deformation of the field line arises from the differential rotation.

The α-effect, on the other hand, is associated with the turbulent motion of the plasma, in particular the upward and downward motions associated with convection. Although this motion is stochastic, its combination with the Coriolis force leads to a turbulent motion which introduces a systematic twist into an originally toroidal field, as shown in Fig. 3.12. The resulting magnetic field coil allows a current parallel to the undisturbed toroidal field.

Inserting (3.129) into (3.131), we obtain

$$\boldsymbol{j}_0 = \sigma\left\{\boldsymbol{E}_0 + (\boldsymbol{u}_0 \times \boldsymbol{B}_0) + \alpha\boldsymbol{B}_0 - \beta(\nabla \times \boldsymbol{B}_0)\right\}\,. \qquad (3.137)$$

The third term on the right-hand side gives, depending on the sign of α, a current parallel or antiparallel to the average magnetic field $\boldsymbol{B}_0$. With Faraday's law (3.44), we can rewrite the last term on the right, and obtain

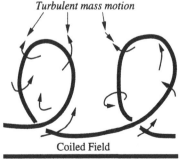

Turbulent mass motion

Coiled Field

Toroidal Field

Fig. 3.12. A combination of the stochastic motion and the Coriolis force leads to turbulent motion (*short twisted arrows*) of the plasma which twists an originally toroidal field (*lower line*) into a coiled field which allows for a current parallel to the original field

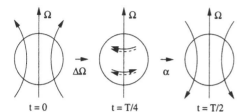

Fig. 3.13. Magnetohydrodynamic dynamo: differential rotation deforms a poloidal magnetic field into a toroidal one. The α-effect allows currents parallel to the field, giving rise to a toroidal magnetic field in the opposite direction

$$j = \sigma_T \left(E_0 + u_0 \times B_0 + \alpha B_0 \right), \tag{3.138}$$

where

$$\frac{1}{\sigma_T} = \frac{1}{\sigma} + \beta \tag{3.139}$$

is the turbulent conductivity. Since β is positive [299], σ_T always is smaller than σ: the turbulence described by the β-term reduces the conductivity. In particular, for $\sigma \to \infty$ turbulent motion would limit the conductivity to a finite value. Fields in a turbulent plasma therefore dissipate faster, and the dissipation time (3.109) becomes a turbulent dissipation time

$$\tau_T \approx \mu_0 \sigma_T L^2 . \tag{3.140}$$

Graphically, σ_T takes into account the fact that the turbulent motion reduces the length scales L of the system.

The combination of the effects of α and Ω allows us to describe the MHD dynamo as sketched in Fig. 3.13. We start with a poloidal field in the Sun at $t = 0$. The differential rotation deforms the magnetic field, leading to a toroidal field ($t = T/4$). The α-effect leads to electromagnetic forces parallel to the field, and thus a toroidal current flows (dashed lines). Although the magnetic field directions are opposite in the two hemispheres, the asymmetry of the Coriolis force leads to an asymmetric α-effect and therefore parallel currents in both hemispheres. This current leads to a magnetic field directed opposite to the original field ($t = T/2$). Half a cycle is now finished. This dynamo is called the $\alpha\Omega$ dynamo because both the α-effect and the differential rotation contribute to the dynamo process. The dynamos inside the Sun and the Earth are based on this principle; their details will be discussed in Sects. 6.6.2 and 8.1.

If the α-effect was not at work, the differential rotation would still transform the poloidal magnetic field into a toroidal one. However, no polarity reversal would occur and, in time, the entire field would dissipate. The differential rotation, on the other hand, is not essential to the MHD dynamo. The α-effect can also work with turbulent motions which, for some reason, have a preferred direction of motion; this is often an upwelling of magnetic flux combined with a particular direction of rotation of the flux tubes [299].

The MHD dynamo requires an initial magnetic field which is amplified by a suitable feedback mechanism. Thus at first glance the MHD dynamo violates Lenz's rule which states that all fields, currents and forces are directed so as to hinder the process that leads to their induction. For instance, an increase in the magnetic field leads to currents which create a magnetic field opposite to the original one. Lenz's rule thus stabilizes the system; it does not allow for the positive feedback required in the MHD dynamo. Were we to build such a dynamo on the basis of one process only, Lenz's rule would be violated. But the MHD dynamo has the remarkable feature that although all individual processes obey Lenz's rule, their sum allows for positive feedback.

3.7 Debye Shielding

So far, we have described a plasma in the context of one-fluid magnetohydrodynamics: the plasma consists of one particle species only and moves with the bulk speed. The thermal motion of the particles is neglected and thus there is no motion of particles relative to each other.

We will now, though in a simple formalism, make use of the stochastic, thermal motion of particles in a two-component plasma consisting of electrons and protons. A local deviation from quasi-neutrality arises from the random thermal motion. Quasi-neutrality depends on the size of the volume under consideration. If the volume is very small, housing only one particle, quasi-neutrality cannot be obtained. But even if we increase the size of the volume, the thermal motion might lead to an excess of particles with one charge sign. Then the shielding of a certain particle with one polarity due to particles of the opposite polarity becomes important. The typical spatial scale for such shielding is the Debye length, already mentioned in the introduction.

The region depleted of electrons due to their random thermal motion is limited in extent because the displaced electrons create an electric field which acts as a restoring force. Consider a sheath of width D depleted of all electrons. Because of their larger mass, the ions are less mobile and stay within this sheath. Within D therefore a positive charge exists while the electrons can be regarded as a surface charge collected at the boundary of the sheath. The electric field is different from zero within the sheath; outside the sheath the field of the positive ions is screened by the surface charge.

The energy in the electric field stems from the kinetic energy of the thermal electron motion. With n_e as the electron density in the undisturbed plasma, the kinetic energy of the electrons in a sheath of thickness D is $n_e k_B T D/2$. If all electrons are removed from this sheath, a restoring force proportional to D acts on them. The energy contained in the electric field created by the charge separation depends on D^2. Thus there is a certain width λ_D at which the energy contained in the field equals the kinetic energy of the electrons originally present within this region: the kinetic energy of the thermal motion of the electrons is converted entirely into field energy.[3]

---

[3] Note that this is different from the discussion of the plasma oscillations in Sect. 4.3.1 because here the sheath depleted of electrons results from their ther-

The electric field created by the ions inside the sheath is $\nabla \cdot \boldsymbol{E} = eZn_i/\varepsilon_0$. Thus the Coulomb potential φ can be written as $\nabla^2\varphi = -eZn_i/\varepsilon_0$. In the one-dimensional case, the potential is

$$\frac{\partial^2\varphi}{\partial x^2} = -eZn_i/\varepsilon_0 = \text{const}. \tag{3.141}$$

The general solution of this equation is $\varphi(x) = -Zen_i x^2/2\varepsilon_0 + C_1 x + C_2$, with the constants determined by the boundary conditions. The coordinate system is fixed such that the potential vanishes at $x = 0$ and C_2 is zero. If the field is symmetric around $x = 0$, then C_1 is zero, too. In addition, the transition between the potential inside and outside the sheath has to be steady. Thus the potential can be written as

$$\varphi = \begin{cases} -Zen_i x^2/2\varepsilon_0 & \text{for } |x| \le D/2 \\ -Zen_i D^2/8\varepsilon_0 & \text{for } |x| \ge D/2 \end{cases}. \tag{3.142}$$

The electric field is

$$E(x) = \begin{cases} Zen_i x/\varepsilon_0 & \text{for } |x| \le D/2 \\ 0 & \text{for } |x| \ge D/2 \end{cases}. \tag{3.143}$$

The energy contained in the electric field can be determined as

$$\int_{-D/2}^{+D/2} \frac{\varepsilon_0 E^2}{2}\,\mathrm{d}x = \frac{e^2 Z^2 n_i^2}{2\varepsilon_0} \int_{-D/2}^{+D/2} x^2\,\mathrm{d}x = \frac{e^2 Z^2 n_i^2}{6\varepsilon_0} D^3. \tag{3.144}$$

We are now looking for the thickness λ_D of a sheath where the field energy equals the kinetic energy of the electrons originally present inside the sheath:

$$\frac{n_e k_B T \lambda_D}{2} = \frac{e^2 Z^2 n_i^2 \lambda_D^3}{6\varepsilon_0}. \tag{3.145}$$

Since the above geometry is one-dimensional, only one degree of freedom is considered in the kinetic energy. Quasi-neutrality requires $n_e = Zn_i$, and thus the Debye length λ_D is given as

$$\lambda_D = \sqrt{\frac{3\varepsilon_0 k_B T}{e^2 n_e}} = \sqrt{\frac{k_B T}{m}}\,\frac{1}{\omega_{pe}}, \tag{3.146}$$

where ω_{pe} is the angular frequency of electron plasma oscillations (Sect. 4.3.1).

Though derived for the one-dimensional case, this equation also holds in the three-dimensional case where the Debye length can be interpreted as the

---

mal motion. In plasma oscillations, instead, a cold plasma is considered and the charge separation results from an external force, e.g. a beam of electrons travelling through the plasma.

maximal radius of a sphere which, due to thermal motion, might be depleted of electrons. In spatial regions small compared with the Debye length, quasineutrality is likely to be violated, while on larger scales the plasma is quasineutral. In the latter case the kinetic energy contained in the thermal motion is not large enough to disturb the particle distribution over the entire region. Figure 1.1 shows typical values of the Debye length and numbers of particles inside a sphere with the Debye radius for different electron densities and temperatures. Some typical plasmas are indicated.

The Debye length also can be used to assess the influence of an instrument on measuring plasma parameters. Let us start from an initially cold plasma, i.e. the motion of the electrons can be ignored. Let us now insert a small positive charge q into this plasma. Immediately, it will be surrounded by a cloud of electrons while the ions are repulsed. The electron cloud screens the additional charge; thus, outside the electron cloud its electric field vanishes. If we now increase the temperature, the electrons gain thermal velocity. Deep inside the electron cloud this velocity will be too small to overcome the attraction of the positive charge. At the outer edges of the cloud, on the other hand, the thermal energy might be large enough to exceed the electrostatic potential of the partly screened charge, allowing an escape from the cloud. The Debye length then can be interpreted as the spatial scale over which the potential of the point charge q has decreased by a characteristic value (see Fig. 3.14): within the Debye length, electrons are influenced by the test charge, while at larger distances the test charge goes unnoticed.

A more general definition would read: only charged particles within a distance of λ_D exert an electrostatic force on each other. This is different from bodies which interact gravitationally, like, for example, interacting stars: gravitation cannot be screened by repulsing forces, it has an indefinite range.

Debye screening also is important in preventing local clusters of charges. For $e\varphi \ll k_B T$ the influence of the electric field on the particle energy is small; thus the particle motion is determined by the thermal speed. The Debye shield arises from small differences in the particle motion: some particles with opposite charge stay close to the test charge slightly longer, while particles with equal charge move away a little bit faster. Charge inhomogeneities in a plasma therefore are balanced on the scale of the thermal propagation time.

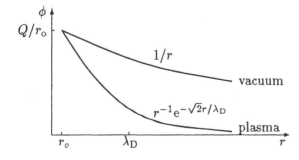

Fig. 3.14. The electric potential of a test charge is reduced by the surrounding plasma

3.8 Summary

Magnetohydrostatics is concerned with the energetics of the electromagnetic field without allowing for the collective motion of the plasma. Basic concepts derived from this approximation are the magnetic pressure and the magnetic tension. Magnetic pressure, graphically interpreted as mutual repulsion of field lines, prevents a magnetic field from being compressed by external forces. If part of a magnetic line of force is displaced from its original position, the magnetic tension creates a restoring force, which graphically can be interpreted as the tendency of a line of force to shorten itself.

Magnetohydrokinematics assumes the energy density of the plasma to be much larger than the energy density of the field, which allows us to ignore the influence of the field on particle motion. The basic concepts are as follows: (a) If the conductivity of the plasma is infinite, the magnetic field lines are frozen-into the plasma. Thus the plasma flow carries away the magnetic field. (b) If the conductivity is finite, the magnetic field is deformed by the plasma flow until diffusion allows the plasma to flow across the field lines. (c) In a stationary plasma, the magnetic field dissipates, with the dissipation time depending on the square of the linear dimensions (small fields dissipate faster) and linearly on the conductivity (if the latter is infinite, the dissipation time is infinite, too, and the field is frozen-in). These concepts are important, e.g. in our understanding of the merging of magnetic field lines, called reconnection, and in dynamo theory.

Exercises and Problems

3.1. Explain the difference between convective and partial derivatives. Find examples to illustrate the differences.

3.2. Recall simple hydrodynamics and give other examples of the momentum balance. Discuss the different forms and compare with the Navier–Stokes equation.

3.3. Derive the hydrostatic equation from the Navier–Stokes equation. Which terms do you need?

3.4. Give a quantitative discussion of the stability of a sunspot (all important numbers are given in Table 6.1).

3.5. Is the filament sketched in Fig. 3.5 realistic? Why does it not dissolve towards the sides (remember, it is a plasma, not a solid body)?

3.6. What is the meaning of viscosity and Reynolds number? What are the formal differences between hydrodynamics and magnetohydrodynamics? What are the differences in substance?

3.7. Explain the consequences of stationary flows parallel and oblique to the magnetic field.

3.8. Why has pressure the units of an energy density?

3.9. Show that (3.76) is a solution of (3.75).

3.10. Show that in an ideal, non-relativistic magnetohydrodynamic plasma the ratio between the electric and the magnetic energies is $(v_\perp/c)^2$.

3.11. Determine the dissipation times for a copper block (side length 10 cm, conductivity 260 A/Vm) and the interstellar medium (linear dimension 10^{21} m, conductivity 2.6 μV/Am). Compare with the age of the universe (about 10^{18} s).

3.12. Determine the Debye length and the number of particles inside a Debye sphere for electrons and protons moving with thermal speeds (see Sect. 5.1.2) in the following fields: (a) the Earth's magnetosphere with $n = 10^4$ cm^{-3}, $T = 10^3$ K, $B = 10^{-2}$ G; (b) the core of the Sun with $n = 10^{26}$ cm^{-3}, $T = 10^{7.2}$ K, $B = 10^6$ G; (c) the solar corona with $n = 10^8$ cm^{-3}, $T = 10^6$ K, $B = 1$ G; and (d) the solar wind with $n = 10$ cm^{-3}, $T = 10^5$ K, $B = 10^{-5}$ G.

4 Plasma Waves

> Roll on, thou deep and dark blue Ocean – roll!
> Ten thousand fleets sweep over thee in vain;
> Man marks the earth with ruin – his control
> Stops with the shore.
> Lord Byron, *Childe Harold's Pilgrimage*

In this chapter we shall catch a glimpse on the vast zoo of plasma waves. The formalism to derive these waves, perturbation theory, briefly will be introduced. We will not derive all types of waves formally; instead, we shall limit ourselves to the magnetohydrodynamic waves, which are Alfvén waves and ion acoustic waves. The derivation of the dispersion relations for other types of waves follows the same scheme, it only differs in the terms considered in the equation of motion, in the assumptions made in the equation of state, and in whether a one-fluid description of the plasma is sufficient or if a two-fluid description is required. Detailed derivations can be found e.g. in [36, 97, 191, 192, 298, 397, 504, 512, 534]. For the purpose of this book, however, it is more important to grasp the nature of the waves than to fiddle around with the mathematical tricks involved in solving the equation of motion.

This chapter is limited to elementary types of waves. First, the geometry always is simple: in a magnetized plasma the waves either propagate parallel or perpendicular to the undisturbed magnetic field – oblique waves are not considered here. Physically more important is the limitation to small disturbances, i.e. small amplitude waves: the basic set of magnetohydrodynamic equations is a set of coupled non-linear partial differential equations. Thus in principle we can expect non-linear couplings between different fluctuating quantities of the wave. If we limit our discussion to small amplitude waves, the equations can be linearized: whenever two oscillating quantities are multiplied, since both are small, we consider this a higher order term and ignore it. If we apply these results to a real situation, we always have to take one step back and justify whether the amplitudes calculated in our real situation are small enough so that the non-linear terms actually are negligible compared with the linear ones.

In a plasma, a large variety of waves exists. A simple phenomenological classification in transversal and longitudinal waves is insufficient. Instead, we have to consider the conditions under which certain types of waves can exist. For instance, in a cold plasma the thermal motion of the particles and therefore the pressure vanishes. Thus elastic waves cannot exist; they can form only in a warm plasma where a pressure gradient can build up. In an isotropic plasma, no magnetic field exists, which allows other modes of propagation than an anisotropic plasma. As a third criterion, we also have to consider which particle species is in motion.

4.1 What is a Wave?

For a start, let us briefly repeat the basic descriptions and properties of waves. A wave is a disturbance propagating through a continuous medium. It gives rise to a periodic motion of the fluid. Even a complex oscillation, at least as long as the amplitude of the disturbance is small, can be decomposed into different sinusoidal waves by the technique of Fourier analysis. Small-amplitude disturbances lead to plane waves with the direction of propagation and the amplitude being the same everywhere.

4.1.1 Wave Parameters

A sinusoidal wave is described by its frequency ω and its wave vector $\boldsymbol{k}$:

$$\boldsymbol{B}(\boldsymbol{r}, t) = \boldsymbol{B}_0 \exp\left\{\mathrm{i}(\boldsymbol{k} \cdot \boldsymbol{r} - \omega t)\right\} . \tag{4.1}$$

The measurable quantity is the real part of this complex expression. The exponent in (4.1) is the phase of the disturbance. The temporal derivative of the phase gives the frequency ω, and the spatial derivative gives the wave vector $\boldsymbol{k}$ that specifies the direction of wave propagation. A surface of constant phase, also called a wave surface, is displaced by the phase velocity v_{ph}, which can be determined from $\mathrm{d}(\boldsymbol{k} \cdot \boldsymbol{r} - \omega t)/\mathrm{d}t$ as $v_{\mathrm{ph}} = \omega/k$ or, in vector form,

$$\boldsymbol{v}_{\mathrm{ph}} = \frac{\omega}{k^2} \boldsymbol{k} . \tag{4.2}$$

If ω/k is positive, the wave moves to the right; for negative ω/k it moves to the left. For electromagnetic waves the refraction index n is defined as the ratio between the speed of light and the phase speed of the wave: $n = c/v_{\mathrm{ph}} = ck/\omega$.

The phase velocity can exceed the speed of light. This is not in contradiction to the theory of relativity because an indefinitely long wave train of constant amplitude does not carry information. Information can be carried by a modulated wave, on which variations in frequency or amplitude are superimposed. We can regard each bit of information in this modulated signal as a wave packet which then moves with the group velocity

$$\boldsymbol{v}_{\mathrm{g}} = \frac{\partial \omega}{\partial \boldsymbol{k}} . \tag{4.3}$$

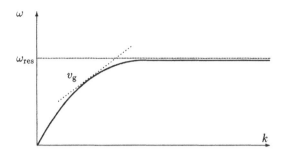

Fig. 4.1. Dispersion relation with a resonance at the frequency ω_{res}. The slope of the curve gives the group velocity v_g, as indicated by the dashed line

The group velocity v_g given by the slope of the dispersion relation (see dotted line in Fig. 4.1). It is always less than the speed of light and determines the velocity with which the energy of the wave is transported.

The most important tool in the description of waves is the dispersion relation $\omega = \omega(k)$, relating frequency and wave vector. From this relation the group and phase velocities of a wave can be determined. The dispersion relation contains the physical parameters of the medium under consideration. If the plot of the dispersion relation shows an asymptotic behavior towards a certain frequency ω_{res}, as depicted in Fig. 4.1, there is a resonance at this frequency: as $\partial\omega/\partial k$ converges towards zero, the wave no longer propagates and all the wave energy is fed into stationary oscillations.

4.1.2 Linearization of the Equations: Perturbation Theory

A wave is a disturbance of the medium with a certain speed, amplitude, and frequency. Thus the parameters of the medium, such as pressure, density, and the electromagnetic field, can be described by an average state, indicated by the index "0", and a superimposed disturbance, indexed as "1":

$$\boldsymbol{B} = \boldsymbol{B}_0 + \boldsymbol{B}_1 , \qquad \boldsymbol{E} = \boldsymbol{E}_0 + \boldsymbol{E}_1 , \qquad \boldsymbol{u} = \boldsymbol{u}_0 + \boldsymbol{u}_1 , \qquad (4.4)$$

$$\boldsymbol{j} = \boldsymbol{j}_0 + \boldsymbol{j}_1 , \qquad \varrho = \varrho_0 + \varrho_1 , \qquad p = p_0 + p_1 , \qquad (4.5)$$

with $\langle \boldsymbol{u}_1 \rangle = \langle \boldsymbol{j}_1 \rangle = \langle \boldsymbol{E}_1 \rangle = \langle \boldsymbol{B}_1 \rangle = \langle \varrho_1 \rangle = \langle p_1 \rangle = 0$. The resulting MHD equations (3.41)–(3.48) are difficult to solve. If we limit ourselves to small disturbances, i.e. $u_1 < u_0$, $B_1 < B_0$, $E_1 < E_0$, $j_1 < j_0$, $\varrho_1 < \varrho_0$, and $p_1 < p_0$, we can derive two sets of equations which are more convenient: the set describing the equilibrium state of the undisturbed medium contains quantities with index "0" only, and a second set for the fluctuating quantities contains fluctuating quantities and products of undisturbed and fluctuating quantities. Products of fluctuating quantities are ignored because they are small in second order.

4.1.3 Reynolds Axioms

We have decomposed the instantaneous quantities $\boldsymbol{x}$ into an average $\boldsymbol{x}_0$ and a fluctuating part $\boldsymbol{x}_1$. By definition, the average of $\boldsymbol{x}$ is $\langle \boldsymbol{x} \rangle = \boldsymbol{x}_0$ and the average of the fluctuations x_1 vanishes: $\langle x_1 \rangle = 0$. All our MHD equations (3.41)–(3.48) are given for the instantaneous quantities. To derive equations for average quantities, the instantaneous quantities have to be expressed in terms of averages and fluctuating parts, and the resulting equation has to be averaged. In doing this, some general rules must be obeyed, described by the Reynolds axioms:

1. The average of the sum of two instantaneous quantities equals the sum of the averages:

$$\langle \boldsymbol{A} + \boldsymbol{B} \rangle = \langle \boldsymbol{A} \rangle + \langle \boldsymbol{B} \rangle = \boldsymbol{A}_0 + \boldsymbol{B}_0 \; . \tag{4.6}$$

2. The average of the product of an average quantity and a fluctuating one vanishes:

$$\begin{aligned}
\langle A_0 B_1 \rangle &= \langle A_0 \rangle \langle B_1 \rangle = A_0\, 0 = 0 \; , \\
\langle \boldsymbol{A}_0 \cdot \boldsymbol{B}_1 \rangle &= \langle \boldsymbol{A}_0 \rangle \cdot \langle \boldsymbol{B}_1 \rangle = \boldsymbol{A}_0 \cdot 0 = 0 \; , \\
\langle \boldsymbol{A}_0 \times \boldsymbol{B}_1 \rangle &= \langle \boldsymbol{A}_0 \rangle \times \langle \boldsymbol{B}_1 \rangle = \boldsymbol{A}_0 \times 0 = 0 \; .
\end{aligned} \tag{4.7}$$

3. The average of the product of two averages is the product of the averages:

$$\begin{aligned}
\langle\, \langle A \rangle \langle B \rangle \,\rangle &= \langle A \rangle \langle B \rangle = A_0 B_0 \; , \\
\langle\, \langle \boldsymbol{A} \rangle \langle \boldsymbol{B} \rangle \,\rangle &= \langle \boldsymbol{A} \rangle \cdot \langle \boldsymbol{B} \rangle = \boldsymbol{A}_0 \cdot \boldsymbol{B}_0 \; , \\
\langle\, \langle \boldsymbol{A} \rangle \times \langle \boldsymbol{B} \rangle \,\rangle &= \langle \boldsymbol{A} \rangle \times \langle \boldsymbol{B} \rangle = \boldsymbol{A}_0 \times \boldsymbol{B}_0 \; .
\end{aligned} \tag{4.8}$$

4. The average of the product of two instantaneous quantities equals the product of the averages plus the average of the product of the fluctuating quantities:

$$\begin{aligned}
\langle AB \rangle &= \langle (A_0 + A_1)(B_0 + B_1) \rangle = \langle A_0 B_0 + A_0 B_1 + A_1 B_0 + A_1 B_1 \rangle \\
&= \langle A_0 B_0 \rangle + \langle A_0 B_1 \rangle + \langle A_1 B_0 \rangle + \langle A_1 B_1 \rangle = A_0 B_0 + \langle A_1 B_1 \rangle \; , \\
\langle \boldsymbol{A} \cdot \boldsymbol{B} \rangle &= \boldsymbol{A}_0 \cdot \boldsymbol{B}_0 + \langle \boldsymbol{A}_1 \cdot \boldsymbol{B}_1 \rangle \; , \\
\langle \boldsymbol{A} \times \boldsymbol{B} \rangle &= \boldsymbol{A}_0 \times \boldsymbol{B}_0 + \langle \boldsymbol{A}_1 \times \boldsymbol{B}_1 \rangle \; .
\end{aligned} \tag{4.9}$$

The last term contains the average of the product of the fluctuating quantities. This is called the covariance or the correlation product. Thus we also have a definition for the correlation product of two quantities $\boldsymbol{x}$ and $\boldsymbol{y}$:

$$\begin{aligned}
\langle x_1 y_1 \rangle &= \langle (x - x_1)(y - y_1) \rangle - x_0 y_0 \; , \\
\langle \boldsymbol{x}_1 \cdot \boldsymbol{y}_1 \rangle &= \langle (\boldsymbol{x} - \boldsymbol{x}_1) \cdot (\boldsymbol{y} - \boldsymbol{y}_1) \rangle - \boldsymbol{x}_0 \cdot \boldsymbol{y}_0 \; , \\
\langle \boldsymbol{x}_1 \times \boldsymbol{y}_1 \rangle &= \langle (\boldsymbol{x} - \boldsymbol{x}_1) \times (\boldsymbol{y} - \boldsymbol{y}_1) \rangle - \boldsymbol{x}_0 \times \boldsymbol{y}_0 \; .
\end{aligned} \tag{4.10}$$

5. The average of the derivative of an instantaneous quantity equals the derivative of its average:

$$\left\langle \frac{\partial \boldsymbol{A}}{\partial \zeta} \right\rangle = \frac{\partial \langle \boldsymbol{A} \rangle}{\partial \zeta} = \frac{\partial \boldsymbol{A}_0}{\partial \zeta} \; . \tag{4.11}$$

6. The average of the integral of an instantaneous quantity is the integral
of the average:

$$\left\langle \int \boldsymbol{A} \, d\zeta \right\rangle = \int \langle \boldsymbol{A} \rangle \, d\zeta = \int \boldsymbol{A}_0 \, d\zeta \, . \tag{4.12}$$

4.1.4 Linearized MHD Equations

The MHD equations (3.41)–(3.48) for the undisturbed quantities are

$$\nabla \times \boldsymbol{B}_0 = \mu_0 \boldsymbol{j}_0 \, , \tag{4.13}$$

$$\nabla \times \boldsymbol{E}_0 = 0 \, , \tag{4.14}$$

$$\nabla \cdot \boldsymbol{B}_0 = 0 \, , \tag{4.15}$$

$$\frac{\boldsymbol{j}_0}{\sigma} = \boldsymbol{E}_0 + \boldsymbol{u}_0 \times \boldsymbol{B}_0 \, , \tag{4.16}$$

$$\varrho_0 \left(\boldsymbol{u}_0 \cdot \nabla \right) \boldsymbol{u}_0 = -\nabla p_0 + \boldsymbol{j}_0 \times \boldsymbol{B}_0 \, , \tag{4.17}$$

$$\nabla \cdot (\varrho_0 \boldsymbol{u}_0) = 0 \, , \tag{4.18}$$

$$p_0 = C \varrho_0^{\gamma_a} \, . \tag{4.19}$$

Note that in the equation of motion friction and gravity have been ignored.
The equation of state describes the plasma as an ideal gas with all changes
occurring adiabatically.

The undisturbed medium is assumed to be homogeneous in pressure, den-
sity, and magnetic field. Since it is assumed to be at rest ($\boldsymbol{u}_0 = 0$), the current
$\boldsymbol{j}_0$ vanishes (4.13) and the undisturbed electric field vanishes too (4.16). Note
that this is just another expression for the high conductivity of the plasma
($\sigma \to \infty$). The equations for the fluctuations then read

$$\nabla \times \boldsymbol{B}_1 = \mu_0 \boldsymbol{j}_1 \, , \tag{4.20}$$

$$\nabla \times \boldsymbol{E}_1 = -\frac{\partial \boldsymbol{B}_1}{\partial t} \, , \tag{4.21}$$

$$\nabla \cdot \boldsymbol{B}_1 = 0 \, , \tag{4.22}$$

$$\boldsymbol{E}_1 = -\boldsymbol{u}_1 \times \boldsymbol{B}_0 \, , \tag{4.23}$$

$$\varrho_0 \frac{\partial \boldsymbol{u}_1}{\partial t} = -\nabla p_1 + \boldsymbol{j}_1 \times \boldsymbol{B}_0 \, , \tag{4.24}$$

$$\frac{\partial \varrho_1}{\partial t} = -\nabla \cdot (\varrho_0 \boldsymbol{u}_1) \, . \tag{4.25}$$

$$\frac{p_1}{p_0} = \gamma_a \frac{\varrho_1}{\varrho_0} \, . \tag{4.26}$$

Now (4.20)–(4.26) is a homogeneous linear system of equations for the fluctuating quantities. Since neither time nor the spatial coordinate are explicit
in one of the equations, the system can be solved by an exponential ansatz.

4.2 Magnetohydrodynamic Waves

Magnetohydrodynamic (MHD) waves are low-frequency waves related to the motion of the plasma's ion component. They can be understood intuitively from the concepts of magnetic pressure (Sect. 3.3.1) and magnetic tension (Sect. 3.3.2): in a magneto-sonic wave, compression of the field lines creates a magnetic pressure pulse which propagates perpendicular to the field in the same way an ordinary pressure pulse propagates through a gas in a sound wave. Magneto-sonic waves therefore are longitudinal waves: the disturbance is parallel to the propagation direction of the wave. The displacement of part of a field line in an Alfvén wave is similar to plucking a string: magnetic tension, like the tension in a string, acts as a restoring force and a transversal wave propagates along the field line.

Despite the simplicity and graphic quality of these concepts, we shall treat these waves formally in a concept which is useful for small disturbances: the linearization of the equations. The MHD waves provide just one example for this concept; it is applied in a more elaborate way also in the quasi-linear theory (QLT) of wave–particle interaction (Sect. 7.3.4).

4.2.1 Alfvén Waves

Alfvén waves are transversal waves propagating parallel to the magnetic field (see Fig. 4.2). The magnetic tension acts as the restoring force. The fluctuating quantities are the electromagnetic field and the current density.

To derive the properties of Alfvén waves, we have to solve the equations for the fluctuating quantities. Let us start with the equation of motion (4.24). The pressure gradient force vanishes because the Alfvén wave is limited to fluctuations in the magnetic field but not in the gas-dynamic pressure. If we express the current j_1 by Ampére's law (4.20), the momentum balance reads

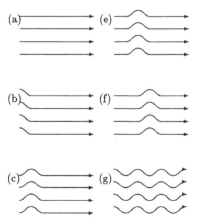

Fig. 4.2. Alfvén waves depicted as an oscillating string or an elastic rope

$$\varrho_0 \frac{\partial \boldsymbol{u}_1}{\partial t} = \boldsymbol{j}_1 \times \boldsymbol{B}_1 = \frac{1}{\mu_0}(\nabla \times \boldsymbol{B}_1) \times \boldsymbol{B}_1 \ . \tag{4.27}$$

Combining Ohm's law (4.23) with Faraday's law (4.21) yields

$$\frac{\partial \boldsymbol{B}_1}{\partial t} = \nabla \times (\boldsymbol{u}_1 \times \boldsymbol{B}_0) \ . \tag{4.28}$$

The remaining equation is Gauss's law for the magnetic field (4.22):

$$\nabla \cdot \boldsymbol{B}_1 = 0 \ . \tag{4.29}$$

The equation of state (4.25) is not considered here because we are not concerned with fluctuations in pressure and density, only in field.

Equations (4.27)–(4.29) can be solved by means of a Fourier transformation. If we assume that the solutions are plane waves, temporal and spatial derivatives can be substituted according to

$$\partial/\partial t \rightarrow -\mathrm{i}\omega \ , \quad \nabla \rightarrow \mathrm{i}\boldsymbol{k} \ , \quad \nabla \cdot \rightarrow \mathrm{i}\boldsymbol{k} \cdot \ , \quad \nabla \times \rightarrow \mathrm{i}\boldsymbol{k} \times \ . \tag{4.30}$$

Equations (4.27)–(4.29) then read

$$-\mathrm{i}\omega\varrho_0\boldsymbol{u}_1 = \frac{\mathrm{i}}{\mu_0}(\boldsymbol{k} \times \boldsymbol{B}_1) \times \boldsymbol{B}_0 \ , \tag{4.31}$$

$$\mathrm{i}\omega\boldsymbol{B}_1 = \mathrm{i}\boldsymbol{k} \times (\boldsymbol{u}_1 \times \boldsymbol{B}_0) \ , \tag{4.32}$$

$$\boldsymbol{k} \cdot \boldsymbol{B}_1 = 0 \ . \tag{4.33}$$

Alfvén waves propagate along the field, thus $\boldsymbol{k} \| \boldsymbol{B}_0$. With $\nabla \cdot \boldsymbol{B}_0 = 0$ the above equations can be reduced to

$$\omega\boldsymbol{u}_1 = \frac{1}{\mu_0\varrho_0}(\boldsymbol{B}_0 \cdot \boldsymbol{B}_1)\boldsymbol{k} \ , \tag{4.34}$$

$$\omega\boldsymbol{B}_1 = (\boldsymbol{k} \cdot \boldsymbol{u}_1)\boldsymbol{B}_0 \ , \tag{4.35}$$

$$\boldsymbol{k} \cdot \boldsymbol{B}_1 = 0 \ . \tag{4.36}$$

Let us now multiply (4.34) by $\boldsymbol{k}$ and (4.35) by $\boldsymbol{B}_0$. Adding both equations gives the dispersion relation for the Alfvén wave:

$$\omega^2 = \frac{B_0^2}{\mu_0\varrho_0}k^2 \ . \tag{4.37}$$

Alfvén waves are non-dispersive waves. Thus group and phase speed are the same. This Alfvén speed

$$v_{\mathrm{A}} = \frac{B_0}{\sqrt{\mu_0\varrho_0}} \tag{4.38}$$

is an important characteristic of a plasma: it is the maximum speed of a disturbance propagating along the magnetic field and can be compared with the speed of sound in a gas: if a disturbance propagates faster than the Alfvén speed, a shock wave develops (Sect. 6.8). Typical Alfvén speeds are some tens of kilometers per second in the interplanetary medium and some 100 km/s in the solar corona.

Example 15. Some average parameters of the solar wind are a number density of 8 cm^{-3}, a magnetic field $B = 7$ nT and a temperature of about 2×10^5 K. The density of the solar wind is then $\varrho = nm_{\mathrm{p}}$ and the Alfvén speed is

$$
v_{\mathrm{A}} = \sqrt{\frac{(7 \text{ nT})^2}{4\pi \times 10^{-7} \text{ (V s/A m)}}} \times 8 \times 1.673 \times 10^{-27} \frac{\text{kg}}{\text{m}^{-3}} = 54 \text{ km/s} .
$$

$$(4.39)$$

Here we have used the fact that the unit T can be expressed as V s/m^2 and that the product of the electrical units volt and ampere gives the watt, which can easily be expressed as a mechanical unit. Table A.4.3 provides some help on units expressed in different forms. □

4.2.2 Magneto-Sonic Waves

A magneto-sonic wave is similar to a sound wave: it is a longitudinal wave parallel to the magnetic field with alternating regions of compression and rarefaction in both the plasma and in the magnetic field (see Fig. 4.3).

Since we allow for a compression of the plasma, we also have to consider the equation of state $p_1 = \gamma_{\mathrm{a}}\varrho_1 p_0/\varrho_0 = v_{\mathrm{s}}^2 \varrho_1$ (4.26), with

$$
v_{\mathrm{s}} = \sqrt{\frac{\gamma_{\mathrm{a}} p_0}{\varrho_0}}
$$

$$(4.40)$$

being the (adiabatic) sound speed. To solve the equation of motion (4.24), we have to express p_1 and $\boldsymbol{j}_1 \times \boldsymbol{B}_1$ by $\boldsymbol{u}_1$.

Fourier transformation of the equation of continuity gives $i\omega\varrho_1 = ik\varrho_0 u_1$ or, combined with the equation of state and (4.40),

$$
p_1 = \frac{v_{\mathrm{s}}^2}{i\omega} ik\varrho_0 u_1 :
$$

$$(4.41)$$

the fluctuating pressure p_1 in the momentum balance can be expressed by $\boldsymbol{u}_1$.

Ohm's law (4.23) yields the dependence of $\boldsymbol{E}_1$ on $\boldsymbol{u}_1$. Combined with Faraday's law (4.21), $\boldsymbol{B}_1$ can be expressed as a function of $\boldsymbol{u}_1$:

$$
\nabla \times \boldsymbol{E}_1 = -\nabla \times (\boldsymbol{u}_1 \times \boldsymbol{B}_0) = -\frac{\partial \boldsymbol{B}_1}{\partial t} .
$$

$$(4.42)$$

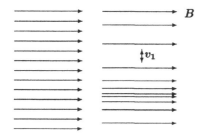

Fig. 4.3. Sketch of a magneto-sonic wave: undisturbed field (*left*) and fluctuating field (*right*)

Transformation yields

$$\mathrm{i}\boldsymbol{k} \times (\boldsymbol{u}_1 \times \boldsymbol{B}_0) = \mathrm{i}\omega\boldsymbol{B}_1 . \tag{4.43}$$

This expression for $\boldsymbol{B}_1$ can be inserted into Ampére's law (4.20):

$$\mu_0\boldsymbol{j}_1 = -\mathrm{i}\boldsymbol{k} \times \left(\frac{\boldsymbol{k}}{\omega} \times \boldsymbol{u}_1 \times \boldsymbol{B}_0\right) . \tag{4.44}$$

With (A.26), this can be simplified to

$$-\mu_0\omega\mathrm{i}\boldsymbol{j}_1 = \boldsymbol{k} \times \boldsymbol{u}_1 \cdot \boldsymbol{k} \times \boldsymbol{B}_0 . \tag{4.45}$$

Here we have used $\boldsymbol{k} \perp \boldsymbol{B}_0$ and therefore $\boldsymbol{k} \cdot \boldsymbol{B}_0 = 0$ and $\boldsymbol{k} \times \boldsymbol{B}_0 = kB_0$. The vector product of (4.45) and $\boldsymbol{B}_0$ yields

$$\boldsymbol{j}_1 \times \boldsymbol{B}_0 = \frac{1}{\mu_0\mathrm{i}\omega}k^2B_0^2\boldsymbol{u}_1 . \tag{4.46}$$

Thus the equation of motion (4.24) combined with (4.41) and (4.46) can be written as

$$-\mathrm{i}\omega\varrho_0\boldsymbol{u}_1 = -v_\mathrm{s}^2\frac{\varrho_0\mathrm{i}k^2}{\omega}\boldsymbol{u}_1 - \frac{B_0^2\mathrm{i}k^2}{\mu_0\omega}\boldsymbol{u}_1 , \tag{4.47}$$

giving the dispersion relation for the magneto-sonic wave. The phase speed is determined by the squared sound and Alfvén speeds:

$$v_\mathrm{ms}^2 = \frac{\omega^2}{k^2} = v_\mathrm{s}^2 + v_\mathrm{A}^2 . \tag{4.48}$$

It is independent of frequency or wave number: the wave is dispersion-free. If the magnetic field vanishes, the Alfvén speed approaches zero and the phase speed of the magneto-sonic wave converges towards the speed of sound. This wave is called the slow magneto-sonic wave. If the magnetic field is strong, the phase speed of the magneto-sonic wave becomes the Alfvén speed. But the wave still behaves differently because it propagates perpendicular to the field instead of parallel to it. This latter kind of wave is also called the compressive Alfvén wave. Since the phase speed of the magneto-sonic wave exceeds the Alfvén speed, it often is called the fast magneto-sonic wave or just the fast MHD wave.

Example 16. Let us briefly return to example 15, where we have already determined the Alfvén speed in the solar wind. To determine the speed of the magneto-sonic wave, we also need the sound speed. From (4.40) we obtain $v_\mathrm{s} = \sqrt{\gamma_a nk_\mathrm{B}T/(nm_\mathrm{p})} = \sqrt{\gamma_a k_\mathrm{B}T/m_\mathrm{p}} = 70$ km/s. For the speed of the magneto-sonic wave, we obtain $v_\mathrm{ms} = 88.7$ km/s from (4.48).   □

4.2.3 MHD Waves Oblique to the Field

So far, we have considered two special geometries: a transversal wave propagating parallel to the field and a longitudinal wave propagating perpendicular to it. But MHD waves can propagate at any angle relative to the field. For this general case, we have to solve the whole set of equations (4.13)–(4.19) for the undisturbed and (4.20)–(4.26) for the fluctuating quantities; neither one of the equations nor a term in one of them can be ignored. Again, a solution can be obtained by combining the equations into one equation for the desired quantity, the speed of the fluctuations. The dispersion relation obtained from this equation yields solutions for the above wave types with phase speeds depending on the angle θ between the wave vector and the undisturbed field. The speed of the MHD wave is determined by

$$u^4 - (v_A^2 + v_s^2)u^2 + v_A^2 v_s^2 \cos^2 \theta = 0 . \tag{4.49}$$

For $\theta = 90°$, the fast magneto-sonic wave (4.48) results. For $\theta = 0°$, two solutions exist: the Alfvén wave with $u = v_A$ and the sound wave with $u = v_s$.

The different solutions can be represented in a hodograph or Friedrichs diagram (see Fig. 4.4). The hodograph is a polar diagram of velocities with the polar angle θ relative to the undisturbed magnetic field direction and the wave's phase speed as distance from the origin. The wave front then propagates perpendicular to the velocity vector. Depending on whether the Alfvén speed is greater or less than the sound speed, two sets of solutions arise (see Fig. 4.4). For almost any direction we find three different phase speeds, corresponding to an Alfvén wave and a slow and a fast mode magnetosonic wave. For $\theta = 90°$ or $270°$ only the fast magneto-sonic wave exists. As can be seen from the hodographs, the Alfvén speed $u_A = v_A \cos \theta$ always

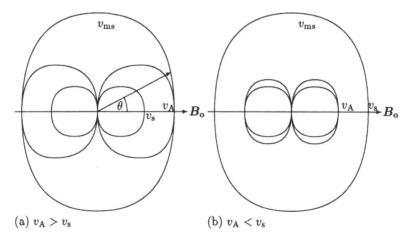

(a) $v_A > v_s$ (b) $v_A < v_s$

Fig. 4.4. Friedrichs diagram (hodograph) representing the different types of MHD waves for Alfvén speeds greater than the speed of sound (**a**) or less (**b**)

is between the phase speeds of the fast (outer curve) and the slow (inner curve) magneto-sonic wave. For waves propagating parallel to the field, the Alfvén speed becomes identical to the fast magneto-sonic speed for $v_A > v_s$. This does not contradict (4.48) because the latter had been derived for waves propagating perpendicular to the field, the direction in which the magneto-sonic speed becomes maximal. The speeds of the Alfvén wave and the slow mode magneto-acoustic wave, on the other hand, are maximal in the direction parallel to the field. The Alfvén speed of waves parallel to the field equals the slow mode speed for $v_A < v_s$.

4.3 Electrostatic Waves in Non-magnetic Plasmas

Electrostatic waves start from a charge imbalance in an initially quasi-neutral fluid element. This charge imbalance accelerates the electrons and ions in its neighborhood, resulting in charges oscillating back and forth. Since these oscillations only involve the electric field, they are defined as electrostatic waves. The oscillating magnetic field is zero. Electrostatic waves can occur in non-magnetic plasmas (this section) or in magnetized plasma (Sect. 4.4). Fourier transformation of Faraday's law (4.21) for the fluctuating quantities of the wave yields $i\boldsymbol{k} \times \boldsymbol{E}_1 = i\omega \boldsymbol{B}_1 = 0$, thus the fluctuating electric field is parallel to the wave vector $\boldsymbol{k}$.

4.3.1 Plasma Oscillations

Plasma oscillations, also called Langmuir oscillations, are a prime example of a plasma phenomenon that requires consideration of both types of charges. Nonetheless, this does not automatically imply that plasma oscillations can be described only in the framework of two-fluid theory. Since the ions are assumed to be stationary, we do not have to consider their equation of motion. Thus a one-fluid description of the electrons is sufficient as long as the electric field created by the ions is considered.

Let us assume that ions and electrons are distributed equally in space. Thus quasi-neutrality is fulfilled even in rather small volumes. In addition, we assume that the thermal motion of the particles vanishes; the fluid is treated as a cold plasma. Such a fluid can be disturbed by the displacement of part of the electrons, as indicated in Fig. 4.5. This displacement creates an electric field that pulls the electrons back to their initial rest positions while the heavier ions stay in their positions. Since the electrons are accelerated along the field, they gain kinetic energy which in turn drives them behind their initial position, creating an electric field in the opposite direction. This field slows down the electrons, eventually driving them back. The period of the resulting electron oscillation around their rest position is the electron plasma frequency ω_{ep}. It can be derived from the equation of motion

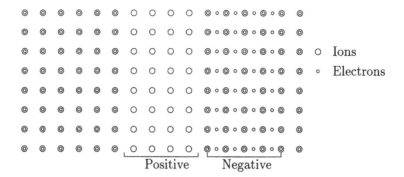

Fig. 4.5. The displacement of electrons in a cold plasma leads to plasma oscillations

$$m_{\mathrm{e}}\frac{\partial^2 x}{\partial t^2} = -eE , \qquad (4.50)$$

where x is the direction parallel to the electric field. E can be determined by applying Gauss's law to a closed surface along the boundary between the positive and negative charges, and extending at least for the displacement x of the charges: $E = 4\pi n_{\mathrm{e}}\,e\,x/\varepsilon_0$. The equation of motion then is

$$\frac{\partial^2 x}{\partial t^2} = -\frac{n_{\mathrm{e}}e^2}{\varepsilon_0 m_{\mathrm{e}}}x = -\omega_{\mathrm{pe}}x . \qquad (4.51)$$

This second order differential equation describes a wave and can be solved with an ansatz $x = x_0\mathrm{e}^{\mathrm{i}\omega t}$. It describes a harmonic oscillator with the electron plasma frequency

$$\omega_{\mathrm{pe}} = \sqrt{\frac{n_{\mathrm{e}}e^2}{\varepsilon_0 m_{\mathrm{e}}}} . \qquad (4.52)$$

The electrons therefore oscillate with a frequency which neither depends on the wave-length nor on the amplitude of the disturbance responsible for their initial displacement.

Plasma oscillations are an important tool in plasma diagnostics because they allow us to measure the electron density n_{e}.

Example 17. Solar radio bursts (Sect. 6.7.1) are an important tool for the diagnosis of coronal disturbances. For the lower corona, we can use a simplified density model $n \sim n_0 \exp(-r/r_0)$ with a scale height of $r_0 = 0.1r_\odot$ and a density $n_0 = 10^{15}$ m^3 at $r = 1r_\odot$ (more correctly, this applies at the base of the corona 2000 km above the solar surface, but for the numerical exercise this 2000 km can be ignored). The radio emission observed from the ground is in the frequency range 10 MHz to about 200 MHz. According to (4.52), this corresponds to coronal heights between $2r_0$ and $0.8r_0$ above the surface. The main mechanisms for the generation of solar radio bursts are electron beams with speeds of about $c/3$ (type III burst) and coronal shock waves

with speeds of the order of 1000 km/s (type II bursts). The height range over which these exciters propagate is $1.2r_0 = 0.12r_\odot = 8.4 \times 10^4$ km. The electrons with speed $c/3$ travel this distance within about 0.3 s, and thus the type III burst shows an extremely fast frequency drift of the order of 750 MHz/s. The shock wave, on the other hand, needs almost 9 s to travel this distance, corresponding to an average frequency drift of about 20 MHz/s. Thus both types of bursts can be easily identified by their frequency drift in radio spectrograms.   □

4.3.2 Electron Plasma Waves (Langmuir Waves)

Plasma oscillations have been derived for a cold plasma: the group velocity equals zero and the disturbance does not propagate. In a warm plasma, the situation is different. The thermal motion of the electrons carries information about a disturbance into the undisturbed ambient plasma. The disturbance then propagates as a wave. Formally, one can derive the dispersion relation for this wave by adding the pressure gradient force $-\nabla p$ to the equation of motion (4.50). If the plasma behaves adiabatically, the dispersion relation for the plasma wave in the one-dimensional case ($\gamma_a = 3$) reads

$$\omega^2 = \omega_{\mathrm{pe}}^2 + \tfrac{3}{2}k^2 v_{\mathrm{th}}^2 \,, \tag{4.53}$$

or in the three-dimensional case

$$\omega^2 = \omega_{\mathrm{pe}}^2 + \tfrac{5}{3}k^2 v_{\mathrm{th}}^2 \,, \tag{4.54}$$

with the thermal velocity $v_{\mathrm{th}} = \sqrt{2k_B T/m_e}$ (see Fig. 4.6). Both equations are based on the assumption that locally a Maxwell equilibrium is established. Thus the plasma must allow for frequent collisions. In space plasmas, in general this is not the case. Here the correct dispersion relation is

$$\omega^2 = \omega_{\mathrm{pe}}^2 + 3k^2 v_{\mathrm{th}}^2 \,. \tag{4.55}$$

This equation is called the Bohm–Gross equation [53]. From (4.55) the group velocity of plasma waves can be determined as

$$v_{\mathrm{g}} = \frac{\mathrm{d}\omega}{\mathrm{d}k} = \frac{3k}{\omega}v_{th}^2 = \frac{3v_{\mathrm{th}}^2}{v_{\mathrm{ph}}} \,. \tag{4.56}$$

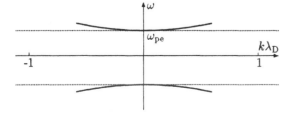

Fig. 4.6. Dispersion relation for electrostatic electron waves in a warm unmagnetized plasma (*Langmuir waves*)

As expected, the group velocity vanishes as the thermal energy of the plasma approaches zero. The group velocity always is significantly smaller than the thermal speed and thus also much smaller than the speed of light. For large wave numbers k, the information propagates with the thermal velocity. For small wave numbers, the information travels slower than v_{th} because the density gradient decreases for large wavelengths and therefore the net flux of momentum into adjacent layers becomes small.

4.3.3 Ion-Acoustic Waves (Ion Waves)

In Langmuir waves, as in plasma oscillations, the ions are assumed to be indefinitely massive; they stay fixed at their position. Langmuir waves thus are high-frequency waves. If we allow for ion motion, the properties of the wave change. The inertia of the ions requires rather slow oscillations. Therefore, ion waves are low-frequency waves. In ordinary fluids, sound waves are the counterpart of the ion wave. They can be derived from the Navier–Stokes equation with the pressure gradient force being the only term on the right-hand side. The dispersion relation than reads $v_s = \omega/k = \sqrt{\gamma_a p_0/\varrho_0} = \sqrt{\gamma_a k_B T/m}$, with v_s being the speed of sound.

Sound waves are pressure waves. They transport momentum from one layer to the next due to collisions between molecules or atoms. Despite its often low density, in a plasma a similar phenomenon exists. Here the momentum is transported by Coulomb collisions; thus information is contained in the charges and the fields. Since we have to consider the motion of both electrons and ions, the ion wave can be derived in the framework of two-fluid theory only. In the equation of motion, we have to consider the pressure gradient force and the force exerted by the electromagnetic field. The dispersion relation than can be derived as

$$\frac{\omega^2}{k^2} = \frac{\gamma_e k_B T_e + \gamma_i k_B T_i}{m_i} = v_s \ . \tag{4.57}$$

Note that $\gamma_e \approx 1$ because the electrons are fast compared with the waves and an isothermal distribution is established easily. The ions, on the other hand, experience a one-dimensional compression in the plane wave, and $\gamma_i = 3$. The group velocity of the ion wave is independent of the wave number k.

If the ion population can be considered as cold, $T_i \to 0$, and the wave length is small, $k\lambda_D \gg 1$, the ions can oscillate with the ion plasma frequency

$$\omega^2 = \omega_{pi}^2 = \frac{n_i Z^2 e^2}{\varepsilon_0 m_i} \ . \tag{4.58}$$

The dispersion relation for ion waves is shown in Fig. 4.7. The ion plasma frequency as asymptote for short wavelengths is indicated in the right part of the figure. There are fundamental differences in the dispersion relation for electron and ion waves. Electron plasma waves are waves with constant

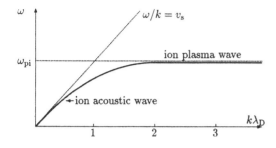

Fig. 4.7. Dispersion relation for ion waves in a warm unmagnetized plasma. The dashed lines are asymptotes for small and large wave numbers

frequency (see Fig. 4.6), the thermal motion only adds a small correction. In contrast, ion waves are waves with a constant speed and require thermal motion. The phase speed and group speed are identical. The difference between these two types of waves is due to the behavior of the second particle component: in the electron plasma wave, the electrons oscillate while the ions stay fixed. In the ion wave, the ions oscillate but the electrons are in motion too. In particular, they can be carried along by the ions, screening part of the electric field resulting from the ion oscillation.

In space plasmas, ion acoustic waves are observed upstream of planetary bow shocks where they are generated by suprathermal particles streaming away from the shock front (Sect. 7.6.4).

Example 18. The proton density upstream of a shock is $n_i = 8$ cm^{-3}, and for the proton $Z_i = 1$. Ion acoustic waves (upstream waves) excited by particles streaming away from the shock then have a ion acoustic frequency $\omega_{pi} = 3.7 \times 10^3$ Hz. $\square$

4.4 Electrostatic Waves in Magnetized Plasmas

Let us now consider electrostatic waves in a magnetized plasma. They can be divided into waves with k parallel or perpendicular to B_0. The terms longitudinal or transversal refer to the direction of the wave vector k relative to the fluctuating electric field E_1. Only longitudinal waves are electrostatic because $k \times E_1 = \omega B_1$ vanishes. In a transverse wave, B_1 is finite and the wave is an electromagnetic one. Waves propagating oblique to the field can be regarded as superposition of longitudinal and transversal waves.

4.4.1 Electron Oscillations Perpendicular to B
(Upper Hybrid Frequency)

As for electron oscillations in an unmagnetized plasma, in the upper hybrid oscillations the ions stay fixed in space, creating a positively charged, uniform background. The plasma is cold; the thermal motion of the electrons can be ignored. The equation of motion then contains only the forces exerted by the electric and magnetic fields.

For longitudinal waves, the dispersion relation reads

$$\omega_{uh}^2 = \omega_{pe}^2 + \omega_{ce}^2 \,, \tag{4.59}$$

with ω_{pe} being the frequency of electron plasma oscillations and ω_{ce} being the electron cyclotron frequency. Only electrostatic waves perpendicular to B have this upper hybrid frequency. Disturbances parallel to B oscillate with the plasma frequency ω_{pe}: particles moving parallel to the magnetic field do not gyrate and therefore ω_{ce} vanishes.

We can understand these waves as the superposition of two motions. In the plane wave, the electrons exhibit regions of compression and rarefaction as in ordinary plasma oscillations. The magnetic field, which is perpendicular to the direction of electron motion, forces the electrons into elliptical orbits instead of oscillations along a straight line. As in simple plasma oscillations, the electric field accelerates the electrons displaced from their rest position. As the electron speed increases, the Lorentz force exerted by the magnetic field increases, too, reversing the direction of electron motion. The electrons therefore move against the electric field, losing energy. Thus two restoring forces act on the electrons: the electrostatic force resulting from the electron displacement and the Lorentz force. This additional restoring force leads to a higher frequency than in simple plasma oscillations. If the magnetic field vanishes, the cyclotron frequency vanishes, too, leaving us with ordinary plasma oscillations. If the plasma density decreases, the plasma frequency decreases, too. For vanishing plasma density, the remaining motion is a gyration around the magnetic field line.

4.4.2 Electrostatic Ion Waves Perpendicular to B (Ion Cyclotron Waves)

The upper hybrid wave is a high-frequency wave with ω much larger than both the plasma and cyclotron frequencies. It results from the motion of electrons in a magnetized plasma. Electrostatic ion waves, like ion acoustic waves, are low-frequency waves. Let us now consider an ion acoustic wave with k *nearly* perpendicular to B. In the equation of motion only the forces of the electromagnetic field are considered. The dispersion relation then is

$$\omega_{pi}^2 = \omega_{ci}^2 + k^2 v_s^2 \,, \tag{4.60}$$

with ω_{ci} being the ion cyclotron frequency. An electrostatic ion wave parallel to the magnetic field has the same frequency $\omega^2 = k^2 v_s^2$ as an ion acoustic wave because the ions do not gyrate around the field and ω_{ci} vanishes.

The physical explanation is the same as in the upper hybrid wave: the Lorentz force provides an additional restoring force leading to an elliptical path and a higher frequency, described by the additional term ω_{ci} in (4.60).

4.4.3 Lower Hybrid Frequency

In the ion cyclotron wave we have assumed k to be nearly perpendicular to B. The small remaining component of the wave vector and therefore the particle motion parallel to B is essential because it allows the electrons to travel freely along B to obtain thermal equilibrium as was required in the derivation of the ion acoustic wave. If this cannot be archived, the equation of motion for the electrons must be solved differently, though that for the ions still is valid. In this case, the lower hybrid wave results with

$$\omega_{lh} = \sqrt{\omega_{ce}\,\omega_{ci}} \; . \tag{4.61}$$

The motion is maintained in the perpendicular direction only. Thus charge neutrality can be maintained in that direction. From this charge neutrality we can understand the frequency given in (4.61). The fluctuating electric field E_1 of the wave is perpendicular to B_0. The ions move along E_1 while the Lorentz force acting on them is rather small. The ion displacement along E_1 is limited because the electric field oscillates; the maximum displacement is about $\Delta x_i = eE_1/m_i\omega_{ci}^2$. The electrons gyrate around the magnetic field. In addition, they experience an $E \times B$ drift perpendicular to both fields. Their displacement parallel to E_1 is given roughly as $\Delta x_e = E_1/B_0\omega_{ce}$. Charge neutrality requires $\Delta x_e = \Delta x_i$ and therefore $\omega = \omega_{lh}$.

Lower hybrid waves are of great importance in the auroral regions where they may be responsible for ion heating.

4.5 Electromagnetic Waves in Non-magnetized Plasmas

Electromagnetic waves consist of both a fluctuating electric and a fluctuating magnetic field. In this section we shall consider electromagnetic waves in an unmagnetized plasma; thus the background field B_0 vanishes and then $B = B_1$. Electromagnetic waves are high-frequency waves. Thus because of their large inertia the ions do not follow the fluctuating field. Electromagnetic waves therefore can be treated within the framework of one-fluid theory. In the equation of motion we have to consider the pressure gradient force and the forces exerted by the electromagnetic field. The dispersion relation for electromagnetic waves then reads

$$\omega^2 = \omega_{pe}^2 + k^2 c^2 \; , \tag{4.62}$$

where $c = 1/\sqrt{\varepsilon_0\mu_0}$ is the speed of light in vacuum.

The dispersion relation (4.62) is shown in Fig. 4.8. For plasma frequencies much smaller than the frequency of the waves we get light waves with $\omega = kc$. The index of refraction for these waves is $n = c/v_{ph} = ck/\omega$ or, taking into consideration the electron plasma frequency,

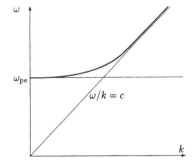

Fig. 4.8. Dispersion relation for electromagnetic waves in a cold unmagnetized plasma. For small wave numbers the group velocity approaches zero and a plasma oscillation results. For large wave numbers both the group and phase speeds converge towards the speed of light

$$n = \sqrt{1 - \frac{\omega_{pe}^2}{\omega^2}} \ . \tag{4.63}$$

Waves can propagate through a medium only if n^2 is larger than zero. Thus electromagnetic waves can exist only if $\omega > \omega_{pe}$. For transverse electromagnetic waves, the ordinary light waves, the cutoff frequency is the electron plasma frequency, the plasma is opaque at lower frequencies.

For $\omega < \omega_{pe}$ an imaginary refraction index results. For a real frequency an imaginary wave vector $\boldsymbol{k}$ would result. Such a wave would not propagate but decay. The second solution, $-\mathrm{i}k$, formally could be interpreted as wave growth. Physically, however, this cannot happen: because we are considering a cold, stationary plasma, the energy required for wave growth cannot be drawn from the plasma. The cold plasma only can absorb the energy of a decaying wave and cannot support wave growth.

Example 19. Electromagnetic waves can be used for plasma diagnostics in space or in the ionosphere. For instance, the density of a space plasma can be determined by detecting the radio signal of a satellite from another satellite or a ground station. As the frequency of the radio signal is varied, absorption sets in at the plasma frequency of the medium; thus the density can be determined.

Another application is the measurement of the electron density in the Earth's ionosphere. Radio waves of variable frequency are sent from a transmitter A to a receiver B, see Fig. 4.9. As the wave enters the ionosphere, it is refracted according to Snell's law. Thus from the travel time of a radio pulse between transmitter and receiver we can determine the height at which the ray path is bent down towards Earth again. From the geometry, we also

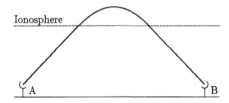

Fig. 4.9. The electron density in the ionosphere can be determined from the reflection of electromagnetic waves

can determine the angle at which the electromagnetic wave enters the ionosphere and, from Snell's law, the refraction index. The latter gives us the
plasma frequency and also the electron density. Ionospheric reflection is used
in short-wave communications because it allows us to send signals around
the Earth. With a maximum ionospheric electron density of about 10^{12} m^{-3},
the critical frequency for reflection is about 10 MHz.

The experiment sketched in Fig. 4.9 can be simplified to a vertical geometry in which the transmitter and receiver are located at the same position.
Total reflection of the pulse then occurs when the emitted frequency equals
the local plasma frequency, i.e. $n = 0$. The travel time of the signal gives the
height of reflection. Emitting at different frequencies over a broad frequency
band, the height profile of the electrons in the ionosphere can be determined.
In principle, this sounding experiment is very simple because reflection occurs
exactly at the point where the signal frequency is equal to the local plasma
frequency. But a small error remains: the electron density varies continuously with height, leading to changes in the propagation speed of the wave.
Therefore, the estimate of the reflection height from the propagation time
is not exact. In reality, the situation becomes even more complex because
the electromagnetic wave propagates into a magnetized plasma and is split
into ordinary and extraordinary modes, both having different propagation
speeds.   □

Example 20. The electron density in the quiet ionosphere at 120 km height is
about 2×10^5 cm^{-3} (see Fig. 8.18). Reflection occurs for $n = 0$, that is, when
the local plasma frequency ω_{pe} equals the wave frequency. From (4.52) we
find $f_{pe} = 2.4$ MHz. During a sudden ionospheric disturbance, the electron
density is increased by a factor of 3. Now the critical frequency for reflection
at 120 km height becomes 7 MHz. The 2.4 MHz wave reflected originally at
this height is now reflected at a lower height, and consequently its range of
propagation is reduced and the signal is not received where it was supposed
to be received: communication is inhibited.   □

4.6 Electromagnetic Waves in Magnetized Plasmas

As in the case of electrostatic waves, electromagnetic waves in magnetized
plasmas are characterized by the direction of the wave vector relative to the
background magnetic field B_0 and the fluctuating electric field E_1.

4.6.1 Electromagnetic Waves Perpendicular to B_0

Let us start with an electromagnetic wave perpendicular to the undisturbed
magnetic field. The wave is assumed to be transversal, and thus $k \perp E_1$.
Two modes are possible: either the fluctuating electric field is parallel to B_0
(ordinary wave) or it is perpendicular to B_0 (extraordinary wave).

Ordinary Waves ($E_1 \| B_0$). For $E_1 \| B_0$ the wave equation takes the same form as in an unmagnetized plasma, leading to the dispersion relation

$$\omega^2 = \omega_{\text{pe}}^2 + c^2 k^2 . \tag{4.64}$$

The propagation of ordinary or O-waves therefore is not influenced by the magnetic field: in an ordinary wave, the fluctuating electric field is parallel to B_0 and therefore the magnetic field does not influence the wave dynamics.

Extraordinary Waves ($E_1 \perp B_0$). If the fluctuating electric field is perpendicular to B_0, the electron motion is influenced by the magnetic field and the dispersion relation has to be modified accordingly. As the wave propagates perpendicular to B_0 and E_1 oscillates perpendicular to B_0, the wave vector k has components parallel and perpendicular to E_1. The wave therefore is a mixed mode of both a transversal and a longitudinal component. While the ordinary wave is linearly polarized (E_1 only along B_0), the extraordinary wave is elliptically polarized. The dispersion relation for the extraordinary wave can be written as

$$\frac{c^2 k^2}{\omega^2} = 1 - \frac{\omega_{\text{pe}}^2}{\omega^2} \frac{\omega^2 - \omega_{\text{pe}}^2}{\omega^2 - \omega_{\text{uh}}^2} = n^2 . \tag{4.65}$$

The influence of the magnetic field is contained in the upper hybrid frequency (4.59). The dispersion relation for both ordinary and extraordinary electromagnetic waves is shown in Fig. 4.10. The hatched area indicates the stop band separating two different modes of the extraordinary or X-wave.

A closer look at the refraction index (4.65) helps us to understand these two modes. The refraction index becomes zero for

$$\omega_{(\text{L,R})}^{\text{X}} = \pm \frac{\omega_{\text{ce}}}{2} + \sqrt{\omega_{\text{pe}}^2 + \frac{\omega_{\text{ce}}^2}{4}} , \tag{4.66}$$

with 'R' and 'L' indicating right-hand and left-hand polarized waves.

The frequencies defined by (4.66) are cutoff frequencies: for lower frequencies the refraction index becomes imaginary and the wave vanishes. The ordinary wave has its cutoff at the electron plasma frequency. The refraction index also can go towards infinity. This happens for very large wave numbers (or very small wavelengths). The corresponding frequency can be found by letting k go towards infinity. This occurs when ω goes towards the upper hybrid frequency ω_{uh} in the denominator of (4.65). As the extraordinary wave approaches this resonance, its phase and group velocities approach zero and the electromagnetic energy is converted into electrostatic oscillations. Therefore, in Fig. 4.10 between the upper hybrid frequency ω_{uh} and the cutoff frequency ω_{R} a stop band results where, because $n < 0$, the extraordinary wave cannot propagate.

The existence of a stop band has an interesting application. Magnetized plasmas are emitters of radio waves. Therefore, all planets emit radio signals. In the solar system, the largest radio source is Jupiter. While it is easy

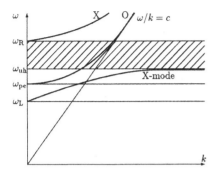

Fig. 4.10. Dispersion relation for ordinary (O) and extraordinary (X) electromagnetic waves in a cold magnetized plasma. The hatched region between ω_{uh} and ω_R separates the two modes of the extraordinary electromagnetic wave

to detect these waves, their interpretation in terms of sources and excitation mechanisms is difficult. The stop band, however, allows us to exclude one possible mechanism. The radio waves are excited close to the planet where both the electron plasma frequency as well as the electron cyclotron frequency are large. As the waves propagate outward, their frequency decreases. Extraordinary waves excited at a frequency below the upper hybrid frequency will encounter a stop band where ω_{uh} has decreased below the frequency of the wave. These waves cannot escape from the vicinity of the planet. Planetary radio waves observed at large distances therefore have not been excited as extraordinary waves below the local upper hybrid frequency.

4.6.2 Waves Parallel to the Magnetic Field: Whistler (R-Waves) and L-Waves

Let us now consider electromagnetic waves propagating parallel to $\boldsymbol{B}_0$. These waves are of particular importance in the magnetosphere because they can propagate along $\boldsymbol{B}$ towards the ground. Again, two cases can be distinguished: the right-hand and the left-hand waves. The dispersion relation becomes

$$\frac{\omega^2 - k^2 c^2}{\omega_{\mathrm{pe}}^2}\left(1 \pm \frac{\omega_{\mathrm{ce}}}{\omega}\right) = 1 \tag{4.67}$$

and the refraction index is

$$n^2 = \frac{c^2 k^2}{\omega^2} = 1 - \frac{\omega_e^2/\omega^2}{1 \pm \omega_{\mathrm{ce}}/\omega} \ . \tag{4.68}$$

In both equations the '+' sign refers to the right-hand or R-wave. The electric field of an R-wave rotates in the same way as an ordinary screw rotates: if the thumb of the right-hand points in the direction of $\boldsymbol{k}$, the curved fingers point in the rotation direction of the electric field. Thus the R-wave rotates in the same direction as an electron gyrates. The L-wave rotates in the opposite direction, following the ion gyration. Thus both waves have a resonance. The

R-wave has a resonance at $\omega = \omega_{ce}$. The electron gyrating around the magnetic field then experiences a constant electric field which, depending on the phase between the electron and the field, either accelerates or decelerates it continuously. This resonance is called cyclotron resonance. At this resonance the phase velocity is zero and the wave does not propagate. The L-wave does not resonate with the electrons because the wave field rotates in the direction opposite to the electron gyration. Thus the L-wave has no resonance at high frequencies, and its resonance is $\omega = \omega_{ci}$. The cutoff frequencies for L- and R-waves are the same as for the extraordinary waves:

$$\omega_{(L,R)}^{cutoff} = \pm\frac{\omega_{ce}}{2} \pm \sqrt{\omega_{ce}^2 + \frac{\omega_{pe}^2}{4}} \, . \tag{4.69}$$

The dispersion relation for electromagnetic waves propagating parallel to the magnetic field is shown in Fig. 4.11. Two cases can be distinguished: $\omega_{pe} < \omega_{ce}$ (right panel) and $\omega_{pe} > \omega_{ce}$ (left panel). The main difference between these cases is the cutoff for the left-hand mode: for $\omega_{pe} < \omega_{ce}$ the cutoff is below the electron cyclotron frequency, while in the opposite case it is above. For a constant cyclotron frequency these waves approach the $n^2 = 1$ line for high frequencies. The R-wave has a stop band between ω_R and ω_{ce}. The R-waves in the lower frequency range also are called Whistler waves and are important for propagation studies in the magnetosphere. Whistler waves can be excited by lightning discharge. Thus the source has a short duration but creates a wide spectrum of different frequencies. Since the propagation time depends on the group speed, the first waves arriving at a distant observer have higher frequencies than those arriving later, rather like a whistle with decreasing pitch. The rate of frequency change contains information about the change in plasma density along the propagation path.

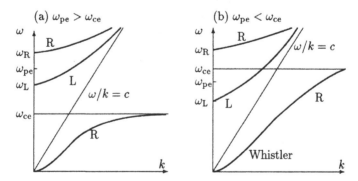

Fig. 4.11. Dispersion relation for electromagnetic waves propagating parallel to the magnetic field in a cold magnetized plasma

4.7 Summary

Table 4.1 summarizes the waves discussed in this chapter. In electrostatic waves only the electric field fluctuates while the magnetic field either is static or zero. In electromagnetic waves, both the electric and magnetic fields oscillate. The waves are further divided into electron and ion waves. In electron waves the electrons oscillate while the ions create a uniform background. Therefore, electron waves are high-frequency waves. In ion waves, both ions and electrons oscillate. Because of the large inertia of the ions, ion waves are low-frequency waves. The propagation of the wave depends on the orientation of the wave vector k relative to the background magnetic field and relative to the fluctuating electric field.

We should note that this set of dispersion relations is greatly simplified in so far as it considers only waves propagating into the principal directions perpendicular or parallel to the field. Waves propagating oblique to the field are more difficult; however, often they can be understood as the superposition of the two modes parallel and perpendicular to the field.

Table 4.1. Different types of plasma waves

Wave	Geometry	Dispersion relation Equation
Electron Waves (Electrostatic)		
Langmuir waves	$B_0 = 0$ or $k \| B_0$	$\omega^2 = \omega_{pe}^2 + 3k^2 v_{th}^2/2$ (4.55)
Upper hybrid waves	$k \perp B_0$	$\omega_{uh} = \omega_{pe}^2 + \omega_{ce}^2$ (4.59)
Ion Waves (Electrostatic)		
Ion acoustic waves	$B_0 = 0$ or $k \| B$	$\omega^2 = k^2 v_s^2$ (4.57)
Ion cyclotron waves	$k \perp B_0$	$\omega^2 = \omega_{ci}^2 + k^2 v_s^2$ (4.60)
Lower hybrid waves	$K \perp B_0$	$\omega_{lh}^2 = \omega_{ci}\,\omega_{ce}$ (4.61)
Electron Waves (Electromagnetic)		
Light waves	$B_0 = 0$	$\omega^2 = \omega_{pe}^2 + k^2 c^2$ (4.62)
O-waves	$k \perp B_0,$ $E_1 \| B_0$	$\omega^2 = c^2 k^2 + \omega_{pe}^2$ (4.64)
X-waves	$k \perp B_0,$ $E_1 \perp B_0$	$\omega^2 = c^2 k^2 +$ $+\omega_{pe}^2\,(\omega^2 - \omega_{pe}^2)/(\omega^2 - \omega_h^2)$ (4.65)
Whistler (R-waves)	$k \| B_0$	$\omega^2 = c^2 k^2 - \omega_{pe}^2/[1 - (\omega_{ce}/\omega)]$ (4.67)
L-waves	$k \| B_0$	$\omega^2 = c^2 k^2 + \omega_{pe}^2/[1 + (\omega_{ce}/\omega)]$ (4.67)
Ion Waves (Electromagnetic)		
Alfvén waves	$k \| B_0$	$\omega^2 = k^2 v_A^2$ (4.37)
Magneto-sonic waves	$k \perp B_0$	$\omega^2 = c^2 k^2\,(v_s^2 + v_A^2)/(c^2 + v_A^2)$ (4.48)

Some of these waves can be used for plasma diagnostics, with the sources of the waves either being natural, such as in Whistler waves to study propagation in the magnetospheric plasma, or artificial, such as radio waves to probe the ionosphere.

Exercises and Problems

4.1. Explain, in your own words, the important quantities characterizing a wave. What are group and phase speeds?

4.2. Show that in an electron plasma wave the energy contained in the electron oscillation exceeds the energy in the ions by the mass ratio $m_i/(Z_i m_e)$.

4.3. On re-entry into the Earth's atmosphere, a spacecraft experiences a radio blackout due to the shock developing in front of the spacecraft. Determine the electron density inside the shock if the transmitter works at 300 MHz.

4.4. Show that the maximum phase speed of a Whistler wave is at a frequency $\omega = \omega_{ce}/2$. Prove that this is below the speed of light.

4.5. Show that if a packet of Whistler waves with a spread in frequency is generated at a given instant, a distant observer will receive the higher frequencies earlier than the lower ones.

4.6. Show that if the finite mass of the ions is included, the frequency of Langmuir waves in a cold plasma is given by $\omega^2 = \omega_{pe}^2 + \omega_{pi}^2$.

4.7. How would you use pulse delay as a function of frequency to measure the average plasma density between the Earth and a distant pulsar?

4.8. Determine the Alfvén speeds and the electron plasma frequencies for the situations described in Problem 3.12.

4.9. Use Fig. 4.4 to describe the properties of magnetohydrodynamic waves propagating parallel to B_0 for $v_A > v_s$ and $v_A < v_s$.

4.10. Show that in an Alfvén wave the average kinetic energy equals the average magnetic energy.

4.11. Discuss an Alfvén wave with $k \| B_0$. (a) Determine the dispersion relation under the assumption of a high but finite conductivity (the displacement current nevertheless can be ignored). (b) Determine the real and the imaginary parts of the wave vector for a real frequency.

5 Kinetic Theory

All animals are equal
but some animals
are more equal
than others.

G. Orwell, *Animal Farm*

Magnetohydrodynamics has treated the plasma as a fluid with all particles moving at the same speed, the bulk speed. The thermal motion is ignored. Space plasmas often are hot plasmas with the thermal speed by far exceeding the flow speed. Thus thermal motion has to be taken into account and the plasma has to be described by a distribution function which considers the positions and velocities of the individual particles. The Boltzmann equation gives the equation of motion for this phase space density.

5.1 The Distribution Function

Kinetic theory starts from the physics of individual particles (the microscopic approach). The macroscopic phenomena then can be described by averaging over a sufficiently large number of particles, an approach which also is used in statistical mechanics. This formalism therefore is called the statistical description of a plasma.

5.1.1 Phase Space and Distribution Function

The mechanical properties of each particle are described completely by its position and momentum. The phase space is a six-dimensional space defined by the three spatial coordinates q_1, q_2, q_3 and the three generalized momenta p_1, p_2, p_3. Each particle is related unambiguously to one point in phase space:

$$\boldsymbol{Q} = (q_1, q_2, q_3; p_1, p_2, p_3) = (\boldsymbol{q}, \boldsymbol{p}) \, . \tag{5.1}$$

The speed of the particle in phase space, i.e. the combined change in its position and momentum in ordinary three-dimensional space, then is

$$C = \frac{\mathrm{d}\boldsymbol{Q}}{\mathrm{d}t} = \left(\frac{\mathrm{d}q_1}{\mathrm{d}t}, \frac{\mathrm{d}q_2}{\mathrm{d}t}, \frac{\mathrm{d}q_3}{\mathrm{d}t}; \frac{\mathrm{d}p_1}{\mathrm{d}t}, \frac{\mathrm{d}p_2}{\mathrm{d}t}, \frac{\mathrm{d}p_3}{\mathrm{d}t}\right) = \left(\frac{\mathrm{d}\boldsymbol{q}}{\mathrm{d}t}, \frac{\mathrm{d}\boldsymbol{p}}{\mathrm{d}t}\right). \qquad (5.2)$$

For a plasma consisting of N particles, in phase space N such points exist. The phase space density $f(\boldsymbol{q}, \boldsymbol{p}, t)$ gives the number of particles inside a volume element $[(q_i, q_i + \mathrm{d}q_i), (p_i, p_i + \mathrm{d}p_i)]$. This density function is also called the distribution function of the plasma. It is related to the particle density in ordinary three-dimensional space by

$$n(\boldsymbol{q}, t) = \int\limits_{-\infty}^{+\infty} f(\boldsymbol{q}, \boldsymbol{p}, t)\, \mathrm{d}^3\boldsymbol{p}\,. \qquad (5.3)$$

The number density is used in the definition of averages which describe the macroscopic properties of the plasma. If $a(\boldsymbol{q}, \boldsymbol{p}, t)$ is a function in phase space, its average is defined as

$$\langle a(\boldsymbol{q}, t)\rangle = \frac{1}{n(\boldsymbol{q}, t)} \int\limits_{-\infty}^{\infty} a(\boldsymbol{q}, \boldsymbol{p}, t)\, f(\boldsymbol{q}, \boldsymbol{p}, t)\, \mathrm{d}^3\boldsymbol{p}\,. \qquad (5.4)$$

The average or bulk speed of the plasma, for instance, is given as

$$\boldsymbol{u}(\boldsymbol{q}, t) = \langle\boldsymbol{v}(\boldsymbol{q}, t)\rangle = \frac{1}{n(\boldsymbol{q}, t)} \int\limits_{-\infty}^{\infty} \boldsymbol{v}(\boldsymbol{p}, \boldsymbol{q}, t)\, f(\boldsymbol{q}, \boldsymbol{p}, t)\, \mathrm{d}^3\boldsymbol{p}\,. \qquad (5.5)$$

Application of this averaging scheme to the equation of motion yields the Vlasov equation (5.23) which is the basic equation of statistical plasma physics and can be used to derive the MHD equations of two-fluid theory.

5.1.2 Maxwell's Velocity Distribution

The average speed $\langle\boldsymbol{v}\rangle$ of the particles in a plasma, as defined by (5.5), is also the plasma flow speed $\boldsymbol{u}$. The speeds $\boldsymbol{v}_i$ of individual particles, however, can be substantially different; in particular, speeds of individual particles can exceed the flow speed by orders of magnitude. A plasma contains different particle species s which all have their own average speed $\boldsymbol{u}_s$. If in an electron–proton plasma the average speeds $\boldsymbol{u}_e$ and $\boldsymbol{u}_p$ are different, a current results.

   To derive the velocity distribution of the particles, let us first determine the kinetic energy contained in a volume element of the plasma. The kinetic energy of the plasma flow is determined by the average speed $\boldsymbol{u}$, while the entire kinetic energy is the sum of the kinetic energies of all particles. Since the latter is larger, there is also kinetic energy contained in the stochastic motion of the particles, which can be described by

$$\left\langle\frac{m}{2}(\boldsymbol{v} - \boldsymbol{u})^2\right\rangle = \frac{\int m(\boldsymbol{v} - \boldsymbol{u})^2 f(\boldsymbol{r}, \boldsymbol{v}, t)\mathrm{d}\boldsymbol{v}}{2 \int f(\boldsymbol{r}, \boldsymbol{v}, t)\mathrm{d}\boldsymbol{v}}\,. \qquad (5.6)$$

This "random" kinetic energy is related to the hydrostatic pressure by

$$\frac{p}{n} = \frac{2}{N} \left\langle \frac{m}{2}(v - u_s)^2 \right\rangle , \tag{5.7}$$

where N is the number of degrees of freedom, normally three.

If the system is in thermal equilibrium, which in a hot plasma is not necessarily the case, the velocity distribution is given by the Maxwell distribution:

$$f(r, v, t) = n \sqrt{\left(\frac{m}{2\pi k_B T}\right)^3} \exp\left\{-\frac{m(v - u)^2}{2k_B T}\right\} \tag{5.8}$$

where T is the plasma temperature and k_B is the Boltzmann constant. According to (5.8), the relative number of particles with large stochastic speeds $|v - u|$ increases with T. The distribution's maximum is at the most probable thermal speed v_{th}:

$$v_{th} = \sqrt{\frac{2k_B T}{m}} . \tag{5.9}$$

In a one-atomic gas in equilibrium, the temperature is related to the kinetic energy of the stochastic motion by

$$\left\langle \tfrac{1}{2}m(v - u)^2 \right\rangle = \tfrac{1}{2}Nk_B T . \tag{5.10}$$

In a plasma, N normally equals 3. Even in a magnetized plasma N equals 3 because the particle speed is described completely by one of the triples $v_x - u_x$, $v_y - u_y$, and $v_z - u_z$ or $v_\parallel$, $|v_\perp - u|$, and ψ_v, the latter describing the direction of the perpendicular speed relative to the gyro-center. Note that the combination of (5.7) and (5.10) yields the ideal gas law $p = nk_B T$.

Occasionally, we are concerned only with particle speeds and not with the direction of motion. This might be the case if the plasma is at rest, i.e. u equals zero. The distribution function (5.8) then is

$$\iint f(r, v, t)\, d\Omega_v\, v^2\, dv = (4\pi f(r, |v|, t)v^2)\, dv = g(r, v)\, dv \tag{5.11}$$

where Ω is the solid angle. Equation (5.11) gives the number of particles inside a volume element with speeds between v and $v + dv$. The function $g(r, v)$ gives the number of particles per velocity unit, again with speeds between v and $v + dv$. For small speeds, this function increases with the square of v while for large speeds it decreases exponentially (see upper panel in Fig. 5.1). If not the speed, but only one component of the velocity is considered, the distribution is symmetric around zero if the plasma is at rest (middle) or symmetric around the flow speed in that particular direction if the plasma is in motion (lower panel).

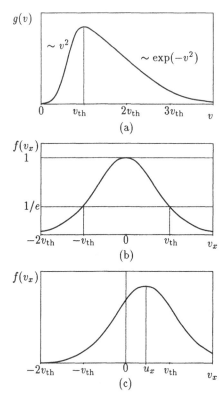

Fig. 5.1. Maxwell distribution. (a) Distribution $g(\boldsymbol{v}, r)$ for the particle speeds in a plasma at rest. (b) Maxwell distribution of one velocity component for a plasma at rest, the other velocity components have been removed by integration. (c) Same as above but for a plasma moving with average speed u_x in the x-direction. Note that an increase in temperature would not affect the position of the maximum in (b) and (c) but would shift the maximum towards the right in (a), combined with an increase of the maximum

5.1.3 Other Distributions

Not all particle distributions can be described by a Maxwellian. If the plasma is not in equilibrium, normally no analytical distribution function exists, although often distribution functions are reasonable approximations.

For instance, if a plasma has a marked difference in the speeds parallel and perpendicular to the magnetic field, the distribution can be approximated by a bi-Maxwellian, that is the product of two Maxwellians, which take into account the different speeds and temperatures of the two motions:

$$
f(\boldsymbol{r}, \boldsymbol{v}, t) = \sqrt{\left(\frac{m}{2\pi k_{\mathrm{B}}}\right)^3} \frac{n}{T_\perp \sqrt{T_\parallel}}
$$
$$
\times \exp\left\{-\frac{m(v_\parallel - u_\parallel)^2}{2k_{\mathrm{B}}T_\parallel}\right\} \exp\left\{-\frac{m(v_\perp - u_\perp)^2}{k_{\mathrm{B}}T_\perp}\right\} . \quad (5.12)
$$

The average kinetic energies parallel and perpendicular to the field are

$$
\left\langle \tfrac{1}{2}m(v_\parallel - u_\parallel)^2 \right\rangle = \tfrac{1}{2}k_{\mathrm{B}}T_{\parallel s} \quad \text{and} \quad \left\langle \tfrac{1}{2}m(v_\perp - u_\perp)^2 \right\rangle = k_{\mathrm{B}}T_\perp . \quad (5.13)
$$

Here the difference in the number of degrees of freedom for the parallel and perpendicular motions becomes obvious: there is one degree of freedom par-

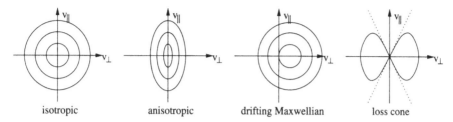

isotropic   anisotropic   drifting Maxwellian   loss cone

Fig. 5.2. Bi-Maxwellian distributions often found in space plasmas

allel to the field while there are two perpendicular to it. Equation (5.13) can also be used to define the pressure perpendicular and parallel to $\boldsymbol{B}$.

The bi-Maxwellian (5.12) is a typical anisotropic distribution function of the form $f(v_\perp, v_\parallel)$. This kind of distribution function is the one most commonly found in space plasmas. It is an essentially two-dimensional and gyrotropic velocity distribution: it does not depend on the phase angle of the gyromotion. Figure 5.2 shows contour-plots for different (bi-)Maxwellians: the isotropic distribution (left) is characterized by circular contours. In the anisotropic Maxwellian, the contours are deformed into elliptical shapes. In this example, the deformation is such that it corresponds to $T_\parallel > T_\perp$. During magnetic pumping (example 9) we have a continuous change from an isotropic to an anisotropic distribution and back. Special distributions arise if either $u_\parallel$ or $u_\perp$ vanishes in (5.12). For instance, a plasma might drift perpendicular to the magnetic field, such as in the case of crossed electric and magnetic fields. Then $u_\parallel$ vanishes and a drifting Maxwellian results (see Fig. 5.2). If the drift velocity is large compared with the thermal velocity, such a distribution is called a streaming distribution. If the distribution drifts along the magnetic field at speed $u_\parallel$, and $u_\perp$ vanishes, it is a parallel-beam distribution. Such distributions are frequently encountered in the auroral regions of the magnetosphere and in the foreshock regions of planetary bow shocks. The last sketch in Fig. 5.2 illustrates a loss cone distribution, where particles with sufficiently small pitch angles are lost from the magnetospheric population.

Occasionally, the particle distribution can be described by a Maxwellian up to a certain energy. At higher energies the particle distribution can be fitted much better by a power law than by the exponential decay of the Maxwell distribution. Here the kappa distribution, sometimes also called Lorentzian distribution,

$$f(\boldsymbol{r}, \boldsymbol{v}, t) = \frac{n_s}{\kappa} \sqrt{\left(\frac{m}{2\pi k_B T}\right)^3} \left[1 + \frac{m(\boldsymbol{v} - \boldsymbol{u})^2}{2\kappa E_T}\right]^{-\kappa-1}, \qquad (5.14)$$

can be used as an approximation. The parameters κ and E_T are characteristics of the distribution, with E_T being closely associated with the temperature T and κ describing the deviation of the distribution from a Maxwellian. For energies $E \gg \kappa E_T$, the distribution decays more slowly than a Maxwellian and can be approximated by a power law in kinetic energy: $f(E) = f_0 E^{-\kappa}$. If κ converges towards infinity, the distribution is a Maxwellian with temperature $kT = E_T$. Note that for the kappa distribution, as well as for all other non-Maxwellian distributions, (5.10) cannot be applied.

5.1.4 Distribution Function and Measured Quantities

While the distribution function is important for our theoretical treatment of (space) plasmas, it is a quantity which cannot be measured directly. Instead, observations give the differential flux $J(E, \boldsymbol{\Omega}, \boldsymbol{r}, t)$ of particles within a solid angle $\mathrm{d}\boldsymbol{\Omega}$ and an energy interval $(E, E + \mathrm{d}E)$. Thus the quantity

$$J(E, \boldsymbol{\Omega}, \boldsymbol{r}, t)\,\mathrm{d}\boldsymbol{A}\,\mathrm{d}\boldsymbol{\Omega}\,\mathrm{d}t\,\mathrm{d}E$$

is the number of particles in the energy band from E to $E + \mathrm{d}E$ coming from the direction $\boldsymbol{\Omega}$ within a solid angle $\mathrm{d}\boldsymbol{\Omega}$, going through a surface $\mathrm{d}\boldsymbol{A}$ perpendicular to $\mathrm{d}\boldsymbol{\Omega}$ during the time interval $\mathrm{d}t$. The differential flux therefore can be measured in units of particles per $(\mathrm{m}^2\ \mathrm{sr}\ \mathrm{s}\ \mathrm{MeV})$. Since J depends on $\boldsymbol{\Omega}$, it can also be interpreted as the angular distribution of the particles. On a rotating spacecraft, often the omnidirectional intensity is measured. The latter can be obtained by averaging over all directions:

$$J_{\mathrm{omni}}(E, \boldsymbol{r}, t) = \frac{1}{4\pi} \int J(E, \boldsymbol{\Omega}, \boldsymbol{r}, t)\,\mathrm{d}\boldsymbol{\Omega} \ . \tag{5.15}$$

The number density of particles with velocity v in a phase space element is given as $\mathrm{d}n = fv^2\mathrm{d}v\mathrm{d}\boldsymbol{\Omega}$. Multiplication by v gives the differential flux of particles with velocity v as $f(\boldsymbol{v}, \boldsymbol{p}, t)v^3\mathrm{d}v\mathrm{d}\boldsymbol{\Omega}$. Comparison with the same quantity expressed by the differential flux yields

$$J(E, \boldsymbol{\Omega}, \boldsymbol{r}, t)\,\mathrm{d}E\,\mathrm{d}\boldsymbol{\Omega} = f(\boldsymbol{r}, \boldsymbol{p}, t)v^3\,\mathrm{d}v\,\mathrm{d}\boldsymbol{\Omega} \ . \tag{5.16}$$

Since the energy is related to speed, $\mathrm{d}E$ is related to $\mathrm{d}v$ by $\mathrm{d}E = mv\mathrm{d}v$. The relation between the differential flux and the distribution function therefore can be written as

$$J(E, \boldsymbol{\Omega}, \boldsymbol{r}, t) = \frac{v^2}{m}f(\boldsymbol{r}, \boldsymbol{p}, t) \ . \tag{5.17}$$

5.2 Basic Equations of Kinetic Theory

As mentioned above, the equation of motion in kinetic theory can be derived by applying the averaging scheme (5.4): the Boltzmann equation is the fundamental equation of motion in kinetic theory; the Vlasov equation can be applied if the forces are purely electromagnetic; and the Fokker–Planck equation also considers scattering.

5.2.1 The Boltzmann Equation

The Boltzmann equation is the fundamental equation of motion in phase space. It is not limited to plasmas, and the only assumption inherent in the Boltzmann equation is that only external forces F act on the particles while internal forces vanish: there are no collisions between the particles.

The Boltzmann equation is a direct consequences of the equation of continuity in phase space:

$$\frac{\partial f}{\partial t} + \nabla_6 \cdot (f C) = 0 \quad \text{or} \quad \frac{\partial f}{\partial t} + \nabla_r \cdot (v f) + \nabla_v \cdot (a f) = 0 , \quad (5.18)$$

where ∇_6, ∇_r, and ∇_v are the divergence in phase space, in ordinary space, and in momentum space, respectively, and v and a are velocity and acceleration. In phase space, r and v are independent variables. If we further assume that the acceleration a, and therefore also the force F, is independent of v, (5.18) can be simplified:

$$\frac{\partial f}{\partial t} + v \cdot \nabla_r f + a \cdot \nabla_v f = 0 \quad \text{or} \quad \frac{\partial f}{\partial t} + v \cdot \nabla f + \frac{F}{m} \cdot \frac{\partial f}{\partial v} = 0 . \quad (5.19)$$

Equation (5.19) is called the collisionless Boltzmann equation. It also can be written as

$$\frac{\mathrm{d}f}{\mathrm{d}t} = 0 , \quad (5.20)$$

which states that the convective derivative of the phase space density is always zero for a collisionless assembly of particles. Thus for an observer moving with the flow, the phase space density is constant. Or, in other words: the substrate of points in phase space behaves like an incompressible fluid. This is also called Liouville's theorem.

The general form of the Boltzmann equation can be written as

$$\frac{\partial f}{\partial t} + v \cdot \nabla f + \frac{F}{m} \cdot \frac{\partial f}{\partial v} = \left(\frac{\partial f}{\partial t} \right)_{\text{coll}} , \quad (5.21)$$

where the term on the right-hand side is the rate of change in phase space density due to collisions (see below).

If changes in f due to collisions are small, e.g. in the case of a thermodynamic equilibrium, the reduced Boltzmann equation can be written as

$$\left(\frac{\partial f}{\partial t} \right)_{\text{coll}} = 0 . \quad (5.22)$$

The solution of this equation is the Maxwell distribution.

5.2.2 The Vlasov Equation

The Vlasov equation is the application of the Boltzmann equation to a plasma on which only electromagnetic forces act. These forces are described by the Lorentz force (2.23). In the derivation of (5.19) we have made the assumption of a force independent on v. At first glance the Lorentz force violates this assumption and therefore should not be considered in (5.19). Closer inspection, however, shows that this is not true. Since the Lorentz force contains the cross product of speed and magnetic field, the resulting force is perpendicular to the speed. Thus each individual component of the force does not depend on the same component of the velocity. Since in (5.18) a scalar product is considered, the only derivatives of a are of the form $\partial a_x / \partial v_x$ and therefore vanish. Thus the Lorentz force can be inserted into (5.19):

$$\frac{\partial f}{\partial t} + v \cdot \nabla f + \frac{q}{m} \left(E + \frac{v \times B}{c} \right) \cdot \frac{\partial f}{\partial v} = 0 . \tag{5.23}$$

Equation (5.23) is called the Vlasov equation. Because of it simplicity, this is the equation most commonly studied in kinetic theory.

The Vlasov equation is derived under the assumption of non-interacting particles. On the other hand, interactions are the very essence of a plasma. Thus we have to discuss whether the Vlasov equation can be applied as often as it is. As we shall see, the Vlasov equation is a valid approach. It does not consider collisions in the sense of short-range, local interactions, such as collisions between two billiard balls or Coulomb collisions between two charged particles. Nonetheless, that kind of interaction, which is essential in a plasma, is considered: each particle moves in the average Coulomb field created by thousands of other particles. Thus the fields in the Vlasov equation are due to the rest of the plasma and describe the interaction of the particles. These fields often are called self-consistent fields. External fields can be included in (5.23), too. Since the fields E and B are determined by the rest of the plasma, they depend on the distribution function f.

The Vlasov equation thus is non-linear; analytical solutions in general are not possible. But Jeans' theorem identifies some solutions. It states: any function of the constants of motion is a solution of the Vlasov equation. For instance, if there are no electric fields, the kinetic energy is a constant of motion. Thus any function of $mv^2/2$ is a solution of the Vlasov equation. In particular, the Maxwell distribution (5.8) is a solution.

Jeans' theorem therefore shows the equivalence of kinetic theory and orbit theory. Following an approach given by Boyd and Sanderson [58], this equivalence can be shown easily. The basic equation of orbit theory is Newton's second law $F = md^2r/dt$. This is a second-order differential equation in three dimensions and therefore the general solution must contain six constants of integration, $\gamma_1, ..., \gamma_6$. Thus the solutions of Newton's second law can be written as $r = r(\gamma_1, ..., \gamma_6, t)$ and $v = v(\gamma_1, ..., \gamma_6, t)$. These six scalar equations can be solved in principle to give the γ_i: $\gamma_i = \gamma_i(v, r, t)$. Jeans' theorem then states that each function $f = f(\gamma_1, ..., \gamma_6)$ is a solution of the fundamental equation of kinetic theory, the Boltzmann equation (5.19). This can be seen easily by inserting the γ_i into the Boltzmann equation:

$$\sum_i \left(\frac{\partial \gamma_i}{\partial t} + \boldsymbol{v} \cdot \nabla \gamma_i + \frac{\boldsymbol{F}}{m} \cdot \frac{\partial \gamma_i}{\partial \boldsymbol{v}} \right) = \sum_i \frac{\partial f}{\partial \gamma_i} \frac{\mathrm{d}\gamma_i}{\mathrm{d}t} = 0 \,. \qquad (5.24)$$

The result is identically zero because the γ_i are constants.

5.2.3 The Fokker–Planck Equation

In contrast to the Vlasov equation, the Fokker–Planck equation considers the short-range, local interactions between particles. Collisions arise from many small Coulomb interactions between charged particles (see Sect. 5.3.2). The collision term has its mechanical analogy in the Brownian motion of particles in a gas; however, both are not equivalent as will be discussed below.

Collisions are not a deterministic but a stochastic process. Thus for a given particle, although we might know its momentary position and velocity, we cannot determine its future motion. Only for an assembly of particles the collective behavior can be determined. This can be done by means of probabilities. Let $\psi(\boldsymbol{v}, \Delta\boldsymbol{v})$ be the probability that a particle with velocity $\boldsymbol{v}$ after many small collisions during a time interval $\mathrm{d}t$ has changed its velocity to $\boldsymbol{v} + \Delta\boldsymbol{v}$. The phase space density $f(\boldsymbol{r}, \boldsymbol{v}, t)$ also is a probability function. At a time t it can be written as the product of the phase space density at an earlier time $t - \Delta t$ multiplied by the probability of changes during this time interval and integrated over all possible velocity changes $\Delta\boldsymbol{v}$:

$$f(\boldsymbol{r}, \boldsymbol{v}, t) = \int f(\boldsymbol{r}, \boldsymbol{v} - \Delta\boldsymbol{v}, t - \Delta t)\, \psi(\boldsymbol{v} - \Delta\boldsymbol{v}, \Delta\boldsymbol{v})\, \mathrm{d}(\Delta\boldsymbol{v}) \,. \qquad (5.25)$$

Since we only consider scattering by small angles, i.e. $|\Delta\boldsymbol{v}| \ll |\boldsymbol{v}|$, Taylor expansion to second order of the product $f\psi$ yields

$$f(\boldsymbol{r}, \boldsymbol{v}, t) = \int \left[f(\boldsymbol{r}, \boldsymbol{v}, t - \Delta t)\, \psi(\boldsymbol{v}, \Delta\boldsymbol{v}) - \Delta\boldsymbol{v} \cdot \frac{\partial (f\psi)}{\partial \boldsymbol{v}} \right] \mathrm{d}(\Delta\boldsymbol{v})$$
$$+ \int \left[\frac{\Delta\boldsymbol{v}\Delta\boldsymbol{v}}{2} \odot \frac{\partial^2 (f\psi)}{\partial \boldsymbol{v} \partial \boldsymbol{v}} \right] \mathrm{d}(\Delta\boldsymbol{v}) \,. \qquad (5.26)$$

Note that the $\odot$ in the last term indicates a product between two tensors.[1] The resulting matrix is the Hess matrix.

Because some interactions always take place, the probability can be normalized to $\int \psi \mathrm{d}(\Delta\boldsymbol{v}) = 1$. Equation (5.26) than can be simplified to

$$f(\boldsymbol{r}, \boldsymbol{v}, t) = f(\boldsymbol{r}, \boldsymbol{v}, t - \Delta t) - \frac{\partial (f\langle \Delta\boldsymbol{v} \rangle)}{\partial \boldsymbol{v}} + \frac{1}{2} \frac{\partial}{\partial \boldsymbol{v} \partial \boldsymbol{v}} \odot (f\langle \Delta\boldsymbol{v}\Delta\boldsymbol{v} \rangle) \,, \qquad (5.27)$$

---

[1] The product $\mathsf{S} \odot \mathsf{T}$ of two tensors S and T itself is a tensor and can be obtained by application of the rules of matrix multiplication.

with

$$\langle \Delta v \rangle = \int \psi \Delta v \, d(\Delta v) \quad \text{and} \quad \langle \Delta v \Delta v \rangle = \int \psi \Delta v \Delta v \, d(\Delta v) \, . \tag{5.28}$$

By definition, the collision term as written down in (5.21) is

$$\left(\frac{\partial f}{\partial t} \right)_{\text{coll}} = \frac{f(\boldsymbol{r}, \boldsymbol{v}, t) - f(\boldsymbol{r}, \boldsymbol{v}, t - \Delta t)}{\Delta t} \, . \tag{5.29}$$

Thus the Fokker–Planck equation can be written as

$$\left(\frac{\partial f}{\partial t} \right)_{\text{coll}} \Delta t = -\frac{\partial}{\partial \boldsymbol{v}} \cdot (f \langle \Delta v \rangle) + \frac{1}{2} \frac{\partial^2}{\partial \boldsymbol{v} \partial \boldsymbol{v}} \odot (f \langle \Delta v \Delta v \rangle) \, . \tag{5.30}$$

The first term on the right basically contains $\langle \Delta v \rangle / \Delta t$, which is an accelera-tion. Thus the term describes the frictional forces leading to an acceleration of the slower and a deceleration of the faster particles, which tends to equalize the speeds. The negative divergence in velocity space describes this narrowing of the distribution function. The second term, $\langle \Delta v \Delta v \rangle / \Delta t$, is a diffusion in velocity space. This term describes the broadening of a narrow velocity dis-tribution, e.g. a beam, as a result of the collisions. The two terms therefore operate in the opposite sense. They are in balance in an equilibrium distri-bution, e.g. the Maxwell distribution. The physics of the collision processes is contained in the probability function ψ.

Equation (5.30) also can be written as

$$\left(\frac{\partial f}{\partial t} \right)_{\text{coll}} \Delta t = -\nabla_{\text{v}} \cdot (\mathsf{D} \cdot \nabla_{\text{v}} \cdot f) \tag{5.31}$$

with the diffusion tensor D derived from the first- and second-order fluctua-tions of the particle velocity.

Figure 5.3 shows the evolution of a suprathermal distribution of particles. At time t_0 the particle distribution is a parallel beam of uniform speed. Since the particles are much faster than the plasma, at early times (t_1 and t_2) pitch angle scattering dominates, and the distribution spreads towards larger pitch

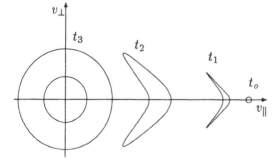

Fig. 5.3. Evolution of a suprathermal distribution of particles from a monoener-getic beam at time t_0 towards an isotropic ring distribution at later times t_3

angles, leading finally to an isotropic ring distribution around the origin (t_3). With increasing time, friction and energy losses become important and the ring distribution contracts towards the origin. This is an example of a more general rule of thumb: the faster a particle moves through a plasma, the smaller is its frictional drag.[2] This longevity property allows the existence of suprathermal particles as a distinct population in some cosmic plasmas. For instance, galactic cosmic rays or solar energetic particles both are long-lived, distinct particle populations in the solar wind; the radiation belt particles are a distinct population in the terrestrial plasmasphere. Obviously, these plasmas can not be described by a Maxwellian. Instead, the kappa distribution gives a quite reasonable description.

5.3 Collisions

We have mentioned collisions twice in this chapter. In Sect. 5.1.2 we introduced the Maxwell distribution. The basic requirement for such a distribution is thermal equilibrium, which requires collisions between the particles. If we have a distribution with a suprathermal tail, such as the kappa distribution, in time collisions will transform it into a Maxwellian. This time scale depends, of course, on the time scales of the collisions. We have also mentioned collisions in connection with the Fokker–Planck equation. We have even mentioned that the collisions should lead to small changes in speed only. But we did not talk explicitly about the nature of these collisions. This section is supplementary, briefly introducing some of the basics of collisions.

Collisions are also important in the energy transfer between different components in a plasma: imagine a plasma which also contains neutral particles. The charged particles might be accelerated by an electric field. In time, collisions between charged and neutral particles will equalize the two distributions, leading to an acceleration of the neutrals.

As in a neutral gas, collisions in a plasma change the path of the individual particle. In a magnetized plasma, collisions between a charged particle and a neutral can shift the gyro-center of a particle onto another field line (see Fig. 5.4). Collisions between charged particles can also lead to a shift in the gyro-center and/or changes in pitch angle.

5.3.1 Collisions Between Neutrals

Collisions between neutral particles give rise to the Brownian motion in a gas. The individual process is a collision between two hard spheres. The hard

---

[2] Graphically this can be understood from the fact that with increasing particle speed the time available for interaction between the energetic particle and a particle of the background plasma decreases; a more formal explanation is given in Sect. 5.4.

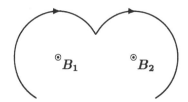

Fig. 5.4. Collisions in a plasma can shift the gyro-center of a particle onto another field line, here illustrated for a collision of a charged particle with a neutral one

sphere model is a simple and quite useful approximation. The full treatment of the collisions between neutrals requires a quantum-mechanical approach considering the attracting van der Waals forces and the repulsing Coulomb forces of the electron shells of the two atoms. The van der Waals potential varies roughly with r^{-6}, the repulsing potential with r^{-12} [569]. Combination of both leads to an extremely steep potential surrounded by a very shallow potential depression in the order of meV compared with the typical eV range in molecule formation. A hard sphere therefore is a reasonable approach to describe the collision of neutrals.

The basic equations are the conservation of momentum and the conservation of energy. The changes in momentum and direction depend on the masses and speeds of the particles and on the angle between their velocities. The change in momentum is largest in a head-on collision: when mass is equal, the particle loses twice its initial momentum as its velocity is reversed. Thus scattering by a large angle up to 180° is possible in collisions between neutrals.

The relevant parameters to describe the scattering process are the mean free path and the scattering cross section. The particle mean free path λ is defined as the average distance travelled by a particle between two subsequent collisions. If we could follow a smoke particle in air, we would detect a path similar to the one depicted on the left-hand side of Fig. 5.5. The statistical motion is composed of many straight lines with different length L. The right-hand side of Fig. 5.5 shows the distribution of these L. This probability distribution can be described by a function $p(L) = a \exp(-L/\lambda)$ with a

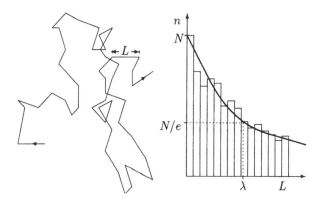

Fig. 5.5. Statistical path of a particle under the influence of collisions (*left*) with other particles (Brownian motion) and distribution of the path length between two collisions (*right*)

being a constant depending on the total number of collisions and λ being the particle mean free path.

The number of collisions, and therefore the mean free path, depends on the number density of particles and on their 'size' as described by the scattering cross section. Consider a fast particle with radius r_1 moving in a gas of slow particles with radii r_2. A collision happens if the distance between the two particles has decreased below $r_1 + r_2$. Alternatively, we can assume the fast particle to be a mass point. Then we have to attribute a radius of $r_1 + r_2$ to the gas molecules. Thus for the fast particle, a gas molecule is equivalent to a disk with the scattering cross section $\sigma = \pi(r_1 + r_2)^2$.

Now consider a beam of particles incident on a slab of area A and thickness dx. The number density of molecules in this slab is n_s, the total number of molecules in the slab is $n_s A dx$. The fraction of the slab blocked by atoms therefore is $\sigma n_s A dx / A = \sigma n_s dx$. Out of N particles incident on the slab, $\Delta N = N n_s \sigma dx$ will experience a collision, leading to a reduction in N according to $dN/N = -\sigma n_s\, dx$. Integration yields

$$N(x) = N_0 \exp(-\sigma n_s x) = N_0 \exp(-x/\lambda) , \qquad (5.32)$$

where the mean free path λ is defined as

$$\lambda = \frac{1}{n_s \sigma} . \qquad (5.33)$$

Thus, the mean free path can also be interpreted as the distance over which the number of particles decreases to $1/e$ of its initial value. After travelling a distance λ, the particle will have a high probability of colliding. The average time between two collisions is

$$\tau = \frac{\lambda}{\langle v \rangle} = \frac{1}{n_s \sigma \langle v \rangle} . \qquad (5.34)$$

This also can be written as a collision frequency ν_c:

$$\nu_c = n_s \sigma \langle v \rangle . \qquad (5.35)$$

The formalism for interactions between a charged particle and a neutral is the same as for a collision between two neutral particles.

5.3.2 Collisions Between Charged Particles

Collisions between charged particles do not require a direct contact, instead the interaction takes place as each particle is deflected in the electric field of the other one. Since the Coulomb force has a long range such an interaction leads to a gradual deflection. Nonetheless, one can derive a kind of cross section for this process. Following the attempt given in Chen [97], we shall only make an order-of-magnitude estimate.

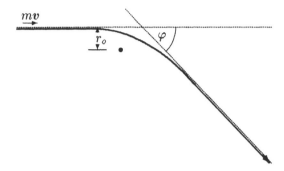

Fig. 5.6. Coulomb scattering: orbit of an electron in the Coulomb field of an ion

The geometry of a Coulomb collision is shown in Fig. 5.6: an electron with velocity v approaches an ion with charge e. If no Coulomb force acts, the electron will pass the ion at a distance r_0, the impact parameter. But the Coulomb force $F = -e^2/(4\pi r^2)$ leads to a deflection of the electron from its original direction by an angle φ. The force acts on the electron for a time $T \approx r_0/v$ when it is in the vicinity of the ion. The change in electron momentum then can be approximated by $\Delta p = |FT| \approx e^2/(4\pi r_0 v)$. For a 90° collision, the change in momentum is of the order of mv. Thus it is $\Delta p \cong mv \cong e^2/(4\pi r_0 v)$. A deflection by 90° results for an impact parameter $r_{90°} = e^2/(4\pi m v^2)$. The cross section for a deflection of at least 90° therefore can be written as

$$\sigma_{>90°} = \pi r_0^2 = \frac{e^4}{16\pi m^2 v^4} \,, \tag{5.36}$$

leading to a collision frequency of

$$\nu_{\text{ei},>90°} = n\sigma v = \frac{n e^4}{16\pi m^2 v^3} \,. \tag{5.37}$$

In a real plasma the situation is more complex. Let us consider the motion of one particle, a test particle, in the field created by the other particles, the field particles. The fields of these particles add to a stochastic field that changes continuously in time and space. Therefore, the test particle will not move in a hyperbolic orbit as in the interaction between two charged particles, instead it basically follows its original direction of motion, though not on a straight line but on a jittery trajectory. Because of the stochastic nature of the collisions, test particles with nearly identical start conditions will diverge in space and velocity. Most of these collisions result in small changes in the particle path only. Occasionally, also large deviations of the original direction result. These are called large-angle collisions.

To understand the different types of collisions, we have to consider the typical spatial scales. One characteristic scale is the Debye length λ_D (3.146): the test particle is screened from the electric field of the charges outside the Debye sphere. The Debye length can be interpreted as the range of microscopic electric fields and separates the field particles into two groups: (i) particles at distances larger than the Debye length can influence the test particle by the macroscopic fields only, leading to gyration, drift, and oscillations; and (ii) particles inside the Debye sphere create a microscopic field leading to the stochastic motion of the test particle.

The other spatial scale is related to the scattering angle, which in turn is related to the impact parameter. In the derivation of the Fokker–Planck equation we have assumed scattering by small angles only. The deflection angle will be small if the kinetic energy $m_T v_T^2/2$ of the test particle is large compared with the electrostatic potential $Z_T Z_F e^2/r_0$, where r_0 is the impact parameter (see Fig. 5.6). The test particle will be deflected by a small angle only if the impact parameter r_0 fulfills the condition $r_{90°} < r_0 < \lambda_D$, with $r_{90°}$ being the impact parameter for a deflection by 90° defined as $r_{90°} = Z_T Z_F e^2/(m_T v_T^2)$. The ratio for deflections by small and large angles can be determined from the ratio of cross sections for both processes:

$$\frac{\lambda_D^2 - r_{90°}^2}{r_{90°}^2} \leq \frac{\lambda_D^2}{r_{90°}^2} \approx \left(\frac{9}{Z_T Z_F}\right)^2 \left(\frac{4\pi}{3} n \lambda_D^3\right)^2 =: \Lambda^2 . \tag{5.38}$$

The expression inside the second parentheses is the number of particles inside a Debye sphere. If we assume this number to be large, (5.38) states that collisions leading to deflections by a small angle by far outnumber the collisions with large-angle deflection. A careful calculation shows that small-angle interactions are about two orders of magnitude more efficient in the deflection of test particles than the few large-angle interactions [42]. Thus, the Fokker–Planck formalism can be applied to the Coulomb collisions in a plasma too. The logarithm of the above quantity, $\lambda_c = \ln \Lambda$, is called the Coulomb logarithm.

5.4 Collisions Between Charged Particles: Formally

We will now have a closer look at the formal treatment of collisions, following [281].

5.4.1 Coulomb Collisions: Unscreened Potential

Again we assume a test particle (this time not necessarily an electron) with speed v_T, mass m_T, and charge q_T approaching a stationary background particle of charge q_B and mass m_B. The geometry is as in Fig. 5.6, and the impact parameter is r_0. The potential energy of the test particle is $W_{C,T} = q_T \Phi_C$, where Φ_C is the undisturbed potential of the background charge. Then we can also write for the potential energy

$$W_{C,T} = \frac{\alpha}{r} , \quad \text{where} \quad \alpha = \frac{q_T q_B}{4\pi\varepsilon_0} . \tag{5.39}$$

The angle of deflection φ is given by

$$\tan\frac{\varphi}{2} = \frac{r_{90°}}{r}, \quad \text{where} \quad r_{90°} = \frac{|\alpha|}{m_T v_T^2} \tag{5.40}$$

is the impact parameter for a deflection by $90°$. The minimum distance $d_{\min}$ of the particle from the background particle and its speed $v_{\min}$ at this position can be derived from the conservation of angular momentum,

$$m_T v_T r_0 = m_T v_{\min} d_{\min}, \tag{5.41}$$

and the conservation of energy,

$$\frac{m_T v_T^2}{2} = \frac{m_T v_{\min}^2}{2} + W_{C,T}(d_{\min}), \tag{5.42}$$

to be

$$d_{\min} = \frac{\alpha + \sqrt{\alpha^2 + m_T^2 r_0^2 v_T^4}}{m_T v_T^2}. \tag{5.43}$$

Depending on the charge of the background particle, α can be positive or negative, and we obtain the following for the minimum distance $d_{\min}^{90°}$ in the case of a deflection by $90°$:

$$d_{\min}^{90°} = \begin{cases} (\sqrt{2}+1)r_{90°} & \text{for } \alpha > 0 \\ (\sqrt{2}-1)r_{90°} & \text{for } \alpha < 0 \end{cases}. \tag{5.44}$$

The corresponding potential energies are

$$W_{C,T}(d_{\min}^{90°}) = \begin{cases} m_T v_T^2/(\sqrt{2}+1) & \text{for } \alpha > 0 \\ m_T v_T^2/(\sqrt{2}-1) & \text{for } \alpha < 0 \end{cases}. \tag{5.45}$$

If we assume that the potential energy is equal to the thermal energy W_{th}, the minimum distance between charged particles in a plasma is

$$d_{\min} = \frac{q_T q_B}{4\pi\varepsilon_0 W_{th}}. \tag{5.46}$$

In an ideal plasma, this distance is

$$d_{\min} \ll n^{-1/3} \ll \lambda_D. \tag{5.47}$$

The cross section σ defines a circle with the impact parameter r_0 as its radius. The differential cross section $d\sigma$ is related to a deflection by an angle $d\varphi$ or to a deflection into a solid angle $d\Omega = 1\pi \sin\varphi\, d\varphi$. With (5.40), we then obtain the following for the differential cross section for Coulomb scattering:

$$\frac{d\sigma}{d\Omega} = \alpha^2 \left(\frac{1}{2m_T v_T^2 \sin^2(\varphi/2)}\right)^2. \tag{5.48}$$

Note that the differential cross section does not depend on the sign of the charges: the cross section is the same if both have the same sign or the opposite sign. Only the hyperbolic path of the particles is different, not the deflection angle φ.

For a large impact parameter r_0, the deflection angle diverges as φ^{-4}. In addition, the total cross section for interaction tends towards infinity:

$$\sigma_{C,\text{unscr}} = \int_0^\infty \frac{\mathrm{d}\sigma}{\mathrm{d}\Omega} \, \mathrm{d}\Omega = \infty \ . \tag{5.49}$$

This infinite cross section is reasonable for a single charge that deflects another one, in a plasma, however, the screening of the Coulomb potential has to be taken into account.

5.4.2 Decelerating Force Between Particles (Drag)

In addition, the significance of an infinite cross section is not clear: although even at large impact parameters deflection occurs, this might be of minor importance, since the deflection angle is small. Thus we have to find a method to describe the impact of background particles on a beam of test particles.

For a first approximation, we assume only one background particle, with infinite mass and at rest. The particle beam is described by its number density n_T and the particle parameters of velocity v_T, mass m_T, and charge q_T. Owing to symmetry, momentum is transported only in the direction of $\boldsymbol{v}_T$. During elastic scattering at a fixed obstacle, the magnitude of the velocity is conserved, and we obtain the following for the change in velocity $\delta \boldsymbol{v}_T$:

$$|\delta \boldsymbol{v}_T| = 2v_T \sin(\varphi/2) \ . \tag{5.50}$$

In addition, we have (see Fig. 5.7)

$$|\delta \boldsymbol{v}_{t,\|}| = 2v_T \sin(\varphi/2) \cos(\varphi/2) \ , \tag{5.51}$$

$$\delta v_{t,z} = -2v_T \sin^2(\varphi/2) = -2v_T \frac{\tan(\varphi/2)}{\tan(\varphi/2)+1} = -2v_T \frac{r_{90°}^2}{r_{90°}^2 + r_0^2} \ , \tag{5.52}$$

$$\delta v_{t,x} = 2v_t \frac{r_0 r_{90°}}{r_{90°}^2 + r_0^2} \cos\varphi \ , \tag{5.53}$$

$$\delta v_{t,y} = 2v_t \frac{r_0 r_{90°}}{r_{90°}^2 + r_0^2} \sin\varphi \ . \tag{5.54}$$

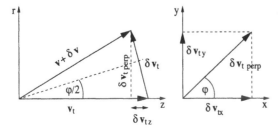

Fig. 5.7. Changes in velocity during a Coulomb collision

The number of particles streaming through a ring with area $d\sigma$ during a time interval dt is $dN_T = n_T v_T \, d\sigma \, dt$. The decelerating force $F_{T,z}$ from the background particle acting on the test particles can now be obtained by integrating over all rings $d\sigma$:

$$F_{z,t} = m_T \int\limits_0^\infty \delta v_{t,z} \frac{dN_T}{d\sigma \, dt} \, d\sigma \ . \tag{5.55}$$

Using the differential cross section (5.48), this can be rewritten as

$$F_{z,t} = -\frac{\pi n_T \alpha^2}{m_T v_T^2} \lim_{\varphi_m \to 0} \int\limits_{\varphi_m}^\pi \frac{\sin\varphi}{\sin(\varphi/2)} \, d\varphi = \frac{4\pi n_T \alpha^2}{m_T v_T^2} \lim_{\varphi_m \to 0} \ln\left(\sin\frac{\varphi_m}{2}\right) \ . \tag{5.56}$$

This quantity is logarithmically divergent for $\varphi_m \to 0$: the drag force due to a background particle in an unscreened Coulomb potential diverges towards infinity.

5.4.3 Coulomb Collisions: Screened Potential

We have already introduced the Debye length in Sect. 3.7 as a parameter that basically describes over what distance a certain charge influences other charges. Using the Debye length, a screened potential or reduced potential Φ_D can be introduced, which is related to the unscreened Coulomb potential Φ_C by

$$\Phi_D(\boldsymbol{r}) = \Phi_C(\boldsymbol{r}) \, e^{-r/\lambda_D} \ . \tag{5.57}$$

Although the screened potential decays faster than the Coulomb potential, it still extends to infinity. To describe the influence of the screened potential we make the following simplification: for scattering with an impact parameter $r_0 < \lambda_D$, the potential is assumed to be Φ_C, while for $r_0 > \lambda_D$, the potential vanishes. In this case, the smallest deflection angle φ_m is given by

$$\tan\frac{\varphi_m}{2} = \frac{r_{90°}}{\lambda_D} \ . \tag{5.58}$$

In an ideal plasma, we therefore have

$$\ln\left(\sin\frac{\varphi_m}{2}\right) = -\ln\frac{\lambda_D \sqrt{1 + r_{90°}^2/\lambda_D^2}}{r_{90°}} \approx -\ln\frac{\lambda_D}{r_{90°}} = -\ln\Lambda \ , \tag{5.59}$$

where $\ln\Lambda$ is the Coulomb logarithm.

Equation (5.59) can also be derived from (5.56) by choosing the upper integration boundary as $\pi/2$ instead of π: in this case scattering is limited to deflection angles less than or equal to 90°.

We then obtain the following for the drag force:

$$F_{t,z} = -\frac{c_A^{TB} n_T}{m_T v_T^2} ,$$ (5.60)

where

$$c_A^{TB} = \frac{q_T^2 q_B^2}{4\pi\varepsilon_0^2} \ln\Lambda \quad \text{and} \quad \ln\Lambda = \ln\frac{\lambda_D 4\pi\varepsilon_0 m_T v_T^2}{|q_T q_B|} .$$ (5.61)

We have already mentioned that the drag force caused by the background plasma decreases with increasing particle speed. In this case we have a background plasma of particle density n_B containing particles with velocity v_B and mass m_B. A test particle then experiences a drag force that is determined by the reduced mass of the particles

$$m_{TB} = \frac{m_T m_B}{m_T + m_B}$$ (5.62)

and their relative velocity $v_R = v_T - v_B$:

$$F_{t,z} = -\frac{c_A^{TB} n_B v_R}{m_{TB} |v_R|^3} .$$ (5.63)

A useful approximation to this equation is

$$F_{t,z} \approx \frac{1}{m_{TB} v_R^2} .$$ (5.64)

The basic result is, as stated before, that the drag force exerted by the background plasma decreases with increasing particle speed. In addition, the drag force is larger for electron–ion collisions ($m_{TB} \approx m_e$) than for particles of equal mass ($m_{TB} = m/2$).

5.5 Summary

Kinetic theory describes a plasma as an assembly of particles with statistically distributed properties. The basic quantity is the phase space density. In thermal equilibrium it is described by the Maxwell distribution. The equation of continuity for the phase space density allows the derivation of the basic equations of kinetic theory: the Boltzmann equation in its most general form or as collisionless Boltzmann equation. For a pure electromagnetic field, the collisionless Boltzmann equation becomes the Vlasov equation. If collisions lead to small-angle deflections only, the collision term in the Boltzmann equation can be described by the Fokker–Planck equation. Despite the limitation to small-angle collisions, the Fokker–Planck equation can be applied to a plasma: while Coulomb collisions can lead to large-angle deflection, these are outnumbered by the small-angle deflections.

Exercises and Problems

5.1. Describe the meaning of the mean free path. What are the physical and formal differences in a neutral gas and in a plasma?

5.2. The solar wind is a proton gas with a temperature of about 1 million K. Plot the distribution function and determine the most probable speed and energy. Compare with the flow speed of 400 km/s and the kinetic energy contained in the flow.

5.3. A spacecraft measures the proton distribution in the solar wind. Above an energy of about 20 keV, the distribution can be described as a power law in E with E^{-4}. Plot the distribution and compare with the results of Problem 5.2. What kind of distribution is this?

Part II

Space Plasmas

6 Sun and Solar Wind: Plasmas in the Heliosphere

> ... but it is reasonable to hope that in not too distant a future we shall be competent to understand so simple a thing as a star.
> A.S. Eddington, *The Internal Constitution of the Stars*[1]

Plasmas in interplanetary space originate from the Sun, as do most of the disturbances and waves embedded in them. The solar atmosphere, the corona, extends as solar wind far beyond the orbit of the outermost planet, Pluto, filling a cavity in the interstellar medium called the heliosphere. The solar magnetic field, frozen-in into the solar wind, is carried out and wound up to Archimedian spirals by the Sun's rotation. Fluctuations and waves on different scales are superimposed, sometimes steepening to collisionless shock waves. The solar wind and the frozen-in magnetic field change during the solar cycle due to systematic changes in solar properties and transient disturbances related to solar activity. Detailed accounts on solar physics and the physics of the interplanetary medium are given in e.g. [113, 484, 572], and the solar corona and the physics of solar activity are described in e.g. [9, 193, 244, 291, 410, 420, 424].

6.1 The Sun

For an astrophysicist, the Sun is an ordinary star of spectral class 2, also called yellow dwarf, and luminosity V. It consists mainly of hydrogen (about 92% in terms of particle number or 72% in terms of mass) as the fuel for solar energy production and helium (about 8%), partly primordial and partly waste product of the energy generation. But the Sun is also more interesting than other stars: owing to its close proximity, we are able to study not only its electromagnetic radiation but also solar emissions of a different kind – plasmas and energetic particles. These not only can be studied for the quiet or average Sun but also for their dependence on the solar cycle, their relation to certain features on the Sun, such as sunspots and filaments, and

---

[1] Copyright 1926, reprinted with kind permission from Cambridge University Press.

Table 6.1. Properties of the Sun

Radius	$r_\odot = 696\,000$ km
Mass	$M_\odot = 1.99 \times 10^{30}$ kg
Average density	$\varrho_\odot = 1.91$ g/cm^3
Gravity at the surface	$g_\odot = 274$ m/s^2
Escape velocity at the surface	$v_{\mathrm{esc}} = 618$ km/s
Luminosity	$L_\odot = 3.86 \times 10^{23}$ kW
Magnetic field	
polar	1 G
general	some G
protuberance	10–100 G
sunspot	3 000 G
Temperature	
core	15 million K
photosphere	5780 K
sunspot (typical)	4200 K
chromosphere	4400–10 000 K
transition region	10 000–800 000 K
corona	2 million K
Sidereal rotation	
equator	26.8 d
30° latitude	28.2 d
60° latitude	30.8 d
75° latitude	31.8 d

their association with the violent processes of the active Sun. Many of these emissions are of interest to the layman, too, because they shape and influence our environment, from the atmosphere and the weather down to the realms of biology and physiology (Chap. 10). The basic properties of the Sun are summarized in Table 6.1.

Most of the Sun's emission is electromagnetic radiation, amounting to 3.86×10^{23} kW integrated over its surface or 6.3×10^4 kW/m^2. With Stefan–Boltzmann's Law an effective temperature T_{eff} of about 5780 K results. At Earth's orbit, the solar constant, which is the solar power received per unit area, is 1380 W/m^2. The solar radiation can be divided into five frequency bands: (a) X-rays and extreme ultra-violet (EUV) with $\lambda < 180$ nm contributes to about 10^{-3} of the total energy output. It is emitted from the lower corona and the chromosphere, and varies during the solar cycle with enhancements up to orders of magnitude during solar flares. (b) Ultra-violet with wavelength between 180 and 350 nm contributes to about 9% of the solar flux. It is radiated from the photosphere and the corona, its variations are similar to the ones in X-rays, although they are smaller. (c) The visible light between 350 and 740 nm contributes to 40% of the energy flux and does not vary significantly with the solar cycle. Only in extremely strong flares a local brightening can be observed. (d) The maximum energy flux of 51% is in the infrared between 740 nm and 10^7 nm, showing no significant variation with solar activity. As the visible light, it is emitted from the photosphere. (e) Radio-emission above 1 mm, originating from the solar corona, contributes to only $10^{-10}\%$ of the solar energy flux, but can be enhanced significantly during solar flares.

6.1.1 Nuclear Fusion

The source of the Sun's energy is nuclear fusion. This idea goes back to Eddington in 1926 [143] and replaced Lane's concept of 1869 which saw the energy source in the Sun's gravitational contraction.

Inside the Sun the basic process of nuclear fusion is the proton–proton cycle: four protons merge to one ^{4}He-nuclei (α-particle). In the first step, two protons merge to a deuteron ^{2}H, emitting a positron and a neutrino. In the second step, a proton collides and merges with the deuteron, forming a ^{3}He-nuclei under emission of a γ-quant. When two ^{3}He-nuclei collide, they merge to an α-particle, emitting two protons and a γ-quant. The mass difference between the four protons and the α-particle corresponds to an energy of 4.3×10^{-12} J or 26.2 MeV. Formally, the reaction can be written as

$$^1\text{H}(\text{p},\text{e}^+\nu_\text{e})^2\text{D}(\text{p},\gamma)^3\text{He}(^3\text{He},2\text{p}\gamma)^4\text{He} + 26.2 \text{ MeV} . \qquad \text{(PPI)}$$

Under solar conditions, half of the hydrogen initially present is converted into deuteron within 10^{10} years: for two protons to merge, their distance must decrease below a proton radius and one of the protons has to undergo spontaneous β-emission. The life-time of ^{2}H is only a few seconds, it immediately captures another proton. The time scale for the fusion of the resulting ^{3}He-nuclei is about 10^6 years. Thus the time scale of the proton–proton cycle basically is determined by the first step, the fusion of two protons.

The last step in the PPI cycle can be replaced by one of the two reactions:

$$^3\text{He}(\alpha,\gamma)^7\text{Be}(\text{e}^-,\nu_\text{e})^7\text{Li}(\text{p},\gamma)^8\text{Be}(\alpha)^4\text{He} + 25.9\text{MeV} \qquad \text{(PPII)}$$

or

$$^3\text{He}(\alpha,\gamma)^7\text{Be}(\text{p},\gamma)^8\text{B}(\text{e}^+\nu_\text{e})^{8*}\text{Be}(\alpha)^4\text{He} + 19.5\text{MeV} . \qquad \text{(PPIII)}$$

In both cases the ^{3}He-nucleus merges with an α-particle. In the PPII chain, the resulting ^{7}Be is converted into ^{7}Li by electron capture. The ^{7}Li then is converted by proton capture into the unstable isotope ^{8}Be which decays into two α-particles. In the PPIII chain, the ^{7}Be-nucleus captures a proton. The resulting ^{8}B-nucleus emits a neutrino and a positron, leading to an excited ^{8}Be-nucleus which decays into two α-particles. With increasing temperature, the latter two reactions become more important compared with the PPI cycle.

Independent of the details of the reaction, energy is liberated in the form of electromagnetic radiation, positrons, neutrinos and, to a smaller extent, kinetic energy of protons. The energy of γs, positrons, and protons immediately is converted into thermal energy while the neutrinos escape: their scattering mean free path is about 7000 AU, compared with the few cm of a photon. Thus although a photon travels with the speed of light, it needs about 100 000 years to diffuse from the Sun's core to its surface.

6.1.2 Solar Neutrinos

Neutrinos emitted in the various chains of the proton–proton cycle have characteristic energies. Thus their energy spectrum gives information about the processes inside the Sun, allowing a test of our standard model of the Sun. Neutrinos are one of the current problems in solar astrophysics. Compared with the first observations, a larger number of solar neutrinos is being detected today; however, it still is smaller than expected.

Part of the problem of detecting neutrinos and of interpreting the results arises from the measurement principle used: different techniques are sensitive to different neutrino energies and thus also to different parts of the energy production cycle. The first neutrino detector was a tank of 615 ton liquid perchloroethylene located about 1500 m under the Homestake mine [120]. This sensitivity of this instrument is such that the ^{8}B neutrinos are expected to generate a signal of 5.6 SNU[2] with ^{7}Be neutrinos contributing 1.1 SNU. Over the years, the Homestake experiment has measured an average of 2.56 ± 0.23 SNU, compared with 7.6 ± 1.2 SNU expected from the standard solar model (SSM). This discrepancy is well known as the solar neutrino problem [103].

The Japanese experiments Kamiokande [175] and Superkamiokande [176], a 680 ton water tank 1 km underground in the Kamioka mine, have an even higher energy threshold and are only sensitive to the high-energy end of the ^{8}B neutrinos; see Fig. 6.1. These experiments show a neutrino deficit of about 50%.

Gallium experiments, such as SAGE, GALLEX, and GNO, have a much lower energy threshold and thus are able to detect also the pp neutrinos. These experiments give a neutrino flux of 74.7 ± 5.0 SNU, lower than the 128 ± 8 SNU expected from the SSM [211].

Two possible explanations have been offered: either our model of the Sun or our understanding of the neutrino is wrong [19, 21, 218, 293, 366, 368]. Fine-tuning of the solar standard model to fit the neutrino observations appears possible without changing the energy flux at the solar surface, although often variations in one parameter lead to a better agreement between the predictions and one of the experiments while the predictions for the other experiments are still in disagreement with the observations. For instance, a reduction in the core temperature in the SSM would allow one to fit the Super-Kamiokande observations, while there still would be a discrepancy with the Homestake observations.

---

[2] Solar neutrino unit; 1 SNU, equals 10^{-36} captures per target atom per second.

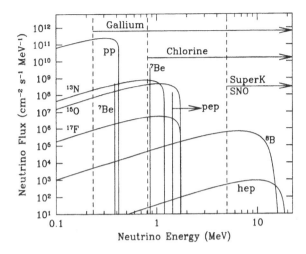

Fig. 6.1. Energy spectra of neutrinos emitted in the various nuclear reactions inside the sun. The threshold energies of various experiments are indicated by *vertical lines*. Figure from S.M. Chitre [100] based on [20], in *Lectures on solar physics* (eds. H.M. Antia, A. Bhatnagar, P. Ulmschneider), Copyright 2003, Springer-Verlag

The other explanation is of great importance to elementary-particle physics: it suggests that neutrinos, contrary to our current understanding, have a mass and can oscillate between different flavors. It has been proposed that the electron neutrinos created in the proton–proton chains are partly converted into τ- or μ-neutrinos during their travel time from the Sun to the Earth. Since only the electron neutrinos can be detected by the above experiments, a lower neutrino flux than would otherwise be expected results.

A more recent experiment, the Sudbury Neutrino Observatory (SNO), suggests that indeed solar neutrinos change their flavor during their journey from the Sun to the Earth [2]. SNO consists of 1000 tons of heavy water (D_2O) at a depth of over 6000 m water equivalent. Owing to the use of deuterium instead of ordinary hydrogen, not only can electron neutrinos be detected but it is also possible to measure the total neutrino flux. Thus electron neutrinos can be distinguished from other neutrino flavors. The results of SNO provide evidence that some electron neutrinos are converted during their travel time to a different neutrino flavor [3] at a rate which gives corrected fluxes in agreement with the SSM. In addition, the KamLAND experiment [147] has detected oscillations in antineutrinos produced in nuclear reactors.

In sum, these results increase our confidence in the SSM because one of the major objections against it, the solar neutrino problem, appears to have its origin in elementary-particle physics rather than in solar physics.

6.1.3 Structure of the Sun

Nuclear fusion takes place in the Sun's core, which is about 0.3 $r_{\odot}$ in radius (see Fig. 6.2). It is surrounded by the radiative core or radiation zone, where energy is transported by radiation. In the surrounding convection zone energy is transported by convection. The top of the convection zone is the

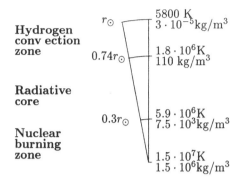

Hydrogen convection zone

$r_\odot$ — 5800 K $3 \cdot 10^{-5} \mathrm{kg/m^3}$

$0.74 r_\odot$ — $1.8 \cdot 10^6 \mathrm{K}$ $110 \mathrm{\ kg/m^3}$

Radiative core

$0.3 r_\odot$ — $5.9 \cdot 10^6 \mathrm{K}$ $7.5 \cdot 10^3 \mathrm{kg/m^3}$

Nuclear burning zone

$1.5 \cdot 10^7 \mathrm{K}$ $1.5 \cdot 10^6 \mathrm{kg/m^3}$

Fig. 6.2. Internal structure of the Sun with typical densities and temperatures

photosphere, the visible surface of the Sun. Here most of the visible light is emitted. The convection cells can be seen as granulation of the photosphere. From the core to the photosphere, the density decreases by more than ten orders of magnitude, and the temperature decreases by a factor of 3000.

The photosphere is optically too thick to receive any electromagnetic radiation emitted from deeper layers of the Sun. Thus solar neutrinos are the only messengers escaping directly from the core of the Sun. All other information about the internal structure of the Sun is obtained indirectly. The most important tool is helioseismology; for a recent review see [7]. Seismology on the Earth is mainly concerned with sudden bursts of mechanical waves during either earthquakes or explosions initiated by humans to probe the terrestrial interior. The goal of helioseismology is similar: to use mechanical vibrations observed on the solar surface to obtain information about the solar interior. But helioseismology is different from seismology on the Earth because oscillations are always present on the Sun. They can be identified from the Doppler shift of spectral lines. These mechanical vibrations of the solar surface are centered around a period of about 5 min [318]. These oscillations have been identified as a superposition of millions of standing waves with amplitudes of the order of a few meters and speeds of the order of a few cm/s [317, 529].

Detailed analysis of these observations allows us to infer parameters such as the sound speed, density, temperature, and chemical composition in the solar interior [10, 203, 294]. This information is used to confirm and refine the standard solar model.

From the viewpoint of plasma physics, a second set of results of helioseismology is more important: it allows us to infer the rotation rates in the solar interior. This is of particular importance for the understanding of the solar dynamo process. The main results are the following: (a) The surface differential rotation persists through the convection zone, while the radiative transfer zone appears to rotate relatively uniformly [465, 516]. However, there appears to be a shear layer beneath the solar surface extending down to about 0.94 solar radii. (b) The transition region, also called the tachocline, is located near the base of the convection zone [33]. The tachocline also seems to be the seat of the solar dynamo. Both features are summarized in Fig. 6.3.

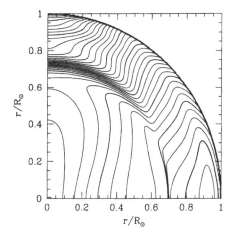

Fig. 6.3. Contours of constant solar rotation rate inferred from helioseismology; figure taken from H.M. Antia [7], in *Lectures on solar physics* (eds. H.M. Antia, A. Bhatnagar, and P. Ulmschneider), Copyright 2003, Springer-Verlag

(c) There is a large-scale flow in the north–south direction, called the meridional flow [217]. The flow speed is of the order of 20 m/s, that is, about two orders of magnitude smaller than the rotation speed. This meridional flow is believed to play an important role in the solar dynamo process [373].

With the good coverage of helioseismological observations since the mid-1990s, an analysis of temporal variations has become possible. So far, no significant temporal variations have been detected in the internal structure of the Sun. But there are some solar-cycle-related modifications close to the surface. For instance, the 5-min oscillations are shifted by up to $0.4\,\mu\text{Hz}$ with higher frequencies during solar maximum [237], probably because of some solar-cycle-related disturbances in the outer layers of the Sun.

In addition, the rotation rate varies with time: on the surface there exist zonal bands of slow and fast rotation which migrate slowly from high to low latitudes during the solar cycle. These flow bands are correlated with migrating magnetic-activity bands already known from the butterfly diagrams [235, 495]. Helioseismology has refined this picture: the bands move towards the equator at low latitudes, while at latitudes above $50°$ they move polewards [8]. This corresponds to the migration seen in magnetic patterns (Fig. 6.28), and thus the poleward migration might be crucial in the understanding of the solar dynamo and, in particular, the polarity reversal [109].

One of the most recent instruments used to study solar oscillations is the MDI (Michelson Doppler Imager) on board SOHO. Details about the instrument, a movie showing a solar quake, and many results from this instrument can be found at `sohowww.nascom.nasa.gov/gallery/MDI/`.

6.1.4 The Solar Atmosphere

Above the photosphere the solar atmosphere consists of three layers: the chromosphere, the transition region, and the corona. The corona can be seen

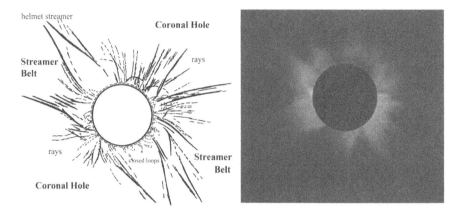

Fig. 6.4. (*Left*) Coronal structure during the total eclipse of 11 July 1991. Based on a sketch by S. Koutchmy in K. Lang [308], *Sun, earth and sky*, Copyright 1995, Springer Verlag; (*Right*) image of the total eclipse on 16 February 1980 (source: www.hao.ucar.edu)

in visible light during a solar eclipse as a structured, irregular ring of rays around the solar disk. Its structure and extent vary with the solar cycle. Figure 6.4 shows on the left-hand side a sketch of the coronal structure during the total eclipse of 11 July 1991 and on the right-hand side a photo taken during the eclipse of 16 February 1980. Since both have been taken during the solar maximum, the corona is highly structured and extends far outwards. During the solar minimum, only few structures are visible and the corona appears smaller. However, the corona does not have a sharp outer boundary, but instead shows structures which extend into different heights and then fade into the background.

The charge states of heavier elements such as O, Si, Mg, and Fe indicate coronal temperatures of about 1 million K. Nonetheless, the corona does not radiate like a black body because it is too thin. Its temperature is roughly independent of height, but it is about a factor of 200 higher than the photospheric and chromospheric temperature. The fastest increase in temperature from about 25 000 to 500 000 K occurs in the transition region, which is only a few hundred kilometers thick and separates the corona from the chromosphere (chromos = color) below. The latter has a height of about 2000 km and can be seen during a total eclipse as a thin red ring around the solar disk, giving the appearance of small flames. The chromospheric emission is weak compared with the photospheric emission because the density of the chromosphere of about 10^{-12} g/cm^3 is about five orders of magnitude smaller than the photospheric density. However, owing to the higher temperature, the maximum of the chromospheric emission is in the UV. Here the chromospheric emission exceeds the photospheric emission: a UV instrument thus "sees" the chromosphere but does not look down to the photosphere. Depending on the wavelength under consideration, different layers of the atmosphere are seen by a telescope. The first observations at different wavelengths in the UV by OSO-4 in 1967 showed particularly well the appearance of the polar coronal hole at larger heights. An example of a modern instrument is EIT (Extreme ultraviolet Imaging Telescope) on board SOHO; a description of the instrument, results, and a picture gallery can be found at umbra.nascom.nasa.gov/eit/.

Excursion 5. *Thermal emission* Chromospheric emission is thermal emission, as is the photospheric emission. Some details of this emission can be approximated under the assumption that the Sun is a black body. In this case the spectrum of the emitted radiation can be described by Planck's law: the energy per unit interval of wavelength emitted by a unit surface area of a black body into a unit solid angle is given by

$$
\begin{aligned}
B_\lambda(T) &= \frac{2hc^2}{\lambda^5} \frac{1}{\exp\left(\frac{hc^2}{k\lambda T}\right) - 1} \, , \\
B_\nu(T) &= \frac{2h\nu^3}{c^2} \frac{1}{\exp\left(\frac{h\nu}{kT}\right) - 1} \, .
\end{aligned}
\tag{6.1}
$$

The total radiation emitted by the black body can be obtained by integration over all wavelengths:

$$
q = \pi F = \pi \int_0^\infty B_\lambda(T)\, \mathrm{d}\lambda = \sigma T^4 \, ,
\tag{6.2}
$$

where $\sigma = 8.26 \times 10^{-11}\ \mathrm{cal\ cm^{-2}\ min^{-1}\ K^{-4}} = 5.6708 \times 10^{-8}\ \mathrm{J\,m^{-2}\,s^{-1}\,K^{-1}}$. This is the Stefan–Boltzmann law. We have already encountered this law when talking about the temperature of the photosphere. The wavelength of the maximum of Planck's curve can be obtained by setting the first derivative of (6.1) equal to zero. We then obtain Wien's law,

$$
\lambda(\mathrm{max})\, T = \mathrm{const} = 2884\ \mu\mathrm{m\,K} \, .
\tag{6.3}
$$

This can be used to determine the wavelength of the maximum of the emission of a black body.

For instance, from (6.3), the photospheric emission (5780 K) has its maximum at 500 nm, well inside the visible. The wavelengths of maximum emission at the bottom (25 000 K) and top (500 000 K) of the transition region are 115 nm, which is in the UV, and 6 nm, which is in soft X-rays, respectively. Thus looking at the Sun in different frequency ranges means looking at different layers of the atmosphere. The soft X-ray emission, since it is viewing greater heights, reveals the coronal holes as dark patches particularly well, but also shows the active regions as bright spots. An example is shown in Fig. 6.5, which shows the evolution of the chromosphere and lower corona during one solar cycle, starting at the maximum in 1990 on the left and continuing to the

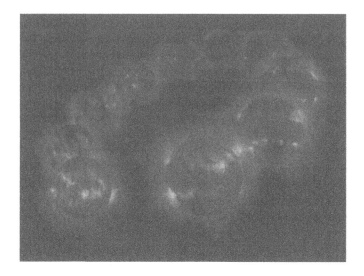

Fig. 6.5. Variation of the Sun in soft X-rays during the solar cycle as observed by Yohkoh; source: `solar.physics.montana.edu/mckenzie/Images/The_Solar_Cycle_XRay_hi.jpg`

maximum in 1999 on the right. A nice movie showing the relation between wavelength and height can be found at `sohowww.nascom.nasa.gov/` as "Five same-day images of Sun in different wavelengths".   □

The coronal emission consists of three components, the emission line or E-corona, the continuum or K-corona (K for the German word *Kontinuum*), and the Fraunhofer or F-corona. The E-corona was first observed during the 1868 solar eclipse. But only in the 1940s did scientists understand the sources of the spectral lines, since these lines were not known from laboratory experiments, because they required extremely high ionization states; for instance, the 530.3 nm green line is from Fe XIV, the 637.4 nm red line is from Fe X, and the 569.4 nm yellow line is from Ca XV. These charge states indicate temperatures of more than one million K. These high temperatures also allow us to understand the great height of the corona. The photospheric temperature of 5780 K would lead to a scale height of about 150 km. At a distance of one solar radius above the photosphere, the density would have dropped by a factor of $\exp(-696\,000/150)$, which is almost zero. With a coronal temperature of two million degrees, a scale height of 10^5 km results, allowing the large extent of the corona.

The main visible coronal emission, the K-corona, is not a real emission but is photospheric light scattered from coronal electrons. Therefore the coronal electron density can be inferred from the intensity of the K-corona. Thus Fig. 6.4 also can be interpreted as an electron density distribution. The K-corona is linearly polarized because the electrons are aligned in the coronal magnetic field.

Typical coronal features include bright arcades, which can be interpreted as closed magnetic loops where electrons are stored, and helmet streamers, which are arcades from which a thin beam extends upwards. The ray-like structures suggest open field lines with electrons streaming away from the Sun. The rather dark regions are depleted of electrons. These coronal holes can also be seen as dark patches in soft X-ray images and are the source regions of the fast solar wind.

Depending on the underlying structures, the electron density at any given height can vary by more than three orders of magnitude [297]. In the lower corona, the intensity gradient is very steep, with a scale height of about $0.1 r_\odot$. Thus, even with an extremely good radiometric resolution, it is not possible to create images of the corona from just above the photosphere out to ten or twenty solar radii. Instead, different instruments have to be combined, as has been done, for example, with the LASCO coronograph on the SOHO spacecraft [161]. Examples and further details can be found at `sohowww.nascom.nasa.gov/gallery/LASCO/` or `lasco-www.nrl.navy.mil/lasco.html`.

The last component, the F-corona, or Fraunhofer corona, results from scattering by slow-moving dust particles. It extends into the interplanetary medium, where it is observed as zodiacal light. The F-corona is not part of the solar atmosphere.

6.1.5 The Coronal Magnetic Field

Figure 6.6 is a very simplified schematic, showing only the most important features, namely coronal holes and helmet streamers. The latter develop over active regions, the legs of the helmet streamer connecting regions of opposite magnetic field polarity. Electrons are captured inside these loops, thus the helmet streamer is a bright feature. The coronal holes on the other hand are regions with open field lines, allowing for a fast electron escape. Thus they appear as dark features, often with rays indicating the direction of the field.

Over the poles of the Sun coronal holes are dominant. Occasionally, they can extend down to the solar equator or even into the opposite hemisphere.

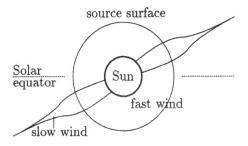

Fig. 6.6. Sketch of the solar corona. Two helmet streamers are shown together with the coronal holes, the solar equator, and the nominal location of the source surface

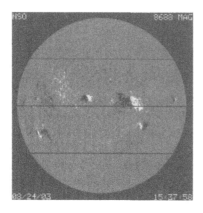

Fig. 6.7. Photospheric magnetic-field measurements from National Solar Observatory (source: www.nso.noao.edu/synoptic/)

Streamer-like configurations are confined to the streamer belt, which is associated with the active regions. Thus the streamer belt's extension and its inclination relative to the solar equator varies over the solar cycle. The resulting magnetic field pattern with different magnetic field polarities on both sides of the streamer belt are carried out into interplanetary space by the solar wind and lead to a sector structure as described in Sect. 6.3.2.

While the coronal magnetic field as described above is rather orderly, spectral line observations of the photosphere reveal a complex magnetic field pattern associated with sunspots and active regions. Figure 6.7 shows an example for the structure of the photospheric magnetic field. The bright and dark spots indicate regions of strong magnetic fields, the color gives the field's polarity. Daily observations of the visible hemisphere of the Sun can be combined to give a map of the photospheric magnetic field such as shown at the top of Fig. 6.8.

 Excursion 6. *Zeeman effect.* The basic tool used to determine the photospheric magnetic field is the Zeeman effect, that is, the splitting of a spectral line λ_0 in a magnetic field into a triplet of lines, with one member at wavelength λ_0, also the π-component, and two members at $\lambda_0 \pm \delta\lambda$, where

$$\delta\lambda = \frac{\pi e}{m_e} \frac{\lambda_0^2 g B}{c} = 4.7 \times 10^{-13} \lambda_0^2 g B \,, \qquad (6.4)$$

g being the Landé g-factor. The split lines are linearly polarized: the π-component is polarized parallel to the field, and the side lines, also called σ_V and σ_R, are polarized perpendicular to it. Thus, when viewed parallel to the magnetic field (longitudinal field), only the σ_V and σ_R, are visible as circularly polarized lines (see Fig. 6.9). Viewed perpendicular to the field (transverse field), all three components are visible, as linearly polarized lines. Thus the Zeeman effect allows the measurement of both the magnetic field strength and its direction.

It should be noted that the photospheric magnetic field is not the only factor influencing the spectral lines. Photospheric motions lead to shifts in

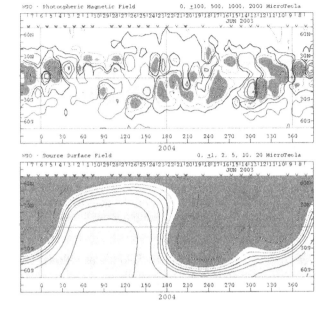

Fig. 6.8. Reconstructed photospheric magnetic field for Carrington rotation 2004 (source: `wso.stanford.edu/synoptic.html`) and calculated photospheric field for the same rotation (source: `quake.stanford.edu/~wso/coronal.html`)

frequency owing to the Doppler effect. Thus a careful analysis of the Fraunhofer lines provide a wealth of information about the photosphere.   □

Coronal loops, structures such as filaments and prominences, and in situ observations in interplanetary space suggest that many of the small-scale photospheric structures form closed loops within less than two solar radii (see Fig. 6.16). Thus a solar source surface can be defined: the small-scale structures are closed below it and the resulting overall field pattern is carried outwards by the solar wind. The source surface is at a height of about 2.5 solar radii and can be determined from the photospheric field pattern using

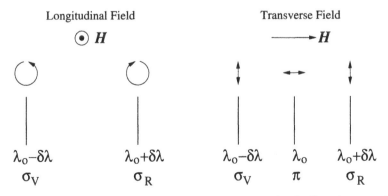

Fig. 6.9. Zeeman splitting when viewed into the fiel (*left*) and perpendicular to it (*right*)

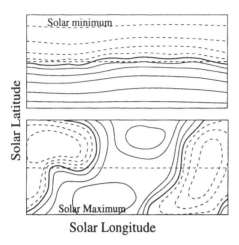

Solar Longitude

Fig. 6.10. Magnetic field pattern on the source surface for solar-minimum (*upper panel*) and solar-maximum conditions (*lower panel*). The *thick line* is the neutral line separating the fields of opposite polarity; the *thin lines* are equipotential lines

potential theory [225,457,571]. The constraints on the field are as follows: (a) the magnetic field at the source surface is directed radially, and (b) currents either vanish or are horizontal in the corona. The resulting map is shown at the bottom of Fig. 6.8.

Figure 6.10 shows source surface maps for solar minimum and maximum conditions. A distinctive feature is the neutral line (thick line), separating the two magnetic field polarities. At the neutral line, the radial magnetic field vanishes. During solar minimum (upper panel) it is roughly aligned with the solar equator while with increasing solar activity (lower panel) the neutral line becomes wavy and extends to higher solar latitudes. During solar maxima, when the Sun changes polarity, the inclination of the neutral line is maximal. Since the neutral line separates magnetic fields of opposite polarity, a current must flow inside it: the neutral line is a current sheet. In Fig. 6.6 it would extend outwards through the tips of the helmet streamers: its extension into interplanetary space is called the heliospheric current sheet (HCS).

The above description gives the impression of a rather static transition from the photospheric to the coronal magnetic field. Recent observation with the Transition Region And Coronal Explorer (TRACE), launched in 1998, revealed a highly filamented corona filled with flows and other dynamic processes [194]. Variability and motions are observed at all spatial locations in the atmosphere and on very short time scales. With the greatly improved spatial resolution of TRACE, a number of new properties in the corona have been identified, as follows. (a) Fine structures: the corona in active regions consists of numerous threads of emitting plasma, which are all clearly separated from each other and in continuous motion. If these threads interact, reconnection can occur. (b) "Moss" is an intricate, dynamic fine structure near the base of active regions. (c) Dynamic structures that change their overall large-scale topology seem to indicate that new magnetic flux emerges in these regions. (d) Bundles of long, nearly linear structures emanating from active regions in the vicinity of sunspots are related to steady outflows of hot material at roughly the local sound speed. (e) The cool, absorbing material embedded within the hot corona close to active regions is also in a highly dynamic state. Examples and the most recent results can be found at vestige.lmsal.com/TRACE/.

6.2 The Solar Wind

The corona does not show a well-defined outer boundary but ragged structures blending into the background. Thus how far does it extend?

The Earth's atmosphere is stationary, shaped by an equilibrium between incoming solar radiation and outgoing terrestrial radiation. On the Sun, the situation is different. Here the temperature is much higher and the solar atmosphere is not stable but blown away as solar wind, filling the entire heliosphere. The first direct measurements of the solar wind [204] started in 1960. However, a particle flow from the Sun towards the Earth had already been suggested at the beginning of the twentieth century. To explain the relationship between aurorae and sunspots, in 1908 Birkeland [50] proposed a continuous particle flow out of these spots. Alternatively, Chapman [91] and Chapman and Ferraro [95] suggested the emission of clouds of ionized particles during flares only. Except for these plasma clouds, interplanetary space was assumed to be empty. Evidence to the contrary came from an entirely different source, i.e. the observation of comet tails. The tail of a comet neither follows the path of the comet nor is directed exactly radially away from the Sun. Instead, its direction deviates several degrees from the radial direction. Hoffmeister [226, 227] suggested that solar particles and not the solar light pressure shape the comet tails. Biermann [48] noted that the fainter dust tails of the comets are indeed directed radially and most probably shaped by light pressure, especially since their spectra resemble the solar spectrum. To explain the shape of the main tail, he too invoked a continuous solar particle radiation.

6.2.1 Properties

The high variability of the solar wind in space and time reflects the underlying coronal structures. The most important features can be summarized as follows [183, 473, 500]: the solar wind is a continuous flow of charged particles. It is supersonic with a speed of about 400 km/s, which is 40 times the sound speed in the solar wind. A plasma parcel travels from the Sun to the Earth within roughly four days. The solar wind carries the solar magnetic field out into the heliosphere, the magnetic field strength amounting to some nanoteslas at the Earth's orbit. The most recent observations are by WIND (see web.mit.edu/afs/athena/org/s/space/www/wind.html for details) and by the plasma instrument on board ACE (see www.srl.caltech.edu/ACE/ for the

instrumentation and general description and www.sel.noaa.gov/ace/ for
the solar wind data).

Two distinct types of plasma flow are observed – the fast and the slow
wind, see, for example, [449, 473, 541]. The fast solar wind originates in the
coronal holes, the dark parts of the corona dominated by open field lines.
Fast solar wind streams are often stable over a long time period (some solar
rotations) and variations from one stream to another are small. The fast
solar wind has flow speeds between 400 km/s and 800 km/s, the average
density is low, about 3 ions/cm^3 at 1 AU. About 4% of the ions are helium.
This ratio is very stable over different fast streams. The average particle flux
is about 2×10^{12} m^{-2} s^{-1}, implying a total particle loss from the Sun of
about 1.3×10^{31}/s. The proton temperature is about 2×10^5 K, the electron
temperature is about 1×10^5 K.

The slow solar wind has lower speeds ranging between 250 km/s and
400 km/s. Its density is about 8 ions/cm^3 at 1 AU, and the flux density is
about twice as large as that in the fast solar wind. During solar minimum
the slow solar wind originates from regions close to the current sheet at the
heliomagnetic equator. The relative amount of helium is highly variable, its
average is about 2%. During solar maximum the slow solar wind originates
above the active regions in the streamer belt, and its helium content is about
4%. Compared with the fast solar wind, it is highly variable and turbulent,
often containing large-scale structures such as magnetic clouds or shocks.
The proton temperatures are markedly lower, about 3×10^4 K, while the
ion temperatures are similar. As in the fast wind, the temperature is always
higher parallel to the magnetic field than perpendicular to it. On the average
it is $T_\parallel \approx 2T_\perp$.

Despite their differences, fast and slow solar wind streams also have similarities. For instance, the momentum flux $M = n_p m_p v_p^2$ on average is similar.
The same is true for the total energy flux, despite the fact that its individual
components, such as kinetic energy, potential energy, thermal energy, electron
and proton heat fluxes, and wave energy flux, are different.

Figure 6.11 shows hourly averages of solar wind parameters during solar
minimum conditions (Carrington rotation 1896). From top to bottom, the
panels give the angles between the solar wind and the Sun–Earth line in the
north–south and east–west directions, the thermal speed of the solar wind,
its density, and the bulk speed of the solar wind. Two fast solar wind streams
start at DOY 143 and 150; a third, albeit slower stream starts late at DOY
160. All three show up as steep increases in solar wind speed and thermal
speed. The density before the arrival of the fast stream increases as solar
wind is swept up, and it decreases abruptly as the spacecraft enters the fast
stream. Prior to the arrival, the flow is slightly from the east, and with the
arrival it turns abruptly to a flow slightly from the west.

A special feature of the solar wind is coronal mass ejections (CMEs). Here
the bulk speed is in the range 400 km/s to 2000 km/s and the composition is

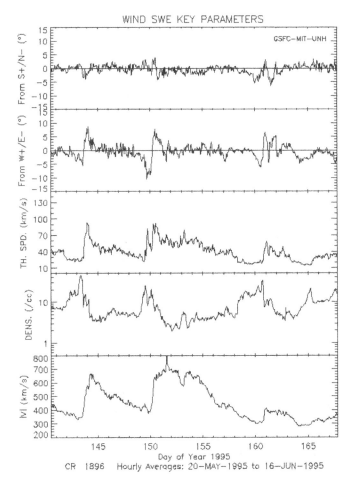

Fig. 6.11. Solar wind under solar minimum conditions, WIND measurements (source: web.mit.edu/afs/athena/org/s/space/www/wind/wind_figures/wind_95may20.gif)

significantly different; in particular, up to 30% of the ions can be α-particles, and even Fe^{16+} or He^+ can be observed occasionally.

Is Slow Solar Wind Always the Same? Observations with the LASCO coronograph on SOHO suggest small-scale density inhomogeneities in the solar wind [477], originating from the tips of helmet streamers. These density variations are assumed to travel along the heliospheric current sheet where they had been detected in situ much earlier as localized maxima in proton density associated with the passage of a sector boundary [55]. Most likely these blobs form when a small plasmoid is disconnected from the helmet streamer by reconnection [280,546]. Thus, the slow solar wind seems to have two distinct sources: plasmoids forming in the streamer belt and strongly overexpanding flux tubes at the boundaries of the coronal holes.

6.2.2 Solar Wind Models

The charge states of heavy ions indicate temperatures of about 10^6 K in the corona, independent of height. Thus hydrogen is completely ionized and the corona basically can be described as an electron–proton gas with small admixtures of heavier elements. In the lower corona, the electron density is about 10^8 to 10^9 cm^{-3} and it decreases with a scale height of about $0.1\,r_\odot$. One of the basic questions in understanding the corona and the solar wind is related to heating: because the photosphere only has a temperature of about 5800 K, how can the corona be heated up to a million Kelvin?

Chapman's Hydrostatic Corona. One of the first models of the corona, introduced by Chapman in 1957 [92], avoided this question and simply described the corona as a static atmosphere, an equilibrium between the pressure gradient force and gravitation:

$$\frac{\mathrm{d}p}{\mathrm{d}r} = -\varrho\frac{GM_\odot}{r^2} \; , \tag{6.5}$$

with G being the universal constant of gravitation and $M_\odot$ the Sun's mass.

Owing to some unknown mechanism, heat is supplied continuously from the photosphere to the corona. Thus thermal energy has to be transported outward through the corona by heat conduction with a heat flow $Q_H = -4\pi r^2 \chi \mathrm{d}T/\mathrm{d}r$. Because the electrons are the more mobile part in the electron–proton gas, the heat basically is transported by electron motion. From the electron density distribution one expects a weakly height-dependent heat conduction coefficient χ. For a completely ionized corona, the variation of density and temperature with height then is

$$\frac{n}{n_0} = \frac{r^{2/7}}{r_0}\exp\left\{-\frac{7r_0}{5H_0}\left[1 - \left(\frac{r}{r_0}\right)^{-5/7}\right]\right\} \tag{6.6}$$

and

$$T = T_0\left(\frac{r}{r_0}\right)^{2/7} \tag{6.7}$$

with the scale height

$$H_0 = \frac{2k_B T_0}{mM_\odot G/r^2} \; . \tag{6.8}$$

Under coronal conditions, heat conduction is about a factor of 20 more efficient than in copper. Thus the temperature decreases only weakly with height (see (6.7)), with the consequence that "... the coronal gas surrounding the Earth may be expected to have a temperature of order of 100 000 K. This is

... consistent with my main inference – that the Earth is surrounded by very hot coronal gas, which greatly distends our outer atmosphere and that heat must flow from it by conduction into our atmosphere" [93], p. 477.

Parker's Hydrodynamic Corona. But the continuous particle flow inferred from the comet tails is in contrast to a static atmosphere extending far behind Earth's orbit. Parker [393] argued that the high temperatures do not allow for a stationary corona and that heat is transported by particle streaming. Thus the solar atmosphere and the continuous particle radiation from the Sun both are the same. Parker used a hydrodynamic approach. Thus the hydrostatic equation has to be complemented by a term describing the fluid motion, leading to Bernoulli's equation. However, Parker did not solve the heating and heat transport problems.

In a simple approach, only protons are considered because they are the dominant ion species and carry virtually all of the mass of the solar wind. The momentum balance then is $\varrho(\boldsymbol{u} \cdot \nabla)\boldsymbol{u} = -\nabla p - \varrho M_\odot G/r^2$ or, in the one-dimensional case for a spherically symmetric corona,

$$u_\mathrm{r} \frac{\mathrm{d}u_\mathrm{r}}{\mathrm{d}r} = \frac{1}{nm} \frac{\mathrm{d}}{\mathrm{d}r}(2nk_\mathrm{B}T) - \frac{GM_\odot}{r^2} . \tag{6.9}$$

The factor 2 in the pressure term $nk_\mathrm{B}T$ considers that both electrons and protons contribute a factor of $nk_\mathrm{B}T$ to the pressure. With the equation of continuity $n(r)\,u_\mathrm{r}(r)\,r^2 = n_0\,u_{\mathrm{r}_0}\,r_0^2$, (6.9) can be written as

$$\frac{\mathrm{d}u_\mathrm{r}}{\mathrm{d}r}\left[u_\mathrm{r} - \frac{2k_\mathrm{B}T}{mv_\mathrm{r}}\right] = \frac{2k_\mathrm{B}r^2}{m}\frac{\mathrm{d}}{\mathrm{d}r}\frac{T}{r^2} - \frac{GM_\odot}{r^2} . \tag{6.10}$$

To describe the temperature gradient, Parker assumed an isothermal corona above about $1.4r_\odot$ as suggested by the charge states of the heavier ions.

Solutions of (6.10) are shown in Fig. 6.12. There are two curves representing special solutions. These curves intersect at the critical point $(u_\mathrm{c}, r_\mathrm{c})$ with

$$r_\mathrm{c} = \frac{GM_\odot m}{4k_\mathrm{B}T} \quad \text{and} \quad u_\mathrm{c} = \sqrt{\frac{2k_\mathrm{B}T_0}{m}} . \tag{6.11}$$

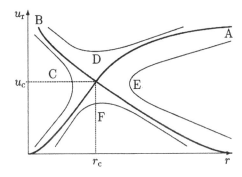

Fig. 6.12. Topology of different solutions for the solar wind equation (6.10)

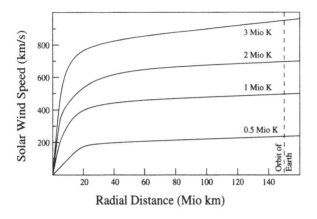

Fig. 6.13. Radial dependence of the solar wind speed for different coronal temperatures, based on [395]

The solution 'A' is the observed solar wind: it starts as a subsonic flow in the lower corona and accelerates with increasing radius. At the critical point r_c, the solar wind becomes supersonic. The solution than takes the form

$$\frac{u_r^2}{u_c^2} - \ln\frac{u_r^2}{u_c^2} = -3 + 4\ln\frac{2u_c^2 r}{wr_0} + \frac{2w^2 r_0}{u_c^2 r} \ , \qquad (6.12)$$

with $w = \sqrt{GM_\odot/r}$. For large distances, (6.12) can be approximated as $u_r \sim 2u_c\sqrt{\ln(r/r_0)}$. Figure 6.13 shows the radial dependence of the solar wind speed for different coronal temperatures. For a coronal density of about $2 \times 10^8 \mathrm{cm}^{-3}$ and a temperature of one million Kelvin, the critical point is at about $6r_\odot$. The solar wind than accelerates up to about $40r_\odot$, afterwards propagating at a nearly constant speed of 500 km/s. The supersonic flow does not extend indefinitely, its density decreases during expansion. At a certain radial distance, most likely beyond 70 AU, the solar wind pressure will become too small to further support a supersonic flow. Where the flow is slowed down to subsonic speed, a termination shock forms. The solar wind then continues as a subsonic flow until the pressure of the interstellar gas becomes larger than the combined pressure of the solar wind and the frozen-in magnetic field. This is the heliopause, the boundary of the heliosphere which is expected beyond 100 AU. In front of the heliopause, a bow shock might develop where the interstellar gas is slowed down by the obstacle heliosphere.

Solutions 'C' and 'F' of the solar wind equation also start as subsonic flows in the lower corona. In solution 'F' the speed increases only weakly with height and the critical velocity is not acquired at the critical radius. The flow continues to propagate radially outward, but then slows down and can be regarded as a solar breeze only. In 'C' the flow has accelerated too fast and has become supersonic before reaching the critical height. It then turns around and flows back towards the Sun as a supersonic flow.

The other curve going through the critical point, 'B', starts as a supersonic flow in the lower corona and becomes subsonic at the critical point. This flow would continue to propagate outwards at subsonic speed. If the flow had decelerated less, as in curve 'D', it still would be supersonic at the critical point, where it would accelerate again, leaving the Sun as a supersonic flow. Solution 'E' is entirely different. It starts as an inward flow blowing subsonically towards the Sun from infinity. The flow accelerates as it approaches the Sun, becoming supersonic at some distance larger than v_c. At that point the flow turns back and propagates outwards as a supersonic flow.

Only one of the mathematically possible solutions of (6.10), curve 'A' in Fig. 6.12, is an approximation of the solar wind. But how good is this model? The assumption of an isotropic pressure p is valid near the Sun where isotropy can be maintained by collisions. As the plasma moves farther out, collisions become less frequent and the pressure parallel to the magnetic field is twice the perpendicular one. The temperature is assumed to be isotropic, too. Again, this might be true close to the Sun but not at the orbit of Earth. In addition, electron and proton temperatures are not the same, as is assumed in the model. These differences do not change the general character of the solution but modify the numbers. Another limitation is the consideration of only one particle species, namely protons. Since α-particles are four times heavier than protons, even the 2% to 4% of He in the solar wind contribute significantly to the momentum transport. Thus an additional set of equations should be considered, leading to a reduction of the flow speed.

A more severe limitation concerns the fields. In the derivation of the hydrodynamic flow, no effects of the magnetic and electric fields were considered: the electromagnetic forces in the momentum balance were ignored. A more elaborate model should consider fields, too. In such a magnetohydrodynamic (MHD) model the critical point is lower in the corona, for average conditions at a height of about two solar radii, which is close to the height of the (fictitious) source surface. The general character of the solution nonetheless is the same as in the hydrodynamic model, for a comparison of such a model with data see e.g. [532].

Parker's hydrodynamic solar wind model nevertheless is a valid approximation of the solar wind in quiet conditions. A comparison of model results with the solar wind observations, however, reveals a rather puzzling fact: the hydrodynamic model is more appropriate to describe the slow wind originating in the streamer belt with its complex magnetic field structures than the fast solar wind blowing directly out of the coronal holes, a situation which is much closer to the assumptions inherent in the model.

Overexpansion of the Solar Wind. Observations suggest that the solar wind and its sources might be even more complicated. In particular, it appears that the solar wind does not expand radially. An expansion factor can be defined as the ratio between the cross-sections of a flux tube at the source surface and in interplanetary space. High-speed solar wind streams from coronal holes then are associated with small expansion factors while the low-latitude slow streams are associated with large expansion factors [545].

These slow streams originate at the boundaries of coronal holes, their strong expansion then causes them to fill the space above low-latitude closed magnetic field structures. Since the expansion factor can be inferred from the magnetic field strength and distribution on the source surface, magnetograms can be used to predict the solar wind speed at the orbit of the Earth.

6.2.3 Coronal Heating and Solar Wind Acceleration

 Although the hydrodynamic description of the solar wind is a reasonable and valuable approach, one fundamental problem has been neglected: the heating of the corona. Basically, two lines of thought have evolved: heating by MHD waves and turbulence or, alternatively, small-scale impulsive energy releases due to reconnection, sometimes called nano-flares. A recent review about heating mechanisms is given in [141,530].

Waves and Turbulence. Of the basic MHD waves, the fast and slow magneto-acoustic waves are compressive, while the Alfvén wave is a noncompressive propagation of fluctuations along the field. In a collisionless plasma, the Alfvén wave propagates undamped, whereas the magneto-acoustic waves undergo Landau damping. Wave energy is then converted into thermal energy, mainly of the ion component in the plasma. Under solar conditions, the slow mode is damped very strongly, while the fast mode can propagate up to a distance of about $20r_\odot$. On the basis of this, Barnes et al. [31] developed a model of the solar wind with coronal heating by fast magneto-acoustic waves. While the Alfvén wave is not damped, it nonetheless contributes to momentum transport and can be interpreted as a radiation pressure, accelerating the plasma. Although non-thermal broadening of some spectral lines indicates the existence of waves or turbulence in the lower corona, it is not completely understood which waves these are, how they propagate outward, and whether the observations really are indicative of wave fields or, rather, of turbulence. A brief review can be found in [30,333]; recent developments are described in [526].

Excursion 7. *Landau damping.* Landau damping is a characteristic feature of collisionless plasmas: waves are damped without energy dissipation by collisions. In Landau damping, a propagating wave accelerates gas particles contained in a distribution function that happen to have a similar direction and speed to the wave: Landau damping is therefore a resonance effect or resonant damping. Landau damping is therefore an example of resonant wave–particle interaction. Chen [97] compares this process to a surfer riding an ocean wave: when surfing, a surfer launches him/herself in the propagation direction into a steepening part of an incoming wave and is further accelerated by this wave. Because a distribution function normally contains many more slower than faster particles, the wave loses energy by accelerating the slower particles. Thus the original distribution function (dashed line in

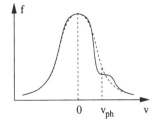

Fig. 6.14. Landau damping: around the phase speed the wave modifies the original particle distribution (*dashed*) by accelerating particles

Fig. 6.14) is deformed, with particles being removed from lower to higher speeds around the phase speed of the wave: the distribution function flattens near the phase velocity, where particles are in resonance with the wave. Interaction between the particles might lead to a redistribution of this energy gain, tending to reestablish the original distribution.

Impulsive Energy Release: Reconnection So far, the corona has been treated only hydrodynamically and the magnetic field has been ignored. Even for coronal heating by MHD waves, the field has been considered as only a carrier for the waves, while its energy content has been neglected. The conversion of field energy into thermal energy has therefore not been considered as a heating mechanism, although this concept is widely applied in space physics on larger scales: models for the acceleration of particles in the magnetosphere's tail or solar flares often involve reconnection because the magnetic field is the only source of energy available.

As we saw in Sect. 3.5, reconnection requires fields of opposite polarity. The photosphere, as the top of the convection zone, is in continuous motion with bubbles rising and falling and plasma flowing in and out, as can be inferred from the Doppler shift of spectral lines and even seen directly in the TRACE data [194]. The plasma motion also shuffles the magnetic field around: magnetic threads emerge in the intergranular lanes between the granulation cells. While the latter are associated with an upwelling flow, a downflow of plasma is observed in the former. A magnetic flux tube therefore sits in the center of a tornado of downflowing gas. Thus, on a small scale, magnetic field configurations suitable for reconnection will form frequently [214], eventually converting field energy into thermal energy [422]. Observational evidence for such small-scale impulsive energy releases is found in electromagnetic radiation, in particular in so-called bright X-ray points [195] and small-scale exploding EUV events. The recent TRACE observations reveal that X-ray bright points are not really points, but can be resolved into highly dynamic loops with distinct features in their footpoints [194]. These observations lend additional support to magnetic reconnection as a mechanism for coronal heating.

But heating by microflares still provides grounds for debate. One problem is related to the energy provided by these flares: it might not be enough to provide the required heating [12], although more recent studies using SOHO/EIT [44] data and TRACE [398] data tend to find larger energy releases. It is also suggested that, since reconnection can generate Alfvén waves, the two models might be connected [479].

6.3 The Interplanetary Magnetic Field (IMF)

The photospheric magnetic field was discovered by Hale in 1902. The splitting of spectral lines due to the Zeeman effect suggests a photospheric field in the order of 10^{-4} T or 1 G (G: gauss) outside and 3000 G to 4000 G inside sunspots. Within less than 2 solar radii, this complex and highly variable field is reduced to a rather simple, radially directed one. Since the conductivity of the solar wind is high, the magnetic field is frozen into it and carried out into interplanetary space. The Sun's rotation winds up these field lines to Archimedian spirals. Thus with increasing radial distance, the originally radial magnetic field becomes more and more toroidal.

6.3.1 Spiral Structure

The Sun rotates with a sidereal rotation period of 27 days. The solar wind flows radially away from the Sun, carrying the frozen-in magnetic field. While the solar wind propagates outward, the base of the field line frozen into the plasma parcel is carried westward, forming an Archimedian spiral,[3] as shown in Fig. 6.15. A similar effect can be observed with a rotating sprinkler; thus the deformation of the field lines also is called the garden-hose effect.

The equation of the Archimedian spiral can be derived from the displacements Δr and $\Delta\varphi$. Initial conditions of the plasma parcel on the Sun are a source longitude φ_0 and a source radius r_0. At a time t the parcel then can

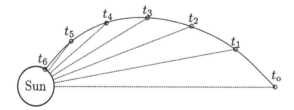

Fig. 6.15. Deformation of a magnetic field line due to the combination of a radial plasma flow and the Sun's rotation. The numbers indicate consecutive times after a plasma parcel has left the Sun at time t_0

---

[3] An Archimedian spiral is defined as a curve resulting from a motion v of a point along an axis that rotates with constant angular speed ω around the origin: $r = a\varphi$ with $\varphi = v/\omega$.

be found at the position $\varphi(t) = \omega_\odot t + \varphi_0$ and $r(t) = u_{\text{sowi}}t + r_0$. Eliminating the time yields the equation for the Archimedian spiral:

$$r = u_{\text{sowi}} \frac{\varphi - \varphi_0}{\omega_\odot} + r_0 . \tag{6.13}$$

With $\tan\psi = \omega_\odot r/u_{\text{sowi}}$, the path length s along the spiral is given as

$$s = \frac{1}{2} \frac{u_{\text{sowi}}}{\omega_\odot} \left(\psi \sqrt{\psi^2 + 1} + \ln\left\{ \psi + \sqrt{\psi^2 + 1} \right\} \right) . \tag{6.14}$$

The magnetic field in the equatorial plane can be expressed in polar coordinates $\boldsymbol{B} = (B_r, B_\varphi)$. The magnitude of $\boldsymbol{B}$ depends on radial distance only: $|\boldsymbol{B}| = B(r)$. Gauss's law in spherical coordinates (see A.3.3) yields

$$\nabla \cdot \boldsymbol{B} = \frac{1}{r^2} \frac{\partial}{\partial r}(r^2 B_r) = 0 \tag{6.15}$$

or $r^2 B_r = r_0^2 B_{r_0}$. Thus the magnetic flux through spherical shells is conserved and the radial component of the field decreases as

$$B_r = B_0 \left(\frac{r_0}{r}\right)^2 . \tag{6.16}$$

Since the magnetic field is constant, it is $\partial B/\partial t = 0$. From the frozen-in condition (3.84) we then get $\nabla \times (\boldsymbol{u} \times \boldsymbol{B}) = 0$, or in spherical coordinates

$$\frac{1}{r} \frac{\partial}{\partial r} (r(u_\varphi B_r - u_r B_\varphi)) = 0 . \tag{6.17}$$

Thus we have $r(u_\varphi B_r - u_r B_\varphi) = \text{const.}$ Let us assume r_0 to be at the source surface. There $\boldsymbol{B}$ is radial and we get

$$r u_\varphi B_r - r u_r B_\varphi = r_0 u_{\varphi_0} B_0 = r_0^2 \omega_\odot B_0 . \tag{6.18}$$

In the last step, the angular speed of the Sun is used to describe the azimuthal component of the solar wind speed at the source surface. From (6.18) the azimuthal component of the magnetic field is

$$B_\varphi = \frac{r u_\varphi B_0 - r_0^2 \omega_\odot B_0}{r u_r} = \frac{u_\varphi - r\omega_\odot}{u_r} B_r . \tag{6.19}$$

For large distances, $r\omega_\odot > u_\varphi$, (6.19) is approximately $B_\varphi = -r\omega_\odot B_r/u_r$. The azimuthal component therefore decreases with $1/r$ while the radial component decreases as $1/r^2$. The field strength decreases with r as

$$B(r) = \frac{B_0 r_0}{r^2} \sqrt{1 + \left(\frac{\omega_\odot r}{u_r}\right)^2} . \tag{6.20}$$

The angle ψ between the magnetic field direction and the radius vector from the Sun is $\tan\psi = B_\varphi/B_r$. For large distances this reduces to $\tan\psi = \omega_\odot r/u_r$. At the Earth's orbit, $\tan\psi$ is about 1 for typical solar wind conditions, and thus the field line is inclined by $45°$ with respect to the radial direction.

The current in the heliospheric current sheet is related to the magnetic field by Ampére's law (2.7). In spherical coordinates the current density in the plane of the ecliptic then is

$$j_r = j_0 \cdot \frac{r_0}{r} \quad \text{and} \quad j_\varphi = \frac{B_\varphi}{B_r} j_r = j_0 \frac{B_\varphi}{B_r} \frac{r_0}{r} . \tag{6.21}$$

6.3.2 Sector Structure

So far we have considered only the shape of the interplanetary magnetic field lines but not their direction. The first long-term observations of the IMF by IMP 1 in 1963 over a couple of solar rotations revealed a sector pattern as shown in Fig. 6.16: the magnetic field polarity is uniform over large angular regions and then abruptly changes polarity. These magnetic field sectors are stable over many solar rotations. At most times either two or four sectors can be observed: if the neutral line is tilted without any wiggles as indicated in the upper panel of Fig. 6.10, a pattern of two magnetic field sectors arises; a wavy neutral line leads to four or more sectors. During solar maximum

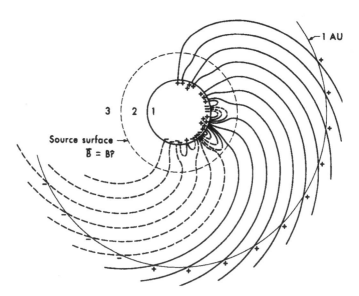

Fig. 6.16. Relationship between the photospheric, source surface and interplanetary magnetic field. Dashed and solid lines indicate negative and positive magnetic field polarities, respectively. Reprinted from K.-H. Schatten et al. [457], *Solar Physics* **6**, Copyright 1969, with kind permission from Kluwer Academic Publishers

the sector structure is complex and distorted by a large number of transient disturbances.

6.3.3 The Ballerina Model

The sector boundaries are the extension of the neutral line into the interplanetary medium, the so-called heliospheric current sheet (HCS). Figure 6.17 shows the HCS extending far into the interplanetary medium. The inclination of the neutral line defines the width of a cone inside which an observer in space alternately sees different polarities of the coronal/interplanetary magnetic field. The maximum inclination of the neutral line at each time is called the tilt angle. It can be used as an alternative measure of solar activity and is an indicator of the range over which both field polarities can be observed.

Figure 6.18 shows a three-dimensional sketch of the wavy current sheet with some field lines for solar minimum conditions. During solar maximum a more complex and more wavy structure would be observed and the neutral line would be bent towards high latitudes, as also suggested in the cross-section shown in Fig. 6.17. Figure 6.18 anticipates the overexpansion of the solar wind: field lines from the borders of the coronal holes expand such that they overlay the closed magnetic field regions on the Sun, even extending down to the heliospheric current sheet.

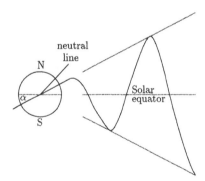

Fig. 6.17. Heliospheric current sheet and definition of the tilt angle α as inclination of the neutral line in the tilted dipole model proposed in [413]. The width of the cone is twice the tilt angle α

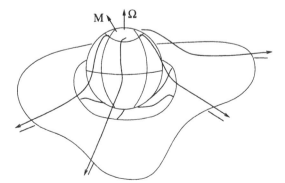

Fig. 6.18. Current sheet in the inner heliosphere in the Ballerina model. The thick lines indicate the magnetic field lines. Based on [491]

6.3.4 Corotating Interaction Regions

Fast and slow solar wind streams originate on the Sun. While these streams propagate outward, the frozen-in magnetic field is wound up to Archimedian spirals. In a slow stream, the field line is curved more strongly than in a fast one. Since field lines are not allowed to intersect, at a certain distance from the Sun an interaction region develops between the two streams. It was soon realizes that "the collision of these plasmas will lead to the formations of two shock waves and a tangential velocity discontinuity between them" [132]. Because this structure rotates with the Sun, it is called a corotating interaction region (CIR). Often the source locations of the fast and slow solar winds are rather stable and an observer in space sees the CIR again during the following solar rotations. In this case, it is called a recurrent corotating interaction region.

Figure 6.19 shows an idealized sketch of the evolution of a CIR in the inner heliosphere. On the Sun, there is an abrupt change in solar wind speeds from fast to slow. As these streams propagate outward, flow compression and deflection on both sides of the interface tend to smoothen the jump, leading to a continuous increase in plasma speed. The region of compressed plasma at the transition between the fast and slow stream at 1 AU typically extends over about 30° while the plasma might originate from a coronal region as

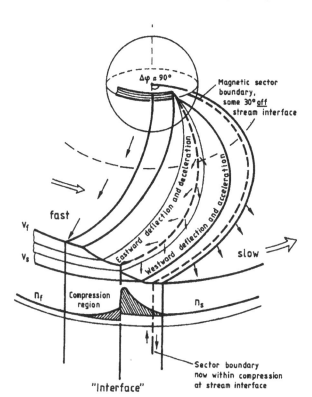

Fig. 6.19. Idealized view of a corotating interaction region (CIR) in the inner heliosphere. Reprinted from R. Schwenn [473], in *Physics of the inner heliosphere, vol. I* (eds. R. Schwenn and E. Marsch), Copyright 1990, Springer-Verlag

wide as 90° or more. Thus magnetic field sector boundaries often found close to the compression region are not necessarily related to the interface but might originate in a coronal region far from the boundary between the fast and slow streams. This is also evident from observations of the corona and the photosphere: the boundaries of the coronal holes are not related to the neutral line of the photospheric field. In particular, the polar coronal hole can extend into the opposite hemisphere, crossing the solar equator as well as the current sheet.

With increasing distance from the Sun, the characteristic propagation speeds, which are the sound and Alfvén speeds, decrease. At some distance between 2 and 3 AU, the density gradient on both sides of the compression region becomes too large and a shock pair develops, propagating away from the interface [247,490]. The shock propagating into the slow wind is called the forward shock, the one propagating into the fast wind is the reverse shock.

Corotating interaction regions tend to distort or even destroy all small-scale fluctuations and disturbances propagating outward from the Sun. In the outer heliosphere, the magnetic field and therefore also the shock fronts are more azimuthally aligned, sometimes extending around the entire Sun. Thus finally the spoke-like structure of different solar wind streams close to the Sun is converted into a shell of concentric shock waves propagating outward like waves from a stone thrown into water. When CIRs or CIRs and travelling interplanetary shocks interact, merged interaction regions result which play a crucial role in the modulation of the galactic cosmic radiation. A summary of the plasma physical properties of CIRs and their consequences for different particle populations in the three-dimensional heliosphere is given in [27].

6.3.5 The Heliosphere During the Solar Cycle

Sunspots strongly modify the photospheric magnetic field, subsequently changing the field on the source surface and the tilt angle of the neutral line. These modifications are transported outwards even to the borders of the heliopause and manifest themselves in spatial and temporal variations in solar wind and magnetic field parameters.

The most dramatic variation of the heliospheric structure during the solar cycle is related to the neutral line of the coronal magnetic field and its interplanetary continuation as the heliospheric current sheet. The waviness of the current sheet, as described by the tilt angle, increases towards the solar maximum. Thus the current sheet is rather flat during the solar minimum, as shown in the left panel in Fig. 6.20, while it extends to much higher latitudes during solar maximum. During solar minimum the CIRs are confined to the vicinity of the equatorial plane while during solar maximum conditions stream interactions also can be observed at higher latitudes.

In the solar wind parameters, the solar cycle variations are less pronounced [183,473]. In general, any long-term variations apparent in the data are small compared with short-term variations. In addition, each solar cycle

Solar Minimum   Solar Maximum

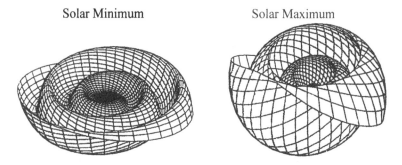

Fig. 6.20. Waviness of the heliospheric current sheet during solar minimum and maximum conditions, based on a sketch by R. Jokipii, University of Arizona

seems to be slightly different, and thus a parameter might be related to solar activity quite well in one cycle but not in another. Despite the uncertainties involved, there are some correlations which can be understood easily; for instance: (a) The average solar wind speed is higher during solar minimum than during solar maximum because high-speed solar wind streams are observed more frequently and for longer times during solar minimum. (b) The average solar wind densities are roughly constant except for individual time periods when exclusively slow solar wind streams were observed. (c) The momentum flux is modulated by $\pm28\%$ with a well-defined minimum at solar maximum, which just is a combination of (a) and (b). For the same reason, the kinetic energy flux is modulated by $\pm40\%$, again with its minimum around solar maximum. (d) The relative amount of helium has a minimum of 2.8% around solar minimum and a maximum of about 4% around solar maximum. Again, this reflects the fact that during solar minimum conditions the fast solar wind is observed more frequently.

6.4 Plasma Waves in Interplanetary Space

The interplanetary magnetic field is highly variable on different temporal and spatial scales. For instance, fast and slow solar wind streams form interaction regions. Coronal mass ejections, sometimes driving an interplanetary shock, are transient disturbances, and on a smaller scale, waves and turbulence are superposed on the average field.

Figure 6.21 illustrates the variability of the magnetic field on time scales of minutes to hours. It shows the magnetic field azimuth (angle between magnetic field line and radial direction), the elevation (inclination of the field with respect to the plane of ecliptic), and the flux density for a time period of 7 hours during unusually quiet (left) and turbulent (right) interplanetary conditions. Magnetic field fluctuations are more pronounced in direction than in flux density and are quite irregular in amplitude and frequency.

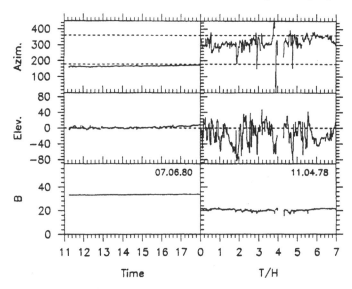

Fig. 6.21. Fluctuations in the magnetic field azimuth, the elevation, and the flux density for extremely quiet conditions (*left*) and during a turbulent phase (*right*). The different azimuth angles indicate an outward direction of the field in the left panel and an inward direction in the right panel. Data from the Helios magnetometer, University of Braunschweig

6.4.1 Power-Density Spectrum

The magnetic field fluctuations can be described by a power-density spectrum [35, 40]:

$$f(k_{\|}) = C \cdot k_{\|}^{-q} . \tag{6.22}$$

Here $k_{\|}$ is the wave number parallel to the field, q the slope, and C a constant describing the level of the turbulence.

The power-density spectrum in Fig. 6.22 shows magnetic field fluctuations on different scales. Its slope is distinct in different parts of the spectrum, indicating different sources and modes of turbulence. Basically, four regimes can be distinguished:

- Large-scale structures lasting a few days up to a solar rotation are related to the stream structure of the solar wind and to solar wind expansion. Both processes are the sources of turbulence on smaller scales; the frequencies of the large-scale turbulence are below 5×10^{-6} Hz.
- Meso-scale fluctuations are associated with the flux-tube structure of the interplanetary medium which originates in the photospheric supergranulation. Frequencies range between 5×10^{-6} and about 10^{-4} Hz.
- In the inertial range, mainly Alfvén waves with periods between some 20 min and more than 15 h are found, corresponding to frequencies between 10^{-4} and 1 Hz. The slope q varies between -1.5 and -1.9. Magnetic

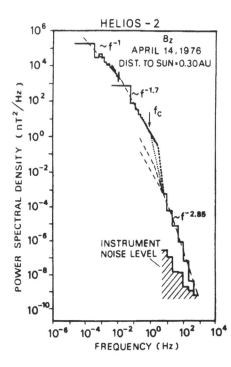

Fig. 6.22. Magnetic field power-density spectrum. Reprinted from K.U. Denskat et al. [129], *J. Geophys. Res.* **87**, Copyright 1983, American Geophysical Union

field fluctuations in the inertial range seem to be responsible for the scattering of protons in interplanetary space (Sect. 7.2).

- The smallest scales are in the dissipation range above 1 Hz. Here the spectrum is steeper with a slope close to −3. The observed fluctuations can be attributed to ion cyclotron waves, ion acoustic waves, and Whistlers.

6.4.2 Waves or Turbulence?

So far we have described the fluctuations in terms of waves. But a single observer in interplanetary space cannot decide whether the fluctuations carried across him by the solar wind are waves or turbulence because he is not able to distinguish between spatial and temporal variations. Thus the question of whether the magnetic field fluctuations should be interpreted in terms of waves or turbulence has led to a long and sometimes fruitless controversy [332, 525]. Only the modern concepts of MHD turbulence, e.g. [365], allowed a kind of unification of both approaches: dynamical MHD turbulence is not the simple superposition of different waves, but rather consists of wave-packets which can interact with each other or can decay and excite new waves.

We will not go into the details of this debate, but only introduce the concept of Alfvénic turbulence or Alfvénicity because it offers a helpful tool in the description of magnetic field turbulence. Alfvén waves are transverse

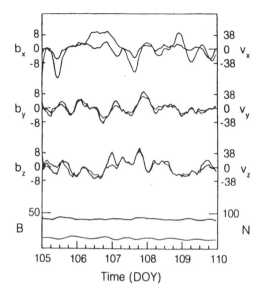

Fig. 6.23. Hourly averages of magnetic field and plasma speed components. The good correlation between the plasma and field fluctuations characteristic of Alfvénic fluctuations is evident. Reprinted from R. Bruno et al. [64], *J. Geophys. Res.* **90**, Copyright 1985, American Geophysical Union

waves propagating along the magnetic field line with the Alfvén speed $v_A = B_0/\sqrt{\mu_0 \varrho}$ (see Sect. 4.2.1). Fluctuations are Alfvénic if the fluctuations δu_{sowi} in flow speed and δB in flux density obey the relation

$$\delta u_{\mathrm{sowi}} = \pm \frac{\delta B}{\sqrt{\mu_0 \varrho}} \,. \tag{6.23}$$

In the plasma and field data shown in Fig. 6.23, the good correlation between these fluctuations is evident. Fluctuations are classified as Alfvénic if the correlation coefficient is larger than 0.6. Obviously, this is true for Alfvén waves. But there is also a large number of other fluctuations which fulfill (6.23). In particular, structures with variable $|B|$ can also fulfill (6.23), as can many of the static structures in the solar wind. Alfvén waves contribute only a small amount to the Alfvénic fluctuations.

The Alfvénicity of fluctuations is useful in the description of the evolution of turbulence from an orderly state (high Alfvénicity) to an entirely stochastic one [525]. For instance, the Alfvénicity is larger close to the Sun than at the Earth's orbit, indicating that in the inner heliosphere most of the fluctuations are of coronal origin. As these fluctuations decay, the Alfvénicity decreases and the slope of the power-density spectrum evolves towards $-5/3$, which is the Kolmogoroff spectrum of random, uncorrelated turbulence. The Alfvénicity is larger in fast solar wind streams than in slower ones; thus in the fast wind an orderly state is preserved over larger spatial scales. If fast and slow streams interact, the Alfvénicity decreases and the spectrum takes the slope of the Kolmogoroff spectrum. The Alfvénicity can also be different on both sides of the heliospheric current sheet.

The solar wind and its magnetic field therefore have to be understood as a dynamically evolving, inhomogeneous, anisotropic, turbulent magneto-fluid. With increasing distance, the fluctuations embedded in this fluid evolve from Alfvénic turbulence close to the Sun towards a Kolmogoroff spectrum.

6.5 The Three-Dimensional Heliosphere

Until the early 1990s our knowledge of the heliosphere had been limited to the plane of ecliptic. The main goal of the Ulysses mission, launched in October 1990, is the study of the heliosphere's third dimension, i.e. plasmas, particles, and fields in the polar regions of the Sun.

The Ulysses trajectory for the prime mission is shown in Fig. 6.24. A swing-by at Jupiter allowed Ulysses to escape out of the plane of ecliptic into an elliptical orbit around the Sun. This orbit is inclined by 80° relative to the solar equator; the orbital period is 6.3 years. Ulysses flew below the Sun's south pole in autumn 1994 and above her north pole in summer 1995, both polar passes were made during solar minimum. The mission will continue for another two orbits, allowing for observations over the poles at solar maximum and during the following minimum. The most important results of the first polar pass are summarized in a series of papers in *Science* **268**.

Plasma and field observations offered some surprises for the scientists. For instance, the radial component of the magnetic field, which is most easily related to the global solar magnetic field, failed to show any latitudinal gradient [26], although the photospheric magnetic field clearly reveals a dipole-like pattern. Thus magnetic flux is removed from the poles towards the equatorial regions, as had been suggested in the sketch in Fig. 6.18 and is also suggested in the (empirical) concept of overexpansion. This observation has consequences for solar wind acceleration models, where the resulting

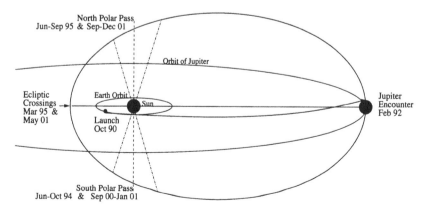

Fig. 6.24. Flight path of Ulysses. After a swing-by at Jupiter in February 1992, the spacecraft left the plane of ecliptic

stress of the magnetic field thus far has not been considered (except for the newer works which also consider the overexpansion), and for models describing the modulation of galactic cosmic rays. In addition, the magnetic flux in the southern and northern hemisphere is different with an increase in radial magnetic field strength of about 30% in the southern hemisphere [160, 493]. This suggests an offset of the heliomagnetic equator by about 7° to the south: the solar magnetic field therefore is not symmetrical about the heliographic equator.

The plasma measurements showed a pronounced latitudinal variation of the solar wind speed. As shown in Fig. 6.25, it increases from about 450 km/s in the equatorial plane to about 750 km/s above the poles. Up to a latitude of about 30°S, which corresponds to the tilt angle at that time, there is a strong variation between fast and slow streams with a period of about 26 days, resulting in a recurrent CIR. At latitudes higher then 50°S, only the fast solar wind streaming out of the coronal hole is observed [409], in agreement with our expectations. The composition of the solar wind, on the other hand, offered some surprises and also a big challenge for theory: the compositions of the fast and slow streams were markedly different, but the abundances were not what would be expected for ions accelerated in the hot corona. Instead, they were more representative of ions formed in the lower temperatures of the chromosphere [184].

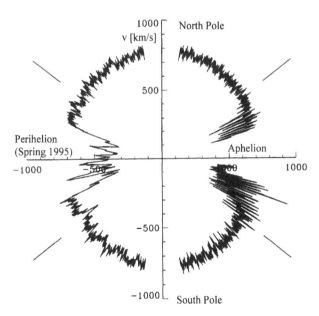

Fig. 6.25. Polar plot of solar wind speed as a function of heliolongitude for the out-of ecliptic phase of the Ulysses mission from February 1992 through January 1997. Reprinted from D.J. McComas et al. [342], *J. Geophys. Res.* **103**, Copyright 1998, American Geophysical Union

A comparison between plasma and magnetic field fluctuations inside the coronal hole revealed the existence of large-amplitude, long-period Alfvén waves propagating outward from the Sun [26]. Thus it appears that the fluctuations, which originate close to the Sun in the acceleration region of the solar wind, are less likely to decay in the uniform fast solar wind flow than they are in the complex and interacting flows in the plane of ecliptic.

6.6 The Active Sun

The variability of the Sun is most obvious in the number and spatial distribution of sunspots. But other properties, such as the electromagnetic radiation, the solar wind, and the solar and interplanetary magnetic fields, change too. And, of course, during times of high solar activity there are the phenomena of violent releases of energy and matter, i.e. flares and coronal mass ejections.

We have already considered a simple model of a sunspot in example 11. Figure 6.26 shows a white light image of a sunspot and its surroundings. The dark spot, also called the umbra, is surrounded by a penumbra which consists of radially oriented filaments. Outside the penumbra, the photospheric granulation cells are visible. The darker the umbra, the lower the temperature and the greater the magnetic field strength. By use of the Stefan–Boltzmann law, the temperature T_{spot} inside the spot can be calculated from the ratio of the intensity I_{spot} of the electromagnetic radiation inside the spot to that for the photosphere I_{photo}:

$$\frac{I_{\mathrm{spot}}}{I_{\mathrm{photo}}} = \left(\frac{T_{\mathrm{spot}}}{T_{\mathrm{photo}}}\right)^4 . \tag{6.24}$$

This temperature difference can be used to calculate the difference in gas-dynamic pressure and thus also the magnetic field strength, as demonstrated in example 11.

6.6.1 The Solar Cycle

The first records of sunspot, the prime indicators for solar activity, date back to the fourth century BC when the Greek astronomers noted dark spots on the

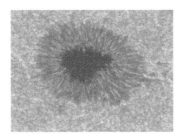

Fig. 6.26. Sunspot with surrounding granules, white light image. Source: www.uni-sw.gwdg.de/research/exp_solar/eflecken.html

solar surface. Ancient Chinese astronomers knew about sunspots, as did the pre-Spanish Peruvians. In Europe, isolated records can be traced back to the ninth century, but systematic observations started only with the development of the telescope early in the seventeenth century. The variability of shape, location, and number of sunspots was recognized early; however, owing to the Maunder minimum, a period of very low solar activity in the seventeenth century, the solar cycle was recognized only in 1843 by H. Schwabe.

Figure 6.27 shows two different representations of the solar cycle. In the top panel, a butterfly diagram gives the latitudinal distribution of sunspots, and in the lower panel, the sunspot number is shown. Alternatively, a sunspot relative number or Wolf number can be used which considers the sunspot size and its relation to other spots or an active region: $R = k(10g + f)$ with g being the number of sunspot groups, f the number of single spots, and k a normalization factor to standardize observations (e.g. corrections for visibility). At the solar minimum, the Sun is almost spotless. Then spots start to appear at latitudes around $30°$. These spots are relatively stable and often can be observed over some solar rotations. They move towards the solar equator while at higher latitudes new spots emerge. The number of sunspots increases until solar maximum. Afterwards, only a few new sunspots appear on the disk while the sunspots at low latitudes dissolve. The total number of sunspots decreases. Just after the solar minimum new sunspots begin to

DAILY SUNSPOT AREA AVERAGED OVER INDIVIDUAL SOLAR ROTATIONS

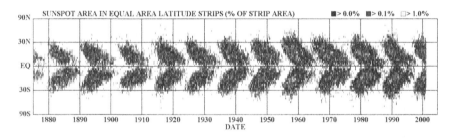

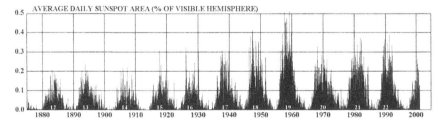

Fig. 6.27. Two different representations of the solar cycle. (*Top*) Butterfly diagram of the latitudinal distribution of sunspots, with each bar marking a sunspot. (*Bottom*) Sunspot number versus time for the same time period (source: science.mfsc.nasa.gov/ssl/pad/solar/images/bfly.gif)

emerge at higher latitudes. The average duration of such a cycle is 11 years with variations between 7 and 15 years.

From the lower panel in Fig. 6.27, it is evident that solar activity is highly variable between different solar cycles. For instance, the sunspot number in the 1958 solar maximum was about twice as large as that in the 1855 maximum. The highest number of sunspots for a month observed so far was 254 in October 1957. Over longer periods, variations by up to a factor of 4 have been observed; at some times, e.g. during the Maunder Minimum of 1650–1710, solar activity and sunspot numbers can be even smaller.

The magnetic cycle is twice as long: within the 11 years of a sunspot cycle, the solar field reverses its polarity once. This is evident in Fig. 6.28, where the magnetic flux is plotted in a butterfly-diagramm. Thus only after 22 years the original polarity pattern is restored. This 22-year cycle is called the Hale cycle or the solar magnetic cycle.

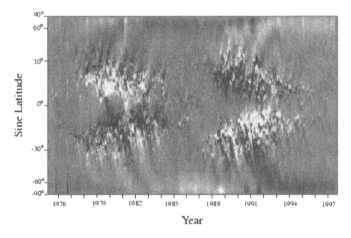

Fig. 6.28. Butterfly diagram of net magnetic flux (constructed from the NSO/KP synoptic rotation magnetic maps) from April 1975 to August 1997. Note the dominant opposite polarities of the magnetic flux poleward and equatorward in the butterfly pattern, reversed between the two hemispheres and between cycles 21, 22, and 23. Large-scale patterns of monopolar magnetic flux extend poleward from the activity belts in several "streams". Those with the polarity of the active-region follower fields ultimately result in the reversal of the polar fields. The change in the tilt of the Sun's axis introduces an annual short-term variation in the polar fields. (These NSO/Kitt Peak data were produced cooperatively by NSF/NOAO, NASA/GSFC, and NOAA/SEC.) Source of figure and caption: www.hao.ucar.edu/smi/SMI_plate1.html; see [324]

6.6.2 A Simple Model of the Solar Cycle

Evidently, the clue to understanding the solar cycle lies in the magnetic field and its reversal. Helpful information, in particular for a proof of models, can be found in the details of sunspots:

1. Sunspots emerge at relatively high latitudes and move towards the equator (Spörer's law). During the solar cycle the latitude of emergence also moves towards the equator.
2. Sunspots are observed in bipolar groups with the leading spot (in the direction of apparent motion and closest to the equator) having the same polarity as the hemisphere it appeared in while the following spot has the opposite polarity (Hale's polarity law). The bipolar groups in opposite hemispheres have opposite magnetic orientation and this orientation reverses in each new solar cycle.
3. The tilt angle of the active regions is proportional to the latitude (Joy's law), which is actually a small deviation from Hale's law.

The motion of the sunspots reveals another important property of the Sun, namely its differential rotation: the Sun rotates faster at its equator and slower at the poles with a sidereal rotation time of 26.8°/day at the equator and 31.8° at 75° latitude (see Table 6.1).

The Solar Dynamo – Basic Idea. The source of solar activity is a MHD dynamo, as first proposed by Babcock [17]. While the details of this process are not completely understood, the underlying principle seems to be valid. The dynamo process works within or at the bottom of the convection zone, most probably at the tachocline where most of the shear is concentrated. During the solar minimum the Sun's magnetic field is poloidal. Differential rotation winds it up to a toroidal field, as already described (Sect. 3.6). The magnetic field is then concentrated in flux tubes with radii of a few hundred kilometers and magnetic field strengths between a few hundred and about 2000 G. The bulk motion in the convection zone twists the field lines, locally increasing the magnetic field strength to up to some thousands of gauss. This high flux density leads to magnetic buoyancy driving up magnetic flux ropes through the photosphere (see Fig. 6.29).

The increased magnetic flux inhibits convection; thus less heat is transported towards the photosphere and the regions of high magnetic flux are cooler. Where the flux tubes intersect the photosphere, bipolar sunspots emerge with the polarity pattern required by Hale's law. The latitude of the first appearance of sunspots is determined by the interplay of differential rotation and magnetic field strength. With the emergence of sunspots the magnetic pressure locally is reduced and the process continues at lower latitudes, leading to the motion of sunspots towards the equator during the solar cycle. Meridional flows in the convection zone combined with magnetic field diffusion and dispersion drive the leading spot towards the equator while the

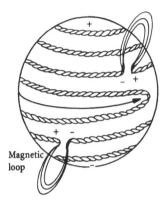

Magnetic
loop

Fig. 6.29. Increased magnetic buoyancy drives
flux tubes through the photosphere, creating
two sunspots of opposite polarity. Reprinted
from K. Lang [308], *Sun, earth, and sky*, Copyright 1995, Springer-Verlag

following spot stays behind. The leading spots of the opposite hemispheres
converge at the equator and dissolve by reconnection. The following spots
undergo reconnection with the polar fields. Since the polar field is slightly
smaller than the field accumulated in the following spots, the polarity of
these spots eventually takes over and the magnetic field is reversed. The possibility of this process is supported by the lanes of different than the prevalent
polarity of the hemisphere at higher latitudes, which can be seen in Fig. 6.28.

This last step, i.e. the pole reversal, is understood least. An alternative
explanation for the polarity reversal is the α-effect discussed in Sect. 3.6 that
creates a toroidal current which in turn gives rise to a poloidal magnetic field
of opposite polarity.

The Solar Dynamo – Details. Two important details in the solar cycle
discussed above are the emerging of magnetic flux in an active region and
the treatment of the α-effect in different dynamo models.

The rise of a flux tube from the tachocline through the photosphere into
the chromosphere or even the corona is influenced predominately by an interplay between magnetic buoyancy, aerodynamic drag, and the Coriolis force.
For an isolated horizontal flux, the pressure balance is

$$p_i + \frac{B^2}{2\mu_0} = p_e \,, \tag{6.25}$$

p_i and p_e are the internal and external gas pressures and $B^2/2\mu_0$ is the
magnetic pressure. Equation (6.25) implies $p_i < p_e$. Expressing the pressure
by the gas law, we can rewrite (6.25) and obtain

$$\varrho_i RT + \frac{B^2}{2\mu_0} = \varrho_e RT \tag{6.26}$$

or, after rearrangement,

$$\frac{\varrho_e - \varrho_i}{\varrho_e} = \frac{B^2}{2\mu_0 p_e} \,. \tag{6.27}$$

As a consequence, the fluid in the flux tube has a lower density than its surroundings and thus the flux tube must be buoyant. This effect is termed magnetic buoyancy. The combination of magnetic buoyancy and aerodynamic drag then determines the details of the rise and, later, the emergence of the flux tube. The Coriolis force does not influence the lifting of the tube but modifies its shape: it twists the rising flux tube from a plane loop into a three-dimensional structure where the area enclosed by the loop has an S-shape in the vertical direction. The inclusion of the Coriolis force allows us to understand Hale's and Joy's laws.

As mentioned in Sect. 3.6, differential rotation and magnetic buoyancy lead to the poloidal field and emerging flux in active regions, while the field reversal requires an additional process to work, the α-effect. Models using the mean-field theory differ in their treatment of this α-effect. These models include cyclic convection, in which α is positive in the unstable layer and negative in the tachocline or overshoot layer below; magnetostrophic waves, which imply a negative α; flux loops which correspond to a positive α; and unstable global-scale Rossby waves.[4] The various models can be summarized as follows [406]:

- Overshoot layer models (OL dynamos), also called co-spatial wave models, combined with $\alpha < 0$, give the correct migration direction and thus are able to model the butterfly diagrams, but tend to give cycle periods that are too short.
- Distributed wave models (IF dynamos) require an abrupt spatial change in diffusivity to excite dynamo waves. So far, strong toroidal fields can be produced in these models.
- Co-spatial transport models (CP dynamos) can describe the field migration in terms of density pumping or advection of the magnetic field but do not address the origin of the deep toroidal field.
- Distributed transport models (BL dynamos) start from an entirely different position: not a dynamo wave but a conveyor belt mechanism is responsible for the emergence and evolution of sunspots as seen in the butterfly diagram. These models can reproduce the confinement of active regions to low latitudes and describe the migration patterns of sunspots; however, present models require an unrealistically low turbulent diffusivity.

If we disregard the many problems with the details of the process, however, we can nonetheless give a simple formal approximation to the solution suggested by Parker [390] that outlines the main features required to describe

---

[4] A Rossby wave is a standing wave that slowly drifts in a fluid layer in a rotating body. The Rossby wave results from an interplay of a thermal or pressure gradient that drives a convection cell, and the Coriolis force. A prominent example of a Rossby wave is the undulating polar jet in the terrestrial atmosphere that guides the pressure regimes and thus is responsible for the weather. A simple laboratory experiment to produce a plane Rossby wave uses a rotating cylindrical tank with a temperature gradient between the outer wall and the axis.

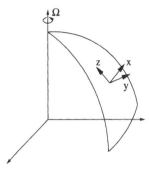

Fig. 6.30. Local Cartesian coordinate system at a point in the northern hemisphere of a spherical rotating body. Note that the coordinate system rotates with the body, and thus the unit vectors vary with time

the dynamo and the solar cycle, following [101]. The starting point is (3.135). Since we shall deal with average quantities only, we shall omit all brackets indicating averages, and $\boldsymbol{B}$ and $\boldsymbol{u}$ are meant to be average quantities. In addition, we assume that viscosity is determined by the turbulent viscosity. Rearrangement of (3.135) then gives

$$\frac{\partial \boldsymbol{B}}{\partial t} = \nabla \times (\boldsymbol{u} \times \boldsymbol{v}) + \nabla \times (\alpha \boldsymbol{B}) + \beta \nabla^2 \boldsymbol{B} \ . \tag{6.28}$$

We choose a Cartesian coordinate system fixed on the surface of the Sun as sketched in Fig. 6.30: the y-axis points radially outwards, the x-axis points in the toroidal direction (east–west direction), and the z-axis points in the direction of increasing latitude.

To be in agreement with the observed butterfly diagram, we need an equatorwards-propagating wave, that is, a wave propagating in the negative z-direction. In addition, the solution should be symmetric with respect to the rotation axis. In a local Cartesian system, this implies $\partial / \partial y = 0$.

The toroidal magnetic field is then simply $B_y \boldsymbol{e}_y$, while the poloidal field lies in the xz-plane. Since this field is solenoidal, it has zero divergence and can be written as the rotation of a scalar field $A(x, z)$:

$$\boldsymbol{B} = B_y(x, z)\boldsymbol{e}_y + \nabla \times [A(x, z)\boldsymbol{e}_y] \ . \tag{6.29}$$

The mean velocity field results from the differential rotation and thus has a component in the y-direction: $\boldsymbol{v} = v_y(x)\,\boldsymbol{e}_y$. The velocity shear then is $G = \partial v_y / \partial x$. Equation (6.28) can then be written as

$$\frac{\partial B_y}{\partial t} = G B_x - \alpha \nabla^2 A + \beta \nabla^2 B_y \ , \tag{6.30}$$

where α is constant. The x- and z- components of (6.28) can be written as

$$\nabla \times \left(\frac{\partial A}{\partial t}\boldsymbol{e}_y - \alpha B_y \boldsymbol{e}_y - \beta \nabla^2 A \boldsymbol{e}_y\right) = 0 \ . \tag{6.31}$$

This equation can be solved is the following condition is satisfied:

$$\frac{\partial A}{\partial t} = \alpha B_y + \beta \nabla^2 A \ . \tag{6.32}$$

Thus, for a field of the form (6.29), the dynamo equation (6.28) is satisfied if A and B_y satisfy (6.30) and (6.32). Thus instead of the dynamo equation (6.28), we can solve the two equations (6.30) and (6.32) simultaneously.

Equation (6.32) describes the evolution of the poloidal field. Without the α-term this would be a simple diffusion equation: any poloidal field would just diffuse away. The α-term works as a source, generating new poloidal field out of the turbulence, since, as discussed in Sect. 3.6, α is a measure of the turbulent motion. Equation (6.30) describes the evolution of the toroidal field. It has two sources: the first results from the velocity shear of the differential rotation, and the second stems from the turbulent motion (as evident from the α contained in this term), because just as the helical motion can twist the toroidal field to produce a poloidal field, it can also twist a poloidal field to produce a toroidal one. If the differential rotation is strong, this second term can be neglected and (6.30) reduces to

$$\frac{\partial B_y}{\partial t} = G\frac{\partial A}{\partial z} + \beta\nabla^2 B_y \ . \tag{6.33}$$

This equation describes the $\alpha\Omega$ dynamo and is a reasonable approximation for the solar dynamo. If differential rotation were weak, the other source term in (6.30) would be dominant and we would get an α^2 dynamo.

The $\alpha\Omega$ dynamo is thus defined by (6.32) and (6.33). These equations can be solved by wave-like solutions. The ansatz is

$$A = A_0 \exp(\omega t + \mathrm{i}kz) \quad \text{and} \quad B_y = B_0 \exp(\omega t + \mathrm{i}kz) \ . \tag{6.34}$$

Substituting in (6.32) and (6.33), we obtain

$$(\omega + \beta k^2)A_0 = \alpha B_0 \quad \text{and} \quad (\omega + \beta bk^2)B_0 = -\mathrm{i}kGA_0 \tag{6.35}$$

or, combined,

$$(\omega + \beta k^2)^2 = -\mathrm{i}k\alpha G \ , \tag{6.36}$$

which has the solution

$$\omega = -\beta k^2 \pm \left(\frac{\mathrm{i}-1}{\sqrt{2}}\right)\sqrt{k\alpha G} \ . \tag{6.37}$$

For maintenance of the dynamo process, the real part of ω must be larger than zero: $\Re(\omega) > 0$. In addition, k is taken to be positive. The crucial term is the product αG, which gives the combined effects of helical motion and differential rotation. For $\alpha G > 0$ and $\Re(\omega) > 0$, we obtain from (6.37)

$$\omega = -\beta k^2 + \sqrt{\frac{k\alpha G}{2}} - \mathrm{i}\sqrt{\frac{k\alpha G}{2}} \quad \text{or} \quad \Re(\omega) = -\beta k^2 + \sqrt{\frac{k\alpha G}{2}} \ . \tag{6.38}$$

We can now define a dynamo parameter

$$N_{\mathrm{d}} = \frac{|\alpha G|}{\beta^2 k^3} \; . \tag{6.39}$$

From (6.38), we see that the condition for dynamo growth is $N_{\mathrm{d}} \geq 2$. The eigenmodes of the marginally stable ($N_{\mathrm{d}} = 2$) dynamo,

$$A, B_y \sim \exp\left\{ -\mathrm{i}\sqrt{\frac{k\alpha G}{2}}\, t + \mathrm{i}kz \right\} , \tag{6.40}$$

correspond to waves propagating in the positive z-direction, that is, polewards.

For $\alpha G < 0$,we obtain from (6.37)

$$\omega = -\beta k^2 + \sqrt{\frac{k|\alpha G|}{2}} - \mathrm{i}\sqrt{\frac{k|\alpha G|}{2}} \quad \text{or} \quad \Re(\omega) = -\beta k^2 + \sqrt{\frac{k|\alpha G|}{2}} \; . \tag{6.41}$$

Dynamo growth is again described by the dynamo parameter (6.39), but the marginally stable solutions now become

$$A, B_y \sim \exp\left\{ \mathrm{i}\sqrt{\frac{k|\alpha G|}{2}}\, t + \mathrm{i}kz \right\} , \tag{6.42}$$

which correspond to an equatorwards-propagating wave. Thus for $\alpha G < 0$, we obtain a solution of the dynamo equation that accounts for both the periodicity and the equatorwards propagation of the solar magnetic field.

Nonetheless, we should be aware that these solutions have been obtained under simplified conditions. In particular, α, β, and G are assumed to be constant, which is certainly not true under realistic conditions. In a realistic scenario, one would have to solve the dynamo equation in a finite region with suitable boundary conditions. In addition, we have assumed that the mean velocity $\boldsymbol{u}$ is purely in the toroidal direction. However, with a suitable velocity field, an equatorwards-propagating wave can be obtained even if $\alpha G > 0$ [102].

6.6.3 Stellar Activity

Owing to its close proximity, the Sun is the only star whose surface we can see directly. Thus we are able to identify such features as sunspots, filaments, flares, and solar cycles, all indicating that the sun is a magnetically active star. Magnetic activity is not unique to the Sun; it is not even strong on the Sun. Other stars can show much stronger magnetic activity. Although spots and flares cannot be observed directly, the Ca II line, which is a strong indicator of the magnetic network on the Sun, can be studied in other stars, too. This has led to the identification of magnetic activity in other stars [25, 429]. Cyclic variations and also magnetically flat stars could be identified. Since many of these stars have shorter star cycles than the Sun, it could be shown that the variations in the solar cycle are small compared with the variations in the magnetic cycles of comparable stars. The other active stars, therefore, are often used to infer information about long-term variations in solar activity [326].

A large number of articles on various aspects of solar and stellar activity can be found in [389, 466, 564].

6.7 Flares and Coronal Mass Ejections

Flares and coronal mass ejections (CMEs) are violent manifestations of solar activity. Both are related to the solar magnetic field, sunspots, and filaments. The energy released in these processes had been stored in the field.

The first record of a solar flare dates back to Carrington in 1859, who observed a sudden brightening of a sunspot in white light. After this "explosion", the sunspot structure had changed and about a day later a violent geomagnetic storm with strong auroral activity was observed. Although Carrington himself noted "a swallow does not make a summer", this was the first direct link between solar activity and its influence on Earth. Coronal mass ejections only have been observed since the early 1970 with the advance of space-borne coronographs. Today, one of the most controversial topics is the relation between flares and CMEs.

Flares normally are associated with active regions and sunspots. Nonetheless, even during sunspot minimum on a spotless Sun sporadic flare events of large magnitude can occur. Such flares observed during solar minimum in general are associated with erupting filaments rather than with sunspots and active regions. In addition, there tend to be fewer flares per solar rotation at times of the highest sunspot number during a solar maximum [302]. For a review on the magnetic nature of solar flares see [423].

In some cases, flares occur repeatedly in the same location and display similar spatial structures during the evolution of the active region. Such flares are called homologous [179, 566]: only part of the energy stored in the magnetic field is released and the overall topology of the field is retained. These flares are not necessarily small because the energy release depends on the available magnetic energy and magnetic stability. If enough energy is stored in the field even the release of only a relatively small fraction of it can lead to a considerable flare.

Some major flares also can trigger flares in their neighbourhood or even in remote active regions. Such flares are called sympathetic, see e.g. [361, 570]. Sympathetic coronal mass ejections also can be observed [362].

6.7.1 Electromagnetic Radiation

A flare is the result of a sudden violent outburst of energy, with energies of up to 10^{25} J being released over a time period of some minutes. White-light flares, as the one observed by Carrington, are rare because even for the largest flares the brightness is less than 1% of the total luminosity of the photosphere. In certain frequency ranges, e.g. at the wings of the black body, the intensity of the electromagnetic radiation can increase by orders of magnitude during a flare. While the flare is defined as an outburst in electromagnetic radiation, often it also is associated with the emission of energetic particles (Sect. 7.2) and huge plasma clouds, the CMEs.

The electromagnetic radiation released in a flare in different frequency ranges shows typical time profiles (see Fig. 6.31). These profiles can be used to define the phases of a flare during which distinct physical processes occur.

In a large flare, the electromagnetic emission can be divided into three parts. In the preflare phase, also called the precursor, the flare site weakly brightens in soft X-rays and Hα. This phase lasts for some minutes; it is observed in very large flares only, and indicates a heating of the flare site. During the impulsive or flash phase, most of the flare energy is released and the harder parts of the electromagnetic spectrum, such as hard X-rays and γ-rays, are most abundant. This phase lasts for a few minutes and can be followed by a gradual or extended phase during which emission mainly occurs in Hα and soft X-rays but microwave and radio emission also can continue. This latter phase can last for some tens of minutes, and occasionally even for a few hours. It is present in the larger events only. Note that most flares are rather small, consisting of an impulsive phase only.

The impulsive phase is related to an impulsive energy release, probably reconnection, inside a closed magnetic field loop in the corona, heating the coronal plasma and accelerating particles. The heated plasma emits soft X-rays. Accelerated electrons generate microwaves and hard X-rays at the top of the loop or hard X-rays, γ-rays, UV emission, and part of the Hα emission at its footpoints, while accelerated ions generate γ-ray line emission. Streaming electrons also generate radio emission. Thus the electromagnetic radiation also provides information about the acceleration and propagation of particles accelerated in a flare.

Soft X-Rays and Hα. In a solar flare most of the electromagnetic radiation is emitted as soft X-rays with wavelength between 0.1 and 10 nm. Soft X-rays originate as thermal emission in hot plasmas with temperatures of about 10^7 K. Most of the radiation is continuum emission; lines of highly ionized O, Ca, and Fe are also present. In a large flare, soft X-rays are emitted during all three phases; the strong increase at the beginning of the impulsive phase is related to an abrupt increase in the temperature at the flare site up to about 5×10^7 K. The Hα emission also is thermal emission, its time profile closely follows the soft X-ray profile.

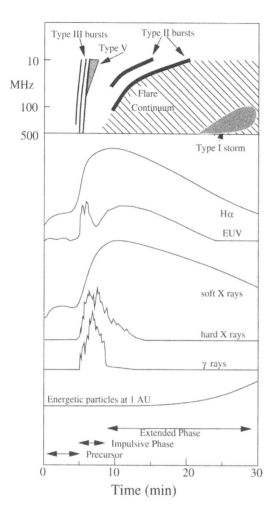

Fig. 6.31. Solar electromagnetic radiation during the different phases of a flare and frequency spectrum of radio bursts, based on [278]

Hard X-Rays. Hard X-rays are photons with energies between a few tens of kiloelectronvolts and a few hundred kiloelectronvolts, generated as bremsstrahlung of electrons with slightly higher energies [279, 304, 535]. Only a very small amount of the total electron energy (about 1 out of 10^5) is converted into hard X-rays.

Microwaves. Solar microwave emission is generated by the same electron population as the hard X-rays, as can be deduced from the similarities in the time profiles [34, 264, 379]. As in hard X-rays, during the impulsive phase the emission often consists of individual spikes, the 'elementary flare bursts'. Microwave emission is gyro-synchrotron radiation of electrons with energies between some 10 keV and some 100 keV [430, 515]. When these mildly relativistic particles gyrate in the coronal magnetic field (about 20–100 G), they emit radiation with frequencies between 10 and 100 times their gyro-frequency. This dependence on the magnetic field strength has a remarkable consequence: as the magnetic field decreases with height, microwave emission at a certain frequency, say 17 GHz, is generated by electrons with energies above 200 keV in low lying flares or small loops, while in a loop extending high into the corona an electron energy of more than 1 MeV is required [296]. While in the impulsive phase the elementary bursts give evidence for individual, isolated energy releases, in the gradual phase the microwave emission most probably is thermal emission.

γ-**Rays.** γ-ray emission indicates the presence of energetic particles. The spectrum can be divided into three parts: (a) Bremsstrahlung of relativistic electrons and, to a lesser extent, the Doppler broadening of closely neighbored γ-ray lines leads to a γ-ray continuum. (b) Nuclear radiation of excited CNO-nuclei leads to a γ-ray line spectrum in the range 4 to 7 MeV. (c) Decaying pions lead to γ-ray continuum emission above 25 MeV. The details of these mechanisms are described in [370, 371, 431]

The most important γ-ray lines are at 2.23 MeV and 4.43 MeV. The 2.23 MeV line is due to neutron capture in the photosphere: ^{4}He-nuclei decay in p/α or α/p interactions in the corona, emitting neutrons. These reactions require particle energies of at least 30 MeV/nucl. The neutrons can propagate independently of the coronal magnetic field. Elastic collisions decelerate neutrons penetrating into the denser regions of the chromosphere or photosphere. Eventually, the neutron is slowed down to thermal energies and can be captured by ^{1}H or ^{3}He. Neutron capture by ^{1}H leads to the emission of a γ-quant. The 4.43 MeV line results from the transition of a ^{12}C nucleus from an excited into a lower state, with the excitation being either due to nuclear decay or inelastic collisions with energetic particles.

Radio Emission. Electrons streaming through the coronal plasma excite Langmuir oscillations. Solar radio bursts are metric bursts; their wavelengths are in the meter range. In interplanetary space, radio bursts are kilometric bursts. The bursts are classified depending on their frequency drift [43, 351]. The type I radio burst is a continuous radio emission from the Sun, basically the normal solar radio noise but enhanced during the late phase of the flare. The other four types of bursts can be divided in fast and slow drifting bursts or continua (see the upper panel in Fig. 6.31).

The type III radio burst starts early in the impulsive phase and shows a fast drift towards lower frequencies. Since the frequency of a Langmuir oscillation depends on the density of the plasma (see (4.52) and example 17), the radial speed of the radio source can be determined from this frequency drift using a density model of the corona. The speed of the type III burst is about $c/3$, it is interpreted as a stream of electrons propagating along open field lines into interplanetary space. Impulsive peaks in the hard X-ray emission can be related to individual type III bursts, indicating individual energy releases. Occasionally, the drift of the type III burst is suddenly reversed, indicating electrons captured in a closed magnetic field loop: as the electrons propagate upward, the burst shows the normal frequency drift which is reversed as the electrons propagate downward on the other leg of the loop.

In the metric type II burst, the frequency drift is much slower, indicating a radial propagation speed of its source of about 1000 km/s. It is interpreted as evidence of a shock propagating through the corona. Nonetheless, it is not the shock itself that generates the type II burst but the shock-accelerated electrons. As these electrons stream away from the shock, they generate small, type III-like structures, giving the burst the appearance of a herringbone (herringbone burst) in the frequency time diagram with the type II as the backbone and the type III structures as fish-bones. The type II burst is split into two parallel frequency bands, interpreted as forward and reverse shocks.

The metric type IV and V bursts are continuous emission directly following the type II and type III bursts, respectively. The type IV burst is generated by gyro-synchrotron emission of electrons with energies of about 100 keV. It consists of two components: a non-drifting part generated by electrons captured in closed magnetic field loops low in the corona, and a propagating type IV burst generated by electrons moving in the higher corona. The type V burst is a similar burst following the type III burst. But in contrast to the type IV burst it is stationary, showing no frequency drift. Most likely, it is radiation of the plasma itself.

Kilometric radio bursts in interplanetary space are interpreted in the same way: type III bursts show a fast frequency drift, indicating electrons streaming along a magnetic field line. If the location of the radio source can be identified, this kilometric type III burst can be used to trace the shape of the interplanetary magnetic field line [282]. The kilometric type II burst gives evidence for a shock propagating through interplanetary space [83].

Solar Quakes Produced by Large Flares. An only recently discovered by-product of flares is a circular wave packet emanating from the flare site. This wave was first observed by the Michelson Doppler Imager (MDI) on board SOHO during the large flare of June 1996 [295]. This magneto-acoustic wave can be interpreted as a kind of solar quake, containing about four orders of magnitude more energy than the 1906 San Francisco earthquakes. The waves of this quake were similar to surface waves on a pond produced by a stone. The waves accelerated from 10 km/s to about 115 km/s during their outward propagation until they finally disappeared in the photosphere.

These solar quakes should not be confused with Moreton waves. The latter are the chromospheric component, seen in $H\alpha$ radiation, of a solar-flare-induced wave that propagates away from the flare site at a roughly constant speed of about 1000 km/s. Moreton waves are attributed to fast-mode MHD shocks generated in the impulsive phase of a flare.

6.7.2 Classes of Flares

Not all the phases indicated in Fig. 6.31 can be observed in all flares. Instead, flares differ in their electromagnetic radiation, in the acceleration of energetic

Table 6.2. Classes of solar flares

	Impulsive	Gradual	
Duration of soft X-rays	< 1 h	>1 h	[387]
Decay constant of soft X-rays	< 10 min	>10 min	[104, 387]
Height in corona	$\leq 10^4$ km	$\sim 5 \cdot 10^4$ km	[387]
Volume	$10^{26} - 10^{27}$ cm^3	$10^{28} - 10^{29}$ cm^3	[387]
Energy density	high	low	[387]
Size in Hα	small	large	[22]
Duration of hard X-rays	<10 min	>10 min	[383]
Duration of microwaves	<5 min	>5 min	[119]
Metric type II burst	75%	always	[84]
Metric type III burst	always	50%	[84]
Metric type IV burst	rare	always	[84, 292]
Coronal mass ejection	rare	always	[84]

particles, and the association with a coronal mass ejection. Standard classifications based on the magnetic structure and the energy release are given in [271, 435, 542].

A useful, although frequently modified, classification scheme for solar flares goes back to Pallavicini et al. [387] who used Skylab soft X-ray images of the Sun, combined with intensity–time profiles. If a flare is observed on the solar limb, the height profile of the electromagnetic emission can be inferred. These limb flares can be divided into three distinct groups: (a) point-like flares, (b) flares in small and compact loop structures, and (c) flares in large systems of more diffuse loops. Flares of classes (a) and (b) are associated with a short duration of the soft X-ray emission, less than one hour, while in flares of class (c) the soft X-ray emission can last for some hours. Therefore, the compact and point-like flares are called impulsive, and the flares in the large diffuse loop are called gradual flares.

The classification scheme, originally introduced for the soft X-rays only, over the years has been extended to other ranges of electromagnetic radiation, as summarized in Table 6.2. These schemes do not always agree. On the basis of the times scales, a flare might appear gradual in soft X-rays but impulsive in hard X-rays or vice versa. These phenomenological criteria provide no sharp separation into two classes but rather a continuous transition from more impulsive to more gradual flares. A better criterion, also pointing to the physical difference, is the occurrence of a coronal mass ejection, leading to an unambiguous classification of flares into confined (corresponding to impulsive) and eruptive (corresponding to gradual).

We have to be careful not to confuse the classes of flares, i.e. impulsive and gradual, with the phases of flares bearing the same name [23]. An impulsive flare appears to be rather simple in so far as it always has an impulsive phase. However, in some small events, which are observed in interplanetary space as so-called ³He-rich events [239], even the impulsive phase is rather small with the Hα flare often too small to be detected although hard X-ray and/or radio emission is observed [437]. Some large impulsive events can also show a small gradual phase. In the larger flares (longer duration, larger volumina) the situation is even more complicated. While the gradual phase is well developed, an impulsive phase is not always present, only the largest gradual flares show all three phases. The gradual flares with a gradual phase only in general are small in electromagnetic emission, their main characteristic is the coronal mass ejection. These solar events are often called disappearing filaments because the expelled matter is their main signature while classic flare emission is weak or absent.

6.7.3 Coronal Mass Ejections

With the aid of a coronograph, the corona can be observed continuously. Basically, a coronograph is a telescope with an occulter screening off the direct photospheric emission. Ground-based coronographs have been in use since the 1930s, observing only selected coronal emission lines. Space-based coronographs, on the other hand, observe the light scattered by the corona. The first space-based coronograph was used on Skylab in 1973/74; the most advanced coronograph is on SOHO [63, 161]. While the older coronographs had a field of view from a height of about 1.5 $r_\odot$ out to 5 or 10 $r_\odot$, the combination of different telescopes in the LASCO coronograph on SOHO has a field of view from 1.1 $r_\odot$ out to about 30 $r_\odot$. In addition, its resolution is much better, thus smaller and fainter mass ejections have been detected. Examples and further details can be found at sohowww.nascom.nasa.gov/gallery/LASCO/ or lasco-www.nrl.navy.mil/lasco.html.

The most striking feature visible in a record of coronograph images is the coronal mass ejection (CME). A coronal mass ejection is a bright structure propagating outward through the corona, as shown in Fig. 6.32. Large data bases on CMEs exist from the Solwind-Coronograph on P78, the HAO-coronograph on Solar-Maximum Mission (SMM), and the LASCO coronograph on SOHO. The basic features of CMEs can be summarized as follows [69, 236, 246, 369, 501]:

Solar Cycle Dependence. During solar maxima, about two CMEs are observed daily, whereas during solar minima one CME is observed per week [551]. This is not too surprising because CMEs are related to flares and filaments which both are more frequent during solar maximum. With the better spatial resolution of the LASCO coronograph these numbers increase, however, the ratio between solar maximum and minimum does not change.

Latitude Distribution. Coronal mass ejections are distributed evenly on both hemispheres, the average latitude is 1.5°N. Their distribution is flat within ±30° and decreases fast towards higher latitudes. The maximum in the ±30° region reflects the latitudinal distribution of sunspots and flares. During solar minimum, the CMEs cluster within ±10° around the equator.

Width. The projected widths of CMEs show a distribution with an average at 46° and a median at 42°. CMEs smaller than 20° or larger than 60° are rare; however, the largest angular extent is more than 120°. The width of CMEs seems to be independent of the solar cycle.

Speeds. CME speeds range from less than 10 km/s up to greater than 2000 km/s with an average of 350 km/s and a median at 285 km/s. The speeds of CMEs do not depend on the solar cycle. Fast and slow CMEs seem to reveal different patterns of energy release: a slow CME can accelerate in the coronograph's field of view, indicating a continuous energy release. A fast CME, on the other hand, does not show evidence for acceleration. Here the energy release appears to be more explosive.

Kinetics. During a CME between 2×10^{14} g and 4×10^{16} g coronal material is ejected with a kinetic energy content between 10^{22} J and 6×10^{24} J. This is comparable with the energy released as electromagnetic radiation in a flare. The combined potential and kinetic energy of the CME is at least comparable with the entire energy released in a flare.

Structure. Textbook coronal mass ejection are loop-like structures as shown in Fig. 6.32. While these are the most impressive and also the most energetic CMEs, other morphologies exist, too, with such picturesque names as spikes, multiple-spikes, clouds, fans, or streamer blow-outs [236]. The basic difference is a smaller extent and a structure distinct from a closed loop. One example will be discussed in connection with Fig. 6.36.

Note that all the geometrical quantities discussed above are apparent quantities only: while the coronal mass ejection is a three-dimensional structure, its image is only a two-dimensional projection into a plane perpendicular to the Sun–Earth axis. Thus sizes and speeds might be underestimated. In particular, if the CME propagates directly towards the observer, its extent and speed cannot be determined. It is even more difficult to detect because it only becomes visible as a halo around the Sun after it has spread far enough.

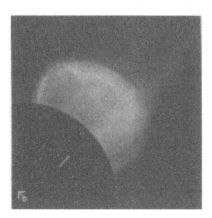

Fig. 6.32. Image of the 4 May 1986 CME, taken by the HAO coronograph. The dark disk in the lower left corner is the occulting disk inside the coronograph, and the dashed circle gives the photosphere. The arrow in the center of the Sun points towards the north. Note that this CME is observed during solar minimum conditions. Reprinted from S.W. Kahler and A.J. Hundhausen [263], *J. Geophys. Res.* **97**, Copyright 1992, American Geophysical Union

Owing to this projection effect, the three-dimensional structure of a CME is debatable: it is not clear whether the leading bright structure is a loop or a bubble. Here observations from two spacecraft from different positions are required, such as planned with the STEREO mission. For details of the mission see `stp.gsfc.nasa.gov/missions/stereo/stereo.htm`, details of the coronograph can be found at `wwwsolar.nrl.navy.mil/STEREO/index.html`.

6.7.4 Coronal Mass Ejections, Flares, and Coronal Shocks

The speeds of coronal mass ejections are highly variable. In Fig. 6.33 the range of CME speeds is compared with the Alfvén and the sound speed. In the corona, most CMEs are too slow to drive a fast MHD shock wave. Nonetheless, because many of the CMEs are still faster than the sound speed, they might drive slow or intermediate MHD shocks. In at least one CME the curvature of the loop suggests a slow shock [248].

In Sect. 6.7.1 we have learned about the metric type II burst, interpreted as a shock wave propagating through the corona. The relationship between type II bursts and CMEs is ambiguous. From a statistical study of CMEs and metric type II bursts [476], it is evident that about two-thirds of the metric type II bursts are accompanied by a fast CME. However, there are also metric type II bursts without CMEs (one-third) as well as CMEs without metric type II burst (three-fifth). In particular, half of the CMEs without type II bursts are fast CMEs, with speeds above 450 km/s. The situation becomes even worse if CMEs and type II radio bursts are compared in individual events. In some events, the CME's radial speed is markedly lower than the speed

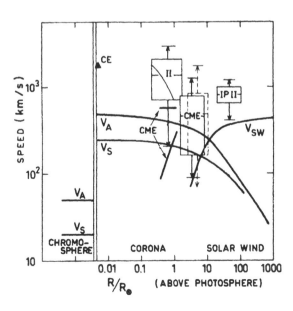

Fig. 6.33. Alfvén and sound speed in the corona and in interplanetary space compared with the solar wind speed v_{SW} and speeds of coronal mass ejections (CMEs), coronal type II bursts, and interplanetary type II bursts. Reprinted from J.-L. Bougeret [57], in *Collisionless shocks in the heliosphere: reviews of current research* (eds. B.T. Tsurutani and R.G. Stone), Copyright 1985, American Geophysical Union

of the radio burst; in other events the source of the metric type II emission is located behind the CME, occasionally overtaking it at later times. Thus a CME is neither a necessary nor a sufficient condition for a coronal shock. Nonetheless, the more energetic CMEs most likely drive a coronal shock.

The relationship between flares and coronal mass ejections also is a hotly debated topic. Traditionally, the CME has been viewed as a phenomenon accompanying large flares. Today, it is suggested that the CME is the primary energy release while the flare is just a secondary process [200]. Both models have their pros and cons, currently we probably should follow the suggestion in [152] that the flare neither is the cause nor the consequence of the CME but that both are triggered by a common mechanism, probably an instability (see below). The observations leading to this statement are as follows.

Flares and CMEs can occur together; however, both also can occur separately: about 90% of the flares are not accompanied by a CME, while about 60% of the CMEs go without a flare. The combined flare and CME events are the most energetic events in both groups. In these events the flare, which is small compared with the angular extent of the CME, is not necessarily centered under the CME but more likely is shifted towards one of its legs.

The prime argument for the CME being the cause and the flare the consequence is based on energetics: the energy released in the CME is larger than the one released in the flare. But the mechanism of the energy release is different, too: if a CME is accompanied by a flare, it has a high and constant speed, indicative of an explosive energy release. A CME without flare, on the contrary, often accelerates, indicating that energy is released continuously.

Timing is another crucial factor in this discussion: in about 65% of the combined CME/flare events, the CME leads, while in the other 35% the flare starts before the CME.

Combined, all these observations suggest that it might be difficult to view flares and CMEs in terms of cause and consequence and they favour a picture of a common trigger.

6.7.5 Models of Coronal Mass Ejections (CMEs)

Many CMEs originate in filaments, and the magnetic field pattern of the filament can even be recognized if the CME is detected in interplanetary space, see, for example, [56].

So far we have learned that magnetic pressure prevents the filament from "falling down" to the photosphere. But how and why can it suddenly be blown out so violently to form a CME? A detailed analysis of the magnetic field pattern of the filament and the photosphere reveals two different configurations. In the normal configuration (Kippenhahn and Schlüter [286], K-S configuration), the magnetic field inside the filament has the same direction as the photospheric field below it (left panel in Fig. 6.34). We have already used this model in example 13. The inverse configuration (Raadu and Kuperus [427, 428], R-K configuration), which is shown in the right panel, is

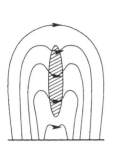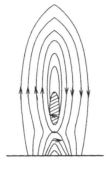

Fig. 6.34. Magnetic field configurations in solar filaments: (*left*) normal configuration, also called K-S configuration, (*right*) inverse configuration, also called R-K configuration

more complex: here the magnetic field inside the filament is directed opposite to the one in the photosphere. In particular, in the large, high rising filaments, which tend to give rise to CMEs, the R-K configuration seems to be dominant.

The crucial feature is obvious in the R-K configuration: below the filament there is an X-point configuration where magnetic fields of opposite polarity can be found in close proximity. Such a location is favorable for reconnection. The configuration might have been stable for a long time; however, the motion of the magnetic field lines anchoring the filament or a slow rise of the filament due to increased buoyancy can lead to the sudden onset of reconnection at the X-point. Then the magnetic field energy is converted into thermal energy and flow energy, leading to a further rise of the filament. At neighboring X-points, reconnection sets in, too, ripping off the filament from its anchoring structure and blowing it out as a CME. A filament of the K-S type can be expelled by the same mechanism, only the forces acting on the anchoring field lines must be larger to create a X-point configuration below the filament.

In the classical model of a filament, reconnection always takes place between the two legs of each field line. The resulting field configuration therefore is a closed loop below the filament and a toroidal field line around it. A different situation arises if the legs of neighboring field lines merge, as sketched in Fig. 6.35. As the filament (gray area) lifts owing to some instability, reconnection sets in. Since neighboring anchoring field lines reconnect, the filament is surrounded by a helical magnetic field. The plasma in front of the erupting structure is compressed. As it flows towards the trailing edge of the eruption, vortices are generated behind the arcade core. The vortices drive the plasma inwards and compress the current sheet below the filament. Therefore the current density increases to a value where kinetic plasma instabilities are excited. The increased resistivity leads to a higher magnetic diffusivity and a new reconnection line below the filament, leading to the formation of a secondary plasmoid below the original filament.

In this three-dimensional reconnection [202, 325], open field lines extending into interplanetary space can also be involved. Thus, particles accelerated in the reconnection region can easily escape into interplanetary space. In ad-

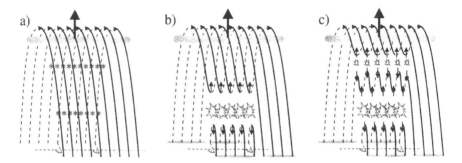

Fig. 6.35. 3-D reconnection below an idealized filament. Figure taken from B. Vrsnak [542], in *Lectures on solar physics* (eds. H.M. Antia, A. Bhatnagar, and P. Ulmschneider), Copyright 2003, Springer-Verlag

dition, the three-dimensional reconnection might evolve along the filament more slowly, leading a slow CME and a more continuous energy release.

The expulsion of a filament is suitable for explaining the loop-shaped CMEs. Other types of CMEs, however, require different geometries. One example is shown in Fig. 6.36: reconnection in the tip of a helmet streamer. Its outer portion consists of opposing magnetic fields. As somewhere on the Sun new magnetic flux emerges, the coronal magnetic field is deformed and at the current sheet reconnection sets in, blowing out an open magnetic field structure along the streamer. In the coronograph image, such a CME is seen as a spike- or jet-like structure. Although operating on a larger scale, this mechanism is the same as in the blobs of high-density slow solar wind discussed above.

6.7.6 Models of Flares

We can expand the above model of a loop-like CME to accommodate a flare. By definition, this would be an eruptive or gradual flare. Figure 6.37 shows again a filament in the inverse configuration. As reconnection sets in at the X-point, three different phenomena occur: (a) The plasma is heated, leading to thermal emission in the soft X-ray and visible ranges. (b) Particles are accelerated, either streaming upward along field lines extending into interplanetary space (solar energetic particles, Sect. 7.2) or downward producing the hard electromagnetic radiation. (c) The filament breaks loose and is ejected as a CME. If the CME is fast enough, it drives a shock wave. Particles accelerated at this shock escape into interplanetary space along open field lines. Because the reconnection occurs on all anchoring field lines, the flare occupies a large volume which extends rather high into the corona, thus fulfilling the criteria of a gradual flare as summarized in Table 6.2.

Large confined flares might be explained by the same model, the only difference would be in point (c): although reconnection sets in at the X-

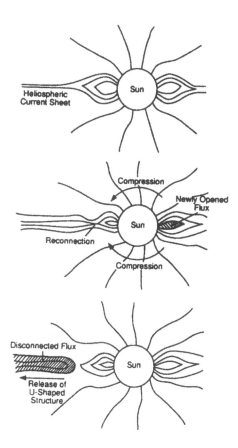

Fig. 6.36. Disconnection event: a CME with an open magnetic field structure is expelled along a helmet streamer. Reprinted from D.J. Mc-Cormas [343], *Geophys. Res. Lett.* **18**, Copyright 1991, American Geophysical Union

point of one or a few anchoring fieldlines, the energy release is not large enough to break all connections between filament and photosphere. Thus the filament is not ejected and the flare is confined to a smaller volume. Such a configuration in which the filament lifts but fails to detach and the overall magnetic topology is retained is a suitable candidate for a series of homologous flares.

In the point-like flares, which also are confined or impulsive flares, the observations of energetic particles in interplanetary space (Sect. 7.2) suggest a different scenario, as sketched in Fig. 6.38. Particles are accelerated inside a closed loop, giving rise to electromagnetic emission. The loop is very stable, preventing particles from escaping into interplanetary space. As the particles bounce back and forth in the closed loop, they excite electromagnetic waves which can propagate in all directions, interacting with the ambient plasma, even accelerating particles. If these "secondary" particles are accelerated on open field lines, they can escape into interplanetary space. Since the acceleration requires particles and waves to be in resonance, different particles are accelerated by different types of waves. If a particle species is common in

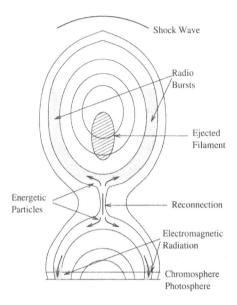

Fig. 6.37. Simplified model of a large eruptive flare. The energy release occurs in the reconnection region below the filament, leading to heating, particle acceleration, and a CME. The heated plasma and the energetic particles give rise to electromagnetic emission in various frequency ranges

the corona, such as H and ^{4}He, the waves in resonance with these particles are absorbed more or less immediately; thus these particles predominately are accelerated inside the closed loop and therefore do not escape into interplanetary space. Other waves, however, travel larger distances before being absorbed by the minor constituents, such as ^{3}He and the heavy ions, and are thus more likely to accelerate these species on open field lines. Since the escaping particle component is selectively enriched in these minor species, this acceleration process is called selective heating.

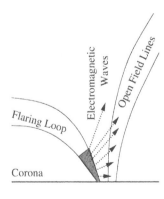

Fig. 6.38. Model of an impulsive ^{3}He-rich flare. Particles accelerated in a closed loop excite waves which in turn accelerate particles on open field lines. These latter can escape into interplanetary space

6.7.7 Magnetic Clouds: CMEs in Interplanetary Space

Coronal mass ejections in interplanetary space still carry the magnetic field patterns from their parent filament. These closed magnetic field structures, also called magnetic clouds, have features different from the ambient medium.

Figure 6.39 shows the magnetic field and plasma data for a typical magnetic cloud. Vertical lines indicate the boundaries of the cloud and the shock wave driven by it. The typical signatures of a magnetic cloud can be summarized as follows: (a) a decrease in magnetic field strength inside the cloud, (b) a rotation of the magnetic field vector, in particular its elevation, (c) decreases in plasma density, plasma speed, plasma temperature, and therefore plasma-β, and (d) a bi-directional streaming of suprathermal electrons back and forth along the length of the cloud (not shown in the figure).

The magnetic field configuration of such clouds can be inferred from the variation in magnetic field elevation: at the beginning of the cloud the magnetic field is almost perpendicular to the plane of ecliptic. Inside the cloud the elevation decreases until at the end the magnetic field vector is almost opposite to the one at the beginning. This is indicative of a magnetic field wrapped around the ejecta as sketched in Fig. 6.40. In this picture the magnetic cloud is sketched as a bundle of twisted magnetic field lines. The direction of field rotation varies from cloud to cloud, reflecting the field configuration of the parent filament. Just how long the magnetic cloud stays connected to the Sun

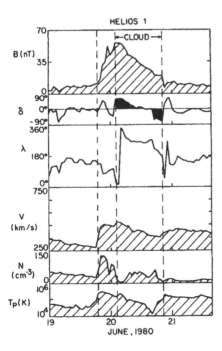

Fig. 6.39. Field and plasma data for an interplanetary shock and the magnetic cloud driving the shock. From top to bottom: magnetic field flux density, elevation, azimuth, solar wind speed, plasma density, and proton plasma temperature. The vertical lines indicate the shock and the boundaries of the magnetic cloud. From L.F. Burlaga [70], in *Physics of the inner heliosphere, vol. II* (eds. R. Schwenn and E. Marsch), Copyright 1991, Springer-Verlag, Berlin

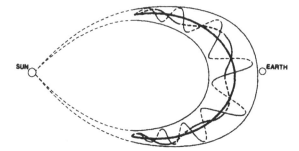

Fig. 6.40. Proposed topology of a magnetic cloud in interplanetary space. Reprinted from L.F. Burlaga [70], in *Physics of the inner heliosphere, vol. II* (eds. R. Schwenn and E. Marsch), Copyright 1991, Springer-Verlag

is still an open question, so therefore the dashed lines in Fig.6.40 indicate the possibility but not the necessity of such a continued connection.

Magnetic clouds are the main cause for geomagnetic disturbances, their geomagnetic effectiveness depends on whether the field at the leading edge has a strong northward or southward component (Sect. 8.5.2).

6.7.8 Interplanetary Shocks

In the event in Fig. 6.39 a shock has been observed in front of the magnetic cloud. But only about one-third of the CMEs in space drive an interplanetary shock [199], while apparently all travelling interplanetary shocks are driven by CMEs, although the magnetic cloud is not necessarily detected if the observer is located at the flank of the shock. Most shocks are observed around the solar maximum.

Interplanetary shocks are identified by characteristic changes in the plasma and field parameters, in particular a sudden increase in plasma density, speed, and temperature, and a jump in the magnetic field strength. Figure 6.41 shows two examples of shocks in the Helios data. From top to bottom, plasma temperature, plasma density, solar wind speed, and magnetic field strength are shown. Both examples can be identified best by the sudden jump in magnetic field strength. Here it is also obvious that each of the shocks consists of a forward shock (marked by a dashed line) and a reverse shock (marked by an arrow). Owing to the rather poor temporal resolution the shocks are more difficult to identify in plasma data. However, in the example on the right, the jumps in the plasma parameters at the forward shock are obvious. This is a fast, strong shock with a local shock speed of 1181 km/s. The example on the left, although the jump in magnetic field is by the same factor, is a slow, rather weak shock with a speed of 508 km/s, only slightly above the solar wind speed.

The properties of interplanetary shocks are highly variable. Between 0.3 AU and 1 AU, the basic characteristics are as follows:

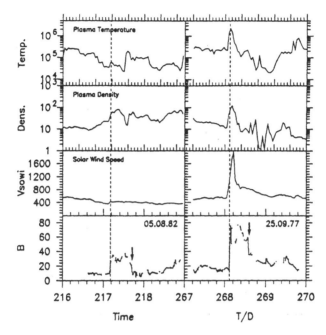

Fig. 6.41. Two examples for shocks observed by Helios 1. *Vertical dashed lines* mark the arrival of the forward shock; the *arrows* mark the reverse shock. Solar wind data from the MPAe Lindau experiment on board Helios, magnetic field data from the University of Braunschweig magnetometer on board Helios

- The compression ratio varies between 1 and 8 with an average close to 2.
- The magnetic compression (ratio between the upstream and downstream magnetic field strengths) varies between 1 and 7 with an average at 1.9.
- Shock speeds in the laboratory frame vary between 300 km/s and 700 km/s with an average of about 600 km/s. Occasionally, shock speeds above 2000 km/s can be observed. Obviously, shocks with speeds only slightly above 300 km/s can be observed in very slow solar wind streams only. Since the shock speed is lower towards the flanks, the tongue-like shape of the shock front closely resembles the shape of the leading edge of the magnetic cloud, as shown in Fig. 6.40.
- The angular extent of the shock varies between a few tens of degrees and up to 180°; the shock is always wider than the driving CME.
- The Alfvén Mach number is between 1 and 13 with an average at 1.7.

The shock parameters, of course, are related to the properties of the CME, such as speed, angular extent, and total energy released.

An interplanetary shock is a disturbance propagating into the expanding solar wind. The shock should develop absolutely because it expands, and also relative to the ambient medium as the latter expands differently. In particular, the expansion of the shock leads to a decrease in the plasma and magnetic

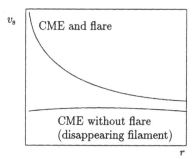

Fig. 6.42. Radial variation of the shock speed from the Sun to 1 AU. If the CME is very fast close to the Sun, shock and CME slow down during propagation. This is generally the case when the CME is accompanied by a flare. If the CME is slow close to the Sun, as in case of a disappearing filament, the propagation speed is roughly constant

flux densities. Thus the energy density also decreases. But the latter decreases not only because of the shock's expansion: turbulence created in the wake of the shock and particles accelerated at the shock front (Sect. 7.6) also reduce the shock's energy.

Changes in shock parameters with radial distance can be quite different from one shock to another. As an example, in Fig. 6.42 the radial variation of the shock speed is shown. Two extreme cases can be distinguished. If a shock is very fast close to the Sun (with CME speeds above 1000 km/s), it is likely to decelerate in interplanetary space. On the other hand, shocks that are rather slow on the Sun do not decelerate but propagate at roughly constant speed. Possible interpretations can be found in the energy release mechanism: it is more explosive in fast shocks and CMEs, which in general are also accompanied by a flare, compared with rather continuous in the slower ones. In addition, the faster shocks in general tend to be more efficient particle accelerators, and thus part of the shock's kinetic energy is converted into kinetic energy of particles. In some sense this relates to the different speed characteristics of the CMEs and to the conversion of shock kinetic energy into particle energy.

6.8 Shock Waves

A shock is a discontinuity separating two different regimes in an otherwise continuous medium. It is associated with something moving faster than the signal speed in the medium: a shock front separates the Mach cone of a supersonic jet from the ambient, undisturbed air. Here the disturbance and the shock are moving, and thus the shock is called a travelling shock. Standing shocks also form: in a river, a shock forms in front of the bridge pier where the fast stream suddenly is slowed down. In space plasmas, both kinds of shocks exist: mass ejections propagating from the Sun through interplanetary space drive travelling shocks. The supersonic solar wind is slowed down at planetary magnetospheres, forming the bow shock, a standing shock wave. At these discontinuities the properties of the medium change dramatically. We can define a shock as follows:

1. The disturbance propagates with a speed faster than the signal speed. In a gas, the signal speed is the speed of sound; in space plasmas, it depends on the Alfvén speed and the sound speed.
2. At the shock front, the properties of the medium change abruptly. In a hydrodynamic shock, pressure and density increase; in a magnetohydrodynamic shock, plasma density and magnetic field strength increase.
3. Behind the shock front a transition back to the properties of the undisturbed medium must occur. Behind a gas-dynamic shock, density and pressure decrease; behind a magnetohydrodynamic shock, plasma density and magnetic field strength decrease. If this decrease is fast, a reverse shock develops.

Shock waves can be stable for long times (in the solar system up to some months) and can propagate even out to the boundary of the solar system.

While the study of gas-dynamic shocks started in the late nineteenth century and had its heyday in the 1940s, the study of plasma shocks started only in the fifties as an interest in fusion plasmas and the consequences of nuclear explosions in the atmosphere awoke. At that time, also a certain kind of shocks in a plasma has been detected that differed strongly from the gas-dynamic shock: in these collisionless shocks, densities are too low to allow for collision between individual atoms or molecules. Instead, the collective effects of the electrical and magnetic properties of the plasma allow for frequent interactions and the formation of a shock wave.

Collisionless shocks therefore are different from gas-dynamic shocks. Nonetheless, the concepts about the fundamental nature of shocks are the same as in a gas-dynamic shock, as are the basic conservation laws.

6.8.1 Information, Dissipation, and Non-linearity

A shock is a non-linear wave of "permanent" form propagating faster than the signal speed. Thus an understanding of a shock has to invoke the concepts of information, dissipation, and non-linearity.

Information can be transferred by a propagating disturbance; the sound wave is the simplest example. It can transfer information either as a continuous stream of different waves (as in language or music) or as a rather sharp pulse like the "bang" following an explosion or the "clap" in hand-clapping. Though the information contained in these latter signals might be more difficult to decipher, they are more useful for the explanation of shock formation because they are wave parcels with a well-defined onset. These sounds travel as pressure pulses through the air. Their distance from the source defines the information horizon: inside the information horizon, the signal has already been received, outside it has not yet been detected.

A sound wave is a compressional wave: the density increases with increasing pressure. Sound waves are "simple" waves. The compression is assumed to be adiabatic: the gas is compressed such that on expansion it returns to its original state. This is possible because the compression is fast enough to prevent thermal conduction from removing heat. Thus the compression is isentropic, the entropy does not change. Furthermore, we assume that the disturbance is small and therefore viscosity, friction, and heat conduction are negligible, and that the gas behaves according to the ideal gas law. We then can calculate the sound speed according to (4.40). It is independent of frequency but depends on the gas parameters, for instance the temperature.

The creation of a supersonic disturbance can then be viewed from different perspectives: we can ask ourselves whether information can travel faster than the speed of sound, we can study a disturbance moving faster than sound, or we can study a large-amplitude disturbance.

Let us start with the latter case. A large-amplitude pressure pulse can be created by an explosion. In air, close to the site of the blast, the sound speed can increase up to about 1000 km/s. But the signal does not propagate as a harmonic wave: in the compressional phase the pressure amplitude can exceed the atmospheric pressure. During decompression, however, the pressure cannot drop below zero. Thus the amplitudes of the positive and negative half-waves are different.[5] In addition, in a large-amplitude wave the change in sound speed during compression and decompression becomes significant: during compression the temperature increases and the wave can propagate faster, while during the other half-wave the temperature decreases, leading to a slower propagation. Thus the wave front steepens in time, similar to a water wave running into shallow water. If the steepening eventually leads to a jump in density and pressure, a shock wave has formed. This kind of shock is called a blast wave shock. Type II radio burst in the solar corona probably indicate blast waves (Sect. 6.7.1). Blast wave shocks are often used in the numerical simulation of interplanetary shocks [138, 245, 494], although today it is realized that interplanetary shocks are driven shocks.

A driven shock is associated with an object moving faster than a sound wave. In this case, even a small-amplitude disturbance can lead to the pressure and density jump that defines a shock. An object moving through air transfers momentum and energy to the ambient molecules. The motion of the object also requires motion of the air: molecules in front of the object must give way to it by streaming around the object and again collecting behind it. But to do this, the molecules must first receive information about the approaching object. The pressure pulse in front of a subsonic vehicle provides this information. For a supersonic vehicle, the information also has to be provided, even if only for a short distance ahead. Thus the medium has to be changed to allow for the faster propagation of information. As the

---

[5] The same argument holds in an even more expressive example, the shallow water wave which steepens until it finally breaks and spills over. Although the water wave as a surface wave requires a slightly different description it should be kept in mind as an illustrations because it is more likely to be related to everyday experience than an acoustical blast wave shock.

motion starts to become supersonic, the initially small-amplitude pressure pulses pile up in front of the vehicle, leading to a large-amplitude disturbance. This large-amplitude disturbance is able to change the properties of the medium irreversibly: after the disturbance has passed by, the medium will be in a different state. Since the disturbance will change the temperature of the medium it also will change the sound speed.

6.8.2 The Shock's Rest Frame

The simplest description of a gas-dynamic shock uses the shock's rest frame (see Fig. 6.43): gas with a speed larger than the signal speed is flowing into the shock from the upstream medium which so far has not received any information about the approaching shock. Since this side is not modified by the shock, it is also called the low-entropy side. At the shock front, irreversible processes lead to the compression of the gas and a change in speed: across the shock front, mass conservation is required, and thus the amount of matter flowing through each element of the shock surface has to be constant. The flow out of the shock front into the downstream medium is subsonic and the density is increased. Thus we can define a shock as an entropy-increasing or irreversible wave that causes a transition from subsonic to supersonic flow.

Theoretically, the disturbance driving the shock can propagate at any speed. Therefore, the shock's propagation speed can increase without any limit. The Mach number M is defined as the ratio between the shock speed in the upstream medium and the sound speed. It is always determined in the rest frame of the shock. In the upstream medium, M is larger than 1, in the downstream medium it is smaller. Thus in the downstream medium the plasma leaves the shock with a speed smaller than the sound speed: any disturbance in the downstream medium can propagate away from the shock.

We can now easily understand that a travelling shock and a standing shock are identical: in a travelling shock a supersonic disturbance propagates through the medium while in a standing shock the object is at rest and the flow is supersonic. Thus the difference between standing and travelling shocks depends on the frame of reference only; the systems are Galilean invariant.

SHOCK FRONT

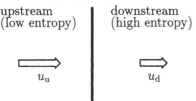

upstream
(low entropy)

downstream
(high entropy)

u_u

u_d

Fig. 6.43. Frame of reference for the description of a shock. The shock front is at rest, plasma flows with a high speed u_u from the upstream medium into the shock front and leaves it with a lower speed u_d into the downstream medium

6.8.3 Collisionless Shock Waves

In a gas-dynamic shock, the important process is the collision between molecules: they establish a temperature distribution, temperatures of different species are equalized, density and temperature fluctuations can propagate, and the viscous forces associated with them lead to dissipation.

Space plasmas are rarefied, and thus collisions are rare. These shocks are called collisionless shocks. The lack of collisions has some implications, for instance: electrons and protons can have different temperatures, their distributions can be very different from a Maxwellian making the classical concept of temperature obsolete, the presence of a magnetic field might even lead to highly anisotropic particle distributions, and processes of dissipation involve complex interactions between particles and fields.

Nonetheless, shocks are frequently observed in space plasmas. While the coupling between the particles due to collisions is negligible, the magnetic field acts as a coupling device, binding the particles together. We have already used this assumption in magnetohydrodynamics (MHD) where we have described a magnetized plasma by concepts such as pressure, density, and bulk velocity. These concepts also prove helpful in the description of the plasmas upstream and downstream of the shock. The details of the shock front and the plasma immediately around it, however, cannot be covered within this framework because MHD does not consider the motions of and the kinetic effects due to individual particles. While we are still far from understanding the details of these processes, observations indicate that the collective behavior of the plasma is mainly due to wave–particle interactions. Thus collisionless shocks are an example of a macroscopic flow phenomenon regulated by microscopic kinetic processes. A popular account can be found in [455], a discussion of laboratory experiments to produce collisionless shocks is given in [135].

6.8.4 Shock Conservation Laws

Plasma properties in the upstream and the downstream media are different in parameters such as bulk flow speed u, magnetic field B, plasma density ϱ, and pressure p. The relationship between these two sets of parameters is established by basic conservation laws, the Rankine–Hugoniot equations.

Rankine–Hugoniot Equations in Ordinary Shocks. A very clear description of the Rankine–Hugoniot equations and their application to hydro- and aerodynamic shocks is given in [108]. These equations describe the conservation of mass, energy, and momentum through the shock front, and can even be applied to simple shocks in space plasmas if the distributions are isotropic Maxwellians and the magnetic field is roughly parallel to the flow.

In these conservation laws the shock is assumed to be infinitesimally thin. In optics, a boundary is thin with respect to the wavelength and thick with respect to the spacing of the molecules in the crystal structure. Analogously, in MHD a boundary is thin with respect to the scale length of the fluid parameters (if waves are involved, as in a shock, it is thin with respect to the wavelength) but thick with respect to the Debye length and the ion gyro-radius, both being characteristic for the collective behavior of the plasma.

In the remainder of this section the abbreviation $[X] = X_u - X_d$ gives the difference of a quantity X in the upstream and the downstream media. The Rankine–Hugoniot equations for a gas-dynamic shock then are:

- conservation of mass:

$$[mu_n] = [\varrho u_n] = 0 ; \qquad (6.43)$$

- conservation of momentum normal to the shock:

$$[\varrho u_n^2 + p] = 0 ; \qquad (6.44)$$

- conservation of momentum tangential to the shock:

$$[\varrho u_n u_t] = 0 ; \qquad (6.45)$$

- conservation of energy normal to the shock:

$$\left[\left(\frac{\varrho u^2}{2} + \frac{\gamma_a}{\gamma_a - 1} p \right) u_n \right] = 0 . \qquad (6.46)$$

Here u is the flow speed, u_n (u_t) the flow speed normal (tangential) to the shock, ϱ the density, p the pressure, and γ_a the specific heat ratio, all measured in the shock's rest frame. The conservation of energy considers both kinetic flow energy and internal energy. Combining (6.43) and (6.44) yields $[u_t = 0]$: the tangential component of the flow is continuous. Therefore, we can choose a coordinate system moving along the shock front with the speed u_t. In this normal incidence frame (see left panel in Fig. 6.44), u equals u_n.

The mass conservation (6.43) can be used to estimate the local shock speed. Making a Galilean transformation into the laboratory system, it can be written as $[\varrho(v_s - u_n)] = 0$. Rearrangement gives the shock speed

$$v_s = \frac{\varrho_d u_{n,d} - \varrho_u u_{n,u}}{\varrho_d - \varrho_u} . \qquad (6.47)$$

In applying (6.47) to shocks in space plasmas, in particular travelling interplanetary shocks, we should be aware of its limitations. First, the magnetic field is neglected; the shock is a simple gas-dynamic one. Second, the shock is assumed to be spherically symmetric with the flow perpendicular to the shock surface. For an interplanetary shock, this is an oversimplification [244, 472] and the speed estimated from (6.47) only gives the radial component of the shock speed. It therefore can be used as a lower limit only. However, observations suggest that the deviation of the shock normal from the radial direction is often less than 20° [89], and thus (6.47) is a reasonable approximation.

Note that the shock speed alone is not indicative of the energetics of the shock. The shock speed becomes large if the denominator in (6.47) is small. Thus a small increase in density across the shock can often be associated with a high shock speed, while the total energy in terms of compression and mass motion is rather small.

Example 21. For the shocks in Fig. 6.41, the following plasma parameters can be determined from the figure: for DOYs 217 and 268, we obtain upstream densites 30 cm^{-3} and 14 cm^{-3}, downstream densities 60 cm^{-3} and 105 cm^{-3}, upstream speeds 360 km/s and 602 km/s, and downstream speeds 420 km/s and 1101 km/s, respectively. From (6.47) we then obtain local shock speeds of 480 km/s and 1204 km/s; the compression ratios r_n are 2 and 7.3, respectively. Both shock speeds are good approximations to the more accurate values given in the figure.   □

Rankine–Hugoniot Equations in MHD Shocks. The crucial difference between a MHD shock and an ordinary shock is the magnetic field. Thus we have to expand the conservation laws to also accommodate the field. In addition, the geometry becomes more complex because the flow is not necessarily parallel to the field.

Often a special rest frame is used, the de Hoffmann–Teller frame (right panel in Fig. 6.44). In the normal incidence frame, the upstream plasma flow is normal to the shock and oblique to the magnetic field. The downstream flow is oblique to both the magnetic field and shock normal. In the de Hoffmann–Teller frame [127], the plasma flow is parallel to the magnetic field on both sides of the shock and the $u \times B$ induction field in the shock front vanishes: the reference frame moves parallel to the shock front with the de Hoffmann–Teller speed $v_{\mathrm{HT}} \times B = -E$.

For a MHD shock, the Rankine–Hugoniot relations can be inferred in the same way as in a gas-dynamic shock [58, 68, 121]. With n being the unit vector along the shock normal, the Rankine–Hugoniot equations are:

- the mass balance, which is the same as for the ordinary shock,

$$[\varrho u \cdot n] = 0 \; ; \qquad\qquad (6.48)$$

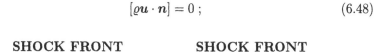

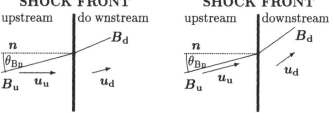

Fig. 6.44. Frames of reference for MHD shocks: normal incidence frame (*left*) and de Hoffmann–Teller frame (*right*)

- momentum balance, where the additional terms describe the magnetic pressure perpendicular and normal to the shock front,

$$\left[\varrho u \left(u \cdot n \right) + \left(p + \frac{B^2}{2\mu_0} \right) n - \frac{(B \cdot n)B}{\mu_0} \right] = 0 ; \qquad (6.49)$$

- energy balance, where the additional terms describe the electromagnetic energy flux $E \times B/\mu_0$ with the electric field expressed by $E = -v \times B$,

$$\left[u \cdot n \left(\frac{\varrho u}{2} + \frac{\gamma}{\gamma - 1} p + \frac{B^2}{\mu_0} \right) - \frac{(B \cdot n)(B \cdot u)}{\mu_0} \right] = 0 ; \qquad (6.50)$$

- Maxwell's equations

$$[B \cdot n] = 0 , \qquad (6.51)$$

which follows from $\nabla B = 0$, and states that the normal component of the magnetic field is continuous ($B_n = $ const), and

$$[n \times (u \times B)] = 0 , \qquad (6.52)$$

which states that the tangential component of the electric field must be continuous.

The Rankine–Hugoniot equations are a set of five equations for the unknown quantities ϱ, u, p, B_n, and B_t.

6.8.5 Jump Conditions and Discontinuities

The Rankine–Hugoniot equations allow the calculation of the downstream plasma parameters from the knowledge of the upstream parameters of a MHD shock. But these conservation equations are more general; the solutions of (6.48)–(6.52) are not necessarily shocks, instead a multitude of different discontinuities can be described, too.

A contact discontinuity does not allow for a plasma flow across it, and thus it is $u_n = 0$. It is associated with an arbitrary density jump while all other quantities remain unchanged. The magnetic field has a component normal to the discontinuity ($B_n \neq 0$), and thus the two sides of the discontinuity are not completely decoupled but tied together by the field such that they flow together at the same tangential speed u_t.

A tangential discontinuity separates two plasma regions completely from each other. There is no flux across the boundary ($u_n = 0$ and $B_n = 0$) and the tangential components of both quantities change ($[u_t] \neq 0$ and $[B_t] \neq 0$). Plasma and field change arbitrarily across the boundary but a static pressure balance is maintained: $[p + B^2/2\mu_0] = 0$. Tangential discontinuities thus are examples for pressure balanced structures. Typical changes in plasma and field parameters are sketched in the left panel in Fig. 6.45.

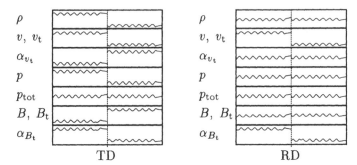

Fig. 6.45. Changes in magnetic field and plasma parameters across a tangential discontinuity (*left*) and a rotational discontinuity (*right*). Here α_x indicates a change in the direction of the quantity x. p_{tot} gives the total pressure, the sum of kinetic, plasma, and magnetic pressures, based on [36]

A rotational discontinuity can be viewed as a large-amplitude wave. In an isotropic plasma, the field and the flow change direction but not magnitude. The rotational discontinuity requires pressure equilibrium according to (6.49). Because there is a flux across the boundary, we get $u_n \neq 0$ and $B_n \neq 0$. The normal flow speed is $u_n = B_n/\sqrt{\mu_0 \varrho}$ and the change in tangential flow speed is related to the change in tangential magnetic field: $[u_t] = [B_t/\sqrt{\mu_0 \varrho}]$. Thus normal flow speed and the change in tangential flow speed are directly related to the Alfvén speed and the change in tangential Alfvén speed. The rotational discontinuity therefore is closely related to the transport of magnetic signals across the boundary. The jump conditions for a rotational discontinuity also apply at boundaries suitable for reconnection. Typical changes in plasma and field parameters are sketched in the right panel in Fig. 6.45.

6.8.6 Shock Geometry

One important parameter in the description of a MHD shock is the local geometry, i.e. the angle θ_{Bn} between the magnetic field direction and the shock normal. Shocks can be classified according to θ_{Bn}:

- a perpendicular shock propagates perpendicular to the magnetic field: $\theta_{Bn} = 90°$;
- a parallel shock propagates parallel to the magnetic field: $\theta_{Bn} = 0°$;
- an oblique shock propagates at any θ_{Bn} between $0°$ and $90°$. Oblique shocks can be subdivided into
  - quasi-parallel shocks with $0° < \theta_{Bn} < 45°$ and
  - quasi-perpendicular shocks with $45° < \theta_{Bn} < 90°$.

The shock shown in Fig. 6.44 therefore is a quasi-parallel shock.

6.8.7 Fast and Slow Shocks

A shock differs from the discontinuities described in Sect. 6.8.5 in so far as there is a flow of plasma through the surface ($u_n \neq 0$) combined with compression and changes in flow speed. Note that in a parallel shock B_t equals 0 and the magnetic field is unchanged by the shock, while in a perpendicular shock the normal component of the magnetic field vanishes, $B_n = 0$, and both plasma pressure and field strength increase at the shock. The parallel shock therefore behaves like a gas-dynamic shock – except for the fact that the collective behavior of the plasma is regulated by the magnetic field and not by collisions.

In a plasma, different modes of MHD waves exist which can steepen to form a shock: fast, slow, and intermediate waves (see Fig. 4.4). Of these waves, only the fast and the slow waves are compressive. The intermediate wave is purely transverse with the velocity perturbation perpendicular to both k_0 and B_0. The intermediate shock, sometimes also called an Alfvén shock, only exists in an anisotropic medium. In an isotropic plasma, such as the solar wind, it is not a shock but a rotational discontinuity: there is a rotation of the magnetic field by 180° in the plane of the shock but no density jump across the shock. Thus there is a flow across the boundary, but without compression or dissipation. In addition, the planes defined by the magnetic field and the plasma flow direction in the upstream and downstream media are not parallel but oblique. In an intermediate shock the propagation speed parallel to the magnetic field equals the Alfvén speed, i.e. $v_{int} = v_A \cos \theta_{Bn}$.

Real shocks are formed by fast and slow magneto-sonic waves only. In both modes, the plasma density and pressure change across the shock. The phase speed of these modes is (see (4.49))

$$2v_{fast,slow}^2 = (v_s^2 + v_A^2) \pm \sqrt{(v_s^2 + v_A^2)^2 - 4v_s^2 v_A^2 \cos^2 \theta} \,, \qquad (6.53)$$

with the $+$ sign referring to the fast and the $-$ sign to the slow mode. If these waves propagate perpendicular to B, then $v_{inter} = v_{slow} = 0$ and $v_{fast} = \sqrt{v_A^2 + v_s^2}$. For propagation parallel to B either v_{fast} equals v_{inter} for $v_A > v_s$ or v_{inter} equals v_{slow} for $v_A < v_s$ (see Fig. 4.4). For the different modes, different Mach numbers can be introduced: M_A is the Alfvén Mach number, M_{cs} the sonic Mach number (the same as the Mach number in a simple gas-dynamic shock), and M_{sl} and M_f the slow and fast Mach numbers, respectively.

The change in the magnetic field is different in fast and slow shocks: in a fast shock the magnetic field increases and is bent away from shock normal because the normal component of the field is constant. The normal component of the upstream (downstream) flow speed is larger (smaller) than the propagation speed of a fast MHD wave and both upstream and downstream flow speeds exceed the Alfvén speed. In a slow shock, the upstream speed

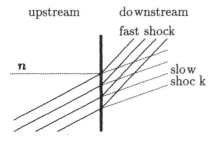

Fig. 6.46. Change in magnetic field direction across a fast and a slow MHD shock

exceeds the sound speed but not the Alfvén speed. In addition, the magnetic field strength decreases across the shock and the field therefore is bent towards the shock normal (see Figs. 6.46 and 6.47).

Travelling interplanetary shocks in general and planetary bow shocks in particular always are fast MHD shocks. So far, only a few slow shocks have been observed *in situ* in the solar system [90, 441]. In the solar corona, however, slow [248] and intermediate [502] shocks might be more common because both the Alfvén and sound speed are much higher (see Fig. 6.33).

For many aspects, in particular shock formation and particle acceleration at the shock, the crucial quantity is not the shock speed but its component $v_{s\parallel} = v_s \sec\theta_{Bn}$ parallel to the magnetic field. A rather slow disturbance can still have a large propagation speed parallel to $\boldsymbol{B}$ if θ_{Bn} is large enough. To form a shock, a disturbance must propagate with $v_{s\parallel} > v_A$ to catch up with the waves propagating along the field, but it must not necessarily propagate with $v_s > v_A$. This problem becomes evident in the definition of the Alfvén Mach number. In a stationary frame of reference, the Alfvén Mach number is defined as

$$M_A = \frac{v_s - u_u}{v_A} \,, \qquad (6.54)$$

which is the shock speed in an upstream reference system relative to the Alfvén speed. With this definition, even fast MHD shocks occasionally can

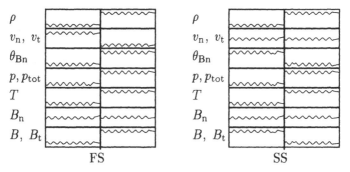

Fig. 6.47. Changes in magnetic field and plasma parameters across a fast (*left*) and a slow (*right*) shock. Based on [36]

have Mach numbers smaller than 1. A better definition, such as the critical Mach number

$$M_{\mathrm{c}} = \frac{v_{\mathrm{s}} - u_{\mathrm{u}}}{v_{\mathrm{A}} \cos \theta_{\mathrm{Bn}}} \, , \tag{6.55}$$

is formally more helpful because $M_{\mathrm{c}} = 1$ exactly gives the intermediate shocks while fast shocks always have $M_{\mathrm{c}} > 1$. But to determine the critical Mach number, the local geometry θ_{Bn} must be known.

6.8.8 The Coplanarity Theorem

In Fig. 6.44 we have tacitly assumed that the shock normal and the magnetic field directions in the upstream and the downstream media all lie in the same plane. This assumption is called the coplanarity theorem and is a consequence of the jump conditions at the shock. It can be expressed as

$$\boldsymbol{n} \cdot (\boldsymbol{B}_{\mathrm{d}} \times \boldsymbol{B}_{\mathrm{u}}) = 0 \, . \tag{6.56}$$

If we only consider the transverse component, the momentum balance (6.49) for an isotropic pressure p can be written as

$$\left[\varrho u_{\mathrm{n}} \boldsymbol{u}_{\mathrm{t}} - \frac{B_{\mathrm{n}}}{2\mu_0} \boldsymbol{B}_{\mathrm{t}} \right] = 0 \, . \tag{6.57}$$

Equation (6.52) can be written as $[u_{\mathrm{n}} \boldsymbol{B}_{\mathrm{t}} - B_{\mathrm{n}} \boldsymbol{u}_{\mathrm{t}}] = 0$. Therefore both $[\boldsymbol{B}_{\mathrm{t}}]$ and $[u_{\mathrm{n}} \boldsymbol{B}_{\mathrm{t}}]$ are parallel to $[\boldsymbol{u}_{\mathrm{t}}]$ and thus also parallel to each other. Then we have $[u_{\mathrm{n}} \boldsymbol{B}_{\mathrm{t}}] \times [u_{\mathrm{n}} \boldsymbol{B}_{\mathrm{t}}] = 0$. Resolving the parentheses gives

$$(u_{\mathrm{n,u}} - u_{\mathrm{n,d}})(\boldsymbol{B}_{\mathrm{t,u}} \times \boldsymbol{B}_{\mathrm{t,d}}) = 0 \, . \tag{6.58}$$

Since $[u_{\mathrm{n}}]$ does not vanish, the upstream and downstream tangential magnetic components must be parallel to each other. Thus the upstream and downstream magnetic field vectors are coplanar with the shock normal vector and the magnetic field across the shock has a two-dimensional geometry. The bulk velocity is coplanar with the shock, too.

6.8.9 The Shock Normal Direction

An application of the coplanarity theorem is the calculation of the shock normal in observational data. If the shock normal is known, the angle θ_{Bn}, which is crucial for shock formation and particle acceleration, can be calculated.

If only magnetic field measurements are available, the coplanarity theorem (6.56) for the magnetic field can be used. Since the magnetic field is divergenceless, we have $(\boldsymbol{B}_{\mathrm{u}} - \boldsymbol{B}_{\mathrm{d}}) \cdot \boldsymbol{n} = 0$. Thus together with (6.56) we have defined two vectors perpendicular to the shock normal. These vectors can be used to calculate the shock normal:

$$\boldsymbol{n} = \frac{(\boldsymbol{B}_{\mathrm{u}} \times \boldsymbol{B}_{\mathrm{d}}) \times (\boldsymbol{B}_{\mathrm{u}} - \boldsymbol{B}_{\mathrm{d}})}{|(\boldsymbol{B}_{\mathrm{u}} \times \boldsymbol{B}_{\mathrm{d}}) \times (\boldsymbol{B}_{\mathrm{u}} - \boldsymbol{B}_{\mathrm{d}})|} \, . \tag{6.59}$$

This method does not work if $\boldsymbol{B}_u$ is parallel to $\boldsymbol{B}_d$. The shock normal derived according to (6.59) is called the coplanarity normal. If also three-dimensional plasma measurements are available, other constraints, in particular the coplanarity theorem for the bulk velocity, can be used to determine the shock normal. In addition, different methods can be combined into one overdetermined solution and solved for a best-fit shock normal [453].

With the known shock normal we are also able to determine the shock speed v_s more accurately than suggested by (6.47):

$$v_s = \frac{\varrho_d \boldsymbol{u}_d - \varrho_u \boldsymbol{u}_u}{\varrho_d - \varrho_u} \cdot \boldsymbol{n} . \tag{6.60}$$

Example 22. Let us briefly return to example 21. From the details of the magnetic field, a colleague has inferred the shock normal and the direction of the upstream and downstream plasma flows relative to it. For simplicity, the shock normal is given here as $(1, 0)$, with the x-component in the direction of shock propagation. For the shock of DOY 217, the direction of the upstream flow is $(0.94, 0.34)$, and that of the downstream flow is $(0.98, 0.17)$; the shock is therefore almost quasi-parallel as expected from the weak compression in the magnetic field. The shock speed then is

$$v_s = \frac{60 \times \begin{pmatrix} 0.98 \\ 0.17 \end{pmatrix} \times 420 - 30 \times \begin{pmatrix} 0.94 \\ 0.34 \end{pmatrix} \times 360}{60 - 30} \begin{pmatrix} 1 \\ 0 \end{pmatrix} \text{km/s} = 485 \text{ km/s} , \tag{6.61}$$

which is slightly above the values of 480 km/s determined in example 21. □

6.9 What I Did Not Tell You

In the previous chapters, we have basically dealt with the "well-defined" topics of physical concepts. In this chapter, we have encountered "real-world" physics in a complex environment. We have used our concepts to describe the observations; however, we should be aware that the basic concepts describe only the general features of the natural phenomena – to understand the details and interpret the observations correctly, we need more advanced models (Chap. 12).

But before we can turn to advanced models, we have to be aware of the limitations of our measurements. For instance, all in situ measurements suffer from one basic problem: interplanetary space is a three-dimensional medium which is highly variable in space and time. Thus, formally, all our variables are fields $\varepsilon(\boldsymbol{r}, t)$ varying in time. What we observe is a time series of the parameter ε at a varying position inside the field because, in general, both the field (for instance, the magnetic field convected outwards with the solar wind) and the observer move. Many of the ideas described in this chapter are therefore based on sparse evidence.

We have already mentioned one example: in the interpretation of magnetic field fluctuations, we are not able to distinguish between waves and turbulence because the observed variations in plasma and magnetic-field parameters are a mixture of temporal and spatial variations. But a one-point observation is not only unable to resolve the nature of the local fluctuations, it is also unable to resolve large-scale structures. For instance, the latitudinal distribution of the solar wind speed shown in Fig. 6.25 is not a snapshot but sampled over almost five years, because it took Ulysses that long to complete one orbit. In addition, the solar wind speeds are measured at radial distances between 1.3 and 5 AU, again a consequence of Ulysses' orbit. This sampling is not necessarily a disadvantage, because in this case all sampling occurred during a solar minimum and thus the figure can be interpreted as a measure for average solar minimum conditions in the intermediate heliosphere between 1 and 5 AU. However, if we want to take a look at shorter time scales, the figure is of limited use because it ignores all the short-term variability. As a consequence, not only do we have to have our data but we also need metadata, that is all relevant information about the data, instrumentation and observational practice.

In this chapter we have also learned about an ambiguous method, that of classification. Classification is useful because it helps us to focus on common aspects of different events/individuals and thus prevents us from getting lost in details. But as long as we do not understand the physical differences between the classes of events, our classification scheme will be based on phenomenological criteria. Since the average height of males exceeds that of females, it might be worthwhile to discuss such a classification approach with a 162 cm male and a 186 cm female to learn about its limitations.

The phenomenological classification scheme has another disadvantage: with each new mission, it is at risk of becoming obsolete. For instance, in this chapter, the occurrence of a CME has been mentioned as a possible physical basis for the classification into impulsive and gradual flares. The observations with the LASCO coronograph, however, showed that there were more CMEs than believed previously and that many of the impulsive events were accompanied by CMEs too. Thus, taking the occurrence of CMEs as a criterion, we would get a different classification scheme, which would not be in agreement with the original classification schemes. But CMEs also show differences among themselves, in particular with respect to their spatial extent, their speed, and their ability to drive a shock. A distinction of CMEs into different groups (again a classification) might help to support the original classification scheme, and thus it is still valid that I have introduced this accepted standard in this book – but you should be aware that this classification is temporary, that it varies, and that, depending on the author's understanding, the terms "impulsive" and "gradual" are sometimes used with slightly different meanings.

6.10 Summary

The heliosphere is structured by the solar wind and the frozen-in magnetic field, which is wound up into Archimedian spirals due to the Sun's rotation. The fast solar wind originates in coronal holes at the poles while the slow wind originates from the streamer belt close to the solar equator where sunspots, filaments, and active regions are located. Where fast and slow solar wind streams meet, corotating interaction regions form. Fluctuations on different scales are superimposed onto the average field. They are related to waves originating in the corona, interactions between different solar wind streams, and transient disturbances such as magnetic clouds and travelling interplanetary shocks. The latter stem from coronal mass ejections, i.e. violent expulsions of huge plasma clouds.

Exercises and Problems

6.1. Explain the basic conservation laws across a shock front. What are the differences between a gas-dynamic and a hydrodynamic shock?

6.2. Explain the differences between a fast and a slow shock.

6.3. Derive (6.14) for the length of the Archimedian spiral.

6.4. Consider a 10 MeV proton in interplanetary space. Determine its gyroradius, its gyration period, and the wave numbers of the Alfvén waves in resonance with the proton (assume three different pitch angles, $10°$, $30°$, and $90°$). Compare with the corresponding values for a 1 MeV electron.

6.5. For an observer on the Earth, calculate the length of the magnetic field line to the Sun and its longitude of origin (connection longitude). Do the same for an observer at 5 AU. (Assume a plane geometry with the field line confined to the plane of ecliptic.)

6.6. Imagine a slow solar wind speed starting on the Sun. $30°$ east of this stream, a fast stream with twice the speed of the slow stream originates. Where would they meet? (Simple assumption of an Archimedian magnetic field spiral.)

6.7. An electron beam with $c/3$ propagates through the interplanetary plasma and excites a radio burst (see example 17). Assume a decrease in plasma density $\sim 1/r^2$. Calculate the frequency drift under the assumption that the electron beam propagates radially. How do these results change if the curvature of the Archimedian field line, along which the electrons propagate, is considered? How would the curvature of the field line influence the frequency drift of a type II burst in front of an interplanetary shock?

6.8. A magnetic loop on the Sun has a parabolic shape with $B = B_0(1 + s^2/H^2)$ with $H = 30\ 000$ m being the height of the loop and s the distance from the top of the loop. Calculate the bounce period of particles with a speed of $2c/3$. As the particles interact with the atmosphere at the mirror points, they create hard X-rays. What is the time interval between two subsequent elementary bursts?

6.9. The plasma instrument on an interplanetary spacecraft detects a sudden increase in plasma density. No other changes in plasma or field are observed. Is this a shock?

6.10. The plasma instrument on an interplanetary spacecraft detects a discontinuity with a jump in plasma density from 4 cm^{-3} to 8 cm^{-3} and a jump in plasma flow speed from 400 km/s to 700 km/s (all quantities in the spacecraft frame). Determine the shock speed. What is the meaning of this speed?

6.11. Assume the following average solar wind properties at the Earth's orbit: proton density 7 cm^{-3}, electron density 7.5 cm^{-3}, He^{2+} density 0.25 cm^{-3}, flow speed 400 km/s almost radial, proton temperature 2×10^5 K, electron temperature 1×10^5 K, and magnetic field 7 nT. Calculate the flux densities and the flux through a sphere of radius 1 AU for the following quantities: protons, mass, radial momentum, kinetic energy, thermal energy, magnetic energy, and radial magnetic flux.

7 Energetic Particles in the Heliosphere

> The space between Heaven and Earth – is it not like a bellow?
> It is empty and yet not depleted;
> Move it and more always comes out.
>
> Lao-Tzu, *Tao Te Ching*

Particles in interplanetary space come from sources as diverse as the Sun, the planets, and the vastness of space. The properties of the different particle populations provide information about the acceleration mechanism(s) and the propagation between source and observer. Thus energetic particles can also be used as probes for the properties of the interplanetary medium. To describe acceleration and propagation, we use concepts such as reconnection, acceleration at shock waves, and wave–particle interactions.

This chapter starts with an overview of the different particle populations and subsequently discusses some of them in detail, in particular solar energetic particles and their propagation, shock-acceleration, particles accelerated at travelling interplanetary shocks and planetary bow shocks, and galactic cosmic rays and their modulation.

7.1 Particle Populations in the Heliosphere

Energetic particles in interplanetary space are observed with energies ranging from the supra-thermal up to 10^{20} eV. The main constituents are protons, α-particles, and electrons; heavier particles up to iron can be found in substantially smaller numbers. The particle populations originate in different sources, all having their typical energy spectrum, temporal development, and spatial extent. Figure 7.1 summarizes the particle populations, Table 7.1 their properties, a recent review can be found in [435].

7.1.1 Populations and Sources

(A) **Galactic Cosmic Rays (GCR)** are the high-energy background population with energies extending up to 10^{20} eV. They are incident upon the

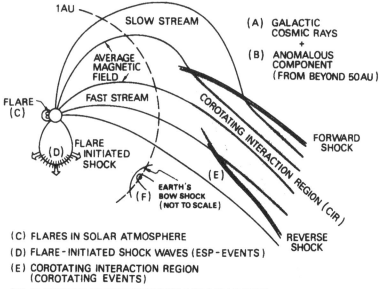

Fig. 7.1. Populations of energetic charged particles in the inner heliosphere. Reprinted from H. Kunow et al. [305], in *Physics of the inner heliosphere, vol. II* (eds. R. Schwenn and E. Marsch), Copyright 1991, Springer-Verlag

heliosphere uniformly and isotropically. In the inner heliosphere, the galactic cosmic radiation is modulated by solar activity: the intensity of GCRs is highest during solar minimum and reduced under solar maximum conditions. For reviews see, for example, [47, 133, 160, 163, 268, 350, 416, 552].

(B) The **Anomalous Cosmic Rays (ACR)**, also called the anomalous component, energetically connect to the lower end of the GCRs but differ from them with respect to composition, charge states, spectrum, and variation with the solar cycle [153, 154, 157, 257]. As neutral particles of the interstellar medium travel through interplanetary space towards the Sun, they become ionized. These charged particles are convected outward with the solar wind and accelerated at the termination shock. Then they propagate towards the inner heliosphere where they are detected as anomalous component.

(C) **Solar Energetic Particles (SEPs)** are accelerated in solar flares, the injection of these particles into the heliosphere thus is point-like in space and time. SEP energies extend up to some tens or a few hundred megaelectronvolts, occasionally even into the gigaelectronvolt range [271, 359]. The latter can be observed with neutron monitors on the ground, the event is called a ground-level event (GLE). Owing to interplanetary scattering, particle events in interplanetary space last between some hours and a few days, depending on the scattering conditions and the observer's distance from the Sun. SEPs

Table 7.1. Characteristics of particle populations in interplanetary space. The letter in the first column is the same as in Fig. 7.1, the subsequent columns give the temporal and spatial scales as well as the typical energy range

	Temporal scales	Spatial scales	Energy range	Acceleration mechanisms
A	continuous	global	GeV-> TeV	diffusive shock
B	continuous	global	10–100 MeV	shock?
C	δ	δ	keV–100 MeV	reconnection, stochastic, selective heating, shock
D	days	extended	keV–10 MeV	diffusive shock, shock-drift, stochastic
E	27 days	large-scale	keV–10 MeV	diffusive shock
F	continuous	local	keV–MeV	diffusive shock, shock drift

not only provide information about the acceleration processes on the Sun but also can be used as probes for the magnetic structure of interplanetary space. Solar energetic particle events show different properties, depending on whether the parent flare is gradual or impulsive [84, 271, 272]. In gradual flares, the solar energetic particles mix with particles accelerated at interplanetary shocks. Although SEPs originate in flares, only a small fraction of all flares leads to enhancements in energetic particles above background.

(D) **Energetic Storm Particles (ESPs)** are particles accelerated at interplanetary shocks. Originally, ESPs were thought to be particle enhancements related to the passage of an interplanetary shock. The name was chosen to reflect their association with the magnetic storm observed as the shock hits the Earth's magnetosphere. Today, we understand the acceleration of particles at the shock, their escape, and the subsequent propagation through interplanetary space as a continuous process, lasting for days to weeks until the shock finally stops accelerating particles. The properties of particles accelerated at interplanetary shocks are summarized for the tens and hundreds of kiloelectronvolt range in [440, 456, 462, 523, 557] and for the megaelectronvolt range in [83, 85, 267]. At very strong shocks, protons can be accelerated up to about 100 MeV or more [434, 528].

(E) **Corotating Interaction Regions (CIR)** also lead to intensity increases. Protons with energies up to about 10 MeV are accelerated at the CIR-shocks [32, 159, 314, 358, 490]. The energetic particles can even be observed remote from the corotating shocks at distances where the shocks have not yet been formed [305, 573] or at higher solar latitudes when a spacecraft is above the streamer belt where the CIRs form [136, 306, 487].

(F) Particles accelerated at **planetary bow shocks** are a local particle component with energies extending up to some 10 keV [107]. An exception is the Jovian magnetosphere where electrons are accelerated up to about 10 MeV.

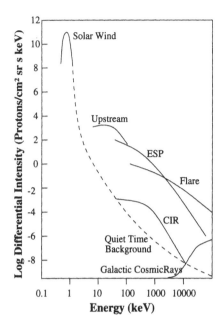

Fig. 7.2. Energy spectra of different ion populations in the heliosphere above a postulated quiescent background, based on [188]

With a suitable magnetic connection between Earth and Jupiter, these Jovian electrons can be observed even at Earth's orbit [98, 106].

Figure 7.2 gives the energy ranges and relative intensities for different ion populations in the heliosphere. The figure is limited to energies below 100 MeV; above that energy the higher energetic end of the SEPs can be found as well as the galactic and anomalous cosmic radiation. The spectrum of the GCR is shown in Fig. 7.29.

7.1.2 Relation to the Solar Cycle

Figure 7.3 shows, from top to bottom, sunspot numbers, intensities of protons above 4 MeV, and intensities of protons above 60 MeV. The particle data were obtained by the University of Chicago instrument on board IMP. The lower proton energies are predominantly influenced by solar energetic-particle events; the higher proton energies reflect the temporal development of galactic cosmic rays, and only the individual events standing out above the background are of solar origin.

The main characteristics of the solar-cycle dependence in the particle components are (letters as in Fig. 7.1 and Table 7.1):

A GCRs have higher intensities during solar minima and lower intensities during solar maxima, as evident from the bottom panel of Fig. 7.3.

B ACRs, which constitute a minority in the lower panel of Fig. 7.3, show the same temporal variation.

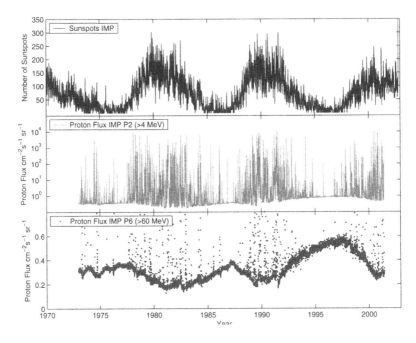

Fig. 7.3. Energetic particles during the solar cycle, IMP measurements; data from Space Physics Interactive Data Resource (SPIDR), `spidr.ngdc.noaa.gov/spidr/index.html`

C SEP events are more frequent during solar minima than during solar maxima (middle panel and spikes in the lower panel): active regions and filaments are more frequent during solar maxima than during solar minima. However, SEP events are also observed during solar minima.

D ESP events are also contained in the middle panel and show the same temporal variation as SEPs.

E CIR events are more frequent during solar minima: a CIR is formed in the interaction of a fast and a slow solar wind stream. The formation of the CIR, the development of the shocks, and the particle acceleration all require time. During solar maximum conditions, the stream pattern is frequently disturbed by coronal mass ejections and magnetic clouds.

F Particle acceleration at planetary bow shocks appears to be rather independent of the solar cycle. The temporal pattern in the observation of Jovian electrons, for instance, is regulated not by solar activity but by the magnetic connection between Jupiter and the observer. Of course, the highly disturbed medium between the source and the observer during solar maxima conditions makes detection more difficult.

In the study of energetic particles, we are interested in their source, in the acceleration mechanisms, and in their propagation. Occasionally, the acceleration mechanism can be studied in situ, for example at travelling and corotating shocks and at planetary bow shocks. For most populations, however, the observer is remote from the source: there is no access to the acceleration sites of solar energetic particles or galactic cosmic rays. Even particles accelerated at shocks in interplanetary space can be observed remotely from the shock. In all these cases, interplanetary propagation has modified the particle populations. Therefore the properties of a particle population observed in interplanetary space reflect in general the combined effects of acceleration and propagation.

7.2 Solar Energetic Particles and Classes of Flares

Particle increases at the Earth's orbit as a consequence of solar flares first were detected in 1942 in neutron monitor records [164]. Because of the small spatial and temporal extent of the particle source, SEPs not only provide information about particle acceleration but also can be used to study interplanetary propagation.

Solar energetic particles are mainly protons, electrons, and α-particles with small admixtures of ^{3}He-nuclei and heavier ions up to Fe. Protons and ions can be accelerated up to some tens or hundreds of MeV/nucl, occasionally even energies in the gigaelectronvolt range can be acquired. The electron acceleration is limited to energies of some megaelectronvolts. The particle increase above background can vary between hardly detectable and some orders of magnitude (see Fig. 7.3). While the flare lasts only for a few minutes, maybe even an hour, a solar energetic particle event at the Earth's orbit typically lasts for a day or two, depending on the number of particles injected into space, scattering conditions, and the presence of a coronal mass ejection fast enough to drive an interplanetary shock.

Solar energetic particle events observed in interplanetary space exhibit different features, depending on whether the parent flare was impulsive or gradual [84, 272]; some of these features are summarized in Table 7.2.

What do the differences in particle populations tell us and how can they be related to the flare scenarios in Sect. 6.7.6? First of all, we can infer that different acceleration mechanisms must be at work. The high charge states of Fe (20) and Si (14) in impulsive events are indicative of acceleration out of a very hot environment with temperatures of about ten million Kelvin [329]. The coronal temperature of about two million Kelvin would imply lower charge states of about 14 for Fe and 10 for Si, such as observed in gradual events. Thus in the impulsive flares, the plasma must be heated while in gradual flares particles can be accelerated out of the corona without significant heating. Particle events following impulsive flares are also enriched in some of the heavier ions, in particular an increase in ^{3}He/^{4}He, Fe/C, and Fe/O is observed [438]. Particle events following gradual flares, on the other hand, show coronal abundances.

Table 7.2. Properties of energetic particle events following impulsive and gradual flares. The numbers refer to particles with energies of a few MeV/nucl

	^{3}He-rich	gradual
Particles	electron-rich	proton-rich
^{3}He/^{4}He	~ 1 (enrichment 2000 times)	~ 0.0005
Fe/O	~ 1.234 (enrichment 8 times)	~ 0.155
H/He	10	100
Q_{Fe}	$\sim +20$	$\sim +14$
Duration	hours	days
Longitudinal cone	$< 30°$	$\leq 180°$
Metric radio bursts	III, V	II, III, IV, V
Coronograph	–	CME
Solar wind	–	ipl. shock
Event rate/a	~ 1000	~ 10

These differences can be understood by the two flare models presented in Sect. 6.7.6. Impulsive flares can be understood in terms of selective heating, while in gradual flares particles are accelerated out of the cool coronal material by the shock in front of the CME. Since the shock accelerates ions, in particular H, more efficiently than electrons, impulsive flares are electron-rich compared with gradual flares and also show a lower H/He ratio.

The shock in front of the CME has another consequence, namely the different longitudinal cones in which energetic particles can be observed. In the model of selective heating, particles are accelerated on a few open field lines adjacent to the closed loop in which the flare occurred. Thus out of the field lines originating on the Sun, only a small bundle is filled with particles. But these particles do not spread in all directions. Since SEPs are charged particles, they propagate along the magnetic field lines but not perpendicular to them. Thus the particles stay on their field lines and an observer in interplanetary space only detects particles if she is on one of these field lines, i.e. if she is magnetically connected to the source region. Owing to the curvature of the interplanetary magnetic field line, under normal solar wind conditions an observer on Earth is magnetically connected to a position at 60°W. Thus for impulsive events, particles are detected on a spacecraft around the orbit of Earth only when the flare occurred in the western hemisphere of the Sun.

In a gradual flare, the situation is different. The shock in front of the CME has a large angular extent. Thus, the particles are not accelerated locally but over a wide region. Therefore, particles are injected into a much larger cone of field lines open to interplanetary space and the particle events therefore can be detected at larger angular distances from the site of the energy release. Thus, close to the Earth, particle events from gradual flares can also be detected if they originated in the eastern hemisphere of the Sun.

So far for a gradual flare we have only considered the particles accelerated at the shock. But in the model of a large flare in Fig. 6.37, the CME and the coronal shock are only one part of the event. The processes occurring below the filament still closely resemble those in a flare without CME. The characteristics of the particles interacting on the Sun can be inferred from the γ-ray line spectrum, and their temporal evolution also from the hard X-ray emission. Corresponding observations indicate that the primary energy release must be quite similar in impulsive and gradual events. In particular, although the electromagnetic emission in gradual flares lasts for a longer time, the elementary energy release in both classes of flares occurs in bursts that are similar. In addition, the compositions of the ions interacting on the Sun are similar and both are indicative of some process similar to selective heating. If the observer is magnetically connected close to the flare site in a gradual event, these flare-accelerated particles can also be detected in interplanetary space. The upper panel of Fig. 7.4 shows the temporal evolution of the intensities of protons, O, and Fe in a gradual flare. In the lower panel the Fe/O ratio is shown for different energies. Early in the event, the Fe/O ratio is high, as expected for an impulsive event. With time, the Fe/O ratio decreases to a value typical of gradual events or the solar wind plasma. Thus early in the event, flare-accelerated particles are present while with increasing time the particles accelerated at the shock become more and more abundant. Note also the differences in the time profiles: the high-energetic Fe channel rises rather fast, and the intensity decreases after its maximum, i.e. a few hours after the flare. This closely resembles a diffusive profile as expected for a short injection on the Sun. In H and O, on the other hand, the intensity stays roughly constant or even increases towards the shock because particles are continuously accelerated and injected from the shock as it propagates towards the observer.

 We should be aware that in most cases a classification scheme is used to order a complex data set rather than to reflect physical processes. Nonetheless, the two classes described in Table 7.2 are distinct kinds of events. But whether there really are only these two classes, or whether the two classes represent extreme cases of a more continuous change from impulsive to gradual, is still under debate. In particular, there are large impulsive flares which are not accompanied by a CME but show characteristics which are in between those given in Table 7.2, while on the other hand there are disappearing-filament events where the particles exhibit features typical of events following gradual flares, although no flare is observed. In addition, even if a flare is accompanied by a coronal mass ejection, the latter does not necessarily drive a shock. This latter point might become more interesting in the future, as observations with the more sensitive SOHO coronograph show that small, slow CMEs are relatively common and can also be observed in rather small impulsive flares. Here the role of the CME is probably not the acceleration of additional particles, because these CMEs tend to be small and slow, but

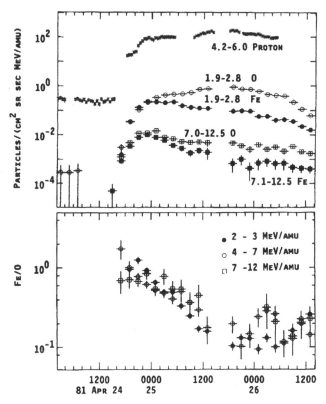

Fig. 7.4. Particle intensities and the development of particle composition in a gradual flare. Reprinted from D.V. Reames et al. [436], *Astrophys. J.* **357**, Copyright 1990, American Astronomical Society

rather the opening of the flare loop in which the initial energy release occurs, allowing the particles to escape into interplanetary space. If the latter scenario turns out to be true, the particles observed in space should exhibit properties in between those given in Table 7.2: although the charge states should be high, the selective enrichment should be diminished because the accelerated particles can escape before the mechanism of selective heating has developed fully. A more detailed discussion of the above classification scheme and the two classes of particle events can be found in [435]; a critical review of that scheme and a discussion of the issues mentioned in this paragraph is given in [271].

7.3 Interplanetary Transport – Theoretical Background

From the long time scales (several hours to some days) of solar particle events compared with the flare duration, in the 1950s Meyer et al. [356]

suggested that interplanetary transport might be diffusive. In this section we shall discuss the basic concepts of diffusion, their physical foundation in particle scattering at magnetic inhomogeneities (resonance scattering), and different transport equations.

7.3.1 Spatial Diffusion

Diffusion is the consequence of frequent, stochastically distributed collisions; thus it is a stochastic process. Therefore, it is not reasonable to discuss the motion of individual particles; instead one has to consider an assembly of particles, described by the distribution function.

But diffusion is not only spatial diffusion. If we carefully drip a drop of ink into a glass of water, in time the drop will spread and eventually ink and water will be mixed completely: the thermal motion leads to collisions between ink and water molecules, distributing both species uniformly. This is spatial diffusion. If we carry out the same experiment with a good drop of cold cream and a cup of steaming hot tea, we find a second consequence of the collisions: after some time, tea and milk have the same temperature, thus thermal energy is transferred from the faster molecules to the slower ones, leading to diffusion in momentum space. The existence of both processes has already been mentioned in connection with Fig. 5.3.

Let us start with spatial diffusion alone. All particles have the same speed and collisions lead to changes in the direction of motion only. To describe the effect of diffusion, we have to keep track of a larger number of small spatial steps for a large number of particles. Because the stochastic aspect is important, we can borrow some concepts from probability calculus and use simple games with coins as illustrations.

Tumbling Drunkards and Tossed Coins. Spatial diffusion, or more correctly, the motion of a particle in spatial diffusion, occasionally is called "drunkards walk". To get the picture, imagine a couple of drunkards, happily lingering around a distiller. As they hear a police siren in the distance, they start to stagger away, everyone in his own direction. They all make steps of equal length but random direction. As a police helicopter arrives at the scene, every drunkard had made N steps of length λ. The spatial distribution of the drunkards, as seen from the helicopter, is shown in Fig. 7.5. None of the drunkards has covered the maximum possible distance $N\lambda$. Instead, they are still relatively close to the distiller. How close, compared with the maximum distance, can be described by a quantity called the expected distance or, mathematically more correct, the average squared distance. This average displacement is $\lambda\sqrt{N}$ as indicated by the circle.

Let us now reduce the problem to the one-dimensional case: the test objects can only move along a straight line, again with constant step length. We can simulate the resulting motion by flipping a coin: a head leads to a step in the positive direction, a tail to a step in the negative one. Let us

Fig. 7.5. Distribution of drunkards staggering away from a distiller D. The circle indicates the average expected distance $\lambda\sqrt{N}$

consider one particle only. At first glance one might expect the expected distance to be close to the starting point. In particular, after a large number of tosses, we would expect the number of heads and tails to be roughly equal and therefore the net displacement to be small. This, however, is a faulty reversion of the law of large numbers, which is often observed in people gambling only occasionally: if the coin has shown tails 9 times in succession, the chance of heads in the next toss is exactly the same as in all previous tosses, 50%, because the coin does not remember the results from the last tosses. Thus in a long series of tosses, there can be quite a large deviation from a deadlock between heads and tails. This has been known since the middle of the seventeenth century when game theory was quite popular. Thus if for a long time one side of the coin can be dominant, as indicated in Fig. 7.6, then a large net gain for the one and a large loss for the other gambler results. Or, in case of one-dimensional motion, the displacement from the starting position can become quite large.

The average squared distance $\langle \Delta x \rangle^2$, or the expected distance for short, can be determined easily. The total squared displacement of the particle is the sum of the displacements dx_i in each individual step:

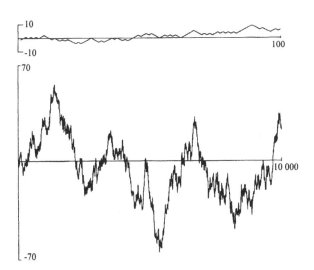

Fig. 7.6. Gain and loss chart for tossing coins. In the upper panel, the result of 100 tosses is shown, in the lower panel 10 000 tosses have been considered

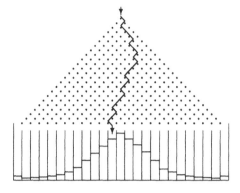

Fig. 7.7. Galton board: many small scatters work together stochastically, forming a bell curve (Gauss's distribution)

$$(\Delta x)^2 = \left(\sum_{i=1}^{N} \mathrm{d}x_i \right)^2 = (\mathrm{d}x_1 + \mathrm{d}x_2 + ... + \mathrm{d}x_N)^2 = \sum_{i=1}^{N} \sum_{j=1}^{N} \mathrm{d}x_i \mathrm{d}x_j \ . \quad (7.1)$$

The individual displacements $\mathrm{d}x_i$ are either $+\lambda$ or $-\lambda$, both with a probability of 0.5. Thus the product $\mathrm{d}x_i \mathrm{d}x_j$ is either λ^2 or $-\lambda^2$. For $i \neq j$, $\mathrm{d}x_i$ and $\mathrm{d}x_j$ are independent and both positive and negative values of the product have the probability 0.5. In the sum (7.1) these terms cancel and only products with $i = j$ remain. They are always $+\lambda^2$ and there are N such products. Equation (7.1) then becomes

$$\langle \Delta x \rangle^2 = N\lambda^2 \ . \quad (7.2)$$

Thus with increasing number N of steps, the average displacement from the starting point increases as $\sqrt{N}$.

If the particle has a speed v, the total distance s travelled during a time t is $s = vt$. If N is the number of direction reversals during this time interval, the distance also can be written as $s = N\lambda$. Therefore in (7.2) we can substitute N by vt/λ:

$$\langle \Delta x \rangle^2 = N\lambda^2 = v\lambda t = 2Dt \ . \quad (7.3)$$

Here D is the diffusion coefficient:

$$D = \tfrac{1}{2} v\lambda \ . \quad (7.4)$$

Note that this diffusion coefficient has been defined for one-dimensional motion. For three-dimensional motion, the diffusion coefficient becomes

$$D = \tfrac{1}{3} v\lambda \ . \quad (7.5)$$

Galton Board and Bell Curve (Gauss's Distribution). The average distance is a statistical term which refers to a large assembly of particles. The individual particles scatter around the starting point. Their distribution can be described by the bell curve (Gauss's distribution).

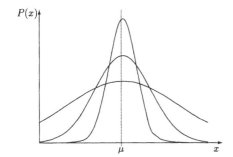

Fig. 7.8. Broadening of a Gauss distribution with increasing standard deviation. Physically, this is equivalent to the diffusive broadening of a distribution with time

The Galton board is a graphical way to derive this distribution. It consists of rows of pins, indicated by dots in Fig. 7.7, and models the scattering the particles experience: as a ball hits a pin, it is deflected either to the left or to the right. Then it hits a pin in the next row, which leads to another deflection and so on. The solid line indicates a sample path. Below the lowest row, the particles are collected in slots. The slot in which a ball finally comes to rest, results from a large number (equal to the number of rows) of stochastic interactions of comparable strength. If we use a large number of balls, the distribution in the slots will be a bell curve or Gauss's distribution:

$$P(x) = \frac{1}{\sqrt{2\pi}\sigma} \exp\left(-\frac{(x - x_0)^2}{2\sigma^2}\right) . \tag{7.6}$$

Here x_0 is the average and σ is the standard deviation. $P(x)$ describes the probability of a ball to be found in the slot at position x. The standard deviation σ defines the width of the distribution: 68.3% of all values will be inside the interval $[x_0 - \sigma, x_0 + \sigma]$ and 95.4% inside $[x_0 - 2\sigma, x_0 + 2\sigma]$. It is given as

$$\sigma^2 = \frac{1}{n} \sum (x - x_0)^2 =: \langle \Delta x \rangle^2 , \tag{7.7}$$

and therefore describes the widening of the particle distribution or the expected displacement from the origin.

We can rewrite (7.6) and (7.7) to find an expression depending on the diffusion coefficient. With (7.3) and (7.4) we find for the standard deviation

$$\sigma = \sqrt{\langle \Delta x \rangle^2} = \sqrt{2Dt} = \sqrt{v\lambda t} , \tag{7.8}$$

and therefore for the bell curve

$$P(x) = \frac{1}{\sqrt{2\pi v\lambda t}} \exp\left(-\frac{(x - x_0)^2}{2v\lambda t}\right) . \tag{7.9}$$

Note that the maximum stays fixed while the distribution broadens with time, as described by (7.3) and shown in Fig. 7.8.

The Diffusion Equation. If collisions happen in a homogeneous gas enclosed in a fixed volume, the relevant quantity to describe the diffusive process is the mean free path. It does not make sense to talk about a diffusion coefficient or an expected displacement because, viewed from the outside, the collisions do not change the properties of the gas, only the individual molecules change positions. This could be depicted by something similar to the Galton board: while in the Galton board the pins are arranged to form a triangle with the input only at the tip of the triangle, the modified board would consist of pins arranged in a rectangle with the input all over the top line. For each input slot, the spatial distribution is the same as for a Galton board. But the superposition of all the different input slots leads to the same number of particles in each output slot. For a gas this implies that on average for each particle leaving a volume element another one enters.

The situation is different if there is a gradient. Then there are more particles of the species under study in one part of the volume than in the other. Accordingly, a random walk carries more particles out of the volume with high density than particles are carried in from the lower density region. Thus a net transport results, reducing the gradient and eventually leading to an equalized distribution. The streaming S of particles can be described as

$$S = -D\nabla U \,, \tag{7.10}$$

with D being the diffusion tensor for anisotropic diffusion and U the particle density. The gradient is the driving force for the flow, a larger gradient leading to a larger flow. The flow also depends on the mobility of the particles, described by the diffusion tensor. If diffusion is isotropic, the diffusion tensor reduces to the diffusion coefficient and the streaming becomes $S = -D\nabla U$. Since the diffusion coefficient depends on the particle speed and on the mean free path, for a given gradient the flow as well as the average displacement are largest for fast particles undergoing only few collisions (thus having a large mean free path) and smallest for slow particles undergoing many collisions.

The equation of continuity (3.35), gives the following for a volume element

$$\frac{\partial N}{\partial t} + \oint_{O(S)} S\mathrm{d}o = 0 \,. \tag{7.11}$$

Here N is the number of particles and S is the flux of particles through the surface o of the volume element V. If U is the particle density, (7.11) yields

$$\frac{\partial}{\partial t} \int_V U\mathrm{d}^3x + \oint_{O(V)} S\mathrm{d}o = 0 \,. \tag{7.12}$$

With Gauss' theorem (A.33) this is

$$\frac{\partial U}{\partial t} + \nabla S = 0 \,. \tag{7.13}$$

With (7.10) we can write the diffusion equation as

$$\frac{\partial U}{\partial t} = \nabla \cdot (\mathsf{D}\, \nabla U) . \tag{7.14}$$

If the diffusion is independent of the direction (isotropic diffusion), we can use the diffusion coefficient (7.4) instead of the diffusion tensor and get

$$\frac{\partial U}{\partial t} = \nabla \cdot (D\, \nabla U) . \tag{7.15}$$

If the diffusion coefficient is also independent of the spatial coordinate, as for example in a homogeneous medium, the equation can be reduced further:

$$\frac{\partial U}{\partial t} = D\Delta U . \tag{7.16}$$

We have already encountered the one-dimensional form of this equation in Sect. 3.4.2 in Eqs. (3.106) and (3.107).

Solutions of the Diffusion Equation. The solution of the diffusion equation depends on the boundary conditions. In the general case we shall consider propagation from the source at a position r_0. Thus we have to consider a source Q in the diffusion equation:

$$\frac{\partial U}{\partial t} - D\Delta U = Q(r_0, t) . \tag{7.17}$$

For a spherical symmetric geometry this can be written as

$$\frac{\partial U}{\partial t} - \frac{1}{r^2} \frac{\partial}{\partial r} \left(r^2 D_{\mathrm{r}} \frac{\partial U}{\partial r} \right) = Q(r_0, t) \tag{7.18}$$

with D_{r} being the radial diffusion coefficient, which describes the diffusion between different radial shells [380, 381].

The simplest case is a pulse-like injection of N_0 particles at the position $r_0 = 0$ at time $t_0 = 0$. A typical example is the injection of solar flare particles into the interplanetary medium. The solution of the diffusion equation for a radial-symmetric geometry then reads

$$U(r, t) = \frac{N_0}{\sqrt{(4\pi D_{\mathrm{r}} t)^3}} \exp\left(-\frac{r^2}{4 D_{\mathrm{r}} t} \right) . \tag{7.19}$$

Two typical diffusive profiles are shown in Fig. 7.9. The intensity rises fast to a maximum and than decays slowly $\sim t^{-3/2}$. The time of maximum t_{m} can be determined by setting the first temporal derivative to zero:

$$t_{\mathrm{m}}(r) = \frac{r^2}{6 D_{\mathrm{r}}} . \tag{7.20}$$

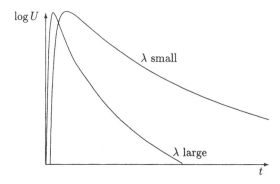

Fig. 7.9. Typical diffusive profiles for small and large λs

The time of the maximum decreases with increasing mean free path and increasing particle speed. That is what we expect from our experience with gases and liquids: the diffusion of a minor constituent is faster with increasing temperature (corresponding to a higher particle speed) and decreasing density (corresponding to an increase in particle mean free path). The time of the maximum increases quadratically with increasing distance. This can be understood easily from (7.8): the average distance increases with $\sqrt{t}$.

Graphically, the time to maximum can be interpreted as follows [560]: if we write (7.20) in the form $t_{\mathrm{m}} = (r/2\lambda)\,(r/v)$, we have r/v as the direct travel time for the distance r and can interpret $r/2\lambda$ as a measure of the number of mean free paths between the origin and the observer at r. The quantity $r/2\lambda$ therefore characterizes the delay due to diffusion.

Inserting (7.20) into (7.19) gives the density at the time of maximum:

$$U(r, t_{\mathrm{m}}) = \frac{N_0}{\sqrt{(4\pi r^2/6)^3}} \exp\left(-\frac{3}{2}\right) \sim \frac{N_0}{r^3} \,. \tag{7.21}$$

The intensity at the time of the maximum thus decreases with increasing radial distance but it is independent of the diffusion coefficient.

Equation (7.20) is used frequently as a simple estimate of the radial mean free path from the time of maximum of a particle profile observed in interplanetary space. Rewriting (7.20) we obtain

$$\lambda_{\mathrm{r}} = \frac{r^2}{2vt_{\mathrm{m}}} \,. \tag{7.22}$$

Solutions of the diffusion equation so far have been obtained for the spheric symmetric case, assuming that particles propagate radially from one shell at r to the next one at $r + \Delta r$. The mean free path λ_{r} then refers to the radial mean free path. In interplanetary space, the geometry is different: particles propagate along the magnetic field line, thus it is more reasonable to use a particle mean free path $\lambda_{\parallel}$ parallel to the magnetic field line. In addition, the field is not radial but Archimedian. The solution, however, is identical to the radial one as long as the relation

$$\lambda_r = \lambda_\parallel \cos^2 \psi \qquad \text{or} \qquad D_r = D_\parallel \cos^2 \psi \qquad (7.23)$$

is obeyed [380, 381]. Here ψ is the spiral angle between the radial direction and the Archimedian magnetic field line. Note that here it is assumed that diffusion perpendicular to the magnetic field is negligible [158, 256, 432].

Diffusion–Convection Equation. So far, we have considered particles being scattered in a medium at rest. A different situation arises if the medium is also moving, for instance an oil spill in a river: the oil is distributed by diffusion but is also carried along with the moving fluid. In interplanetary space, the convection is due to the solar wind: the particles are scattered at inhomogeneities frozen into the solar wind and therefore propagating together with the solar wind. Thus the streaming in (7.10) has to be supplemented by the convective streaming $S_{\text{conv}} = U u$, giving $S = U u - D\nabla U$. Here u is the velocity of the convective flow.

As above, the streaming can be inserted into the equation of continuity (3.34), giving the diffusion–convection equation

$$\frac{\partial U}{\partial t} + \nabla(U u) = \nabla(D\nabla U) . \qquad (7.24)$$

If u and D are independent of the spatial coordinate, (7.24) reads

$$\frac{\partial U}{\partial t} + u\nabla U = D\Delta U . \qquad (7.25)$$

In the radial-symmetric case, the solution for a δ-injection then is

$$U(r,t) = \frac{N_0}{\sqrt{(4\pi D_r t)^3}} \exp\left\{ -\frac{(r - ut)^2}{4 D_r t} \right\} . \qquad (7.26)$$

For small bulk speeds u of the medium the transport equation as well as its solution converge towards the simple diffusion equation.

7.3.2 Pitch Angle Diffusion

Thus far we have not discussed the physical process of scattering. In the graphical description we have tacitly assumed that we are dealing with large-angle interactions: either the particle continues to propagate in its original direction of motion or it is turned around by 180°. In chapter 5, however, we have learned that fast particles in a plasma are more likely to encounter small-angle interactions. Thus to turn a particle around, a large number of interactions is required.

In space plasmas small-angle interactions are not due to Coulomb-scattering in the electric field of the background plasma but due to scattering at plasma waves. The physical processes will be briefly sketched in Sect. 7.3.4; the formal description is similar to the one discussed for spatial diffusion.

Let us assume a magnetized plasma with an energy density exceeding that of the energetic particles: the energetic particles then can be regarded as test particles. Thus the particles gyrate around the lines of force and a pitch angle can be assigned to each particle, often written in the form $\mu = \cos\alpha$. Each interaction leads to a small change in μ, i.e. a diffusion in pitch angle space. We can derive a scattering term strictly analogous to spatial diffusion. Let us start from (7.14). This can be rewritten easily: while the driving force for spatial diffusion is a spatial gradient, the driving force for pitch angle diffusion is a gradient in pitch angle space. Therefore the spatial derivatives have to be replaced by a derivative to μ and the scattering term reads

$$\frac{\partial}{\partial\mu}\left(\kappa(\mu)\frac{\partial f}{\partial\mu}\right) , \qquad (7.27)$$

with κ being the pitch angle diffusion coefficient and f the phase space density. Note that the scattering depends on μ, and thus the scattering can be different for different pitch angles, depending on the waves available for wave–particle interaction. The pitch angle diffusion coefficient can be related to the particle mean free path parallel to the magnetic field [215, 256] by

$$\lambda_\| := \frac{3}{8}v \int\limits_{-1}^{+1} \frac{(1-\mu^2)^2}{\kappa(\mu)}\,\mathrm{d}\mu . \qquad (7.28)$$

Here $\lambda_\|$ does not describe the average distance travelled between two consecutive small-angle scatterings, but the distance travelled before the particles pitch angle has been changed by 90°, i.e. the direction of motion has been reversed. Thus for the overall motion, λ has a meaning comparable to the mean free path in ordinary spatial diffusion.

The term (7.27) also can be used to describe spatial scattering if we also consider the field-parallel motion μv of the particles. Thus as in the diffusion–convection equation we have to consider the streaming of particles with respect to the scattering centers. The transport equation then can be written as

$$\frac{\partial f}{\partial t} + \mu v\frac{\partial f}{\partial s} = \frac{\partial}{\partial\mu}\left(\kappa(\mu)\frac{\partial f}{\partial\mu}\right) . \qquad (7.29)$$

Here $\partial f/\partial s$ is the spatial gradient along the magnetic field line. This dependence is sufficient, because the motion of the guiding center is one-dimensional along the magnetic field line and the particle gyrates around it. We will encounter this equation again as part of the focused transport equation (7.36) for particles in interplanetary space.

7.3.3 Diffusion in Momentum Space

Collisions not only change a particle's direction of motion but also its energy. We had already mentioned this as a basic requirement in the establishment of a thermal distribution. Momentum transfer can happen by collisions between particles as well as by wave–particle interaction. If the energy gain in each interaction is small compared with the particle energy, this can be described as diffusion in momentum space [167,527]. Instead of particle flow, streaming S_p in momentum results in

$$S_p = -D_{pp}\frac{\partial f}{\partial p} + \frac{dp}{dt}f \ . \tag{7.30}$$

Here D_{pp} is the diffusion coefficient in momentum space. The second term describes non-diffusive changes in momentum, such as ionization, and corresponds to the convective term in the spatial diffusion equation. It therefore can also be described as convection in momentum space. Again, the physics of the scattering process is hidden in D_{pp}.

7.3.4 Wave–Particle Interactions

In this section we shall briefly introduce some of the basic processes of wave–particle interactions. While in all previous sections the plasma was regarded as well-behaved, wave–particle interactions are an example of the non-linearity of plasma physics. While for the linear aspects treated before a well-developed mathematical description is available, in the non-linear theory no general algorithms exist. Only few analytical methods are known, most of them relying on approximations. One of them is the limitation to lowest-order perturbations, similar to the approximation described in Sect. 4.2. Let us start this section with a brief introduction to quasi-linear theory.

Quasi-Linear Theory. Quasi-linear theory is based on perturbation theory; interactions between waves and particles are considered to first order only. Thus all terms of second order in the disturbance should be small enough to be ignored. Only weakly turbulent wave–particle interactions can be treated this way: the particle distribution is only weakly affected by the self-excited waves in a random-phase uncorrelated way. This requirement not only corresponds to small disturbances in perturbation theory but even directly results from it as the waves are described in the framework of perturbation theory. The waves generated by the particles will affect the particles in a way which will tend to reduce the waves. Thus we assume the plasma to be a self-stabilizing system: neither indefinite wave growth happens nor are the particles trapped in a wave well.

The basic equation is the Vlasov equation (5.23). We split all quantities into a slowly evolving average part, such as f_0, $E_0 = 0$, and B_0, and a fluctuating part f_1, E_1, and B_1. The long-term averages of the fluctuating quantities vanish: $\langle f_1 \rangle = \langle E_1 \rangle = \langle B_1 \rangle = 0$. Note that in contrast to the ansatz in Sect. 4.1.2 here the quantities with index '0' are not constant background quantities but slowly evolving average properties of the system.

These are the quantities we are interested in – the fluctuating quantities are of interest only in so far as they give rise to the evolution of the phase space density. In Sect. 4.1.2, on the other hand, we were interested in the fluctuating quantities because they gave rise to a new phenomenon, the waves.

With the above ansatz, the average Vlasov equation reads

$$\frac{\partial f_0}{\partial t} + \boldsymbol{v} \cdot \nabla f_0 + \frac{q}{m} \boldsymbol{v} \times \boldsymbol{B_0} \cdot \frac{\partial F_0}{\partial \boldsymbol{v}} = -\frac{q}{m} \left\langle (\boldsymbol{E}_1 + \boldsymbol{v} \times \boldsymbol{B}_1) \cdot \frac{\partial f_1}{\partial \boldsymbol{v}} \right\rangle . \quad (7.31)$$

The term on the right-hand side contains the non-vanishing averages of the fluctuations and describes the interactions between the fluctuating fields and the fluctuating part of the particle distribution. These interactions combined with the slowly evolving fields on the left-hand side of (7.31) lead to the phase space evolution of the slowly varying part of the distribution. Note that we have not made any assumptions about the smallness of the fluctuations, the only limitation is a clear separation between the fluctuating part and the average behavior of the plasma.

Equation (7.31) is the fundamental equation in non-linear plasma physics. Solutions, however, are difficult to obtain because they require an a priori knowledge of the fluctuating fields to calculate the average term on the right-hand side. This term has the nature of a Boltzmann collision term. Note that these collisions are not particle–particle interactions but result from the non-linear coupling between the particles and the fluctuating wave fields.

If the particles and the fluctuating fields are known, the term on the right-hand side can be calculated. It can then be used to derive an expression for the scattering coefficients mentioned above which depends on particle properties, in general the rigidity, and the properties of the waves, in particular their power density spectrum.

Resonance Scattering. The scattering of particles by waves can be described as a random walk process if the individual interactions lead to small-angle scattering only. Thus a reversal of the direction of motion requires a large number of these small-angle scatters. If we assume a particle to be in resonance with the wave, the scattering is more efficient because the small-angle changes all work together into one direction instead of trying to cancel each other. Thus pitch angle scattering will mainly occur from interactions with field fluctuations with wavelengths in resonance with the particle motion along the field. Such a resonance interaction can formally be understood from a simple mechanical or electrical analogy, such as a light torsion pendulum in a turbulent gas or a resonant circuit excited by noise [560]. A full treatment of the theory with application to the scattering of particles in interplanetary space was first given by Jokipii [256] and, with a somewhat different approach, by Hasselmann and Wibberenz [215].

The idea is sketched in Fig. 7.10. Let us assume a simple model of magnetic field fluctuations: the (relevant) waves propagate only along the magnetic field ($\boldsymbol{k} \| \boldsymbol{B}_0$) and the fluctuating quantities are symmetric around the wave vector.

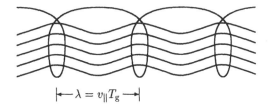

$\leftarrow \lambda = v_\| T_g \rightarrow$

Fig. 7.10. Resonance scattering: scattering is more efficient if the particle interacts only with waves in resonance with its motion parallel to the magnetic field

This assumption is called the slab model. Let us single out a certain wave number k. A particle is in resonance with this wave if it propagates a wave length $\lambda_\|$ along the magnetic field during one gyration: $\lambda_\| = v_\| T_c = \mu v T_c$. With $k_\| = 2\pi/\lambda_\|$ and $T_c = 2\pi/\omega_c$ the resonance condition can be written as

$$k_\| = \frac{\omega_c}{v_\|} = \frac{\omega_c}{\mu v} . \tag{7.32}$$

The amount of scattering a particle with pitch angle α experiences basically depends on the power density $f(k_\|)$ of the waves at the resonance frequency. This dependence clearly shows why the pitch angle diffusion coefficient κ depends on μ: particles with different pitch angles are scattered by different wave numbers with different power densities.

Alfvén Waves and Interplanetary Propagation. Magnetohydrodynamic waves are Alfvén waves with wave vectors parallel and magneto-sonic waves with k perpendicular to the undisturbed field. In interplanetary space, the resonance scattering of energetic particles at Alfvén waves plays an important role in particle propagation. The formalism describing this process is derived in [215, 256] or in the review [156].

The basic process is the interaction of the particle with the fluctuating component of the magnetic field. Since the fluctuations are assumed to be small, $B_1 \ll B_0$, the change in pitch angle during a single gyration is small, too. Scattering at waves in resonance with the particle motion parallel to the field is essential for efficient scattering. The relation between the power density of the magnetic field fluctuations and the pitch angle diffusion coefficient is relatively simple for the limitation to waves with wave vectors parallel to the field and axially symmetric transverse fluctuating components (slab model).

The magnetic field density power spectrum in interplanetary space can be described by the power law (see Sect. 6.4, in particular Fig. 6.22) $f(k_\|) = C k_\|^{-q}$. Here q is the spectral shape, $k_\|$ the wave number parallel to the field, and C the power at a certain frequency. The pitch angle diffusion coefficient is related to this spectrum by

$$\kappa(\mu) = A(1 - \mu^2)|\mu|^{q-1} , \tag{7.33}$$

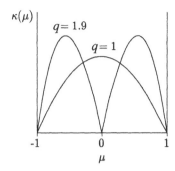

Fig. 7.11. Shape of the pitch angle scattering coefficient for different slopes q of the power density spectrum

with A being a constant related to the level C of the turbulence. The particle mean free path can then be determined by (7.28). The mean free path depends on particle rigidity as $\lambda_\| \sim P^{2-q}$, as long as $q < 2$.

Figure 7.11 shows the shape of the pitch angle scattering coefficient for different slopes q of the magnetic field power density spectrum. The case $q = 1$ corresponds to isotropic scattering, i.e. the strength of the scattering is independent of the particle's pitch angle and the diffusion coefficient is given as $\kappa_{iso} = A(1 - \mu^2)$. With increasing q, a gap develops around $\mu = 0$, the width of the gap increasing with q. Thus for particles with pitch angles close to 90°, scattering decreases as the power spectrum becomes steeper and it becomes increasingly difficult to scatter the particle through 90°, i.e. to turn around its motion along the field line. The physical basis is simple: particles with large pitch angles are in resonance with waves with small wavelength or high wave numbers. Since the power density spectrum decreases towards higher wave numbers, there is not enough power left for efficient scattering. For $q \geq 2$ the gap around $\mu = 0$ becomes too wide to scatter particles through 90°: although pitch angle scattering still occurs, the direction of the particle motion along the field is not reversed. Or, in other words: the small-angle interactions do not sum up to a large-angle interaction. In this case the particle transport into the two hemispheres parallel and anti-parallel to the magnetic field is decoupled.

7.3.5 Electromagnetic Waves

Pitch angle scattering also occurs in the magnetosphere. Here electrons trapped in the radiation belts are scattered into the loss cone by Whistler waves. Since Whistlers are electromagnetic waves, both a fluctuating electric and a fluctuating magnetic field component exist. Whistlers propagate parallel to the field, and they are circularly polarized with the electric field rotating in the same direction as an electron gyrates (see Sect. 4.6.2). Whistler waves therefore have a resonance at the electron cyclotron frequency.

Particles trapped in the radiation belt perform a bounce motion along the field line. Since they are trapped between two magnetic mirrors, they show a typical pitch angle distribution: particles with small pitch angles are missing because their mirror point lies deep inside the atmosphere and thus interaction with the atmosphere is far more likely than reflection. The resulting pitch angle distribution is called a loss cone distribution. The population of particles with large pitch angles, on the other hand, will bounce back and forth between north and south. This distribution would be stable if sufficiently large electron fluxes did not excite Whistlers. These Whistlers propagate on the background plasma. Like the electrons, Whistlers are trapped in the geomagnetic field, bouncing back and forth between the lower hybrid resonance points. These Whistlers lead to pitch angle scattering, scattering the electrons into the loss cone, and providing a mechanism for the depletion of the radiation belts. This scattering is particularly strong if the electron component in the radiation belts is large, e.g. during a substorm, because large electron fluxes are necessary for efficient wave generation. The electrons scattered into the loss cone then precipitate into the ionosphere.

Under these conditions the quasi-linear diffusion equation is a pure pitch angle diffusion equation:

$$\frac{\partial f}{\partial t} = \frac{1}{\sin \alpha} \frac{\partial}{\partial \alpha} \left(D \sin \alpha \frac{\partial f}{\partial \alpha} \right) . \tag{7.34}$$

The diffusion coefficient written in terms of the fluctuating magnetic field then is

$$D = \pi^2 \omega_{\text{thresh}} \sum_k \frac{1}{|k_\parallel|} \left| \frac{B_1(k)}{B_0} \right|^2 \delta \left(v \cos \alpha - \frac{\omega - \omega_{\text{thresh}}}{k_\parallel} \right) , \tag{7.35}$$

with ω_{thresh} being the threshold frequency up to which Whistler waves can be excited by electrons.

7.3.6 Transport Equations

To describe the development of a particle distribution we need a transport equation. So far, we have encountered the diffusion equation (7.14) and the diffusion–convection equation (7.24).

Focused Transport. Diffusion due to scattering at magnetic field irregularities is only one aspect of particle propagation. As we saw in Sect. 6.3, the interplanetary magnetic field is divergent. Since the magnetic moment of a charged particle is constant, the particle's pitch angle decreases as it propagates outward: a particle starting with a pitch angle close to 90° on the Sun will have a pitch angle of only 0.7° at the Earth's orbit. This effect is called focusing.

In the inner heliosphere, i.e. within about 1 AU, particle propagation is influenced mainly by pitch angle scattering and focusing. The resulting transport equation then is the focused transport equation [446]:

$$\frac{\partial f}{\partial t} + \mu v \frac{\partial f}{\partial s} + \frac{1 - \mu^2}{2\zeta} v \frac{\partial f}{\partial \mu} - \frac{\partial}{\partial \mu} \left(\kappa(\mu) \frac{\partial f}{\partial \mu} \right) = Q(r, v, t) . \tag{7.36}$$

Here s is the length along the magnetic field spiral as described by (6.14), and $\zeta = -B(s)/(\partial B/\partial s)$ is the focusing length. The terms from left to right describe the field parallel propagation, focusing in the diverging magnetic field, and pitch angle scattering. The term on the right-hand side describes the particle source. We have already encountered part of this equation in the discussion of pitch angle scattering in Sect. 7.3.2. Since convection is ignored, (7.36) should only be applied to particles markedly faster than the solar wind. Note that (7.36) cannot be solved analytically but only numerically.

From solutions of the transport equation we can determine the phase space density as a function of time, location and pitch angle. Thus, for a fixed location not only can we determine the intensity–time profile but also the temporal evolution of the pitch angle distribution. The latter can be described as an anisotropy: if the anisotropy vanishes, particles are streaming isotropically from all directions, while for a large anisotropy particles predominately come from one direction.

 Focused Transport Including Solar Wind Effects. For low-energy particles in the inner heliosphere, both focusing and convection with the solar wind are strong systematic effects and should be considered in the transport equation. Here an extension of (7.36) has been suggested which also includes convection with the solar wind and adiabatic deceleration [448]. An already simplified form of the equation reads

$$
\frac{\partial F}{\partial t} + \frac{\partial}{\partial s}\left(\left[\mu'v' + \left\{1 - \frac{(\mu'v')^2}{c^2}\right\}v_{\text{sowi}}\sec\psi\right]F\right)
$$
$$
- \frac{\partial}{\partial p'}\left(p'v_{\text{sowi}}\left[\frac{\sec\psi}{2\zeta}(1 - \mu'^2) + \cos\psi\frac{\mathrm{d}}{\mathrm{d}r}\sec\psi\mu'^2\right]F\right)
$$
$$
+ \frac{\partial}{\partial\mu'}\left(v'\frac{1 - \mu'^2}{2\zeta}F - \kappa(s,\mu')\frac{\partial F}{\partial\mu'}\right) = Q(t,s,\mu',p') . \qquad (7.37)
$$

The distribution function $F(t,s,\mu',p')$ depends on time t, distance s along the field line, pitch angle μ' and momentum p'. The primes indicate that the latter two quantities are measured in the solar wind frame. Note that F is not the phase space density but a distribution function. In particular, in contrast to phase space density, it depends on momentum but not energy.

Equation (7.37) considers, in addition to the terms already mentioned in connection with (7.36), the convection with the solar wind (the additional term in the first set of parentheses) and adiabatic deceleration (the $\partial/\partial p'$ term). Adiabatic deceleration is related to solar wind expansion: as the solar wind expands, the distance between the scattering centers frozen into the solar wind increases. Thus, the "cosmic ray gas" expands too, and, therefore, cools. Adiabatic deceleration differs from the other transport processes, insofar as it changes particle momentum. This makes numerical solutions to the transport equation even more complex since momentum is added as an additional dimension [269, 448]. Thus, for application to observational data, in general the focused transport equation (7.36) or the diffusion–convection equation (7.24) are used.

Figure 7.12 demonstrates the influence of the solar wind effects and their variation with energy. Solid lines are solutions of (7.37), dashed ones are solutions of (7.36), both for different energies between 50 keV and 340 MeV, a radial mean free path of 0.1 AU, and an observer at 1 AU. The consideration of solar wind effects leads to an earlier onset in particle intensity (and, therefore, also in anisotropy), an earlier maximum and a faster decrease in intensity. The earlier onset and maximum are mainly due to convection with the solar wind. Therefore, their effect is most pronounced in low particle energies where the average particle speed is comparable to the solar wind speed. With increasing solar wind speed, convection with the solar wind becomes less and less important: at energies above a few MeV onset and maximum are the same, independent of whether solar wind effects are considered or not. The faster decrease in intensity (and anisotropy) is due to adiabatic deceleration: for a certain energy interval more particles are removed to lower energies than are added from higher ones, owing to the shape of the energy

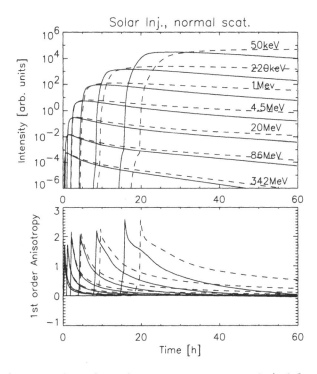

Fig. 7.12. Solution of the focused transport equation with (*solid*) and without (*dashed*) consideration of solar wind effects

spectrum. At energies below a few MeV, this effect is enhanced by convection with the solar wind.

7.4 Interplanetary Propagation – Observations

To infer interplanetary propagation conditions from observations, we can choose from two methods: (a) we can fit a suitable transport equation on intensity– and anisotropy–time profiles; or (b) we analyze the interplanetary magnetic field turbulence. In both cases, we get a pitch angle diffusion coefficient $\kappa(\mu)$ or a mean free path λ.

7.4.1 Fits with a Transport Equation

As an example, Fig. 7.13 shows the intensity–time and anisotropy–time profiles of 0.5 MeV electrons. In the upper panel, the intensity time profile shows the fast rise and slow decay typical of a diffusive profile (see Fig. 7.9). The anisotropy (lower panel) is high early in the event because the first particles always come from the solar direction. It decays as particles are scattered back towards the Sun. The solid lines give a fit of (7.36) on the intensity and anisotropy profile with a mean free path λ of 0.05 AU.

Particle mean free paths are different from event to event and also depend on particle rigidity. Figure 7.14 summarizes the results from fits on a large

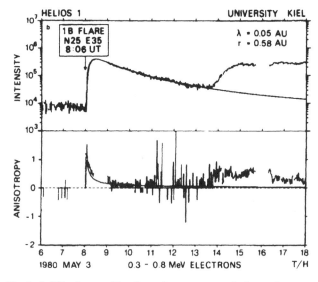

Fig. 7.13. Typical diffusive profile of an electron event in interplanetary space. The solid line gives a fit with the focused transport model (7.36); the second increase starting at about 14 UT is due to a second flare. Reprinted from M.-B. Kallenrode et al. [276], *Astrophys. J.* **391**, Copyright 1992, American Astronomical Society

sample of particle events with the mean free path plotted versus magnetic rigidity. Filled symbols refer to electron observations, open ones to protons. The shaded area gives the Palmer consensus range [388], where most of the events cluster. Here λ is between 0.08 AU and 0.3 AU and, at least in this statistical study, is independent of rigidity. In individual events, however, a small increase in λ with increasing rigidity can be observed.

Over the course of time, a large number of particle events has been analyzed. The main results can be summarized as follows:

- Particle motion predominately is parallel to the field, diffusion perpendicular to the magnetic field can be neglected in the inner heliosphere [158, 256, 300, 432].
- Particle profiles observed in interplanetary space are determined by both the time–development of the particle injection from the Sun and the subsequent interplanetary scattering. If only the intensity is considered, different combinations of injection and diffusion could produce the same profile. This problem can be circumvented by fitting simultaneously intensity– and anisotropy–time profiles [346, 469].
- Particle propagation can range from almost scatter-free (with mean free paths of the order of 1 AU) to strongly diffusive (with λ about 0.01 AU); for most events $\lambda_{\parallel}$ is between 0.08 and 0.3 AU (Palmer consensus range [388]).

7.4.2 Analysis of Magnetic Field Fluctuations

The dotted line in Fig. 7.14 shows the expected rigidity dependence of the particle mean free path determined in an entirely different approach. Since the elementary mechanism is pitch angle scattering by resonant wave–particle interactions, the magnetic field turbulence determines the amount of interplanetary scattering. Thus, an analysis of the magnetic field fluctuations also gives a particle mean free path (see Sect. 7.3.2). The fluctuations are assumed to be small ($\delta B \ll B$), and so quasi-linear theory can be applied. In addition, the wave vectors k of the fluctuating field are assumed to be parallel to the average magnetic field B_0 (slab model). This approach is also called standard quasi-linear theory or standard QLT. As can be seen from Fig. 7.14, the mean free path determined with standard QLT is always below that obtained by fitting. In addition, standard QLT predicts an increase of λ with rigidity while the observations suggest λ to be independent of rigidity. This discrepancy has been known since the early 1970s [216], and it has been called the discrepancy problem. However, as in case of the fits, magnetic field fluctuations and the mean free paths derived from them turned out to be highly variable from event to event (or time span to time span) [547].

7.4.3 Comparison Between Both Approaches

More hints on the discrepancy problem can be gained from the comparison between mean free paths derived from fits and magnetic field fluctuations on

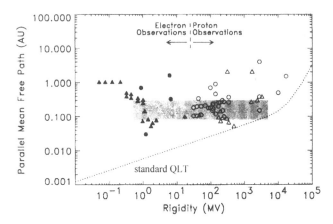

Fig. 7.14. Particle mean free paths in different solar energetic particle events plotted versus the particle's magnetic rigidity. Reprinted from J.W. Bieber et al. [46], *Astrophys. J.* **420**, Copyright 1994, American Astronomical Society

an event to event basis. Figure 7.15 shows the results of such a comparison: (a) the mean free paths determined from magnetic field fluctuations always (with one exception) are smaller than those derived from fits: $\lambda_{QLT} < \lambda_{fit}$; and (b) the discrepancy $\lambda_{QLT}/\lambda_{fit}$ is highly variable from event to event. In addition, the observed rigidity dependence of the mean free path is different from that predicted from quasi-linear theory: for 1 MV electrons and 187 MV protons the ratio between the mean free paths obtained from fits is $\lambda_p/\lambda_e = 1.6 \pm 0.9$ while quasi-linear theory predicts a ratio of 6 [265].

 The main reason for this discrepancy problem appears to be the interpretation of the magnetic field fluctuations. If their scattering power is overrated, the particle mean free path obtained by standard QLT would be too

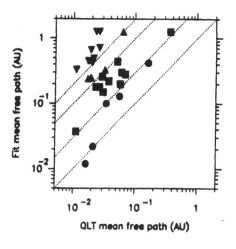

Fig. 7.15. Comparison of mean free paths derived from fits (*vertical axis*) and from magnetic field fluctuations using standard QLT (*horizontal axis*) on an event to event basis [548]. Reprinted from Wanner et al. [548], *Adv. Space Res.* **13**(9), Copyright 1993, Pergamon Press

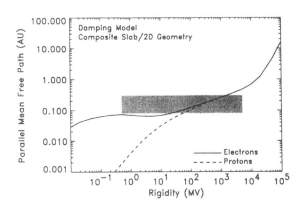

Fig. 7.16. Magnetic field fluctuations as 2D dynamical turbulence. The reduced amount of slab turbulence lets the mean free paths approach the Palmer consensus range (*shaded*). Reprinted from J.W. Bieber et al. [46], *Astrophys. J.* **420**, Copyright 1994, American Astronomical Society

small. Misinterpretation might be easy: observations with one spacecraft give a scalar power density spectrum of magnetic field fluctuations as these are carried across the observer. However, different three-dimensional fluctuations can lead to the same scalar spectrum. In standard QLT it is assumed that all fluctuations arise from waves with $k\|B_0$. Thus, all power contained in the fluctuations also scatters the particles. To reduce the discrepancy, we can interpret the power-density spectrum so as to remove power from the waves parallel to the magnetic field. For instance, if we assume the fluctuations to be Alfvén waves propagating radially away from the Sun, the field-parallel power available for particle scattering would be reduced [254]. This effect would be particularly strong at larger distances where the angle between the interplanetary magnetic field line and the radial direction is large. Since some observations suggest an increase of the discrepancy with increasing distance [549], this ansatz is attractive, although it is problematic because Alfvén waves with large angles with respect to the field are easily dampened.

A different approach interprets the magnetic field fluctuations not as waves but as two-dimensional dynamical turbulence [46]. Observations of the field fluctuations suggest that the correlation function is strongest in the directions perpendicular and parallel to the field [337]. Thus, the fluctuations seem to consist of two components: slab-like Alfvén waves parallel to the magnetic field, and a two-dimensional turbulent component perpendicular to the field. The relative amount of the slab-component is between 12% and 20%, and the two-dimensional turbulence contributes between 80% and 88% to the power-density spectrum. Since only the slab-component scatters the particles, the mean free paths determined with this modified QLT are much closer to the observed ones (see Fig. 7.16). In addition, for low rigidities, electron and proton mean free paths decouple: thus, the observed rigidity (in)dependence can be reproduced with this model, too.

Nonetheless, even with this ansatz it might be too early to declare the discrepancy problem solved and interplanetary transport understood. In our description of particle propagation we always assume that scattering occurs by fluctuations uniformly distributed in space and time. There are no isolated scattering centers. On the other hand, we know that the level of turbulence in interplanetary space can be quite different. We have already seen from Fig. 6.21 that quiet and turbulent periods exist. On shorter time scales of about an hour the fluctuation level can also be highly variable. Again, with one spacecraft it is difficult to interpret this: is the level of fluctuations uniform along a magnetic field line but different on neighboring ones and do the changes in fluctuation level result from the motion of the observer relative to the field lines, or do the magnetic field fluctuations show a completely irregular pattern with variations along the field line as well as between neighboring field lines? The latter case would require an entirely new approach to our understanding of propagation in interplanetary space, in particular because we would have to consider a stochastic distribution of scattering centers instead of the continuous wave fields assumed today.

7.5 Particle Acceleration at Shocks – Theory

Particle acceleration at collisionless shocks can be observed best at planetary bow shocks and travelling interplanetary shocks. Compared with shock waves in other astrophysical objects, such as supernovae remnants or quasars, in these shocks both plasmas and energetic particles can be observed in situ. In this section the fundamentals of shock acceleration will be reviewed. More detailed reviews can be found in [259, 412, 506, 519, 524].

There are different physical mechanisms involved in the particle acceleration at interplanetary shocks:

- the shock drift acceleration (SDA), sometimes also called scatter-free acceleration, in the electric induction field in the shock front;
- the diffusive shock acceleration due to repeated reflections in the plasmas converging at the shock front; and
- the stochastic acceleration in the turbulence behind the shock front.

The relative contributions of these mechanisms depend on the properties of the shock. For instance, shock drift acceleration is important in perpendicular shocks ($\theta_{Bn} = 90°$) where the electric induction field is maximal, but vanishes in parallel shocks. Stochastic acceleration requires a strong enhancement in downstream turbulence to become effective, while diffusive acceleration requires a sufficient amount of scattering in both upstream and downstream media. In addition, shock parameters such as speed, compression ratio (ratio of the densities in the upstream and downstream media), or Mach number, determine the efficiency of the acceleration mechanism.

If we discuss particle acceleration at shocks, we treat the particles as test particles, which do not affect the shock. The shock is thin compared with the Larmor radius, and thus the adiabatic invariants still hold. The jump in field and plasma may be arbitrarily large but both are homogeneous in both upstream and downstream media, and the shock front is planar. Thus effects due to the curvature, in particular drifts, are neglected. Since the total extension of a shock is large compared with the Larmor radius, this approximation is reasonable for a start.

7.5.1 Shock Drift Acceleration (SDA)

Shock drift acceleration (SDA) takes advantage of the electric induction field in the shock front [11,122,134,165]. To allow for a reasonably long drift path, scattering is assumed to be negligible. SDA therefore is also called scatter-free shock acceleration. The calculations by Schatzmann [458] were the first theoretical studies containing an order-of-magnitude estimate of the efficiency of this mechanism. They showed that a very efficient acceleration on short temporal and spatial scales is only possible if there is additional scattering which feeds the particles back into the shock for further acceleration. As we shall see below, this problem of feeding the particles back into the acceleration mechanism is a recurrent problem in the study of shock acceleration, independent of the actual acceleration mechanism.

In shock drift acceleration, a charged particle drifts in the electric induction field in the shock front. In the shock's rest frame, this is

$$\boldsymbol{E} = -\boldsymbol{u}_{\mathrm{u}} \times \boldsymbol{B}_{\mathrm{u}} = -\boldsymbol{u}_{\mathrm{d}} \times \boldsymbol{B}_{\mathrm{d}} . \tag{7.38}$$

This field is directed along the shock front and perpendicular to both magnetic field and bulk flow; it is maximal at a perpendicular shock and vanishes at a parallel shock. In addition, the shock is a discontinuity in magnetic field strength. Thus a particle can drift along the shock front according to (2.54). The direction of the drift depends on the charge of the particle and is always such that the particle gains energy.

Figure 7.17 shows sample trajectories for particles in the rest frame of a quasi-perpendicular shock with $\theta_{\mathrm{Bn}} = 80°$. The abscissa shows the distance from the shock in gyro-radii; the ordinate gives the particle energy in units of the initial energy E_0. The dashed lines indicate the shock; the upstream medium is to the left, the downstream medium to the right. The electric induction field is parallel to the shock front in the upward direction. In the left panel, the motion of the particle starts in the upstream medium. The particle then drifts along the shock front, gaining energy. This energy gain changes the details of the drift path: the velocity component perpendicular to the shock increases, eventually becoming larger than the shock speed. Then the particle separates itself from the shock front. In the left panel, the particle is reflected back into the upstream medium. The other two panels show particles transmitted through the shock, either from the upstream medium to the downstream medium (middle panel) or vice versa (right panel).

The details of the particle trajectory, in particular the question of whether the particle is transmitted or reflected, strongly depends on its initial energy

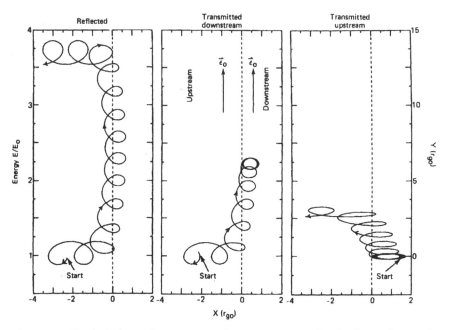

Fig. 7.17. Shock drift acceleration: sample trajectories in the shock rest frame of a quasi-perpendicular shock with $\theta_{Bn} = 80°$. Reprinted from R.B. Decker [123], *Space Science Reviews* **48**, Copyright 1988, with kind permission from Kluwer Academic Publishers

and pitch angle (for additional examples of trajectories see, for example, [11, 123]), leading to a characteristic angular distribution of the accelerated particles: an initially isotropic distribution of particles in the upstream and downstream media is converted to a very strong field-parallel beam in the upstream medium and to a smaller beam roughly perpendicular to the field in the downstream medium.

The energy gain of a particle is largest if the particle can interact with the shock front for a long time. This time depends on the particle's speed perpendicular to the shock. If this is small, the particle sticks to the shock, if it is large, the particle escapes before it has gained a large amount of energy. This particle speed relative to the shock is determined by the particle speed, the shock speed, the pitch angle, and θ_{Bn}. With increasing particle speed the range of the other three parameters, which allow for efficient acceleration, narrows, reducing the chance of efficient acceleration. Thus, acceleration to higher and higher energies becomes increasingly difficult.

Since the magnetic moment is conserved, the momentum perpendicular to both shock and field after an interaction between particle and shock is

$$\frac{p_{2\perp}}{p_{1\perp}} = \frac{B_d}{B_u} = r_B \ . \tag{7.39}$$

The normal component of the momentum is unchanged by the interaction. Thus the change in momentum is determined by the magnetic compression r_B, the ratio between the upstream and downstream magnetic field strengths. Equation (7.39) is valid only for perpendicular shocks. For oblique shocks, Toptygin [519] gives the approximation $\Delta E \sim p u_u / \theta_{Bn}$.

The average energy gain is a factor 1.5–5. Evidently, such an energy gain is much too small to accelerate particles out of the solar wind to energies of some tens of kiloelectronvolts or even some tens of megaelectronvolts. Acceleration up to higher energies by shock drift acceleration would require repeated interactions between particles and shock. If a particle is in the downstream medium, further acceleration is not difficult because the turbulence created by the shock leads to stochastic acceleration or scatters the particle back towards the shock front. But once a particle has escaped into the upstream medium, the shock will not be able to catch up with it again, and thus no further acceleration is possible. This statement is only true for scatter-free conditions as originally assumed in shock drift acceleration. Space plasmas, however, are turbulent plasmas, and particles are scattered depending on the level of turbulence. This allows for repeated interactions between particles and shock and thus a higher energy gain [124–126]. Note that in the presence of scattering the energy gain in each individual interaction between particle and shock in general is smaller because the particle is easily scattered off from its drift path along the shock front. The larger net effect results from the larger number of interactions.

Shock drift acceleration out of the solar wind up to energies of a few kiloelectronvolts to some tens of kiloelectronvolts can be observed at the Earth's bow shock and at interplanetary travelling shocks (see below).

7.5.2 Diffusive Shock Acceleration

Diffusive shock acceleration is the dominant mechanism at quasi-parallel shocks. Here the electric induction field in the shock front is small and shock drift acceleration is negligible. In diffusive shock acceleration, the particle scattering on both sides of the shock is crucial. Since the scattering centers are frozen into the plasma, particle scattering back and forth across the shock can be understood as repeated reflection between converging scattering centers. The concept was developed in the 1970s [16, 39, 51, 137]; for reviews, see, for example, [13, 165, 315, 461, 462, 538].

Figure 7.18 shows the motion of a particle in the rest frame of a quasi-parallel shock. The magnetic fields on both sides of the shock are turbulent; the resulting pitch angle scattering is quantified by the diffusion coefficients D_u and D_d or the mean free paths λ_u and λ_d. Assume that the particle has just traversed the shock front in the upstream direction. The particle then follows a zigzag path and eventually is scattered back to the shock front, traversing it into the downstream medium. Here the same process is repeated, only with the diffusion coefficient characteristic of the downstream

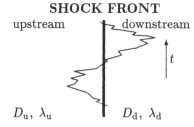

Fig. 7.18. Diffusive shock acceleration: motion of a particle in the shock rest frame. One cycle of motion consists of a crossing from one side of the shock to the other and back again

medium. As the particle is finally scattered back towards the shock front, traversing it into the upstream medium, a new cycle of motion begins.

But where in this cycle does the acceleration occur? Whereas in shock drift acceleration the location of acceleration is well defined, in diffusive shock acceleration the acceleration is given by the sum of all pitch angle scatters within the cycle. The energy gain can be determined from a simplification. Let us reduce the cycle to two isolated collisions, one in the upstream medium and the other in the downstream medium: in the upstream medium the particle gains energy due to a head-on collision with a scattering center; in the downstream medium it loses energy because the scattering center moves in the same direction as the particle. Since the flow speed, and therefore the speed of the scattering center, is larger upstream than downstream, a net gain in particle energy results. The amount of energy gained in each interaction depends on the relative speed between particle and scattering centers. The simplified picture sketched here assumes particle motion parallel to the field. In reality, the particle will have a pitch angle. Then the energy gain depends on the particle velocity parallel to the magnetic field. Thus for a given particle speed the energy gain depends on pitch angle, too.

Diffusive shock acceleration is based on scattering, which is a stochastic process. Therefore we cannot calculate the path of a certain particle. However, from the initial particle distribution, the shock parameters, and the scattering conditions in the upstream and downstream media we can calculate the average behavior of a large number of particles. Diffusive shock acceleration, therefore, must be described in terms of a phase space density and a transport equation. The latter can be written as

$$\frac{\partial f}{\partial t} + U\nabla f - \nabla \cdot (\mathsf{D}\,\nabla f) - \frac{\nabla U}{3}p\frac{\partial f}{\partial p} + \frac{f}{T} + \frac{1}{p^2}\frac{\partial}{\partial p}\left(p^2\left(\frac{\mathrm{d}p}{\mathrm{d}t}\right)f\right)$$
$$= Q(p, r, t)\,. \tag{7.40}$$

Here f is the phase space density, U the plasma speed, D the diffusion tensor, and T the loss time. The terms, from left to right, present: convection of particles with the plasma flow, spatial diffusion, diffusion in momentum space (acceleration), losses due to particle escape from the acceleration region, and convection in momentum space due to processes which affect all particles, such as ionization or Coulomb losses. The term on the right-hand side is a source term describing the injection of particles into the acceleration process. In a first-order approximation losses and convection in momentum can be ignored, reducing the transport equation to

$$\frac{\partial f}{\partial t} + \boldsymbol{U}\nabla f - \nabla\left(D\,\nabla f\right) - \frac{\nabla\cdot\boldsymbol{U}}{3}p\frac{\partial f}{\partial p} = Q(p,r,t)\,. \tag{7.41}$$

Here the terms containing the plasma speed $\boldsymbol{U}$ and the diffusion tensor D describe the acceleration of the particles across the shock front.

Often we are not interested in the evolution of the particle distribution at the shock but only want to know the particle spectrum and the acceleration time in steady state. In this case, $\partial f/\partial t$ equals zero and the transport equation can be solved for suitable boundary conditions, such as steadiness in particle density and in the normal component of the particle flow $S = -4\pi p^2[Up(\partial f/\partial p) + D\cdot\nabla f]$ across the shock front. This steady-state equation leads to predictions of the acceleration time, the particle energy spectrum, and the intensity increase upstream of the shock.

Characteristic Acceleration Time. From the transport equation (7.41) the time required to accelerate particles from momentum p_0 to momentum p can be inferred to be

$$t = \frac{3}{u_{\mathrm{u}} - u_{\mathrm{d}}}\int_{p_0}^{p}\frac{\mathrm{d}p}{p}\left(\frac{D_{\mathrm{u}}}{u_{\mathrm{u}}} + \frac{D_{\mathrm{d}}}{u_{\mathrm{d}}}\right)\,. \tag{7.42}$$

For all momenta below p the particle distribution at the shock is in steady state. Note that here isotropic scattering in a homogeneous medium is assumed, reducing the diffusion tensor D to a diffusion coefficient D.

If we assume the diffusion coefficient D to be independent of momentum p, then (7.42) can be integrated to give a characteristic acceleration time

$$\tau_{\mathrm{a}} = \frac{p}{\mathrm{d}p/\mathrm{d}t} = \frac{3}{u_{\mathrm{u}} - u_{\mathrm{d}}}\left(\frac{D_{\mathrm{u}}}{u_{\mathrm{u}}} + \frac{D_{\mathrm{d}}}{u_{\mathrm{d}}}\right) \tag{7.43}$$

in $p(t) = p_0\exp(t/\tau_{\mathrm{a}})$. Equation (7.43) is often written as

$$\tau_{\mathrm{a}} = \frac{3r}{r-1}\frac{D_{\mathrm{u}}}{u_{\mathrm{u}}^2}\,, \tag{7.44}$$

with $r = u_{\mathrm{u}}/u_{\mathrm{d}}$ being the ratio of the flow speeds in the shock rest frame. For a parallel shock, this ratio equals the compression ratio. In (7.44) $D_{\mathrm{d}}/u_{\mathrm{d}}$ is assumed to be small compared with $D_{\mathrm{u}}/u_{\mathrm{u}}$ because of enhanced downstream turbulence.

Example 23. Let us now assume a shock with an upstream speed of 800 km/s in the shock rest frame and a ratio of flow speeds $r = 3$. With a typical upstream mean free path $\lambda_u = 0.1$ AU for 10 MeV protons, we obtain an upstream diffusion coefficient $D_u = v\lambda/2 = 3.26 \times 10^{17}$ m^2/s. With (7.44), we then obtain a characteristic acceleration time $\tau = 2.3 \times 10^6$ s, that is, almost 27 days or one solar rotation. Even if we were to increase the scattering by a factor of 10, the characteristic acceleration time would be almost 3 days, which is longer than the average travel time of a shock between the Sun and the Earth. However, under these conditions, the mean free path would be rather small (see Fig. 7.15) and the acceleration would still be inefficient, because during the characteristic acceleration time the momentum is increased only by a factor e. The situation is different at a lower particle speed, say 100 keV. In that case, the diffusion coefficient is an order of magnitude smaller and, consequently, the characteristic acceleration time is only a tenth of that for 10 MeV protons, that is, roughly 7 hours (for the optimistic case of strong scattering with $\lambda = 0.01$ AU). During a travel time of 2 days, the particle momentum therefore increases by about three orders of magnitude (under the simplifying assumption of a diffusion coefficient independent of particle energy/momentum).   □

The Energy Spectrum. The energy spectrum expected from diffusive shock acceleration is a power law:

$$J(E) = J_0 \, E^{-\gamma} \, . \tag{7.45}$$

The spectral index γ depends only on r. In the non-relativistic case this is

$$\gamma = \frac{1}{2} \frac{r+2}{r-1} \, . \tag{7.46}$$

In the relativistic case, the spectral index becomes $\gamma_{\rm rel} = 2\gamma$. Again, this holds only for the steady state, i.e. $t \gg \tau_a$.

Why do we get an energy spectrum and not, for instance, a monoenergetic distribution? The energy gain for each particle is determined by its pitch angle and the number of shock crossings. The latter is determined by the stochastic process of scattering. For a high energy gain, the particle must be "lucky" to be scattered back towards the shock again and again. Most particles make a few shock crossings and then escape into the upstream medium. They are still scattered there, but they have escaped too far to be scattered back to the shock. Thus the stochastic nature of diffusion allows high gains for a few particles, while most particles make only small gains.

Example 24. Typical values for the shock compression ratio r are between slightly above 1 and about 8; the average value is close to 2. With this value from (7.46) we obtain a spectral index $\gamma = 2$, which is rather flat compared with the observed spectral indices, which cluster around 3.   □

Intensity Increase Upstream of the Shock. With an infinitely thin shock at $x = 0$ and a continuous particle injection $Q = q\delta(x)\delta(p-p_0)/(4\pi p_0^2)$ at the shock front, the spatial intensity variation around the shock is

$$f(x,p) = f(0,p)\exp\{-\beta_i|x|\}\,,\qquad(7.47)$$

with $\beta_i = u_i/D_i$, the index i indicating the upstream and the downstream medium, respectively. If particle escape is considered (the f/T term in (7.40)), the scale length β_i for the upstream intensity increase is

$$\beta_i = \frac{u_i + \sqrt{u_i^2 + 4D_i/T}}{2D_i}\,.\qquad(7.48)$$

Figure 7.19 shows a typical upstream intensity increase and its dependence on particle energy. If β is spatially constant, the intensity increase is exponential, with its slope depending on D. Because λ increases with energy (Sect. 7.4), the ramp is steeper for lower energies E_1. In addition, the intensity at the shock is higher for lower energies, reflecting the power-law spectrum (7.45). In interplanetary space, the slope of the intensity increase is often used to determine the scattering conditions upstream of the shock.

Example 25. With the parameters given in example 23, the scale length of the upstream increase is 2.5×10^{-12} m^{-1}, or 0.35 AU^{-1}. Let us assume that the shock has reached steady state. In this case this upstream increase would be convected across the observer at the shock speed. If this is 1200 km/s, a scale length of 0.35 AU takes about 54 000 s, or 15 h, to pass the observer. Thus, in the intensity time profile upstream of the shock, we would expect an increase by a factor of e over a time of 15 h. Again the situation is different for 100 keV protons. Here the diffusion coefficient and thus the scale length are an order of magnitude smaller and, consequently, the rise in intensity upstream of the shock will be much steeper: a factor of e in 1.5 h.   □

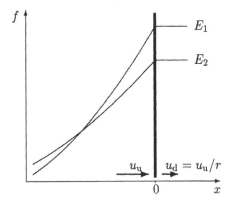

Fig. 7.19. Intensity increase upstream of the shock front for two energies $E_1 < E_2$

7.5.3 Diffusive Shock Acceleration
and Self-Generated Turbulence

Occasionally, (7.44) is used to show that proton acceleration up to mega-electronvolt energies at travelling interplanetary shocks is difficult: "Yet the Fermi process develops so slowly that the protons accelerated by quasi-parallel interplanetary shocks only reach energies of a few hundred thousand electron volts in the one day it takes the shock to travel from the Sun to the Earth" [455]. Note that this conclusion depends strongly on the assumptions about the particle mean free path: the particle mean free paths inferred from solar energetic particle events, lead to acceleration times too high to explain the observations. For higher energies (e.g. in the megaelectronvolt or tens of megaelectronvolt range), the mean free path inferred from solar energetic particle events is so large that a particle would interact with the shock only a few times during its travel time from the Sun to the Earth. This even violates the assumption of frequent interactions inherent in the transport equation.

However, if the turbulence upstream of the shock could be increased, the acceleration would be more efficient. But how can we increase the amount of upstream scattering? As we saw in Chap. 7.3, scattering is a consequence of wave–particle interactions. But waves cannot propagate away from the shock into the upstream medium. Thus the upstream medium does not have any knowledge of the approaching shock (which, of course, is inherent in the definition of a shock). This statement is true from the viewpoint of the plasma. But as we saw in the previous section, the energetic particles do escape from the shock front, as evident in the upstream intensity increase. Whenever energetic particles stream faster than the Alfvén speed, they generate and amplify MHD waves with wavelengths in resonance with the field parallel motion of the particles [29], i.e. the same resonance condition as in Fig. 7.10. These waves grow in response to the intensity gradient of the energetic particles. A hydrodynamic approximation on the wave growth, treating the energetic particles as a fluid, can be found in [349]. Lee has introduced a model in which the diffusion equations describing the particle transport and the wave kinetic equations describing the wave transport are solved self-consistently for the Earth's bow shock [312] and travelling interplanetary shocks [313]. Lee's model allows some predictions regarding the particles and waves upstream of the shock: (i) At or just upstream of the shock the particle spectrum can be described by a power law in momentum $f \sim p^{-\sigma}$ with $\sigma = (1 - u_2/u_1)/3$. The scale-length of the intensity increase just upstream of the shock increases with increasing particle momentum as $p^{\sigma-3}$. (ii) The upstream waves generated by the particles propagate into the upstream direction. The growth rate is largest for waves with wave vectors parallel to the ambient magnetic field. The scale-length for the decay of the upstream waves is about the scale-length for the intensity increase of particles in resonance with these waves. Thus the scale-length of wave decay depends on the wave number. The power density spectrum of the excited upstream waves can be described as $f(k) \sim k^{\sigma-6}$.

(iii) The ratio between the particle energy density ε_p and the magnetic field density of the fluctuating waves $\langle|\delta\boldsymbol{B}|^2\rangle/2\mu_0$ is proportional to the ratio between upstream flow speed and Alfvén speed:

$$\frac{\varepsilon_p}{\langle|\delta\boldsymbol{B}|^2\rangle/2\mu_0} \propto \frac{u_u}{v_A} . \tag{7.49}$$

These predictions have been found to be in fair agreement with the observations in the 11 November 1978 event [283, 315], although some differences in the details between the observed and the predicted turbulence have been found. The energy up to which the accelerated particle population is in steady state is a few hundred kiloelectronvolts.

In sum, Lee's model suggests acceleration at the shock to occur as follows. First, accelerated particles stream away from the shock. As they propagate upstream, the particles amplify low-frequency MHD waves in resonance with them. Particles escaping from the shock at a later time are scattered by these waves and are partly reflected back towards the shock. These latter particles again interact with the shock, gaining additional energy. Thus even more energetic particles escape from the shock, amplifying waves in resonance with these higher energies. The net effect is an equilibrium between particles and waves which in time shifts to higher energies and larger wavelengths.

7.5.4 Stochastic Acceleration

In the previous section we saw that turbulence generated (or amplified) by particles streaming into the upstream medium is important for efficient shock acceleration. In the downstream medium, the situation is different. As the shock passes by, waves are generated with wave vectors in both the upstream and downstream directions. Thus a particle can either gain energy from the wave or supply energy to it. This is a stochastic process, and the resulting particle acceleration is called stochastic acceleration or second-order Fermi acceleration. Fermi originally had suggested particle scattering at uncorrelated magnetic inhomogeneities moving in arbitrary directions. A particle then gains or loses energy, depending on whether it hits the inhomogeneity head-on or not. Because of the Doppler effect, head-on collisions are slightly more frequent, leading to a net gain in energy. The first considerations of the possibility of second-order Fermi acceleration in the turbulence in the downstream medium suggested that this effect is of minor importance [367] compared with the first-order Fermi acceleration described above. Thus applications of stochastic acceleration in general are limited to solar flares or astrophysical objects. At travelling interplanetary shocks stochastic acceleration occasionally seems to lead to intensity increases in the upstream medium which become evident as jumps in the intensity as the shock passes over the observer (Sect. 7.6).

Physically, stochastic acceleration is based on wave–particle interactions. In contrast to spatial scattering here the interaction leads to scattering in

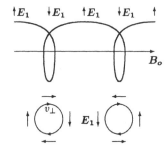

Fig. 7.20. Resonant wave–particle interaction: the particle interacts with the fluctuating electric field of a circularly polarized wave. The particle is in resonance with the wave if the wave frequency equals the gyration frequency. Depending on the phase between wave and particle gyration, the particle is either constantly accelerated or decelerated

momentum space. In pitch angle scattering the particle interacts with the fluctuating magnetic field of the waves. A different kind of interaction can take place if the particle interacts with the fluctuating electric field of a wave, in particular if this field rotates around the magnetic field line as in a circularly polarized wave. A particle is in resonance with the wave if its gyration frequency equals the frequency of the wave. In that case, the particle is either accelerated or decelerated continuously. Figure 7.20 illustrates this principle. In the upper panel the gyration of a particle around B_0 is shown together with the fluctuating electric field. To make the figure readable, the fluctuating magnetic field is not shown. In the lower panel, the two extreme cases are illustrated: depending on the phase between the wave and gyro-orbit, the particle either moves parallel or anti-parallel to the electric field. Thus, either deceleration or acceleration results. In the latter case, the wave energy is converted into particle energy, and in the former case the particle energy is converted into wave energy. The resulting acceleration is called stochastic acceleration because the result (acceleration or deceleration) depends on the random phase between wave and particle.

Since the acceleration or deceleration changes the particle speed perpendicular to the average magnetic field, the particle's pitch angle changes too. Stochastic acceleration can be described as diffusion in momentum space as long as the energy gain is small compared with the particle's energy (see Sect. 7.3.3). Then the accompanying change in pitch angle is small, too.

7.5.5 The Shock as a Non-linear System

Shocks are highly non-linear systems with complex interactions between plasmas, waves, and energetic particles. We often use linear or quasi-linear approximations to describe some aspects of the entire phenomenon, such as shock formation or particle acceleration, but we have to be aware that these are approximations. For instance, particle acceleration and wave generation require energy which, of course, has to be taken from the plasma flow, thus altering the shock.

Figure 7.21 sketches the complexity of the phenomenon shock. The system can be divided into three parts: the plasma, the wave field, and the

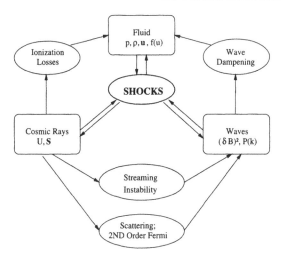

Fig. 7.21. Sketch of the non-linear interactions at a shock front. Arrows indicate the direction of energy flows, based on [13]

energetic particles (all shown as rectangles). The plasma is characterized by its distribution function f which also serves as the injection function into the acceleration process, and by its hydromagnetic parameters pressure p, density ϱ, bulk speed $\boldsymbol{u}$, and magnetic field $\boldsymbol{B}$. The energetic particles are characterized by their density U and their streaming $\boldsymbol{S}$. The wave field can be described by the power spectrum $P(k)$ or the average squared magnetic fluctuations $\langle(\delta B)^2\rangle$. The wave field determines the diffusion coefficient and therefore the coupling between background plasma and energetic particles.

In a linear system, we could determine the relevant quantities for the waves and fields independently and use the resulting diffusion coefficient and velocity field to calculate the acceleration of energetic particles. In a non-linear system, we also have to consider mutual interactions. This can be illustrated by any pair of shock components in Fig. 7.21. Let us start with the wave field and the background plasma: waves can be amplified by the interaction with the shock, they can be dampened by the plasma, feeding wave energy into the plasma, or the wave pressure and energy flux can change the dynamics of the plasma. The non-linear interaction between another pair of shock constituents, the wave field and the energetic particles, has been partially discussed before: energetic particles can excite or amplify waves by streaming instabilities while these waves in turn scatter the particles, leading either to second-order Fermi acceleration in the downstream medium or more efficient first-order Fermi acceleration. Part of the particle's energy therefore can be transferred to the plasma by first converting it to wave energy which is then converted into plasma energy by damping.

A shock therefore is a complex, highly non-linear system. Even if all relevant equations can be written down, it is not possible to solve them self-consistently. Therefore, either approximations can be used, such as the ones presented in this chapter, or the system or part of the system can be studied by means of Monte–Carlo simulations.

Another crucial aspect of shock acceleration not addressed so far is the injection problem: to be accelerated, the particle must be sufficiently energetic to be scattered across all the micro- and macro-structures of the shock to experience the compression between the upstream and the downstream medium. Thus although we assume particle acceleration out of the solar wind, not all solar wind particles can be accelerated because they are not energetic enough. Different scenarios can be developed ranging from the acceleration of particles out of the high energetic tail of the distribution only to some kind of pre-acceleration by direct electric fields or the requirement of a pre-accelerated particle component. A review of this injection problem is given in [568].

7.5.6 Summary Shock Acceleration

Shocks provide an important acceleration mechanism for particles of planetary, solar and even galactic origin. Three acceleration mechanisms can be distinguished which are not mutually exclusive but work together with different relative contributions depending on the properties of the shock:

- In shock drift acceleration (SDA, also called scatter-free acceleration) particles gain energy by drifting along the electric induction field in the shock front. The energy gain is between a factor of 1.5 and 5, and the particle interacts with the shock only once. If particles are scattered in the ambient plasma, the energy gain per interaction is reduced because the drift path is shortened, but particles can interact repeatedly with the shock. This might lead to a higher total energy gain.
- In diffusive shock acceleration the particles gain energy due to repeated scattering in the plasmas converging at the shock front (first-order Fermi acceleration). The time scales of acceleration and the energies acquired depend on the amount of scattering. Diffusive shock acceleration is a slow process, the maximum energy acquired depends on scattering conditions, shock parameters, and the time available for acceleration.
- Diffusive shock acceleration becomes more efficient if the turbulence created by the particles is considered. This self-generated turbulence in the upstream medium scatters the particles back towards the shock front, leading to further acceleration. This is an example for the non-linearity of processes at a shock.
- Stochastic acceleration is a second-order Fermi process during which the particle gains energy due to scattering in the downstream turbulence.

7.6 Particles at Shocks in Interplanetary Space

In Fig. 7.4 we saw a particle event consisting of both a flare-accelerated component and a particle component accelerated at the shock in front of the CME. Let us now take a look at the particle distribution right at an interplanetary travelling shock.

Figure 7.22 shows the energy spectrum between 200 eV and 1.6 MeV for ions in the spacecraft rest frame just downstream of an interplanetary shock. The pronounced peak around 1 keV is the solar wind plasma; the power-law spectrum for higher energies gives the energetic storm particles. The rather smooth transition between these two spectra is in agreement with the assumption that the particles are accelerated out of the solar wind plasma. While the solar wind spectrum is described by a Maxwellian, the combined spectrum can be approximated by a kappa-distribution (5.14). Note the break in the power-law spectrum at energies of a few hundred kiloelectronvolts: here the spectrum is steeper, indicating a less efficient acceleration.

It appears reasonable to discuss energetic particles accelerated at interplanetary shocks in two separate energy bands: (a) a low-energy component

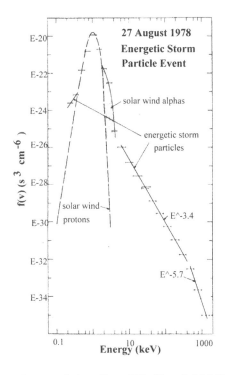

Fig. 7.22. Spectrum of energetic ions from 200 eV to 1.6 MeV at an interplanetary shock in the spacecraft's reference frame. Reprinted from J. Gosling et al. [201], *J. Geophys. Res.* **86**, Copyright 1981, American Geophysical Union

with ion energies up to a few hundred keV/nucl and electron energies up to some tens of kiloelectronvolts; and (b) a high-energy component with ion energies in the megaelectronvolts and tens of MeV/nucl range and electron energies in the hundreds of kiloelectronvolt to megaelectronvolt range. One reason for such a separation is the break in the ion spectrum, as indicated in Fig. 7.22. Another argument stems from the different particle speeds relative to the shock: while protons in the tens of kiloelectronvolt range on average are only slightly faster than the shock and therefore stick close to the shock front, protons in the megaelectronvolt range are much faster and can escape easily from the shock. Thus, an observer in interplanetary space at lower energies detects the shock-accelerated particles mainly around the time of shock passage, while at higher energies particles accelerated at the shock can be detected when the shock is still remote from the observer. In addition, acceleration is more efficient if particles stay close to the shock and interact repeatedly than for particles that escape easily from the vicinity of the shock.

Particle events associated with interplanetary shocks are reviewed, for instance, in [266, 271, 435].

7.6.1 Low-Energy Particles (Tens of keV) at Travelling Shocks

At low energies, three types of particle events can be distinguished, each related to another acceleration mechanism (see Fig. 7.23). Nonetheless, as we saw in Sect. 7.5, combinations of acceleration mechanisms are possible and so too are combinations of different types of events.

In the left panel of Fig. 7.23 an energetic storm particle event is sketched. This is the classical type of particle event at interplanetary shocks. In the particle data, it starts some hours prior to shock arrival at the decaying flank of the preceding solar energetic particle event. The intensity maximum is close to the time of shock passage. Often there is no dip between the first maximum and the maximum at the time of shock passage but a more plateau-like profile or continuous increase as in the protons in Fig. 7.4. In many cases, a few hours after the arrival of the shock, a sudden decrease in intensity is observed as the spacecraft enters the ejecta or magnetic cloud, the former coronal mass ejection, driving the shock. With increasing energy, the intensity increase related to the shock becomes smaller, reflecting the rather steep energy spectrum of ESPs compared with SEPs. These events are connected with quasi-parallel shocks, and thus the particles are accelerated by diffusive shock acceleration. The features of the particle event, for instance the quasi-exponential intensity increase upstream of the shock and the energy spectrum, are in agreement with the predictions of this acceleration mechanism. In addition, the enhanced upstream turbulence is indicative of self-generated turbulence [283, 456].

At quasi-perpendicular shocks, a different type of particle event can be observed, namely the shock spike (middle panel in Fig. 7.23). Here the inten-

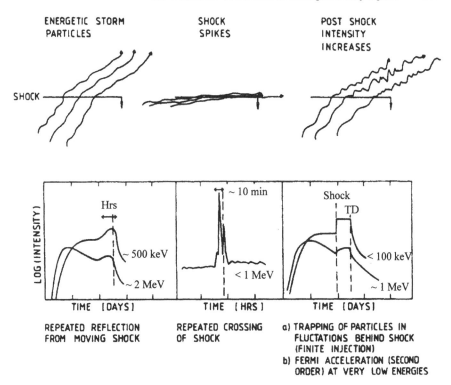

Fig. 7.23. Different types of energetic particle events in the tens to hundreds of kiloelectronvolt range at interplanetary shocks and the acceleration mechanisms. Reprinted from M. Scholer and G. Morfill [464], in *Study of travelling interplanetary phenomena*, 1977, Air Force Geophysics Laboratory

sity increase is limited to a few minutes around the time of the shock passage. Despite its short duration, the intensity increase can be quite large. Shock spikes are explained by shock-drift acceleration. However, to acquire the observed energies, the particles must cross the shock more than once, and thus here shock-drift acceleration works in combination with scattering and not as scatter-free shock acceleration. The details of the particle event (single-spike, multiple-spikes, etc.) are more variable than in the classical ESP events. This structuring, however, is not necessarily related to the acceleration process but might also reflect the differences between neighboring field lines or the variability of the interplanetary medium on short temporal and spatial scales.

The third type of event, the post-shock increase, can occur in isolation, on the decaying flank of a SEP event or in the wake of an ESP event. In the particle data, it can be identified as an abrupt increase in intensity as the shock passes the observer. It often ends abruptly as the observer encounters the ejecta. Post-shock intensity increases are generally associated with strong turbulence in the downstream medium. This turbulence leads to an efficient stochastic acceleration and a storage of the accelerated particles downstream of the shock front. In the low-energy electrons, the post-shock increases are the most common type of event; however, in ions and higher energetic electrons they are less frequent.

While the local geometry, the angle θ_{Bn} between the magnetic field direction and the flare normal, determines the shape of the shock-related particle increase, its size is determined by the compression ratio, the magnetic compression, and the shock speed $v_{SB} = v_s \sec \theta_{Bn}$ parallel to the magnetic field. This latter parameter even seems to distinguish between shocks with little or no particle acceleration ($v_{SB} < 250$ km/s) and shocks that lead to significant particle acceleration [523]. Although v_{SB} is less expressive than the shock speed, its physical significance is more important: as the particles are bound to travel along the magnetic field line, the shock speed parallel to the field determines the ability of a particle to escape from the shock. Even at rather slow shocks, a large number of particles might be kept close to the shock for further acceleration as long as θ_{Bn} is large, i.e. the shock is quasi-perpendicular. The physical significance of the magnetic compression also might be simple: the change in B across the shock front creates a magnetic mirror for particles moving from the upstream medium towards the shock. As these particles are reflected at this mirror, they gain energy. This process is faster and more efficient than the reflection in the turbulence downstream of the shock as assumed in standard diffusive shock acceleration. However, for efficient particle acceleration multiple encounters with the shock are required, and thus again the upstream turbulence is the crucial factor. For one particle event a complete analysis of particles and waves upstream of the shock has confirmed the existence of self-generated turbulence and its good agreement with the theoretical predictions [283].

7.6.2 High-Energetic Particles (MeVs) at Travelling Shocks

In megaelectronvolt protons and ions different intensity profiles can be distinguished, too. However, they do not reflect the local acceleration mechanism as in the low energies but the location of the observer relative to the shock. This is a consequence of the higher speeds of the megaelectronvolt particles which allows them to escape from the shock front. An observer in interplanetary space thus samples all particles that the shock has accelerated on the observer's magnetic field line while it propagates outward. The intensity profile therefore is the superposition of particle injections with different sizes and at different positions.

Figure 7.24 shows sketches of typical intensity profiles of 20 MeV protons for different locations at an interplanetary shock, the latter propagating downward. The intensity of the first strong intensity increase is largest for an observer on the eastern flank. Since these particles arrive very early, they must have been accelerated at a time when the shock still was close to the Sun. This component can be called a solar component because the particles

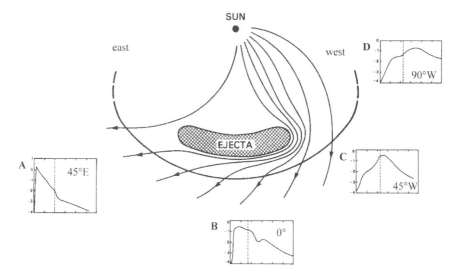

Fig. 7.24. Representative profiles of about 20 MeV proton events for different positions of the observer with respect to the shock. The draping of the field lines around the ejecta is only a suggestion. Reprinted from H.V. Cane et al. [85], *J. Geophys. Res.* **93**, Copyright 1988, American Geophysical Union

are accelerated close to the Sun: either by the flare, by the shock, or by a combination of both. Going towards the central meridian or the western flank of the shock, the intensity of the solar component decreases. The intensity of the shock-accelerated component, on the other hand, is largest at the nose of the shock at the central meridian while it decreases towards the flanks. Note that here shock-accelerated means "accelerated at the shock while it propagates outward".

Now we can easily understand the different profiles. The observer in panel D is located at the eastern flank of the shock and is magnetically connected to the flare site. Thus he sees a strong intensity increase early in the event due to the solar component. Often the flare-accelerated particles might be identified from their charge states and composition (see Fig. 7.4). The local acceleration at the shock, however, is rather meager because the observer is located only on its flank. Going to the east, observer A is located at the nose of the shock but is magnetically connected to a position about 60° west of the flare site. The solar component therefore is smaller; however, the particle component accelerated at the shock is maximal. Whether the first increase is larger than the one close to the shock, as depicted in the figure, or vice versa, depends on the properties of the flare, the shock's acceleration efficiency, and the scattering conditions in interplanetary space. Moving farther to the west, the solar component decreases or becomes undetectable because of an insufficient magnetic connection. The component accelerated at the shock is also smaller than the one at the central meridian; however, it is always larger than the solar component.

Note that Fig. 7.24 is a schematic based on a statistical study. The pattern of the changes in the profiles along the shock front have also been confirmed by multi-spacecraft observations at individual shocks; however, the variations are not necessarily as pronounced as those indicated in Fig. 7.24 [270]. Figure 7.25 shows an example of a shock observed by three spacecraft at different positions: the top panel shows the geometry, the center of the shock prop-

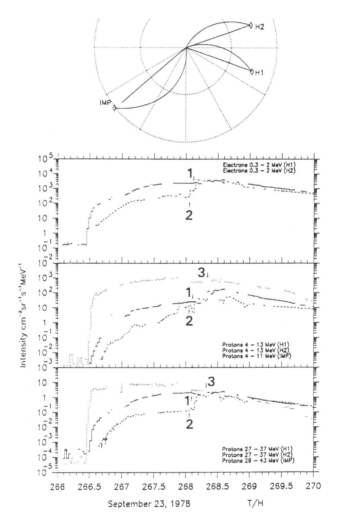

Fig. 7.25. Dependence of the particle profiles on location of the observer from multi-spacecraft observations (Helios and IMP); data from the University of Kiel particle instrument on board Helios and the Goddard Spaceflight particle instrument on IMP

agating straight downwards. The three spacecraft were separated by almost 180°. The particle intensities are given for 0.5 MeV electrons (top, 1 indicating Helios 1 observations, 2 indicating Helios 2), ~7 MeV protons (middle; curve 3 is for IMP), and ~30 MeV protons (bottom); the arrows indicate the arrival of the shock. Although the variations with the location of the observer relative to the shock follow the trend shown in Fig. 7.24, they are much smaller than suggested there. Note the post-shock increase in the Helios 2 data, which is one of the rare occasions of a clear indication of particle storage and acceleration in the turbulence behind the shock front at the rather high energies shown here.

In addition, we should be aware that all shocks and flares exhibit their own features, and therefore the relative intensity between the first intensity increase and the hump at the time of shock passage can vary by orders of magnitude, even at a fixed location relative to the shock. Thus the typical profiles sketched in Fig. 7.24 can also be observed some ten degrees farther to the west or to the east.

As at low energies, the intensity at the time of shock passage is correlated with the magnetic compression and the shock speed [267]. However, these correlations are rather weak, indicating that other parameters might influence the acceleration efficiency too. In addition, we should expect only a weak correlation, because the shock parameters are measured locally at the position of the observer while the particle event is a superimposition of all injections along the observer's magnetic field line, where the shock parameters of the various injections most likely would have been different.

Note that the details of the particle profile, and, in particular, the relative sizes of the solar and shock-accelerated components also vary with energy. Figure 7.26 shows proton intensities in five energy channels between 15 MeV and 850 MeV as observed by GOES during the Bastille Day event of 14 July 2000. The vertical lines indicate the arrival of two shocks, the second shock belonging to the flare and particle event under study. The highest

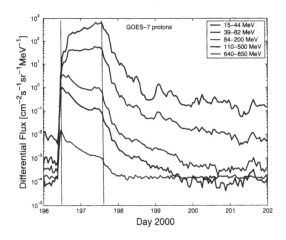

Fig. 7.26. Proton intensities between 15 MeV (*upper curve*) and 850 MeV (*lower curve*) observed by GOES during the Bastille Day event of 14 July 2000. Data courtesy NOAA-SPIDR

energies (bottom curve) closely resemble the profiles of a solar event (see Fig. 7.13), with only a very small increase in intensity around the arrival of the shock: the particles must have been accelerated on or close to the Sun. With decreasing energy (going upwards from curve to curve), the bump of ESP particles becomes more pronounced; at about 20 MeV the intensity continues to rise from the onset until the shock arrives. In the latter case, particle acceleration at the shock in interplanetary space is dominant, and particles accelerated at or close to the Sun play only a minor role. Although the relative sizes of the solar and shock-accelerated particle components are different, this energy dependence is similar in all events.

The observations clearly show that megaelectronvolt protons are accelerated at interplanetary shocks. But how? Of the acceleration mechanisms discussed in Sect. 7.5, only diffusive shock acceleration might to be able to explain the observations. Stochastic acceleration in the turbulence behind the shock appears to be too slow. In addition, it would keep the particles confined in the downstream medium, leading to a post-shock increase. The latter can be observed in megaelectronvolt protons, although not frequently. Shock-drift acceleration also is not very likely because its characteristic feature, the shock spike, is not observed. In addition, it is debatable whether under interplanetary conditions SDA will be able to accelerate protons to megaelectronvolt energies.

But the explanation of the observations by diffusive shock acceleration also is inconsistent. For instance, if we assume typical scattering conditions, the characteristic acceleration time would be much larger than the travel time of the shock from the Sun to the observer. If, on the other hand, we use the upstream intensity increase to determine λ according to (7.47), we find mean free paths much lower than those typically observed in interplanetary space, even lower than those inferred from the magnetic field fluctuations using standard QLT (which always are a lower limit only). The characteristic acceleration times inferred with these λs, on the other hand, are still larger than the shock's travel time. Thus the predictions of diffusive shock acceleration cannot be used to derive a consistent picture. In addition, evidence for self-generated waves upstream of the shock still is missing [270]. Nonetheless, this does not imply that the particles are not accelerated by this mechanism. First of all, all the tests and predictions assume a steady state. Obviously, this is not acquired because it would require acceleration times larger than the shock's travel time. Instead, a steady state is acquired up to a few hundred kiloelectronvolts only, as is also evident in the break in the power-law spectrum in Fig. 7.22. But even if a steady state is acquired only up to a lower energy, some particles have already been accelerated to much higher energies. In addition, the higher energies might require the consideration of additional effects, in particular particle escape from the shock or the curvature of the shock front. Thus the acceleration of megaelectronvolt protons at interplanetary shocks is still not completely understood. One attempt to gain insight into the problem is the simulation of the observed particle profiles with a model that takes into account the propagation of the shock and a continuous, though variable, injection of particles, combined with the subsequent interplanetary propagation of the particles [222, 268, 275]. Currently the shock is assumed to be a black box, although the inclusion of a real acceleration mechanisms, for instance time-dependent diffusive shock acceleration, should be the target. Nonetheless, such a model still yields valuable results, in particular the variation of the particle injection from the shock as the latter propagates outward can be determined, allowing us to determine the change in shock acceleration efficiency with time.

In the above discussion, we have left open one question: how is the solar component accelerated, at the shock or in a flare process? This question is hotly debated. Originally [84, 276] the shock-accelerated particles were considered as a particle component in addition to the particles accelerated in the flare. Later, the current paradigm evolved that in gradual events all particles, and therefore also the solar component, are accelerated at the shock as it propagates outwards [435]. At present, the pendulum appears to be swinging back a little because there are a number of indicators that particles accelerated in the flare also contribute to the event observed in interplanetary space [271]. (1) The particles accelerated in the flare leave clear signatures of their composition and spectra in the form of γ-ray line emission. This is essentially the same in impulsive and gradual flares: "in both impulsive and gradual flares the particles that interact and produce γ-rays are always accelerated by the same mechanism that operates in impulsive flares, namely, stochastic acceleration through gyro-resonant wave particle interactions" [331]. (2) In many particle events, the composition evolves from flare-like to shock-accelerated during the event, as already indicated in Fig. 7.4 [528]. (3) With increasing energy, the charge states more closely resemble those in flares than those in the corona [105, 339, 360].

This discussion is important insofar as it is directly related to the injection problem: if particles from a flare are already present, acceleration by the shock must work on a preaccelerated population. In addition, the shock must only reaccelerate particles, but not accelerate particles out of the solar wind – thus, to explain the observations, an energy gain by a factor of say five or an order of magnitude might be sufficient, instead of the several orders of magnitude required in the case of acceleration out of the solar wind. There is also observational evidence for reacceleration: enrichments in ^{3}He, which are believed to be a typical signature of acceleration in the flare process, of between 3 and 600 times the background ratio, have been found at shocks in EPS events, with the probability for a given particle event at a shock to be ^{3}He-rich increasing with increasing solar activity, that is, when more ^{3}He from earlier flares is in interplanetary space [130]. Thus particles from previous impulsive events are still around as the shock passes by and form a seed population: residual ^{3}He and Fe ions from impulsive solar flares can fill a substantial volume (>50%) of the in-ecliptic interplanetary medium during periods of high solar activity [336].

7.6.3 Particles at CIR Shocks

Particles observed at corotating interaction regions are also accelerated at shocks. As in the case of travelling shocks, particles and shocks can be observed. Thus the similarities between the two particle components are: (a) particles are accelerated at shocks (acceleration process), (b) particles are accelerated out of a seed population (injection problem), and (c) particles are accelerated remotely from the observer (propagation problem). However, there are also some differences between the two populations of shock-accelerated particles, see Table 7.3.

The most important features of particle events in association with CIR shocks are (see [335]): (1) the maximum particle intensities are observed at the reverse shock at ∼4 AU and 20° heliolatitude, (2) the particle flow at 1 AU is mostly sunward with occasional significant non-field-aligned anisotropies, (3) low-energy ions can be observed at 1 AU even if the shock is not observed, (4) the composition of the ions resembles that of the solar wind except for enhancements in He and C relative to O, and (5) the particle ratios He/O, C/O, and Ne/O increase with increasing speed of the high-speed stream. Basically, these observations fit into the same picture as used in the current paradigm of particle acceleration at travelling interplanetary shocks: particles

Table 7.3. Comparison between particles accelerated at travelling shocks and at CIR shocks

Particles accelerated at CME-driven shocks	Particles accelerated at CIR shocks
Particle energies up to some 100 MeV (protons) and some MeV (electrons).	Particle energies limited to about 10 MeV (protons) and 200 keV (electrons).
Particles stream away from the shock (the sign of anisotropy changes as the shock passes by).	At 1 AU, particles stream towards the Sun.
Most efficient acceleration close to the Sun.	Most efficient acceleration at about 4 AU.
At low energies mainly, solar wind composition but there are exceptions (a) for individual events, (b) in energy, (c) with time.	Nearly solar system composition, except for enhanced He and C relative to O.
Particles can be observed even if the shock is not observed.	Low energy ions observed at 1 AU or high latitudes in the absence of shocks.
No evidence for pickup of anomalous cosmic rays.	Singly charged He indicates pickup of ACRs.

are accelerated by the shock (acceleration mechanism) out of the solar wind (source population and injection) and propagate almost scatter-free (propagation) except in the vicinity of the shock, where strong scattering is required for efficient particle acceleration.

The difference in the particle events at CIR shocks and travelling shocks, as outlined in Table 7.3, results from the different properties of the shocks: while travelling shocks are strong, often high-speed shocks with travel times between the Sun and the Earth of about 2 days, CIR shocks are rather weak, slowly moving shocks with lifetimes of many solar rotations. In addition, while the geometry is quasi-parallel in travelling shocks, at least close to the Sun, in CIR shocks it is quasi-perpendicular rather than quasi-parallel. Thus the boundary conditions for the acceleration mechanism in travelling and CIR shocks are quite different. In particular, the main problem for travelling shocks, the small time span available for acceleration, does not apply to CIR shocks. Thus it is even more tempting to assume the paradigm of shock acceleration out of the solar wind to be valid, as discussed in [435].

Nonetheless, as in the case of travelling shocks, recent observations pose challenges to this paradigm. A major challenge is the observation of singly charged ions in the particle population accelerated at CIR shocks. These pickup ions, which have entered the heliosphere as neutrals and are ionized by the Sun's hard electromagnetic radiation on their way towards the inner heliosphere, are picked up for acceleration with much higher efficiency than the ambient solar wind [189, 340].

7.6.4 Particles at Planetary Bow Shocks

Particles and waves can also be observed upstream of planetary bow shocks which form when the supersonic solar wind hits the obstacle magnetosphere and is slowed down to subsonic speed. Basically, the bow shock is parabolic and symmetric around the Sun–Earth line. At the Earth's position, the interplanetary magnetic field spiral has an angle of 45° with respect to the Sun–Earth axis. Thus along the bow shock, the local geometry θ_{Bn} is highly variable (see Fig. 7.27), ranging from quasi-perpendicular close to the nose of the bow shock to quasi-parallel at the dawn side of the magnetosphere.

Upstream of the bow shock, a foreshock region develops which is characterized by energetic particles streaming away from the shock front and by waves excited by these particles. Depending on the local geometry θ_{Bn}, different particle distributions and different types of waves can be observed. Close to the nose of the shock, the subsolar point, the geometry is quasi-perpendicular and shock-drift acceleration leads to a particle distribution in the form of a rather narrow beam of reflected ions, see the left panel of Fig. 7.28. The spiky peak in the middle of the distributions is the solar wind while the broader peaks are reflected and accelerated ions. The wave field consists of low-amplitude waves with frequencies of about 1 Hz.

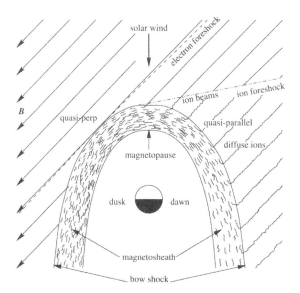

Fig. 7.27. Geometry of the bow shock and waves in the upstream and downstream media. The different ion distributions are indicated

At the dawn side the geometry is more quasi-parallel and particles are accelerated by diffusive shock acceleration. The resulting particle distribution is a diffusive one, forming a ring around the solar wind peak, as is evident in the right panel in Fig. 7.28. The waves excited by these ions again are low-frequency waves with larger amplitudes, occasionally containing shocklets in the sense of discrete wave packets which often are associated with discrete beams of particles. The relationship between particles and waves

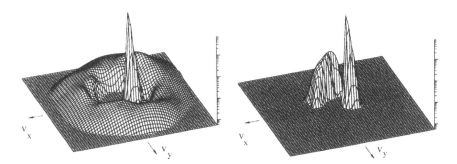

Fig. 7.28. Particle distributions upstream of the Earth's bow shock. In the beam distribution (*left*) particles propagate from the shock front towards the solar wind only while the diffusive population (*right*) covers all pitch angles. In both panels, the sharp peak in the middle is the solar wind incident on the bow shock. Reprinted from G. Paschmann et al. [401], in *J. Geophys. Res.* **86**, Copyright 1981, American Geophysical Union

can be described in the model of coupled MHD wave excitation and ion acceleration [312]; see Sect. 7.5.3.

In between, the local geometry is oblique and both mechanisms contribute to the particle acceleration. The resulting particle distribution is an intermediate distribution in which both the reflected beam and the more diffusive population can be found. The accompanying wave field basically consists of transverse low-frequency waves.

The electrons form a foreshock region, too. It starts close to the nose of the magnetosphere (see Fig. 7.27): at the nose of the bow shock, the geometry is quasi-perpendicular and particles are accelerated by shock-drift acceleration. Since the gyro-radii of the electrons are much smaller than those of the protons, their drift path along the shock front is shorter, leading to an earlier escape and therefore a more extended foreshock region.

Compared with the particle populations observed at interplanetary shocks, the bow shock particles, although significantly more energetic than the solar wind, are still low-energy particles. Electron acceleration is observed only up to a few kiloelectronvolts, ion acceleration up to some tens of kiloelectronvolts. Waves and particles upstream of bow shocks cannot only be found on Earth but also on other planets [386, 451, 454]; see Sect. 9.4.4.

7.7 Galactic Cosmic Rays (GCRs)

Galactic cosmic radiation is incident on the solar system isotropically and constantly. It basically consists of hydrogen, helium, and electrons, but also heavier ions such as C, O, and Fe. Energetically, the galactic cosmic radiation starts where the spectrum in Fig. 7.2 ended, at some tens of MeV/nucl. The energy spectrum has a positive slope, i.e. the intensity increases with increasing energy, up to some hundreds of megaelectronvolts (see Fig. 7.29). At higher energies, the spectrum has a slope of -2.5. Very few of the GCRs can have energies up to 10^{20} eV, corresponding to about 20 J (that is the kinetic energy contained in an apple of 200 g moving at a speed of 50 km/h). GCRs hit the Earth at a rate of about $1000/(m^2 s)$.

7.7.1 Variations

The intensities of galactic cosmic rays vary on different temporal and spatial scales. Some of these variations will be discussed here.

Modulation with the Solar Cycle. At energies below a few gigaelectronvolts, the GCR intensities show a strong dependence on solar activity with a maximum during the solar minimum (see Fig. 7.30). This effect is called modulation. With increasing solar activity, the maximum of the energy spectrum shifts towards higher energies. At proton energies of about 100 MeV the modulation is maximal, while at energies of about 4 GeV the modulation

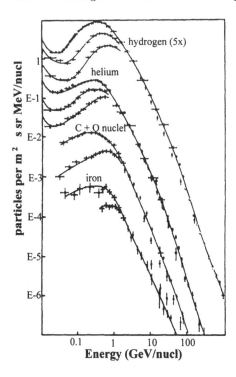

Fig. 7.29. Spectrum of galactic cosmic radiation. At energies below some GeV the three curves for each particle species indicate the spectra for solar minimum (*upper curve*), average (*middle curve*), and solar maximum conditions (*lower curve*). Reprinted from *Physics Today* **27**, P. Meyer et al. [357], Copyright 1974, with kind permission from the American Institute of Physics

only is 15%–20%. Up to about 10 GeV galactic electrons show a spectrum similar to the protons. They also show a modulation with solar activity in the energy range 0.1 to 1 GeV.

Forbush Decreases. Interplanetary shocks not only accelerate energetic particles, they also block part of the galactic cosmic radiation in the hundreds of megaelectronvolt to gigaelectronvolt range. This is called a Forbush decrease. Forbush decreases occur in two steps [28, 162]: first, there is an intensity decrease as the shock passes by, followed by a more pronounced decrease as the observer enters the ejecta driving the shock [86]. Typical values at 500 MeV are about 2% for the shock decrease and about 5% for the decrease related to the arrival of the ejecta. While both decreases are rather abrupt (within minutes), the recovery phase is long and can last for days. Two examples are shown in Fig. 7.31.

Decreases in the galactic cosmic radiation are not only found at travelling interplanetary bow shocks but also at the shocks at corotating interaction regions. Amazingly, the modulation of GCRs by CIRs continues even to latitudes well above the streamer belt where the CIRs form: while the plasma instruments on Ulysses could neither detect shocks nor the changes in solar wind effects, the particle instruments still detected the recurrent modulation of GCRs associated with CIRs that had formed at lower latitudes [485].

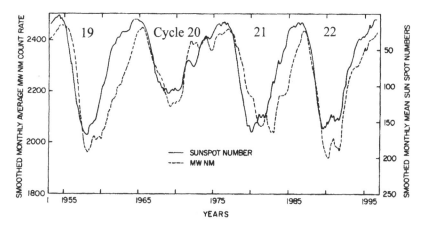

Fig. 7.30. Yearly running averages of Mount Washington neutron monitor GCR intensities (*dashed*) and monthly sunspot numbers (*solid*) from 1954 to 1996 [327]. Note the reversed scale in sunspot numbers: sunspot numbers are high during GCR minima and low during GCR maxima. Reprinted from J.A. Lockwood and W.R. Webber, *J. Geophys. Res.* **102**, Copyright 1997, American Geophysical Union

CR-B Relation. When a corotating interaction region collides with a travelling interplanetary shock or another corotating interaction region, it forms a merged interaction region (MIR). Three types of MIRs can be distin-

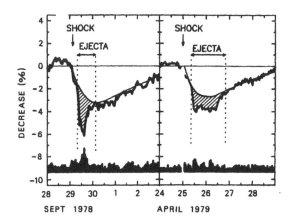

Fig. 7.31. Two-step Forbush decreases. The smooth line suggests the shape of the Forbush decrease after subtraction of the local ejecta effect (*shaded*). The bottom panel gives the standard deviation of counting rates [561]. Reprinted from G. Wibberenz, in L.A. Fisk, J.R. Jokipii, G.M. Simnett, R. von Steiger, and K.-P. Wenzel (eds.), *Cosmic rays in the Heliosphere*, Copyright 1998. Kluwer Academic Publishers

guished [73]: global merged interaction regions (GMIRs), corotating merged interaction regions (CMIRs) and local merged interaction regions (LMIRs).

Global merged interaction regions (GMIRs) are shell-like structures with intense magnetic fields extending around the Sun up to high latitudes. They originate in the interactions of transient and corotating MIRs and produce step-like intensity decreases in galactic cosmic rays throughout the heliosphere which, in turn, produce most of the modulation [404,418]. GMIRs are long-lived (1.5–1.8 years), they might even extend over the poles.

Corotating merged interaction regions (CMIRs) are MIRs with spiral forms associated with the coalescence of two or more corotating interaction regions. CMIRs and rarefaction regions generally produce several successive decreases and increases in GCRs over several months while the background intensity stays roughly constant. They do not lead to an appreciable net modulation.

Local merged interaction regions (LMIRs) are non-corotating MIRs with a limited longitudinal and latitudinal extend. Most likely, they are formed by interactions among transient and perhaps corotating flows. Their effect on galactic cosmic rays is local, comparable to the typical Forbush decrease observed on Earth.

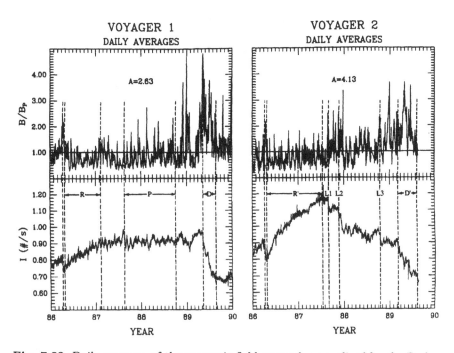

Fig. 7.32. Daily averages of the magnetic field strength normalized by the Parker magnetic field strength B_P and cosmic ray intensity of > 70 MeV protons [73]. Reprinted from L.F. Burlaga, *J. Geophys. Res.* **98**, Copyright 1993, American Geophysical Union

Figure 7.32 shows daily averages of the intensities of > 70 MeV/nucl galactic cosmic rays and the magnetic field from the beginning of 1986 to the end of 1989 for Voyagers 1 and 2. Note that these intensities are measured on an outward propagating spacecraft, thus the long-term variations include effects due to the radial gradient (see below). The most important results are: (1) GCRs decrease when a strong GMIR moves past the spacecraft (time periods D and D′). (2) GCRs tend to increase over periods of several months when MIRs are weak and the strength of the magnetic field is relatively low (R and R′). (3) GCRs fluctuate about a plateau when MIRs are of intermediate strength and are balanced by rarefaction regions (time period P when CMIRs passed Voyager 1). Some local merged interaction regions (L1–L3) produce step-like intensity decreases; however, they are observed locally on one Voyager spacecraft only.

From the close correlation between increasing magnetic field strength and decreasing cosmic ray intensity, Burlaga et al. [72] suggested an empirical relation, the CR-B relation, between the change in GCR intensity J and the magnetic field strength B relative to the Parker value B_P:

$$\frac{\mathrm{d}J}{\mathrm{d}t} = -D\left(\frac{B}{B_P} - 1\right) \qquad \text{for } B > B_P \qquad (7.50)$$

and

$$\frac{\mathrm{d}J}{\mathrm{d}t} = R \qquad \text{for } B < B_P \qquad (7.51)$$

with D and R being constant.

Radial Gradients. Spatial variations in cosmic ray intensity can be represented as

$$\frac{1}{J}\,\mathrm{d}J = g_r\,\mathrm{d}r + g_\lambda\,\mathrm{d}\lambda \qquad (7.52)$$

with the local radial (latitudinal) intensity gradient g_r (g_λ) defined as

$$g_r = \frac{1}{J}\frac{\mathrm{d}J}{\mathrm{d}r} \qquad \text{and} \qquad g_\lambda = \frac{1}{J}\frac{\mathrm{d}J}{\mathrm{d}\lambda}\,. \qquad (7.53)$$

Actual measurements are made between two often widely separated spacecraft, thus only an average, non-local gradient can be determined. Data sampled over successive solar minima by the Pioneer and Voyager spacecraft indicate that the radial gradient might be a function of heliospheric distance [174]

$$g_r = G_0\,r^\alpha\,. \qquad (7.54)$$

Both G_0 and α show a rather complex dependence on time [174]. General features of these variations are: at solar maxima radial gradients are larger than at solar minima and there is a significant change in the radial dependence of g_r, in particular, α might change sign during the solar cycle. At solar minima there is a strong decrease in g_r with increasing r and the magnitude of g_r is appreciably larger in $qA < 0$ (1981) than in $qA > 0$ epochs (1977).

Latitudinal Gradients. Latitudinal gradients are less well studied than radial ones: Ulysses is the first spacecraft to reach heliographic latitudes of 80°, the only other spacecraft outside the ecliptic plane is Voyager 1 at ∼ 34°.

Figure 7.33 shows a comparison of IMP and Ulysses > 106 MeV counting rates between early 1993 and the end of 1996. Ulysses radial distance and heliographic latitude are given at the bottom of the figure, IMP is at 1 AU in the ecliptic plane. From the IMP data, the recovery of GCRs in the declining phase of solar cycle 22 is evident. Owing to the spacecraft orbit, galactic cosmic rays on Ulysses show a more complex time development: at the beginning of the time period under study Ulysses slowly moves inward and to higher latitudes, passing the Sun's south pole in fall 1994 at a maximum southern latitude of 80°S at a radial distance of 2.3 AU. Within 11 months, Ulysses performs a fast latitude scan up to 80°N. Afterwards, Ulysses descends slowly in latitude and moves outwards. During the fast latitude scan Ulysses crosses the ecliptic plane at a radial distance of 1.3 AU. At this time, GCR intensities on both spacecraft agree, their difference is largest when Ulysses is over the poles. The fast latitude scan is most suitable to study latitudinal gradients because the radial distance of Ulysses does not vary strongly (thus results are not affected by radial gradients) and the time period is rather short and GCR intensities at the Earth's orbit are roughly constant, indicating that temporal variations are negligible too.

From the data in Fig. 7.33 a latitudinal gradient of ∼ 0.3%/degree can be determined [221]. However, significant latitudinal effects are only observed when Ulysses is totally embedded in the high-speed solar wind streams of the coronal holes. As long as fast and slow solar wind streams can be observed, the latitudinal gradient vanishes. The latitudinal variation in galactic cosmic

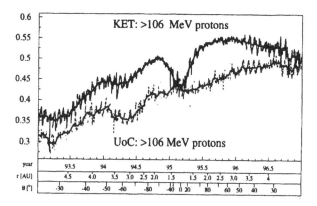

Fig. 7.33. Twenty-six day running mean quiet time counting rates of > 106 MeV proton observed by KET on Ulysses and by the UoC Instrument on IMP 8 from 1993 to the end of 1998 [221]. The fluctuations are caused by CIRs. Reprinted from B. Heber et al., *J. Geophys. Res.* **103**, Copyright 1998, American Geophysical Union

rays is not symmetric around the heliographic equator but has an offset to 7–10°S [220, 488], indicating an offset of the heliospheric current sheet towards the south. This offset is confirmed by the solar wind and magnetic field observations (Sect. 6.5).

Electrons. Most information about galactic cosmic rays are obtained for nuclei, electron observations are rather sparse [148]. In general the modulation of cosmic ray electrons is similar to that of nuclei, although some clear differences exist: (a) the slope of the electron spectrum below 100 MeV is negative because it is dominated by Jovian electrons; and (b) the ratio between electrons and helium strongly depends on the polarity of the solar cycle [181], indicating a charge dependence of modulation.

7.7.2 Modulation Models

The galactic cosmic ray intensity is anti-correlated with the sunspot number and thus solar activity. But how can we understand this behavior? The galactic cosmic rays have to propagate from the heliopause towards the inner heliosphere and it appears that during solar maximum fewer particles manage this task. What is hindering them and which forces do they experience?

Like solar particles, galactic cosmic rays basically experience the Lorentz force: they travel field parallel, they can drift in the large-scale structures of the field, and they are scattered at the magnetic field irregularities. In addition, the particles experience convection and adiabatic deceleration and they are blocked and reflected at transient inhomogeneities such as magnetic clouds and shocks. With increasing solar activity the structure of the interplanetary field changes: first, the tilt angle increases, leading to a wavier heliospheric current sheet (see Fig. 6.20); second, shocks and the ejecta driving them disturb the interplanetary medium.

Diffusion. Galactic cosmic rays propagate inwards into the solar system and we are basically interested in their advance by a piece Δr in radius, although the actual motion is a gyration around the field line. For solar energetic particles observations indicate that diffusion perpendicular to the field is negligible (Sect. 7.3). This partly is due to the fact that the magnetic field line is almost radial close to the Sun. At large heliocentric distances, the field line is tightly wound up and therefore almost circular around the Sun. Thus, even if the diffusion coefficient $\kappa_\perp$ perpendicular to the field is much smaller than that parallel to the field, the net transport in the radial direction might be more efficient for particles crossing from one field line to a neighboring one instead of moving all the way along the field line.

The perpendicular diffusion coefficient depends on the particle speed, the Larmor radius and the perpendicular mean free path $\lambda_\perp$ [166]:

$$\kappa_\perp = \frac{v r_{\mathrm{L}}}{3} \frac{\lambda_\parallel / r_{\mathrm{L}}}{1 + (\lambda_\perp / r_{\mathrm{L}})^2} \tag{7.55}$$

or for $\lambda_\perp \gg r_L$:

$$\kappa_\perp = \frac{v r_L}{3} \frac{r_L}{\lambda_\perp} . \qquad (7.56)$$

Parallel and perpendicular diffusion coefficients can be combined to yield a radial diffusion coefficient κ_{rr}

$$\kappa_{rr} = \kappa_\parallel \cos^2 \psi + \kappa_\perp \sin^2 \psi \qquad (7.57)$$

where ψ is the spiral angle.

Drifts. Owing to their high speeds, galactic cosmic rays have large gyro-radii of the order of AU. Thus, they can drift in the large-scale structure of the heliosphere (see Fig. 7.34). In the heliosphere, the most important drifts are curvature drift (2.55) and gradient drift (2.54). The latter is extremely efficient along the heliospheric current sheet (see Fig. 2.2). Drifts in the mean Archimedian spiral pattern can be characterized by a drift velocity [391]

$$v_D = \frac{cvp}{3q} \left[\nabla \times \frac{\boldsymbol{B}_0}{B_0^2} \right] . \qquad (7.58)$$

Drift then results in a convection of particles with the drift velocity v_D in the transport equation (7.60). Drift speeds can be several times the solar wind speed; thus, drifts can by far exceed convection with the solar wind [258].

In contrast to diffusion and convection with the solar wind, drift depends on the polarity of the magnetic field: drift directions are reversed when the magnetic field polarity is reversed. In addition, drifts are charge-dependent with electrons and nuclei drifting into opposite directions. Thus, the consideration of drift effects in modulation models leads to a charge dependence and different features in cycles with opposite magnetic field polarities.

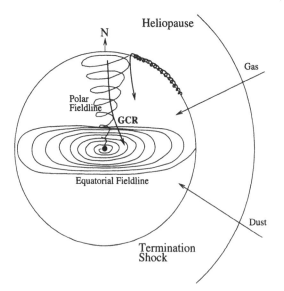

Fig. 7.34. Simplified sketch of the heliosphere and the energetic particles. In reality, the heliosphere has a shape similar to the magnetosphere only in that the deformation is due to the interstellar wind. Note the different shapes of magnetic field lines originating from the equatorial and the polar regions

Formally, the drift also can be included in the antisymmetric terms κ^T of the diffusion tensor

$$\kappa = \begin{pmatrix} \kappa_{\parallel} & 0 & 0 \\ 0 & \kappa_{\perp} & \kappa^T \\ 0 & -\kappa^T & \kappa_{\perp} \end{pmatrix} . \tag{7.59}$$

The drift speed can be obtained as the divergence of the antisymmetric part.

Transport Equation. The development of the cosmic ray density U over time can be described by a transport equation [394, 396]

$$\frac{\partial U}{\partial t} = \nabla \left(\kappa^s \nabla U \right) - \left(v_{\text{sowi}} + v_D \right) \cdot \nabla U + \frac{1}{3} \nabla v_{\text{sowi}} \frac{\text{d}(\alpha T U)}{\text{d}T} . \tag{7.60}$$

Here T is the particle kinetic energy, $\alpha = (T + 2m_0)/(T + m_0)$ with m_0 being the particle rest mass, κ^s the symmetric part of the diffusion tensor, and v_D the drift velocity. The terms on the right-hand side then give the diffusion of particles in the irregular magnetic field, bulk motion due to the outward convection with the solar wind and particle drifts, and adiabatic deceleration resulting from the divergence of the solar wind flow. Note that no effects of transient disturbances are included, thus (7.60) should be applied only during the relatively undisturbed conditions during solar minima.

The Modulation Parameter. The first attempts in modeling modulation reduced (7.60) to a simple diffusion–convection equation. For quasi-stationary conditions $\partial U/\partial t \approx 0$, a roughly isotropic cosmic ray flux and a spherical-symmetric geometry, a modulation parameter Φ could be determined [187]

$$\Phi = \int_r^R \frac{v_{\text{sowi}} \text{d}r}{3\kappa(r, P)} \tag{7.61}$$

with r the radius at which the observer is located, R the outer boundary of the modulation region and $\kappa(r, P)$ the diffusion coefficient. Physically, Φ roughly corresponds to the average energy loss of inward propagating particles due to adiabatic deceleration. This energy loss might be several 100 MeV for particles travelling from the outer boundary to 1 AU. Thus, particles with energies below some hundred MeV/nucl in the interstellar medium are completely excluded from the vicinity of Earth. Or, in other words: near-Earth observations at energies below a few hundred MeV/nucl do not provide any information regarding the spectrum of the local interstellar medium at these energies. The part of the interstellar spectrum blocked by modulation is not negligible, it contains about 1/3 of the GCR pressure or energy density.

The Importance of Drifts. Although drifts were included in the original transport equation (7.60), they were generally neglected until in the late 1970s Jokipii et al. [258] pointed out that the inclusion of drift effects may profoundly alter our picture of modulation.

In the heliosphere, the following drift pattern arises: In an $A > 0$ cycle (the Sun's magnetic field in the northern hemisphere is directed outwards, the configuration in the 1970s and 1990s) positively charged particles drift inwards in the polar regions, downward to the heliomagnetic equator, and outward along the neutral sheet. The sense of drift is reversed for an $A < 0$ cycle. At the termination shock there is a fast drift upward along the shock. Note that drift itself does not cause modulation but only changes the path along which particles enter the heliosphere. Modulation itself can only happen if also transient disturbances and MIRs are present or if the tilt angle changes, which in turn alters the drift path of the particles: with increasing tilt angle, the waviness of the heliospheric current sheet increases and the drift path in the current sheet becomes longer.

The relative roles of drift and diffusion are crucial for our understanding of modulation. For typical conditions, diffusion dominates drift on small time scales. On longer time scales, however, drift effects can accumulate and therefore can become important compared with diffusion. Two conditions have to be fulfilled: (a) noticeable effects of drift can only be expected if the particle spends enough time in the heliosphere to drift at least a significant portion of $\pi/2$ in latitude. (b) Perpendicular diffusion should not be too strong to wash out the drift pattern. If perpendicular diffusion was too strong, particles would not drift along the polar axis or neutral sheet but would spread in latitude. This spread depends on the ratio $\kappa_\perp/\kappa^T$: if $\kappa_\perp \ll \kappa^T$, drift dominates, while for $\kappa^T \ll \kappa_\perp$ diffusion destroys the drift pattern. For intermediate cases, both effects have to be considered.

The inclusion of drift effects in the transport of cosmic rays leads to the following consequences [416, 417]:

- A polarity-dependent 11-year cycle with a pronounced maximum in a $qA < 0$ cycle and a flat plateau-like maximum in a $qA > 0$ cycle, which is observed (see Fig. 7.30).
- A correlation of modulation with the tilt angle in $qA < 0$ cycles only when positively charged particles travel inwards along the heliospheric current sheet and a larger tilt angle automatically implies a longer drift path.
- A charge asymmetry which implies differences in, for instance, electron to helium ratios in different polarity cycles.

Since these features are observed, the importance of drifts is without doubt. The details of the modulation process and the relative importance of the different processes, however, are still subject to debate. A state-of-the-art review with many accompanying papers on detailed problems can be found in [47, 160].

7.8 What I Did Not Tell You

Some of the simplifications we have encountered in this chapter are consequences of problems already mentioned in Sect. 6.9. For instance, we show a sketch of the shock and the ejecta in Fig. 7.24, but our idea is based on one-point observations only: neither have we measured the shock parameters at a fixed time at different positions along the shock front nor have we observed the entire shock at different times. At most, we can have observations from a few points at different times and positions, as suggested in Fig. 7.25. Thus we do not even know whether the shock front is smooth and continuous, as suggested in the figure, or whether it more closely resembles a Swiss cheese, with holes and excursions depending on the varying properties of the upstream interplanetary medium.

And we again encounter the problem of classification. For instance, from Table 7.2 we learn that impulsive events are electron-rich and gradual events are proton-rich, or, in other words, that the electron-to-proton ratio is higher in impulsive than in gradual events. But we should be careful, and look into the data [84, 272]: it is not meant that the e:p ratio in each impulsive event is higher than that in any gradual event. Instead, it is meant that on average the e:p ratio is higher in impulsive than in gradual events. Thus, instead of two distinct distributions for e:p, we find a big overlap in the distributions for the two classes and only the averages are different; see Fig. 7.35. Again this situation is fairly similar to that for the height distribution of males and females: if we pick out one event, we cannot tell from the e:p ratio whether it is impulsive or gradual, just as we cannot tell from the height of an individual whether that person is male or female. And if we pick out one impulsive and one gradual event, the one with the larger e:p ratio might be the gradual one. This not only points to the problem of a classification based on phenomenological criteria, as already described in Sect. 6.9, but also gives a constraint on model development: any model to describe the differences in particle events from impulsive and gradual flares must account not only for the difference in one property but also for the large scatter in the properties in both classes.

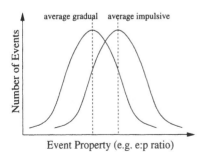

Fig. 7.35. The properties of gradual and impulsive SEP events are different on average, but the distributions for the two classes of events scatter over a broad range and overlap

Like the plasma observations, particle observations are one-point in situ measurements. The plasma is convected over the observer with the solar wind, that is, its motion relative to the observer is rather regular and simple. As we have already seen in the discussion of upstream turbulence generated by energetic particles, particles are fast and, owing to scattering, travel back and forth. That is the reason why particles can be used as probes of the structure of the interplanetary medium, for instance when we study scattering conditions. However, this running ahead also implies that a particle event might be influenced by a structure in the solar wind from a much earlier event. This is not considered in our present methods when interpreting particle events: normally, we relate the intensity profile to a parent flare and the accompanying CME and shock. But this picture might be oversimplified. For instance, as early as in the 1970s Levy et al. [322] suggested that extremely large particle events might result from a fast shock, accompanying the flare and event under consideration, running towards a slower shock from an earlier event. In that picture, particles are trapped between the shocks and are accelerated as the distance between the shocks decreases because of the second adiabatic invariant. Such a scenario can be applied to many of the larger events [273]; modeling now considers even the CMEs [274]. The Bastille Day event in Fig. 7.26 is one likely candidate for such an event: the first shock in the rising phase is from an earlier event and might act as the barrier that prevents particles from escaping to larger distances and might even lead to further acceleration by a Fermi I process. On a shorter time scale and smaller spatial scales, the interaction between CMEs has also been discussed as a possible requirement for efficient particle acceleration: CMEs can catch up with each other even in the field of the coronograph. These cannibalizing CMEs [197] show some peculiarities in the plasma parameters which can be observed in situ in the interplanetary medium, and it appears that cannibalizing CMEs are more efficient in particle acceleration than single CMEs [198], although the detailed mechanisms are not understood yet.

7.9 Summary

Energetic particles in interplanetary space originate from various sources, such as planetary bow shocks, travelling and corotating shocks, and solar flares, or from outside the solar system. The corresponding populations have characteristic spectra, compositions, and time profiles, providing information about the acceleration and propagation mechanism. For almost all particle populations, shock acceleration is important; in solar flares, reconnection and selective heating are also at work. The particles, in particular solar energetic particles, can also be used as probes of the magnetic structure of the interplanetary medium and the superimposed turbulence.

Exercises and Problems

7.1. Determine the Larmor radius, gyro-period, and speed of galactic cosmic rays with an energy of 10 GeV in a 5 nT magnetic field. Compare with the same values for a solar proton with an energy of 10 MeV. Determine the travel time between the Sun and the heliopause at 100 AU for a straight path and a path following an Archimedian magnetic field line.

7.2. Assume a Galton board with n rows of pins. For each pin, the possibility of a deflection to the left or right is 0.5. (a) Give the probability distribution in the nth layer. (b) Show that for large n this distribution converges toward the bell curve. Give the standard deviation. (c) Write a small computer program to simulate a Galton board. Compare the runs of your simulation with the expected result for a different number of rows. Alternatively, simulate the results for a Galton board with 5 rows and 100 balls by tossing a coin. Compare with the expected results.

7.3. Get an idea about changes of time scales in diffusion. Imagine a horde of ants released at time $t_0 = 0$ onto a track in the woods. The speed of the ants is 1 m/min, their mean free path 10 cm. How long do you have to wait until the number of ants passing your observation point at 1 m (10 m, 100 m) is largest? How do your results change if you get faster ants (10 m/min, 50 m/min) or ants moving more erratically (mean free paths reduced to 5 cm, 1 cm). Can you imagine different populations of ants characterized by different speeds and different mean free paths reaching their maximum at the same time at the same place? (More realistic numbers for interplanetary space: particle speeds of 0.1 AU/h, 1 AU/h, and 6 AU/h, distances of 0.3 AU, 1 AU, 5 AU, and mean free paths of 0.01 AU and 0.1 AU).

7.4. In interplanetary space, propagation should be described by the diffusion-convection equation instead of a simple diffusion equation. The flow speed of the solar wind is about 400 km/s. Calculate profiles with the diffusion–convection equation with the numbers given in the parentheses for Problem 7.3. Compare with solutions of the simple diffusion model. Discuss the differences: how do they change with particle speed and mean free paths and why? (Note: Solving this problem you should get an idea about the influence of convection. And this influence is quite similar when additional processes in the transport equation are considered too.)

7.5. Explain the shape of $\kappa(\mu)$ in Fig. 7.11 for isotropic scattering. Why is it not a straight line?

7.6. Shock acceleration is important for many of the particle populations discussed in this chapter. Describe them and find arguments for the differences, in particular the maximum energy gained by the different populations.

7.7. In Fig. 7.4 the composition slowly evolves from one characteristic of flare acceleration to another one characteristic of shock acceleration. Can you explain this slow evolution in terms of a δ-like solar acceleration, a continuous acceleration of particles at the shock, and interplanetary propagation?

7.8. An interplanetary shock propagates with a speed of 800 km/s in the space craft frame into a solar wind with a speed of 400 km/s. The ratio of upstream to downstream flow speed in the shock rest frame is 3, and the upstream diffusion coefficient is 10^{21} cm^2/s. Determine the characteristic acceleration time. Determine the power-law spectral index for times longer than the acceleration time.

7.9. A shock propagates with a speed of 1000 km/s through interplanetary space. The solar wind speed is 400 km/s. The particle instrument on a spacecraft observes an exponential intensity increase by two orders of magnitude starting 3 h prior to shock arrival. Determine the diffusion coefficient in the upstream medium (losses from the shock can be ignored, the shock is assumed to be quasi-parallel).

7.10. Perpendicular transport in modulation: compare the travel paths of a particle at $r = 80$ AU if the particle has to follow an Archimedian spiral around the Sun for one turn of the spiral and if it travels the same distance straight along a radius.

8 The Terrestrial Magnetosphere

> The poet's eye, in a fine frenzy rolling,
> Doth glance from Heaven to Earth, from Earth to Heaven,
> and, as imagination bodies forth
> the *forms of things unknown*, the poet's pen
> turns them to shapes, and gives airy nothing
> a local habitation and a name.
> W. Shakespeare, *Much Ado About Nothing*

A magnetosphere is shaped by the interaction between a planetary magnetic field and the solar wind. The magnetopause is a discontinuity separating both fields, forming a cavity in the solar wind. Since the solar wind is a supersonic flow, a standing shock wave, the bow shock, develops in front of the magnetopause. In the anti-sunward direction, the magnetosphere is stretched by the solar wind, forming the magnetotail. Inside the magnetosphere, different plasma regimes exist, dominated by ionospheric plasma in the plasmasphere, a highly variable mixture of ionospheric and heliospheric plasma in the geosphere, and by the solar wind plasma in the outer magnetosphere. These different regimes are coupled by fields and currents. Inside the plasmasphere energetic particles are trapped in the radiation belts. The inner magnetosphere can be approximated as a slightly distorted dipole field. It is coupled to the ionospheric current system, with energy in the form of particles and waves exchanged between both regimes. Both ionospheric currents and the ring current associated with the radiation belts modify the dipole field.

Particles and energy are fed into the magnetosphere from different sources: (a) the solar wind can penetrate into the magnetosphere due to reconnection at the dayside (flux transfer events), convection above the polar cusps, and and diffusion into the magnetotail, (b) solar energetic particles can penetrate into the magnetosphere at the polar cusps, (c) galactic cosmic rays travelling along Størmer orbits even can penetrate down to ground level, and (d) plasmas are exchanged between the ionosphere and the magnetosphere.

The magnetosphere was recognized as a dynamic phenomenon as soon as the first systematic magnetic field measurements at the ground became available: aside from diurnal variations, strong transient disturbances can be observed. These magnetic storms often are accompanied by aurorae and might influence our technical environment, as evident in disruptions in radio communication or power-line breakdowns. All these phenomena are caused by strong fluctuations or discontinuities in the solar wind and can be related to changes in magnetospheric structure, in particular in the magnetospheric current system and in the plasma sheet inside the magnetotail. This chapter provides an introduction to these phenomena and a supplementary section about the aurora and the history of aurora research.

8.1 The Geomagnetic Field

Magnetic fields either originate in currents or from magnetized bodies. The Earth is not a magnetized body, as can be seen from the variations in the terrestrial field, in particular the pole reversals. Instead, the terrestrial field originates in a dynamo process similar to the one working inside the Sun. Close to the Earth's surface the field can be approximated as a dipole; at higher altitudes or under magnetically disturbed conditions it deviates from the dipole due to currents, the solar wind pressure, and plasma and field exchange with the interplanetary medium. A recent review about these geomagnetic fields and their variability is given in [82, 355].

8.1.1 Description of the Geomagnetic Field

To first order, the Earth can be described as a sphere magnetized uniformly along its dipole axis. This axis intersects the surface in two points, the austral (southern) pole at 78.3°S 111°E close to the Vostok station in Antarctica and the boreal (northern) pole at 78.3°N 69°W close to Thule (Greenland). Both positions are about 800 km from the geographic poles and the magnetic dipole axis is inclined by 11.3° with respect to the axis of rotation. The dipole moment M_E of the Earth is 8×10^{25} G cm^3 or 8×10^{22} A m^2.

Geomagnetic Coordinates. The geomagnetic coordinate system is oriented along the magnetic dipole axis. A plane perpendicular to the dipole axis intersecting the center of the Earth defines the equatorial plane. Its intersection with the Earth's surface marks the geomagnetic equator. The geomagnetic longitude Λ and latitude Φ then are defined analogously to the geographic longitude λ and latitude φ. With $\varphi_0 = 78.3°N$ and $\lambda_0 = 291°E$ as the latitude and longitude of the boreal magnetic pole, the magnetic and geographic coordinates are related by the transformations

$$\sin \Phi = \sin \varphi \sin \varphi_0 + \cos \varphi \cos \varphi_0 \cos(\lambda - \lambda_0) \tag{8.1}$$

and

$$\sin \Lambda = \frac{\cos \varphi \sin(\lambda - \lambda_0)}{\cos \Phi} . \tag{8.2}$$

The magnetic potential at a position r from the Earth's center is

$$V = \frac{\mu_0}{4\pi} \frac{\boldsymbol{M}_E \cdot \boldsymbol{r}}{r^3} = -\frac{\mu_0}{4\pi} \frac{M_E \sin \varphi}{r^2} \ . \tag{8.3}$$

From this, the magnetic field strength $\boldsymbol{B} = -\nabla V$ can be derived:

$$\boldsymbol{B} = \frac{\mu_0}{4\pi} \frac{M_E}{r^3} \left(-2 \sin \Phi \, \boldsymbol{e}_r + \cos \Phi \, \boldsymbol{e}_\Phi \right) \ . \tag{8.4}$$

The flux density

$$B = \sqrt{B_r^2 + B_\Phi^2} = \frac{\mu_0}{4\pi} \frac{M_E}{r^3} \sqrt{1 + 3 \sin^2 \Phi} \tag{8.5}$$

falls off with distance as r^3. At the Earth's surface the magnetic field components can be approximated as

$$B_\Phi = B_E \cos \Phi \qquad \text{and} \qquad B_r = 2 B_E \sin \Phi \tag{8.6}$$

where

$$B_E = \frac{m_0 M_E}{(4\pi R_E)^3} = 3.11 \times 10^{-5} \ \text{T} \tag{8.7}$$

is the equatorial field at the Earth's surface. At the pole B equals B_r while at the equator B equals B_Φ. Thus the magnetic field strength at the pole is twice that at the equator. This ratio does not change with distance.

The geomagnetic field can be described in different systems. In a rectangular Cartesian system, the triple (X, Y, Z) gives the northward, eastward, and vertical components. In a cylindrical system, the triple (D, H, Z) is used with Z as the vertical intensity (that is B_r), H as the horizontal intensity (that is B_Φ), and D as the declination of the field. In a spherical system with Z and X as the axes of reference, the field can be described by the triple (B, I, D) with total intensity B, inclination I, and declination D. Lines with constant declination D are called isogones. Lines with constant inclination I are isoclines. The line with $I = 0°$ is the dip-equator or geomagnetic equator. The magnetic inclination is $\tan I = Z/H = B_r/B_\Phi = -2 \tan \Phi$. Thus, close to the dip-equator the inclination increases twice as fast as the geomagnetic latitude. Figure 8.1 shows the relation between the different systems.

The magnetic field components discussed so far have been intrinsic because their reference direction is the magnetic field itself. The other components are relative ones, their reference direction is the geographic north. The declination D is defined as the angle between the magnetic field direction and the geographic north: $D = Y/X$. The northward and eastward components of the field then are given as $X = H \cos D$ and $Y = H \sin D$.

The equation of a field line $r = r(\Phi)$ can be inferred from (8.6). The magnetic field vector is always tangential to the line of force. The angle α between the radius vector and the magnetic field line is given as

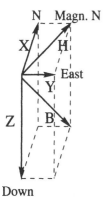

N   Magn. N

X

H

East

Y

Z

B

Down

Fig. 8.1. Components of the geomagnetic field at the Earth's surface

$$\frac{B_\Phi}{B_r} = \frac{r\,\mathrm{d}\Phi}{\mathrm{d}r} = \frac{1}{2\tan\Phi}\ . \tag{8.8}$$

Thus we get

$$\frac{\mathrm{d}r}{r} = 2\mathrm{d}\Phi\tan\Phi = \frac{2\sin\Phi}{\cos\Phi}\mathrm{d}\Phi = 2\frac{\mathrm{d}(\cos\Phi)}{\cos\Phi}\ . \tag{8.9}$$

Integration yields $\ln r = 2\ln(\cos\Phi) + \text{const}$, which can be written as

$$r = r_{\mathrm{eq}}\cos^2\Phi\ , \tag{8.10}$$

where r_{eq} is the distance of the field line from the Earth's center above the equator. This is also the largest distance of a field line; it is used to define the L-shell parameter $L_0 = r_{\mathrm{eq}}/R_E$ (see Fig. 8.2). The magnetic field then is

$$B(L_0,\Phi) = \frac{B_E}{L_0^3}\frac{\sqrt{1+3\sin^2\Phi}}{\cos^6\Phi} \tag{8.11}$$

and the equation of the field line can be written as $L = L_0\cos^2\Phi$. The field line intersects the Earth's surface at a latitude $\cos\Phi_E = 1/\sqrt{L_0}$, where $L = 1$. Physically, the L-shell is the surface traced out by the guiding center of a trapped particle as it drifts around the Earth while oscillating between

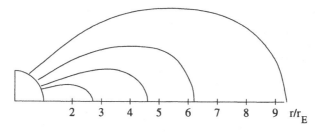

2   3   4   5   6   7   8   9 r/r_E

Fig. 8.2. Shape of magnetic dipole field lines and the definition of the L-shell parameter as the intersection between the field line and the equatorial plane

its northern and southern mirror points. Note that on a given L-shell the particle's physical distance from the Earth's surface might change while its 'magnetic distance' stays constant.

Multipole Expansion. A more correct expression for the geomagnetic field not too high above the Earth's surface is a multipole expansion. In spherical coordinates r, θ and φ, the potential V can be written as

$$V(r) = \sum_{n=1}^{\infty} \frac{r_\mathrm{E}}{r^{n+1}} \sum_{m=0}^{n} \{g_n^m \cos(m\varphi) + h_n^m \sin(m\varphi)\} P_n^m(\cos\theta) . \qquad (8.12)$$

The g_n^m and h_n^m are normalization coefficients and the P_n^m are the Legendre coefficients. For $m > n$, P_n^m equals 0. The quantity n gives the order of the multipole: with $n = 1$, (8.12) describes a dipole field, with $n = 2$ a quadrupole. The magnetic monopole ($n = 0$) is not contained in (8.12). The potential of a dipole field can be inferred from (8.12) as $V = (r_\mathrm{E}/r^2)g_1^0 \cos\theta$.

The coefficients in (8.12) are determined from fits to the measured magnetic field. Because the field changes with time, the coefficients have to be adjusted too. The field determined with the multipole expansion is accurate to 0.5% close to the Earth's surface.

The Surface Field: Shift of the Dipole Axis. Fits on the field measured close to the Earth's surface reveal an offset of the magnetic dipole relative to the center of the Earth by 436 km in the direction of the westerly Pacific. This offset leads to a region of unusually small magnetic flux density in the south Atlantic, just off the coast of Brazil, the South Atlantic Anomaly (SAA); see Fig. 8.3. Since the radiation belts are roughly symmetric around the dipole axis, they come closest to the Earth's surface, too, making the SAA suitable for the study of radiation belts by rockets, but also making it a radiation risk for manned space-flight.

Deviations from the Multipole: The Outer Field. With increasing height, the shifted multipole approximation becomes less efficient. The importance of the higher moments of the multipole decrease, but the field does not become dipole-like. Instead, it is distorted by the influence of electric currents in the ionosphere and magnetosphere and by the direct action of the solar wind. While the solar wind strongly modifies the structure of the outer magnetic field, only the influence of the ring current can be observed at the surface: the ring current results from the opposite drifts of electrons and protons in the radiation belts and gives rise to a magnetic field opposite to the Earth's field, thereby reducing the magnetic flux density.

8.1.2 Variability of the Internal Field

With the beginning of systematic measurements of the terrestrial magnetic field in the middle of the nineteenth century, its variability became evident:

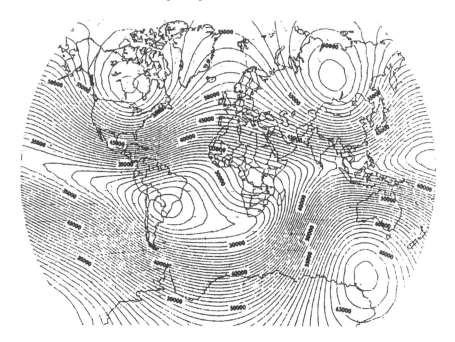

Fig. 8.3. International geomagnetic reference field. Contour intervals are 1000 nT. In a pure centered dipole field the iso-intensity lines would be horizontal. For the most recent list of coefficients see www.agu.org/eos_elec/000441e.html. Reprinted from IAGA [250], *EOS* **67**, Copyright 1985, American Geophysical Union

a systematic daily variation (Sq variation) was occasionally superimposed by much stronger variations, the geomagnetic disturbances. Later it was discovered that the terrestrial magnetic field also varies with the solar cycle. On geological time scales, variations become more pronounced, including the reversal of the field. The origins of these variations are quite different: daily variations result from the asymmetric shape of the magnetosphere. Magnetic storms and variations with the solar cycle reflect the variability of the solar wind and solar activity; variations on even longer time scales are related to the MHD dynamo inside the Earth.

Pole reversals give strong evidence for a variability of the internal field and the dynamo process inside the Earth. Although we have not witnessed such a reversal, the different sheath of lava at the deep-sea trenches have preserved a record of magnetic field polarity (see Fig. 8.4). It appears that the typical cycle for field reversal is about 500 000 years; however, shorter polarity reversals, also called magnetic events, can be observed on time scales between a few thousand years and about 200 000 years. The last polarity reversal occurred about 30 000 years ago, a time when the early humans already had spread across the Earth and the Neanderthals still where alive.

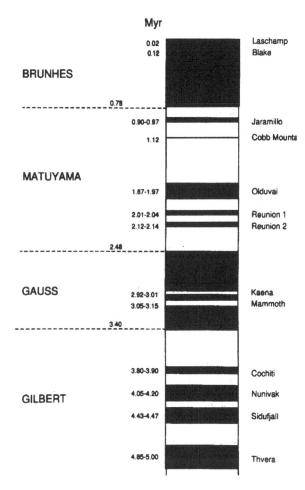

Fig. 8.4. Polarity reversals of the geomagnetic field during the last 5 million years as derived from lava records. Four magnetic field epochs are indicated on the left, and reversals on shorter time scales, magnetic events, are indicated on the right. Reprinted from D. Gubbins [205], *Rev. Geophys.* **32**, Copyright 1994, American Geophysical Union

At the time of polarity reversal, fossil records often indicate the extinction of different plant species [253]. Today it is not clear whether this points to a causal relationship. Four models are discussed. (a) The weak or absent magnetic field at the time of polarity reversal allows the cosmic radiation to penetrate down to the biosphere, causing increased radiation damage in certain plant species and leading to the extinction of at least some of them. Although tempting, this interpretation probably will not hold as most of the cosmic rays are absorbed by the atmosphere well above the biosphere. The field reversal can be seen as an increase in the records of cosmogenic nuclides produced by the interaction between cosmic rays and the atmosphere, such as ^{14}C or ^{10}Be; however, its amplitude is too small to be considered a biological hazard. (b) Certain species in the micro fauna are sensitive to the magnetic field. Thus field reversals might have changed the biochemical processes within these micro-organisms, leading to their extinction, or might have pushed them from their original habitat into a life-threatening one.

Since they are the start of the food chain, the extinction of micro-organisms also can lead to the extinction of other species. (c) Within the framework of catastrophe theory, there is discussion as to whether an external event might have caused both the polarity reversal and the extinction of species. (d) Magnetic field polarity reversals are often associated with climate changes [567]. Although the mechanisms are not yet understood, it appears possible that they invoke an interaction between galactic cosmic rays and the atmosphere, in particular changes in the global circulation patterns in the stratosphere or upper troposphere, in the ozone column, or in cloud cover. Another explanation for a link between pole reversals and climate change might be a modified circulation and heat transfer pattern inside the Earth's core which might lead to both changes in the heat flux through the Earth's surface and a modified dynamo process.

But even if the magnetic field has a certain polarity, it is not necessarily constant. Instead, variations in the dipole moment can be found (see Fig. 8.5). For instance, about 2000 years BP, the magnetic dipole was almost 50% stronger, as can be seen from the dotted curve, while about 25 000 years ago, the magnetic field had only half of its present value.

Figure 8.6 shows the variation of the magnetic moment (top) and the location of the geographic north since 1600. Measurements are sparse in the early part of that period but frequent since the time of Gauss's analysis in 1835. Since that time, the magnetic moment of the dipole has decreased by about 5% per century, while the location of the north magnetic pole has been roughly constant. Extending the time period back to 1600, the decrease in magnetic moment, although slightly weaker, is again prominent, while the drift of the magnetic north is more pronounced: about 0.08° per year to the west and about 0.01° per year to the south.

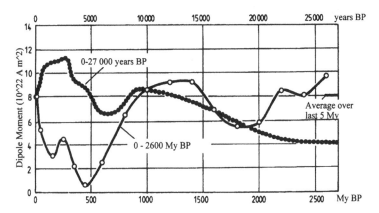

Fig. 8.5. Dipole moment of the terrestrial magnetic field for the last 27 000 years (*dotted line*) and the last 2600 million years (*solid line*). Reprinted from K. Strohbach [507], *Unser Planet Erde*, Copyright 1991, with kind permission from Gebrüder Borntraeger Verlag

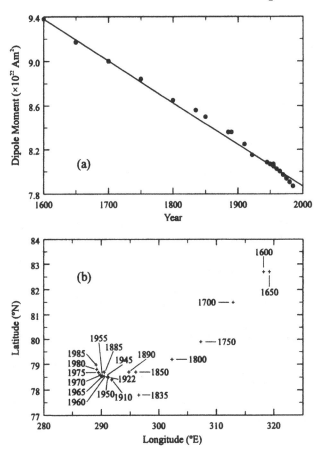

Fig. 8.6. Variation of the magnetic dipole moment (*top*) and the position of the geomagnetic north pole (*bottom*) since 1600. Figure from A.C. Fraser-Smith [170], *Rev. Geophys.* **25**, Copyright 1987, American Geophysical Union

Information about recent work on geomagnetic variations and its terrestrial consequences as well as useful links on this topic, can be found at `www.tu-bs.de/institute/geophysik/spp/index_en.html`. The identification of changes in the terrestrial magnetic field is only possible if accurate measurements are available. While, historically, ground-based observatories have been used, satellite measurements have the advantage of a global coverage. CHAMP (`op.gfz-potsdam.de/champ/index_CHAMP.html`) is one example of such a satellite.

Thus not only the times of polarity reversal but also the strength of the geomagnetic dipole are distributed stochastically. For instance, in the period between 118 million and 83 million years BP, no magnetic field reversal occurred. For the last 5 million years, the statistical distribution of field reversals can be described by an asymmetric random walk distribution [354].

Such distributions are found if the observed process, here the polarity reversal, results from a large number of small individual events following a bell shape distribution. In the case of the terrestrial magnetic field, these small events probably are the patterns of the convection cells in the outer core which play an important role in the dynamo process.

8.1.3 The Terrestrial Dynamo

The principle of a MHD dynamo has already been discussed in Sect. 3.6. The special topology of the terrestrial dynamo is shown in Fig. 8.7. Both panels show the Earth's core only, the combined processes give a complete description of the terrestrial dynamo.

The motion of matter is depicted in the left panel: the liquid inside the outer core rotates but the angular speed decreases with increasing distance from the Earth's center. This differential rotation is the consequence of the vertical transport of angular momentum: convection transports matter upwards from the deeper layers of the outer core, where the linear speed is small, while matter from the higher layers, where linear speeds are high, is transported downwards. As a result, the inner part of the outer core rotates faster than its outer one. In Fig. 8.7, this difference is indicated as the speed v_1 of the deeper layers relative to the outer layers. Let us now assume a magnetic seed field B_1 which is at rest relative to the outer layers and parallel to the axis of rotation. In the inner layers, the relative motion of the fluid with respect to the magnetic field causes an electric induction field E_1 directed towards the axis of rotation. Since the matter inside the core is highly con-

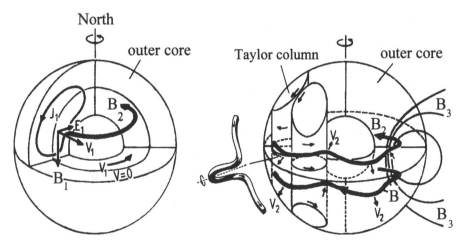

Fig. 8.7. Origin of the terrestrial magnetic field in a dynamo process. Only the core of the Earth is shown. The processes sketched in both panels have to be viewed together. Reprinted from K. Strohbach [507], *Unser Planet Erde*, Copyright 1991, with kind permission from Gebrüder Borntraeger Verlag

ductive, a ring current j_1 results which is directed counter-clockwise in the northern hemisphere and gives rise to a toroidal magnetic field B_2. In the southern hemisphere, the current j_1 and the magnetic field B_2 have opposite directions.

In addition, in the outer core convection takes place in the Taylor columns, that are rotating vertical columns. The toroidal magnetic field B_2 is pushed upward by the convective motion and simultaneously twisted by the Coriolis force, as indicated in the small loop in the lower left of the right panel in Fig. 8.7. The resulting magnetic field B_3 is poloidal and adds to the initial field: the small seed field is amplified.

This dynamo process can explain the basic features of the terrestrial magnetic field. Many details, in particular the details of the polarity reversal, are not yet understood. One important ingredient for the terrestrial dynamo is the core's differential rotation, leading to the poloidal field; the other is the helical twist of the field lines. Thus our description of the geomagnetic dynamo is, as in the case of the solar dynamo, in terms of an $\alpha\Omega$ process. If the differential rotation were absent, no amplification of the seed field would be possible. This appears to be the case in the extremely slowly rotating Venus and the deep-frozen Mars.

8.2 Topology of the Magnetosphere

Now let us put the terrestrial dipole field into a magnetized plasma, the solar wind. How will the dipole be distorted by this flow? Which topology of the magnetosphere arises? How deep into the magnetic field does the influence of the solar wind extend?

8.2.1 Overview

The structure of the magnetosphere is best described in a frame of reference with a fixed Sun–Earth axis. The magnetosphere than stays fixed in space while the Earth rotates inside it. This system divides the magnetosphere into two parts, a dayside directed towards the Sun and a nightside facing the magnetotail. The corresponding directions in the equatorial plane are noon and night, and the direction perpendicular to it dusk and dawn, always referring to local time.

The most important features of the magnetosphere are shown in the noon–midnight cross-section in Fig. 8.8. Typical extensions are about 10 r_E in the solar direction and more than hundred r_E tailwards. The bowshock, the magnetosheath and magnetopause, the cusps and the tail are indicated. These components are strongly determined by the interaction between the terrestrial magnetic field and the solar wind plasma. However, Fig. 8.8 also indicates that the magnetosphere is far from being homogeneous but is highly structured by different plasma and particle components.

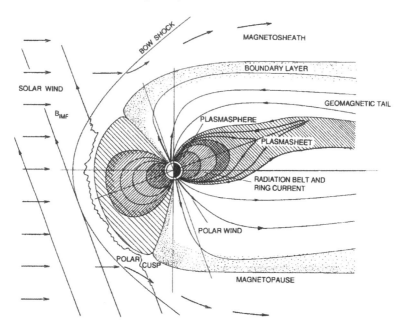

Fig. 8.8. Noon–midnight cross-section through the magnetosphere. Reprinted from G.K. Parks [397], *Physics of Space Plasmas*, Copyright 1991, with kind permission from Addison-Wesley Publishing Company

Starting from the surface of the Earth, on the way to interplanetary space, an astronaut will encounter different regimes of magnetospheric plasma. Relatively low in the atmosphere, at a height of about 70 km, the ionosphere begins. As a conductive layer, it forms the bottom of the magnetosphere. Above it, the plasmasphere, dominated by ionospheric plasma, extends up to a few Earth radii. The radiation belts are embedded in the plasmasphere. The plasmasphere still is relatively symmetric, except for a small bulge on the nightside. The overlying geosphere is more strongly influenced by the interaction between the geomagnetic field and the solar wind: it is highly asymmetric with a larger extension towards the tail. In addition, it is highly variable, at times being dominated by solar wind plasma while at other times ionospheric plasma is more abundant. The outer magnetosphere, the region between the geosphere and the magnetopause, is filled by a plasma of solar wind origin although the magnetic field still is the geomagnetic one.

8.2.2 The Magnetopause

Let us first have a look at a boundary between the solar wind regime and the terrestrial one. This boundary is called the magnetopause and is defined as an equilibrium between the solar wind kinetic pressure and the pressure of the terrestrial magnetic field. The interplanetary magnetic field and the thermal pressure of the solar wind do not contribute significantly to this balance; their combined pressure is less than 1% of the plasma kinetic pressure. For a similar reason, the gas-dynamic pressure inside the magnetosphere does not enter into the balance: it is too small compared with the magnetic pressure.

In the pressure balance we have to consider the geometry of the field and the plasma flow. Since the magnetic pressure is anisotropic, only the tangential magnetic field B_t contributes to the magnetic pressure:

$$p_{\mathrm{mag}} = \frac{B_t^2}{2\mu_0} . \tag{8.13}$$

If we assume the magnetopause to be a perfect boundary between the solar wind and the terrestrial field, B_n equals zero at the magnetopause. If we describe the solar wind as an electron and ion plasma flowing at an angle ψ with respect to the normal direction on the magnetopause, within a second each surface element of the magnetopause is hit by $n_{\mathrm{sowi}} u_{\mathrm{sowi}} \cos \psi$ particles, with n_{sowi} being the number density and u_{sowi} the bulk speed of the solar wind. These particles transfer a momentum $2 m n_{\mathrm{sowi}} u_{\mathrm{sowi}}^2 \cos^2 \varphi = 2 \varrho u_{\mathrm{sowi}}^2 \cos^2 \psi$. The pressure balance at the magnetopause therefore can be written as

$$2 \varrho u_{\mathrm{sowi}}^2 \cos^2 \psi = \frac{B_t^2}{2\mu_0} . \tag{8.14}$$

Note that this is a simplification because we have not considered a slow-down of the solar wind as it passes through the bow shock and part of the flow energy is converted into thermal energy; see Sect. 8.2.5. Nonetheless, (8.14) is still valid as long as the factor 2 in the kinetic pressure is substituted by a factor $K < 2$, which at the terrestrial bow shock is about 0.88.

Let us assume the magnetosphere to be axisymmetric. In cylinder coordinates r is the radial distance from the axis of symmetry and s the distance along the magnetopause, measured from the subsolar point. With $dr/ds = \cos \psi$ we can determine the cosine in (8.14). If B_t were known, the position of the magnetopause could be determined. But B_t is known only as the solution of a potential problem with the boundary conditions defined by (8.13). Thus (8.14) has to be solved iteratively: we make an assumption for B_t which is a solution of (8.13). Then we can determine the electric currents inside the magnetopause, which in turn give a new B_t, which again can be used as input.

As a crude measure for the size of the magnetosphere, the position of its subsolar point on the Sun–Earth line is used. Here the plasma flow is perpendicular to the magnetopause and $\cos \psi$ equals 1. B_t can be approximated from B_0, the magnetic field at the Earth's surface. If we assume a mirror dipole at a distance $2d$, the normal component of the magnetic field would vanish if the tangential field were doubled in d: $B_t = 2B_0/d^3$. Then the distance of the subsolar point can be determined from (8.14):

$$d_{\mathrm{so}} = \sqrt[6]{\frac{4B_0^2}{2\mu_0 K \varrho u_{\mathrm{sowi}}^2}} . \tag{8.15}$$

For typical solar wind conditions, the subsolar point, or stand-off distance, is at $10r_E$.

As can be seen from (8.15), the stand-off distance depends on the solar wind speed, and to a lesser extent, also on its density. The variability of the solar wind conditions thus leads to continuous changes in the size and therefore also in the shape of the magnetopause. Depending on the solar wind speed, the subsolar point is between 4.5 and 20 r_E. Since solar wind speed changes can be quite abrupt, e.g. across a travelling interplanetary shock or when a fast stream is suddenly swept across Earth, the magnetopause has to adjust to this changed environment rather fast. Satellite observations indicate speeds of the magnetopause between a few kilometers per second in response to solar wind fluctuations and up to about 600 km/s in response to discontinuities. The average speed of the magnetopause is about 40 km/s.

Although the magnetopause is defined as a three-dimensional boundary at which an equilibrium between the solar wind and the planetary magnetic field is established, it is not infinitely thin. Instead, it is an extended sheath with a thickness between a few hundred up to thousand kilometers: the solar wind is reflected at the magnetopause only after it has penetrated into the magnetic field and has been turned around by the Lorentz force (see Fig. 8.9). Since the Lorentz force depends on the charge, electrons and ions are deflected in opposite directions, forming the Chapman–Ferraro current inside the magnetopause. This charge separation leads to a pile-up of charges at the flanks of the low-latitude magnetosphere (low-latitude boundary layer (LLBL)) with an excess of positive charges on the dawn side and negative charges on the dusk side. This can also be interpreted as a dawn-to-dusk electric field. Field lines intersecting the LLBL map back this potential pattern towards the high-latitude ionosphere. Relevant aspects concerning the LLBL are summarized in a series of articles in [377].

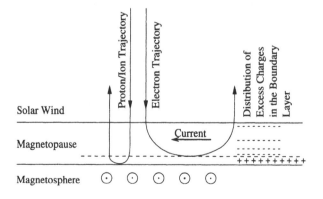

Fig. 8.9. The deflection of solar wind electrons and protons is responsible for both the finite thickness of the magnetopause and the Chapman–Ferraro current inside it. Based on [563]

This deflection can also be described as an extreme case of the grad B drift, with the magnetic field vanishing on one side of the boundary. The drift speed, as given by (2.54), only depends on the particle's Larmor radius which also defines the depth at which the particles penetrate into the magnetopause. Since the Larmor radius depends on $m/|q|$, ions penetrate deeper into the magnetopause than electrons, leading to an excess of negative charges in the outer magnetopause and an excess of ions in the inner one. The resulting electrical polarization field accelerates (decelerates) an electron (ion) on entering the magnetopause and decelerates (accelerates) the particle on leaving it. Thus the particles do not travel along semicircles, but along elliptical orbits as sketched in Fig. 8.9.

The physics of the magnetopause is discussed in details in a series of articles in [497].

8.2.3 Polar Cusps

The polar cusps are two singularities in the dayside magnetosphere: here the magnetic field vanishes and particles and plasma can penetrate freely into the magnetosphere. The polar cusps separate closed field lines on the dayside magnetosphere from open field lines swept to its nightside. The cusps are not located at the dipole axis but at lower geomagnetic latitudes because the higher latitude field, which in the dipole field still would close on the dayside, are convected with the solar wind to the nightside magnetosphere. The magnetic field lines connect the cusps back to geomagnetic latitudes of about 78°; they are the only ones connecting the surface of the Earth to the magnetopause. Thus all field lines of the magnetopause converge at the cusps. The cusps themselves are filled with plasma from the magnetosheath but not from the magnetosphere. Thus at the cusps, plasma of solar wind origin can penetrate deep into the Earth's atmosphere, as can energetic particles.

8.2.4 The Tail and the Polar Caps

Magnetic field lines extending from the cusps to the nightside form the boundary of the magnetotail. Close to the Earth, in addition to these open field lines also closed field lines can be found inside the tail, preserving, at least partly, the dipole character of the inner magnetosphere. As on the dayside magnetosphere, the closed field lines originate in geomagnetic latitudes below 78°. At higher latitudes, all field lines are open and swept into the night side. This region is called the polar cap.

Figure 8.10 shows a sketch of the magnetosphere, drawn to scale to visualize the extent of the magnetotail. The solid lines give the field lines, and the dashed lines are the trajectories of plasma particles, which will be discussed in Sect. 8.3. Two important features are obvious: first, plasma is convected from the plasma mantle at the polar cusps towards the plasma sheet in the

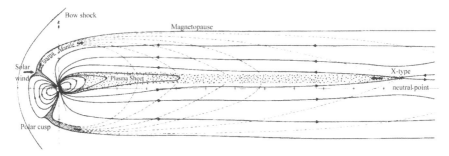

Fig. 8.10. Sketch of the magnetosphere and the magnetotail, drawn to scale. The solid lines are magnetic field lines, and the dashed lines give the trajectories of plasma parcels, filling the plasma sheet from the mantle. X-points inside the plasma sheet are favorable positions for reconnection. Reprinted from W.G. Pilip and G. Morfill [411], *J. Geophys. Res.* **83**, Copyright 1978, American Geophysical Union

magnetotail. Thus a continuous flow of hot solar wind plasma fills part of the magnetosphere. Second, in the equatorial plane there is a plasma sheet separating the oppositely directed magnetic fields of the north and south lobes (fields directed toward and away from the Earth). Inside this plasma sheet, neutral point configurations suitable for reconnection can form. One example is indicated as the X-point. Thus the magnetotail obviously has dynamic aspects: it is filled with plasma from the outside and occasionally reconnection in the neutral sheet leads to an ejection of plasma towards the Earth, where it can create beautiful displays of aurora.

This magnetotail configuration requires currents: the Chapman–Ferraro current in the magnetopause, and the tail current inside the plasma sheet separating the north and south lobes. Both currents are perpendicular to the magnetic field lines and form the closed current system sketched in Fig. 8.11.

The radius of the magnetotail can be estimated by a simple approximation. All field lines of the tail connect back to the polar caps of the corre-

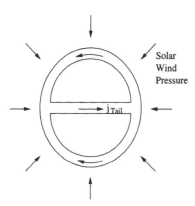

Fig. 8.11. Cross-section of the magnetotail with the currents inside the magnetopause and in the neutral sheet forming a closed loop

sponding hemisphere. The magnetic flux leaving the polar caps is defined by the vertical component of the magnetic field integrated over the polar caps:

$$\Phi_{PC} = 2\pi (r_E \cos\theta_{PC})^2 B_0 . \qquad (8.16)$$

Here θ_{PC} is the latitude of the boundary of the polar cap and B_0 is the equatorial magnetic field strength, which is about half the flux density at the polar cap. This flux is convected outwards into the magnetotail. If we assume the lobe to be semicircular with radius r_{Tail} and magnetic field strength B_{Tail}, the flux inside one lobe is $\Phi_{Tail} = \pi r_{Tail} B_{Tail}/2$. With (8.16) we get

$$\frac{r_{Tail}}{r_E} = \sqrt{\frac{4B_0}{B_{Tail}}} \cos\theta_{PC} . \qquad (8.17)$$

With $\theta_{PC} = 75°$ and $B_0 = 31\,000$ nT, the radius of the magnetotail is 20 r_E for $B_{Tail} = 20$ nT or 29 r_E for $B_{Tail} = 10$ nT. The latter value is typical for the outer magnetosphere.

The current density inside the magnetopause can be determined from the momentum balance at the magnetopause: $\nabla p_{sowi} = \boldsymbol{j} \times \boldsymbol{B}/c$. Alternatively, the cross-tail current density can also be determined from Ampére's law to be $\delta B = 2B_{Tail} = 4\pi j/c$, with δB being the jump in magnetic field strength across the current sheet. With $B_{Tail} = 20$ nT we find $j = 30$ mA/m for the cross-tail current and half this value for the magnetopause current.

How far does the tail extend? Do the tail field lines close far away from the Earth or do they connect, at least partly, to the interplanetary magnetic field? Figure 8.12 sketches such an open magnetosphere: close to the Earth, three types of magnetic field lines can be observed: (i) an interplanetary magnetic field line of solar origin passing by; (ii) closed dipole field lines of planetary origin; and (iii) a merged planetary and interplanetary magnetic field line which connects the surface of the Earth magnetically to the interplanetary medium. The dashed lines give neutral sheets where the interplanetary magnetic field vanishes. A careful discussion of the many aspects of the magnetotail is given in [378].

For orientation, Fig. 8.13 gives the volumes of the magnetosphere predominantly occupied by open (top) and closed (bottom) field lines. The volume with the open field lines contains the tail lobes, including the mantle, the

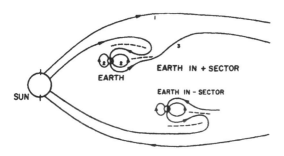

Fig. 8.12. Sketch of possible connections of the magnetospheric field to the interplanetary medium for two different polarities. Based on K.A. Anderson and R.B. Lin [6], *J. Geophys. Res.* **74**, Copyright 1969, American Geophysical Union

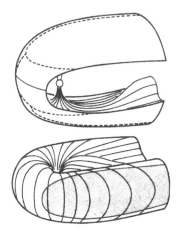

Fig. 8.13. Volumes of the magnetosphere occupied by open (*top*) and closed (*bottom*) field lines, adapted from N. Crooker [114], *J. Geophys. Res.* **82**, Copyright 1977, American Geophysical Union

cusps, and the open portions of the low-latitude boundary layer on the dayside magnetosphere. The volume containing the closed field lines includes the plasma sheet, the quasi-dipolar inner magnetosphere, and the closed part of the low-latitude boundary layer. The shaded area indicates the part of the magnetosphere that abuts closed field lines. The figure is drawn as symmetric with respect to the equatorial plane and to noon. In reality, these symmetries are broken by the tilt of the geomagnetic dipole with respect to the plane of the ecliptic and the direction of the interplanetary magnetic field.

8.2.5 Magnetosheath and Bow Shock

A prominent feature in front of the magnetopause is the bow shock where the supersonic solar wind is slowed down to subsonic speed. The bow shock is about 2 to 3 r_E ahead of the magnetopause, and its upstream medium is characterized by turbulence and energetic particles; see Sect. 7.6.4.

The solar wind flow passes through the bow shock but does not penetrate the magnetopause. Thus the position of the bow shock must be adjusted so as to allow the solar wind to flow around the obstacle magnetopause. At the subsolar point, observational evidence suggests that the ratio between the position of the bow shock and the stand-off distance of the magnetopause is $1.1n$, with n being the density jump at the bow shock. If we assume a gas-dynamic shock, the density jump depends on the Mach number M and on γ_{ad} as $n = [(\gamma_{\mathrm{ad}} - 1)M^2 + 2]/(\gamma_{\mathrm{ad}} + 1)M$. With $M = 8$ and $\gamma_{\mathrm{ad}} = 5/3$, the bow shock is 29% farther out than the stand-off distance of the magnetopause.

In particular, close to the subsolar point the observations are in quite good agreement with these theoretical predictions, as can be seen in Fig. 8.14. Here the positions of the bow shock and magnetopause in the equatorial plane are shown. The solid lines give the calculated magnetopause and bow shock

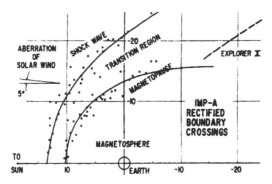

Fig. 8.14. Observed and calculated positions of the bow shock and the magnetopause in the equatorial plane; the scatter in the symbols indicates the variability of the magnetopause and bow shock due to solar wind variations. Reprinted from N.F. Ness et al. [376], *J. Geophys. Res.* **69**, Copyright 1964, American Geophysical Union

positions for average solar wind conditions, and the symbols give observed distances. Their scatter reflects the variability of the solar wind.

In the magnetosheath, the region between the bow shock and the magnetopause, the solar wind plasma is deflected and slowed down. Kinetic flow energy is converted into thermal energy, heating the plasma to about 5 to 10 times the solar wind temperature. Spreiter et al. [499] used a gas-dynamic model to describe the plasma flow inside the magnetosheath. The magnetic field is considered only in so far as it is convected by the solar wind; however, it does not modify the dynamics of the process. Two sample solutions for the geometry symmetric around the Sun–Earth line are shown in Fig. 8.15. In the left panel, the stream lines give the deflection of the plasma flow around the magnetopause and, as the magnetic field is convected with the plasma,

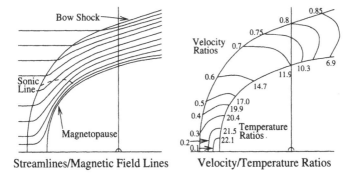

Fig. 8.15. Plasma flow inside the magnetosheath. (*Left*) Streamlines, which are also magnetic field lines, are shown. The dashed curve marks the transition from supersonic to subsonic flow. (*Right*) Velocity and temperature ratios. Based on [499]

also the magnetic field configuration in the magnetosheath. The dotted line marks the transition from supersonic to subsonic flow. The right panel shows the velocity and temperature ratios. Their contours are identical, as can be seen from integration of the energy equation [499] which gives

$$\frac{T}{T_\infty} = 1 + \frac{(\gamma_{ad} - 1)M_\infty^2}{2}\left(1 - \frac{u^2}{u_\infty^2}\right) . \tag{8.18}$$

The increase in plasma temperature can be quite substantial: on the dayside, the solar wind temperature can increase by up to a factor of 20. Since it is the sum of both electron temperature and ion temperature, with the electron temperature often about twice as high as the ion temperature, and since the electron temperature does not change significantly, the increase in ion temperature can be much larger than indicated in Fig. 8.15.

8.3 Plasmas and Currents in the Magnetosphere

We will now follow the path of our hypothetical astronaut from Sect. 8.2.1 in more detail. However, to get the broader scope, we shall start with some basics of atmospheres.

8.3.1 The Atmosphere

The solar electromagnetic radiation determines the structure and dynamics of the atmosphere. Depending on the radiation's wavelength, it interacts with the atmosphere at different altitudes. Since interaction always is associated with heating, a characteristic temperature profile develops which can be used to define atmospheric layers (see Fig. 8.16).

The bottom layer of the atmosphere, the troposphere, has a thickness between about 16 km at the equator and less than 10 km close to the poles. This layer contains more than three-quarters of the atmospheric mass and therefore, energywise, is the most important layer. Its composition is basically 78% N_2, 21% O_2, and a number of trace gases, as well as up to 4% water vapor. The presence of water vapor allows the formation of clouds and precipitation, thus the troposphere is the weather layer of our planet. The combined effects of radiation, convection, and the transport of latent heat cause a negative temperature gradient of about 6.5 K/km. Out of the incident solar radiation, only the visible and infrared penetrate down to the troposphere; the shorter wavelengths are absorbed at higher altitudes. The top of the troposphere is the tropopause, the local minimum in the temperature profile.

The next layer, the stratosphere, is characterized by a positive temperature gradient caused by the absorption of UV in the ozone layer. Its shape stems from the same combination of effects that is responsible for the formation of the ionospheric Chapman layers (Sect. 8.3.2): the incoming electromagnetic radiation increases with height, while the density decreases. Thus

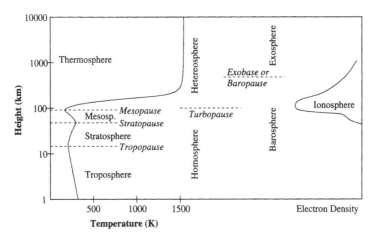

Fig. 8.16. Horizontal structure of the terrestrial atmosphere with different layers defined by extrema in the temperature profile, mixing of the components, the possibility of particle escape, and the ionization

at a certain height, a maximum in the ionization is established: below this height, the intensity of the electromagnetic radiation is too small for efficient ionization, and above this height there are not enough particles left to ionize. The maximum of the ozone layer is at a height of about 25 km. The stratosphere is dry; its water vapor content is almost negligible, although water vapor plays an important role in the ozone chemistry (Sect. 10.4.2). Since the tropopause is a temperature inversion, ideally there would be no exchange of matter across it. Exchange, however, happens, as is evident, for instance, from the influence of the anthropogenic CFCs on the ozone layer and from the deposition of cosmogenic nuclides formed in the stratosphere. Two effects allow such an exchange: violent processes, such as nuclear explosions or erupting volcanoes, and a seasonal slow exchange at the jet streams, where the tropopause is "leaky". The consequences of this slow exchange are twofold: luckily, it is rather difficult for man-made gases to reach the stratosphere, but unfortunately, once such substances have entered it, they will be removed only slowly. The stratosphere extends up to altitudes of about 40–50 km.

In the mesosphere, the temperature gradient is negative again. The mesosphere extends up to a height of about 80 km; it is characterized by photodissociation, ionization (the D-layer of the ionosphere is inside the mesosphere), and a variety of chemical processes. One prominent feature of the mesosphere is noctilucent clouds [178, 287]: for examples see www.meteo.helsinki.fi/~tpnousia/nlcgal/nlcgal.html, www.nlcnet.co.uk/, www.polarimage.fi, or www.iap-kborn.de/optik/nlc/nlc_kb_d.htm.

Above the mesosphere, the temperature increases again because the hard electromagnetic radiation is absorbed, leading to the formation of the main ionospheric layers. Thus the lower thermosphere is characterized by a positive temperature gradient. Above an altitude of 150–200 km the thermosphere is isothermal, with temperatures between about 1300 K (nightside at solar minimum) and 2000 K (dayside at solar maximum). It is characterized by extended circulation systems which vary seasonally and with the solar cycle. It also is the atmospheric layer that is connected directly to the magnetospheric processes: the currents providing the ionosphere–magnetosphere coupling run through the thermosphere, and the particles causing aurorae penetrate down to its bottom.

8.3.2 The Ionosphere

The ionosphere starts in a height of about 70 km as a charged-particle component inside the atmosphere. The ionosphere often is described as the base of the magnetosphere. Because of its high density and the existence of a large neutral component it does not obey the definition of a magnetosphere, namely that particle motion is determined by the magnetic field only. Nonetheless, it is vital for the understanding of the magnetosphere because it provides a highly conducting bottom layer and ionosphere–magnetosphere coupling is important for the energetics of the magnetosphere.

Chapman Layers. The ionosphere is formed due to the ionization of atmospheric constituents by hard electromagnetic radiation in the UV and EUV range. Conveniently it is described as consisting of different layers. To derive the height profile of such a layer, the Chapman profile, we can use a simplified model considering one atomic species and monochromatic electromagnetic radiation only. The variation of density n with height z is described by the barometric height formula

$$n(z) = n_0 \exp\{-z/H\} , \tag{8.19}$$

where $H = k_B T/mg$ is the scale height. Thus the intensity decreases exponentially with increasing height. The intensity I of the ionizing electromagnetic radiation, on the other hand, increases with increasing height: it is maximal at the top of the atmosphere and then is absorbed according to Bougert–Lambert–Beer's law

$$dI/dz = -I_\infty \sigma_a n , \tag{8.20}$$

where σ_a is the absorption cross-section for the particle species and frequency range under study. The intensity at height z then is given by

$$I(z) = I_\infty \exp\left\{ -\frac{1}{\cos\theta} \int_z^\infty \sigma - an(z)\,dz \right\} = I_\infty \exp\left\{ -\frac{\tau}{\cos\theta} \right\} , \tag{8.21}$$

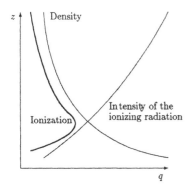

Fig. 8.17. The Chapman profile of an ionospheric layer results from the superimposition of the height dependences of the particle density and the flux of the ionizing electromagnetic radiation

where θ is the Sun's altitude and

$$\tau = \int_z^\infty \sigma_a n(z) \, dz \tag{8.22}$$

the optical depth.

The electromagnetic radiation leads to a height-dependent ionization rate

$$q(z) = n\sigma_i I(z) \tag{8.23}$$

where σ_i is the ionization cross-section. It is $\sigma_a > \sigma_i$ because absorption not necessarily leads to ionization.

The combination of the two profiles gives the Chapman profile (see Fig. 8.17): at a certain height, the ionization, and therefore also the charge density, is highest. Below, it decreases as the intensity of the ionizing radiation decreases. At higher altitudes, although the intensity of the ionizing radiation is higher, the charge density decreases, too, because the density of particles available for ionization is lower. Formally, we can insert (8.21) and (8.19) into (8.23) and get the charge density in the Chapman layer:

$$q(z) = \sigma_i n_0 I_\infty \exp\left\{-\frac{\tau}{\cos\theta} - \frac{z}{H}\right\} . \tag{8.24}$$

If the Sun is in the zenith, the charge density is largest and the maximum of the Chapman layer is at lower altitudes. With decreasing solar altitude, the Chapman layer shrinks and its maximum shifts to higher altitudes.

Equation (8.24) gives the ionization rate. It is thus a good approximation for the number of electrons created at a certain height. The dynamics of an ionospheric layer, however, are not only determined by the ionization but also by losses due to recombination and attachment to neutrals. These loss processes modify the daily variation of the electron density.

Since the atmosphere consists of different particle species and the incoming radiation covers a broad spectrum, for each particle species such a layer forms in the ionosphere at its typical height. Figure 8.18 summarizes these layers: on the right, the densities of the neutrals are shown, and on the left,

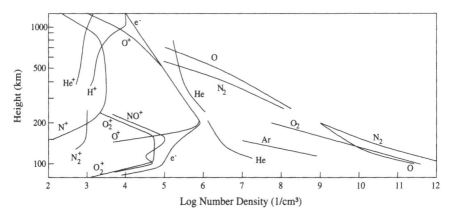

Fig. 8.18. Height dependence of different constituents of the ionosphere and atmosphere for quiet solar conditions. Based on [255]

the ionospheric layers are depicted for the different ion species as well as for the electrons. The electron distribution is the sum of the different ion layers. Note that with increasing height the relative importance of the ionized component increases: while at a height of about 100 km only 10^{-7} of the atoms and molecules are ionized, above a height of about 800 km only ionized particles exist.

Since these layers have a large extension in height and partly overlap, they are not easily identified in observations. Instead, from the observations a more simple scheme for layering in the ionosphere has emerged early in ionospheric studies. The earliest detected layer was the E-layer, named so due to the reflection of electric fields. It is the best studied layer and is dominated by O_2^+ and NO^+. Here we find about one electron for every 10^8 neutral particles. Below the E-region, between 60 and 90 km, is the D-region. It is highly variable with a much smaller electron content. Above the E-region is the F-region, which also contains the maximum in electron concentration. This maximum is typically given at altitudes around 300 km, however, it can shift with solar activity between 200 km and 800 km.

Sudden Ionospheric Disturbances SID. The profiles of the different Chapman layers in Fig. 8.18 are shown for quiet solar conditions. Thus the hard electromagnetic radiation is at a rather low level and no additional radiation is emitted in flares. During solar maximum, the intensity of the hard electromagnetic radiation is higher, leading to a stronger ionization. In addition, during solar flares the hard electromagnetic radiation can be enhanced even further. Then the charge density in the ionosphere can become large enough to absorb electromagnetic waves instead of reflecting them, leading to a break-down in long-wave communication. Such an event is called a sudden ionospheric disturbance (SID).

Ionospheric Conductivity. Currents and plasma flows couple the ionosphere and the magnetosphere. The ionosphere differs from the magnetosphere in so far as collisions of charged particles with the neutrals of the
atmosphere occur frequently; the ionosphere therefore is characterized by a
collision-dominated plasma. Thus the conductivity is finite and the frozen-in
approximation is no longer valid. The conductivity is not only finite but also
highly anisotropic. Three typical conductivities can be defined. The field-
aligned conductivity parallel to B depends on the masses m_e and m_i of the
electrons and ions and on their collision frequencies ν_e and ν_i:

$$\sigma_\| = \left(\frac{1}{m_i \nu_i} + \frac{1}{m_e \nu_e} \right) n e^2 . \tag{8.25}$$

This expression corresponds to the ordinary conductivity.

The Pederson conductivity is concerned with currents parallel to the electric field. With ω_i as the gyro-frequencies we have

$$\sigma_{\mathrm{Ped}} = \left[\frac{\nu_i}{m_i(\nu_i^2 + \omega_i^2)} + \frac{\nu_e}{m_e(\nu_e^2 + \omega_e^2)} \right] n e^2 . \tag{8.26}$$

Pederson currents dissipate energy since $E \cdot j > 0$. The Hall conductivity is
concerned with currents perpendicular to both the electric and the magnetic
fields. It is free of dissipation and can be written as

$$\sigma_{\mathrm{Hall}} = \left[\frac{\omega_i}{m_i(\nu_i^2 + \omega_i^2)} + \frac{\omega_e}{m_e(\nu_e^2 + \omega_e^2)} \right] n e^2 . \tag{8.27}$$

The total current in the ionosphere therefore can be written as

$$j = \sigma_\| E_\| + \sigma_{\mathrm{Ped}} E + \sigma_{\mathrm{Hall}} \frac{B \times E}{B} = \sigma E , \tag{8.28}$$

where

$$\sigma = \begin{pmatrix} \sigma_P & -\sigma_{\mathrm{Hall}} & 0 \\ \sigma_{\mathrm{Hall}} & \sigma_P & 0 \\ 0 & 0 & \sigma_\| \end{pmatrix} \tag{8.29}$$

is the conductivity tensor.

Of these conductivities, the field-aligned one generally is the largest. To
maintain current continuity, the electric field component $E_\|$ parallel to the
magnetic field has to be very small. In particular, at high latitudes the magnetic field is almost perpendicular to an ionospheric layer and therefore allows
for an efficient current (Birkeland current) between the lower ionosphere and
higher altitudes, the basis for magnetosphere-ionosphere coupling.

The Ionospheric Current System. Conductivities are highest in the E-
region, where also strong winds and tidal oscillations can be observed. Owing
to their different masses, ions and electrons are influenced differently by these

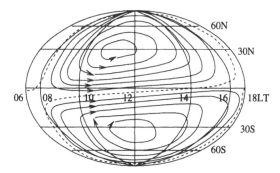

Fig. 8.19. Average Sq currents in the ionosphere

motions of the neutral atmosphere: while the ions are forced to move across the field lines, electrons tend to move perpendicular to both the magnetic field and the neutral wind, albeit at a slower pace. This relative motion causes a charge separation and thus an electric field which, in turn, can affect the currents. Owing to the creation of an electric field by the atmospheric motion, the E-layer also is called the dynamo layer.

The relation between the conductivity and the electric field is described by Ohm's law. However, here we have to add a term which considers the motion imparted by the neutral wind

$$j = \sigma \left(E + v_n \times B \right) \tag{8.30}$$

with v_n being the velocity of the neutral wind. At low latitudes, the dynamo current is mainly driven by the $v_n \times B$ field arising from the ion motion across the B-field while at higher latitudes the contribution from the neutral wind is small and the main driving force is the electric field.

In mid-latitudes, the driving force mainly is provided by atmospheric tides excited by solar heating of the atmosphere. The resulting current system is called the solar quiet or sq-current system. These currents are responsible for the daily sq-variations in magnetometer records. Figure 8.19 gives a sketch of the basic features of the current system viewed from above the ionosphere. Basic features are the two vortices, one in each hemisphere. These systems touch at the equator where they form a strong, jet-like current, the equatorial electrojet. This current is larger than just the sum of the two currents in the vortices because the special geometry of the magnetic field (almost horizontal) and the nearly perpendicular incidence of the solar electromagnetic radiation increase the conductivity.

8.3.3 Magnetosphere–Ionosphere Coupling

The motion of particles, plasmas, and magnetic fields gives rise to currents. Currents in the magnetosphere associated with its large-scale structure are the Chapman–Ferraro current inside the magnetopause and the tail current separating the southern and northern lobes. These currents do not affect

Solar Wind–Magnetosphere Coupling

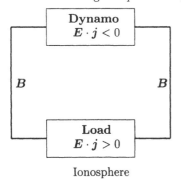

Fig. 8.20. Magnetosphere–ionosphere coupling as a closed circuit

the terrestrial magnetic field, and so their fluctuations are not recorded by ground-based magnetometers. In addition, the drift of charged particles trapped in the radiation belts gives rise to a ring current. These currents are perpendicular to the magnetic field. Another kind of current flows parallel to the magnetic field and therefore can provide coupling between the ionosphere and the magnetosphere. Since the magnetic field is perpendicular to the ionospheric layer only at high latitudes, these field-parallel currents are a phenomenon typical of the polar ionosphere and magnetosphere.

The entire configuration then can be interpreted as a circuit with the solar wind–magnetosphere interaction working as a dynamo ($\boldsymbol{E} \cdot \boldsymbol{j} > 0$) and the ionosphere being a load with dissipative losses ($\boldsymbol{E} \cdot \boldsymbol{j} < 0$). The circuit then is closed by currents parallel to $\boldsymbol{B}$; see Fig. 8.20. The driving force, and thus also the source of energy, is the solar wind. Thus the dynamo has correctly been called the solar wind dynamo.

Birkeland Currents. The field-parallel currents can be observed in the polar ionosphere where the magnetic field lines are almost perpendicular. The average patterns of these Birkeland currents are shown in Fig. 8.21. The currents are plotted versus geomagnetic longitude with the dark areas indicating currents into the ionosphere and the lighter shaded areas indicating currents out of it. At high latitudes, currents are flowing out of the ionosphere in the evening side and into it at the morning side. They are called region 1 currents. At somewhat lower latitudes, the region 2 currents show the opposite pattern: they flow into the ionosphere in the evening side and out of it in the morning side. At very high latitudes around noon, i.e. below the polar cusps, the pattern of the field-parallel current is highly variable and strongly depends on the northwards or southwards component of the interplanetary magnetic field owing to the convection of magnetic field lines across the polar caps (see Sect. 8.4). Around midnight, the currents in regions 1 and 2 overlap without a clear separation.

Most of the field-parallel currents are carried by the electrons. Thus an inward current implies an outward motion of electrons and vice versa. The

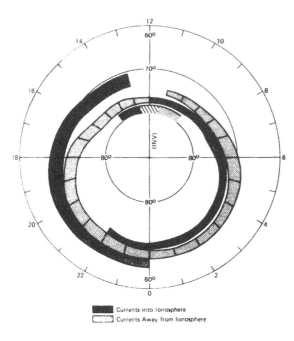

Fig. 8.21. Distribution of the field-parallel Birkeland currents in the polar ionosphere. Note that the numbers refer to the geomagnetic latitude. Reprinted from T. Iijima and T.A. Potemra [249], *J. Geophys. Res.* **81**, Copyright 1976, American Geophysical Union

pattern of Birkeland currents also reflects the spatial distribution of aurorae at geomagnetic quiet periods: aurorae are observed where electrons stream down to the ionosphere, i.e. where the Birkeland current is directed upwards.

The upward and downward Birkeland currents are closed by ionospheric currents. The auroral zone electric field $\boldsymbol{E}_{\mathrm{a}}$ in the ionosphere is directed northwards in the dusk sector and southwards in the dawn sector, i.e. from dusk to dawn. Since the conductivity of the ionosphere is finite, the Birkeland currents will be closed in the ionosphere by field-parallel currents (Pederson currents) northwards in the dusk sector and southwards in the dawn sector. A possible closure of the circuit is sketched in Fig. 8.22. Here a dusk-to-dawn cross-section of the magnetosphere is shown viewed from the tail towards

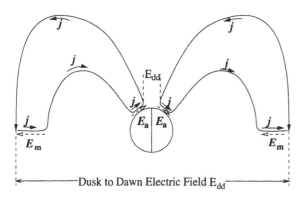

Fig. 8.22. Possible scenario for the closure of the ionosphere–magnetosphere–current system in a cross-section in the dusk-to-dawn plane viewed from the tail towards the Sun. Based on [59]

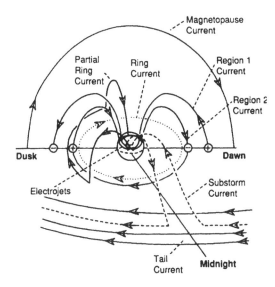

Fig. 8.23. Schematic representation of the various current systems in the magnetosphere and ionosphere. Reprinted from R.L. McPherron [352], in *Introduction to space Physics* (eds. M.G. Kivelson and C.T. Russell), Copyright 1995, with kind permission from Cambridge University Press

the Sun. The closure of the ionosphere–magnetosphere circuit in the equatorial plane is suggested to be radial in the dusk-to-dawn direction. Note that the dusk-to-dawn auroral zone electric field is reversed if mapped outward, in agreement with the dawn-to-dusk electric field driven by the solar wind and also with the cross-tail electric field. Thus in the magnetosphere, the electric field and currents are antiparallel, forming the generator proposed in Fig. 8.20, while they are parallel in the ionospheric load.

Ring Current. Part of the ring current is also involved in the magnetosphere-ionosphere coupling. It flows near dusk in the equatorial magnetosphere and is closed through the ionosphere by field-parallel currents. Part of the tail current is diverted into the ionosphere by field-aligned currents, forming the substorm current. The substorm current, as its name suggests, plays an important role in geomagnetic and auroral activity. These currents, together with the current systems discussed so far, are summarized in Fig. 8.23.

A summary of the entire magnetospheric current system and hints on many open questions is given in the articles in [384].

8.3.4 The Plasmasphere

The plasmasphere is dominated by a dense and cold plasma of ionospheric origin, as is evident from the high O^+/H^+ ratio and the existence of other ion species such as He^+, O^{2+}, N^+, and N^{2+} which cannot be found in the completely ionized solar wind. Spatially, it coexists with the radiation belts, extending up to heights of about 3 to 5 Earth radii. The particles have energies close to 1 eV, and the density varies between 10^4 cm^{-3} at about 1000 km

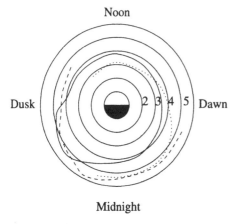

Noon

Dusk   Dawn

Midnight

Fig. 8.24. Cross-section through the plasmasphere in the equatorial plane. The location of the plasmapause, as determined from Whistler observations, is indicated for different times corresponding to different levels of geomagnetic activity. Based on [87]

and 10–100 cm^{-3} at the outer boundary of the plasmasphere. The plasmasphere is filled from the ionosphere by the polar wind. A comprehensive review of the properties of the plasmasphere can be found in [319].

The plasmapause as the relatively sharp outer boundary of the plasmasphere was first proposed from the properties of Whistler waves in the magnetosphere. Figure 8.24 shows the location of the plasmapause in the equatorial plane for three different times corresponding to different levels of geomagnetic activity. The bulge at the duskside is a persistent feature, although it can rotate somewhat in local time, depending on magnetospheric conditions. The plasmasphere – and with it the plasmapause – basically is a field-aligned structure. It can be traced from the equatorial plane down to the ionosphere.

Figure 8.25 shows the plasma density plotted versus the L-shell parameter for the nightside plasmasphere for different levels of geomagnetic activity.

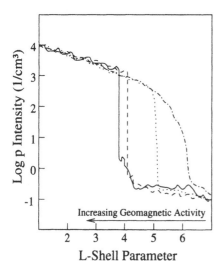

Increasing Geomagnetic Activity

L-Shell Parameter

Fig. 8.25. Variation of the nightside plasmapause with geomagnetic activity. Based on [96]

With increasing geomagnetic activity, the plasmasphere shrinks and its outer boundary becomes more pronounced.

The plasmasphere corotates with the Earth, leading to an electric induction field $\boldsymbol{E}_{\mathrm{corot}} = -(\boldsymbol{\omega} \times \boldsymbol{r}) \times \boldsymbol{B}(\boldsymbol{r})$. This field is smaller than the electric field driving the current system in the magnetotail; the plasmapause as the outer boundary of the plasmasphere separates these two current systems.

8.3.5 The Geosphere

The plasmasphere is embedded in the geosphere, a highly variable region filled with a hot plasma of low density. The plasma inside the geosphere has two sources, the ionosphere and the solar wind, as is evident from the composition. Figure 8.26 shows the size and location of the geosphere together with the three main current systems. In the upper panel, the familiar meridional cross-section is given. The middle panel corresponds to a view from high above the North Pole towards the equatorial plane, and in the lower panel

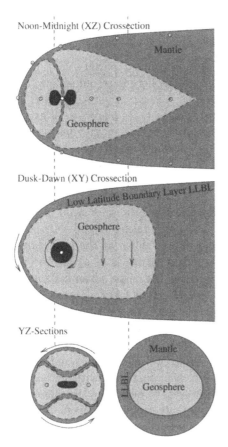

Fig. 8.26. Plasma regimes in the magnetosphere with the darkness of the shading indicating the plasma density. Views from the duskside to the dawn (*upper panel*), from pole to equator (*middle panel*), and from the subsolar point tailwards (*lower panel*). The arrows give currents. Based on [363]

two cross-sections with views from the subsolar point tailwards are shown. The shading indicates the plasma density, the arrows indicate the Chapman–Ferraro current in the magnetopause, the tail current across the tail in the equatorial plane, and the ring current around the Earth associated with the radiation belt particles.

The plasma density in the geosphere is much lower than in the plasmasphere; it is also lower than in the outer magnetosphere. Evidence for a contribution of the ionospheric plasma to the geosphere again comes from the presence of O^+ and other heavier ions. The dominant species, H^+ and He^{2+}, in number are roughly independent of solar and geomagnetic activity, while their energy as well as the relative amount of ionospheric ions dramatically increases with increasing geomagnetic activity. Thus the solar wind ions H^+ and He^{2+} are fed into the geosphere at a roughly constant rate while geomagnetic activity strongly enhances the energy imparted to these particles as well as the density of the ionospheric component. In addition, the ionospheric outflow into the geosphere tends to be larger by up to a factor of 4 if the interplanetary magnetic field has a northward component, that is for a closed magnetosphere (see Sect. 8.4).

8.3.6 The Outer Magnetosphere

The outer magnetosphere is dominated by solar wind plasma. The basic mechanism for feeding plasma into the magnetosphere is reconnection. The magnetopause therefore is not an unpenetrable boundary separating two completely decoupled systems. Instead, plasma transfer takes place almost everywhere along the magnetopause. Since the field-lines of the magnetopause converge at the polar cusps, the high-latitude ionosphere is directly influenced by the solar wind plasma penetrating through the magnetopause.

Even the simplest models of the magnetosphere had predicted an direct access of solar wind at the polar cusps. But the plasma in some properties, in particular in thermal energy, is different from the solar wind plasma: since it comes from the magnetosheath, the temperature is higher than in the solar wind because the latter had been slowed down at the bow shock. This downward plasma flow can be detected at latitudes of about $78°$ for about ± 3 h around local noon, i.e. exactly below the cusps, as an increase in electron density and temperature. While the electrons form a narrow beam penetrating downwards through the cusps, the ions are spread back towards the tail. This is caused by an $E \times B$ drift in the dawn-to-dusk electric field. Since the electrons have a very short travel time, their displacement by this drift is negligible. The ions, on the other hand, are much slower and therefore are affected by the drift. Their displacement increases with increasing travel time, i.e. with decreasing energy. However, to penetrate the magnetosphere efficiently, the particles must already gyrate around the magnetic field lines bordering the cusp. Thus, the process is far more efficient if the field lines are not closed but open: they do not connect the cusp regions of the two hemispheres but connect to the interplanetary magnetic field. Then the magnetosphere is called an open magnetosphere.

8.4 The Open Magnetosphere: Reconnection Applied

In the early days of magnetospheric research two different explanations for the entry of solar wind plasma into the magnetosphere were offered: the concept of an open magnetosphere [139] where plasma and fields are exchanged between the magnetosphere and the interplanetary medium by reconnection and convection as opposed to the viscous interaction model [15] where particles diffuse across the magnetopause of a closed magnetosphere. Since the entrance of solar wind plasma into the magnetosphere strongly depends on the orientation of the interplanetary magnetic field, namely a southward component, evidence is in favor of the open magnetosphere.

In an open magnetosphere plasma transfer across the magnetopause is due to reconnection and thus requires an X-point configuration or neutral line where fields of opposite polarity meet. This is most likely to occur at the dayside magnetopause if the interplanetary magnetic field has a southward component. Then the magnetopause is a rotational discontinuity while at times of a northward interplanetary magnetic field it is a tangential discontinuity completely separating the interplanetary and the planetary plasmas. The corresponding configurations are an open and a closed magnetosphere.

Compared to other astrophysical objects, such as solar flares, the magnetosphere is the only plasma laboratory where reconnection can be studied directly: at the day side magnetosphere flux transfer events give direct evidence for reconnection, in the tail reconnection can be observed in the formation of substorms. The evidence for these reconnection processes is summarized in [463].

8.4.1 Convection of Plasma Into the Magnetosphere

Figure 8.27 sketches the dynamics of an open magnetosphere: solar wind convects the interplanetary magnetic field lines towards the magnetopause. If the interplanetary magnetic field has a southward component (open magnetosphere), interplanetary magnetic field line 1' eventually merges (or reconnects) with the planetary field line 1 in a diffusion region in the dayside magnetosphere. Thus two mixed planetary/interplanetary field lines result (2 and 2') which are convected tailwards with the solar wind, eventually becoming field lines of the geomagnetic tail (5 and 5'). The F-region ionospheric plasma joins this anti-sunward flow since it is still magnetically connected to the convected field lines (see also the small inset in Fig. 8.27). During this anti-sunward motion of field lines, plasma from the magnetosheath is convected into the tail, first filling the mantle and later also moving down to lower latitudes, as indicated by the dashed lines in Fig. 8.10. If this process

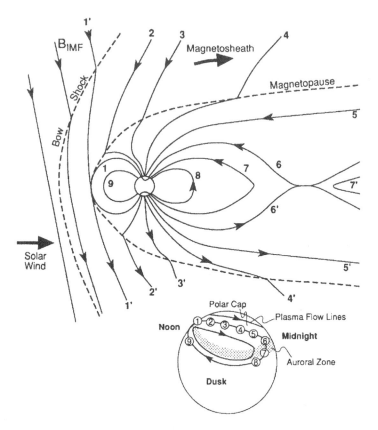

Fig. 8.27. The convection of plasma into the magnetosphere. The numbered field lines show a succession of configurations of an interplanetary magnetic field line 1′ at the front of the magnetosphere. Field lines 6 and 6′ reconnect in the tail and return to the dayside at lower latitudes. The small figure shows the resulting convection pattern of the plasma in the northern high-latitude ionosphere: an anti-sunward flow in the polar cap and a return flow at lower latitudes. The shaded area is the auroral oval. Reprinted from W.J. Hughes [240], in *Introduction to Space Physics*, Copyright 1995, with kind permission from Cambridge University Press

were to continue indefinitely, the entire geomagnetic field would soon be connected with the interplanetary field and magnetic flux would pile up in the tail. Since this is not observed, the magnetic flux must be returned to the closed magnetic field. This is achieved on the nightside magnetosphere: as the mixed field lines are pushed further towards the equatorial plane, a X-point results in the tail's plasmasheet where lines 6 and 6′ meet. Here reconnection sets in, forming a closed geomagnetic field line and a purely interplanetary magnetic field line. Magnetic tension will relax the geomagnetic field lines (7, 8), which in time will return to the dayside magnetosphere at lower altitudes leading to a sunward flow of magnetic flux in the ionosphere. Thus a return flow results. Magnetic tension also allows the interplanetary field line 7′ to shorten and therefore to be pulled outwards through the geomagnetic tail.

This picture is grossly simplified since in reality the entire process will be essentially non-steady. Thus although the time-averaged reconnection rates at the dayside magnetopause and in the tail must be equal, at any given time they can be quite different. Nonetheless, there is observational evidence for such a process to occur. In the late 1950s, it was realized that the plasma flow in the polar and auroral ionospheres must map outward and be related to a magnetospheric flow pattern. Magnetometer measurements indicated a plasma flow over the polar regions from noon to midnight (points 1–6 in the inset in Fig. 8.27) and a flow back towards the dayside at lower latitudes (points 7–9). This process also can be described as a solar wind dynamo and is the driving process in the magnetosphere–ionosphere coupling and thus the vertical exchange of matter and energy. The plasma carried with the convected field lines leads to an electric convection field of the order of 50–100 kV, the dawn-to-dusk field. Combined with the corotating field of the plasmasphere, an asymmetric field results that also contributes to the partial ring current.

The pattern described in Fig. 8.27 is roughly stationary in local time, i.e. in a frame of reference with a fixed Sun–Earth line. An observer on Earth rotates underneath this flow pattern, seeing it as a diurnal magnetic field variation. Since the flow pattern resembles a thermally driven flow cell, it has been termed a convection pattern, although it is not thermally driven. The lower panel in Fig. 8.27 shows this flow pattern on the duskside. Since the magnetic field is frozen into the plasma, we can map back the plasma flow to a pattern of motion of a magnetospheric field line which is swept across the polar cap and then returns to the dayside magnetosphere at lower latitudes, which is exactly the motion of the field line shown in the upper panel of Fig. 8.27.

Note that Fig. 8.27 is meant as a schematic only. The details of the reconnection process strongly depend on the local direction of the interplanetary medium. Also, correspondingly, the results of this process, in particular the amount of field lines convected into the tail and the resulting ionospheric currents, are highly variable. Magnetospheric parameters influenced by the properties of the interplanetary magnetic field, in particular its southward component, include the cross-tail electric field, the ring current and the aurora, all of them smaller or weaker in the case of a non-southward interplanetary magnetic field. Some empirical relations between solar wind conditions and geomagnetic parameters, such as the D_{st} index have been suggested. The most global relation is concerned with the total solar wind power input into the magnetosphere [403, 539],

$$P_{\mathrm{in}} = u_{\mathrm{sowi}} \frac{B_{\mathrm{ip}}^2}{2\mu_0} \sin^4 \frac{\Theta}{2} \, 2\pi r_{\mathrm{mp}} \,, \tag{8.31}$$

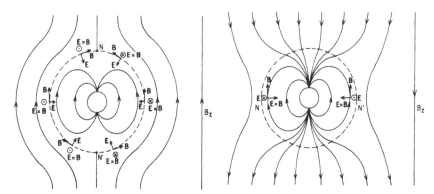

Fig. 8.28. The Earth with an idealized dipole field, in a northward (*left*) and a southward (*right*) magnetic field. Reprinted from A. Brekke, *Physics of the upper polar atmosphere* [59], Copyright 1997, with kind permission from Wiley

where B_{ip} is the strength of the interplanetary magnetic field, Θ its angle with respect to the z-axis, and r_{mp} the magnetopause distance. Equation (8.31) gives the flux density of magnetic energy ($u_{sowi} B_{ip}^2/2\mu_0$) into the approximate surface area of the magnetopause.

In (8.31) a northward-pointing interplanetary magnetic field corresponds to $\Theta = 0$ and thus $P_{in} = 0$: in the case of a northward field, no energy is transferred into the magnetosphere and the magnetosphere is closed. At most times, however, Θ will be different from zero and thus some amount of energy is transferred to the magnetopause.

Figure 8.28 illustrates this energy flow, which depends on the north–south component of the interplanetary magnetic field. The terrestrial field is drawn as an idealized dipole; the interplanetary field has a northward component in the left panel and a southward component in the right panel. For the northward field, the magnetosphere is closed and the energy flux (Poynting vector) $\mathbf{E} \times \mathbf{B} = -(\mathbf{u}_{sowi} \times \mathbf{B}) \times \mathbf{B}$ is parallel to the field lines: no energy enters the magnetosphere. For a southward interplanetary magnetic field (right panel), the Poynting vector points into the magnetosphere and energy enters everywhere.

The magnetic energy is only a small portion of the total solar wind energy because most energy is contained in the flow. The maximum ratio (for $\Theta = 180°$) between the imparted magnetic energy and the bulk kinetic energy of the solar wind can be approximated by

$$\frac{P_{in}}{P_{sowi,kin}} \approx \frac{1}{M_A^2} , \tag{8.32}$$

where M_A is the Alfvénic Mach number. For typical solar wind conditions, $M_A \approx 7$.

The details of the reconnection process at the magnetopause are still being debated; in fact, this problem is at the cutting edge of space physics. For summaries see, for example, [210, 241, 399, 463].

8.4.2 Flux Transfer Events

Direct evidence for reconnection between the planetary and interplanetary magnetic fields is obtained from in situ observations at the dayside magnetopause. These observations are not related to the convection pattern sketched above, but directly confirm the exchange of plasma and field between the magnetosphere and the interplanetary medium. As a satellite passes through the magnetopause, short events in the magnetic-field, plasma, and particle data provide evidence for such flux exchange. These events therefore are termed flux transfer events; for observational evidence see, for example, [210, 277]. The earliest observations were based on magnetic-field data only. A flux transfer event can be identified as two consecutive short excursions of the magnetic field component normal to the magnetopause, first to positive values, later to negative values, and afterwards returning to the undisturbed value around zero, if the satellite is in the northern hemisphere. The sequence is opposite for an observer in the southern hemisphere; see Fig. 8.29. These signatures are interpreted as an isolated magnetic flux tube connected through the magnetopause to the geomagnetic field and convected northwards across the satellite by the solar wind; see Fig. 8.30. The composi-

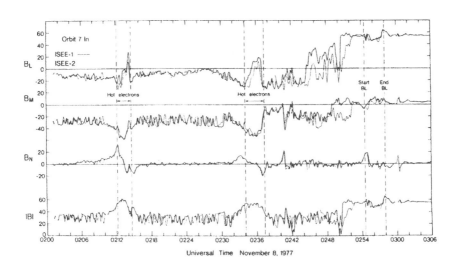

Fig. 8.29. Magnetic field data during a magnetopause crossing, with evidence for flux transfer events indicated by the *dashed vertical lines*. Reprinted from C.T. Russell and R.C. Elphic [450], *Geophys. Res. Lett.* **6**, Copyright 1979, American Geophysical Union

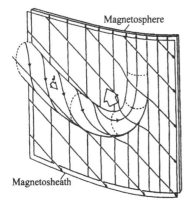

Fig. 8.30. Sketch of a magnetic flux tube through the magnetopause. The motion of such a flux tube leads to the signature of a flux transfer event in the magnetic field data. Reprinted from C.T. Russell and R.C. Elphic [450], *Geophys. Res. Lett.* **6**, Copyright 1979, American Geophysical Union

tion of the plasma and the energetic particles inside the flux tube were both indicative of a magnetospheric origin in the example illustrated here. Since flux transfer events require reconnection at the dayside magnetosphere, these events are observed at times when the interplanetary magnetic field has a southward component.

8.4.3 Release of Accumulated Matter: Substorms

The release of matter convected into the magnetotail's plasmasheet causes magnetic substorms which are seen in magnetic field variations, in particular in the AE index, as well as the aurora. The primary location of energy storage is the tail; the physical mechanism for the energy release is reconnection in the plasma sheet. The basic process can be likened to a dripping fountain: a drop at the outflow forms by the interplay between gravity pulling the water down and surface tension keeping it up. If enough water has been collected at the outflow, gravity takes over and the drop falls. In a magnetospheric substorm, a plasmoid in the magnetotail takes over the role of the drop: solar wind drag pulls on the magnetosphere, feeding plasma and energy into it. As the plasmoid grows, a magnetic neutral point forms close to the Earth. Here reconnection sets in, expelling the plasmoid tailwards and accelerating a small amount of plasma towards the Earth, causing the aurora at the Earth's high-latitude nightside and the accompanying geomagnetic disturbance.

This scenario is sketched in more physical terms in the Hones substorm model as shown in Fig. 8.31. Panel 1 gives the quiet-time configuration of the magnetosphere, with an X-point at a radial distance of about $100r_\mathrm{E}$. The field line at this point is the last closed field line: in the direction of the Earth, all field lines are closed; at larger distances, all field lines are open. As reconnection sets in at this X-point, the energetics of the plasma sheet are changed so that a second X-point forms much closer to the Earth (panel 2). As more energy is fed from the solar wind into the magnetosphere, reconnection at this inner X-point continues and a plasmoid is formed between the inner and

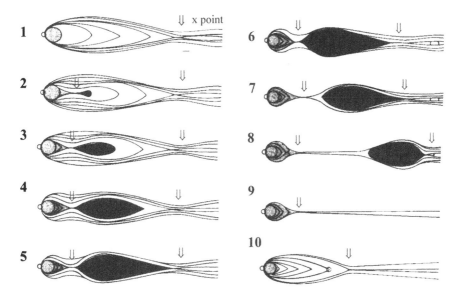

Fig. 8.31. Sequence of events leading to a magnetospheric substorm (see text). Based on E.W. Hones [228], in *Magnetic reconnection* (ed. E.W. Hones), Copyright 1984, American Geophysical Union

outer X-points (panels 3–6), until finally the plasmoid becomes detached at the inner X-point (panel 7) and is accelerated, leaving the magnetotail (panel 8). The bulk of the energy fed from the solar wind into the magnetosphere is contained inside the plasmoid and thus is fed back into the solar wind. Only a small amount is converted to kinetic energy of plasma moving towards the Earth. As this plasma interacts with the high-latitude ionosphere, it causes an aurora and enhances the auroral electrojet. The tail slowly fills with new solar wind plasma (panels 9 and 10), until the initial configuration (panel 1) is restored and the cycle starts anew.

Observational evidence for this scenario, together with a shortened version of the Hones substorm model, is presented in Fig. 8.32. The left panel shows the model and the location of the spacecraft; the right panel shows a superposed epoch analysis of the tailward plasma velocity, the total magnetic field, the north–south excursion B_z of the magnetic field, the flux of >30 keV electrons in geosynchronous orbit, and the auroral-electrojet index AL. The dashed line marks the first occurrence of the high-speed flow. With the arrival of this flow, the spacecraft is engulfed by a region of lower magnetic field strength and the north–south component of the field rotates as the plasmoid travels across the spacecraft. The two lower panels indicate the characteristic consequences of substorms: an increase in electron flux and a sudden depression of the auroral electrojet. The onset of the substorm is about 30 min before the passage of the plasmoid in the tail – the delay is due to the fact

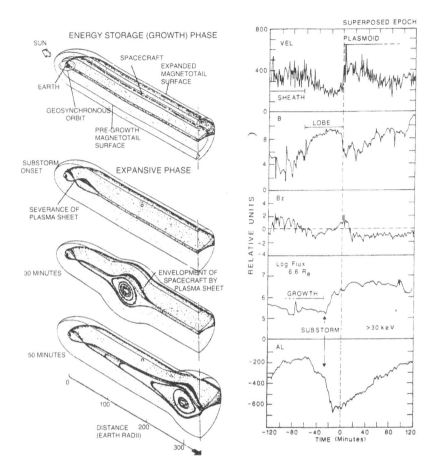

Fig. 8.32. Model and observations during plasmoid formation in the magnetotail. From M. Scholer [463], Copyright 2003, Springer-Verlag, based on [24]

that the spacecraft does not observe the plasmoid during formation but only after it has propagated a considerable distance through the tail.

The plasma travelling towards the Earth during the discharge of the magnetotail leads to the substorm current, which has already been shown in Fig. 8.23. It is associated with the tail current, which during a substorm, is partly diverted as a field-parallel current towards the ionosphere in the evening sector, continues through the ionosphere as an auroral electrojet, and flows back towards the tail as a field-parallel current. The excursion of the tail field happens when, during the onset of reconnection, the tail field collapses. This current system is called a substorm current wedge because a projection of the current system onto the equatorial plane takes the form of a

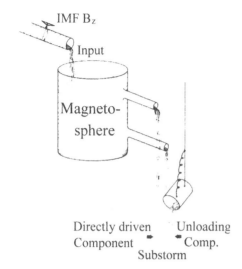

Directly driven   Unloading
Component   Comp.
Substorm

Fig. 8.33. Energy flux from the solar wind into the magnetosphere. Reprinted from S.-I. Akasofu [4], *EOS* **70**, Copyright 1989, American Geophysical Union

wedge. The opening angle of the wedge is typically 70°, the current is about 2×10^6 A, and the wedge extends from the Earth into the tail for about $5r_E$.

Figure 8.33 offers a simple illustration of this dependence of geomagnetic activity on solar wind flow and the southward component of the interplanetary magnetic field in terms of the superimposition of two energy fluxes. The solar wind feeds energy continuously into the magnetosphere. The energy partly is stored as magnetic field energy in the tail (small bucket to the right) and partly is converted to geomagnetic activity (outflow to the left). During weak geomagnetic activity as well as during large storms the solar wind energy is fed directly into the magnetosphere and ionosphere. At times of moderate geomagnetic activity, however, the energy is stored in the magnetosphere before it is released. Solar wind energy is fed more efficiently into the magnetosphere if the latter is open, i.e. if the interplanetary magnetic field has a southward component, indicated by the handle at the outflow. Thus, more energy is available for release in the form of geomagnetic activity. This release, however, is not continuous, but energy is stored over a time period of typically an hour and then liberated abruptly in a substorm.

During the sequence shown in Fig. 8.31, an observer on Earth sees characteristic changes in the aurora. As the plasmoid in the tail grows, the auroral oval slowly expands equatorwards with the aurora still being a quiet arc. As reconnection sets in at the inner X-point, the initial auroral arc brightens, exhibits more structures and fast changing features, and moves rapidly equatorwards. During the recovery, a rather quiet auroral arc or curtain contracts to the size of the initial auroral oval.

More detailed discussions about the magnetosphere, its plasma sources and losses and its variability are given in [234, 242].

8.4.4 Closed, but Only Almost Closed

From the above discussion, it may appear that all the interesting things happen when the interplanetary magnetic field has a southward component and that the magnetosphere is rather boring at times of a northward field. Judged from the aurora as a visible indicator of geomagnetic activity, this is simultaneously true and false: true in that the aurora is rather quiet and limited in space when there is a northward field, and false in that a special kind of aurora, the theta aurora, can be observed.

There is also evidence for reconnection between interplanetary magnetic field lines and the geomagnetic field, and thus the magnetosphere is not entirely closed. Obviously, this reconnection cannot happen at the nose of the magnetosphere, because there the terrestrial and interplanetary magnetic fields are parallel. However, as the interplanetary field lines are convected over the magnetosphere, they encounter configurations suitable for reconnection at the cusps where the fields of the magnetosheath and the lobes are antiparallel; see Fig. 8.34. After reconnection, the poleward portion of the flux tube is convected tailwards by the solar wind. This is similar to the dayside reconnection in a southward interplanetary magnetic field. In the dayside portion of the field line, on the other hand, plasmas from the magnetosphere and the magnetosheath mix and the field line sinks into the magnetosphere. The observational signature of these reconnection events is unidirectional electron streams and an outflow of ionospheric O^+. These observations also suggest that reconnection does not occur simultaneously in both hemispheres.

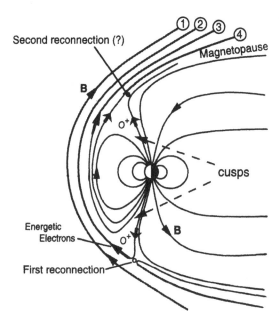

Fig. 8.34. Reconnection in the dayside magnetosphere for a northward interplanetary magnetic field. Reprinted from Fuselier et al. [177], *J. Geophys. Res.* **106**, Copyright 2001, American Geophysical Union

8.5 Geomagnetic Disturbances

8.5.1 Daily Variations

The magnetogram of a geomagnetically quiet day shows systematic variations in all three field components. The most pronounced variation can be observed around local noon. These variations are regular excursions, which are repeated each day. Their direction and magnitude depends on the geomagnetic latitude of the observer. These Sq, or solar quiet, variations are related to the ionospheric sq current system (Fig. 8.19)

Occasionally, the Sq variations are enhanced by a solar flare: since its hard electromagnetic radiation leads to a stronger ionization of the dayside ionosphere, the ionospheric current system is enhanced. In this case the change in magnetic field is called the sfe variation (sfe: solar flare effect).

8.5.2 Geomagnetic Indices

For a quantitative description of the geomagnetic field the K and A indices are used. The basis for these indices are the magnetograms. The K index is a quasi-logarithmic number between 0 and 9 determined at the end of specified 3-h intervals as the maximum deviation of the observed magnetic field from the expected quiet field. It is determined for each of the three magnetic field components separately. The largest of the maxima is converted to a standardized K index taking into account the geomagnetic properties of the observation site. Thus K indices of different stations, in particular of stations at high and low latitudes, can be compared and can be combined to give a planetary K index.

At individual stations, the eight daily K indices are linearized to give an a index and than averaged arithmetically to give the A index describing the daily averaged magnetic activity. The construction of the K and A indices and the global network of observatories is described in [353].

Geomagnetic disturbances generally are described by two indices. Magnetic storms can be quantified by the D_{st} index which gives the excursion of the equatorial H component compared with quiet times. Physically, the D_{st} index is related to the ring current. At high latitudes, an AE index is used, related to the auroral electrojet. It is also determined from the excursion of the H component compared with quiet times. Both indices are determined globally by combining different observatories at comparable geomagnetic latitudes but different longitudes. The index with the longer time record is the AA index which also uses the difference between observed and expected horizontal components but now at mid-latitudes. Owing to its long record, it is often used for correlative studies; however, physically more significant and easier to understand are the AE and D_{st} indices.

On these periodic quiet time variations, irregular disturbances on different time scales are superimposed. Fluctuations with time scales below about

0.2 s are waves, fluctuations with time scales between 0.2 s and 600 s are pulsations of the magnetosphere, and variations with time scales above 10 min are geomagnetic disturbances. These geomagnetic disturbances can be observed worldwide; however, their amplitudes generally are largest at high geomagnetic latitudes and smaller, or even vanishing, at low latitudes.

8.5.3 Geomagnetic Pulsations

Magnetospheric waves and pulsations are phenomena which affect the entire magnetosphere. For instance, magnetic field fluctuations observed from the ground are highly correlated to fluctuations in the electric field observed from a satellite in geostationary orbit [186]. Nonetheless, amplitudes might vary with position. Such fluctuations are called geomagnetic pulsations. Continuous pulsations are grouped from Pc1 to Pc5 according to their periods, with Pc1 starting at periods of 0.2 s and Pc5 ending at 600 s. These waves are ultra-low frequency waves. Geomagnetic pulsations can be quite regular during quiet geomagnetic periods and become quite irregular during geomagnetic storms. Then they are termed Pi1 and Pi2.

Geomagnetic pulsations act as coupling devices between different parts of the magnetosphere and ionosphere because they transport energy and information. Physically, the quiet time pulsations (Pc1–Pc5) best can be interpreted in terms of a cavity vibration of the entire magnetospheric cavity. The irregular pulsations Pi, on the other hand, seem to be Alfvén waves.

Wave generation requires an energy input. The departure from the equilibrium of the plasma and the field that drives the waves at least at quiet times appears to be related to the large scale convective flux. Thus the energy input is at the dayside magnetosphere. The irregular pulsations are driven by sporadic events, for instance the compression of the frontside magnetosphere at the beginning of a sudden commencement also causes an oscillation of the magnetospheric cavity [289].

8.5.4 Geomagnetic Storms

Geomagnetic disturbances are also called magnetic storms or substorms, depending on their temporal and spatial extent. Magnetospheric substorms are the most frequent type of geomagnetic activity. Its most obvious manifestation is the sudden explosion of a quiet auroral arc to more brilliant colors and moving structures. Over a period of an hour, they develop through an orderly sequence that depends on time and location. Simultaneously, a magnetometer on the ground below the aurora will record intense disturbances caused by the electric currents accompanying the aurora. These auroral electrojets are roughly parallel to a geomagnetic parallel circle, flowing at a height of about 120 km in concentrated channels of high conductivity produced by the same particles that generate the auroral emission. The disturbances in the

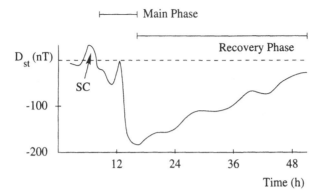

Fig. 8.35. The variation of the D_{st} index for a large magnetic storm

magnetic field are in the range 200 nT–2000 nT, they typically last between 1 h and 3 h, and are most pronounced at high geomagnetic latitudes.

If the coupling of matter and energy from the solar wind into the magnetosphere is stronger and lasts longer, a magnetic storm develops. Its temporal development can best be seen in the D_{st} index. Figure 8.35 shows the variation of the D_{st} index for a large geomagnetic storm. The storm often begins with a sudden increase in the magnetic field. This sudden commencement may last for many hours. This initial phase is followed by a rapid and sometimes highly disturbed decrease in D_{st}, which defines the storm's main phase. Subsequently, D_{st} recovers, first rather quickly, later more slowly. A storm lasts between 1 and 5 days with the initial phase anything up to 1 day, the main phase normally about 1 day, and the recovery phase lasting for several days. The distribution of storm magnitudes obeys a power law: storms with D_{st} between 50 nT and 150 nT occur about once in a month. Disturbances with D_{st} between 150 nT and 300 nT occur several times a year, while only a few storms with D_{st} above 500 nT can be observed over an entire solar cycle.

The phases of a geomagnetic disturbance can be understood as follows: the increase in the magnetic field strength at the beginning of the disturbance can be attributed to the compression of the magnetosphere as the magnetopause is pushed inward by the increased solar wind speed. The decrease in the field strength during the main phase of the storm is due to an increase in the ring current, which creates a magnetic field opposite to the terrestrial one. A typical current density in the undisturbed ring current is about 10^{-8} A m^{-2}, mainly carried by particles with energies between 10 keV and 100 keV at a height between 3 r_E and 6 r_E. During a strong magnetic storm, particles are injected from the plasma sheet in the magnetotail into the radiation belts, enhancing its density by an order of magnitude on time scales as short as 10 min. This enhanced ring current then reduces the magnetic field measured on the ground.

Magnetic storms can be caused by fast solar wind streams or also by transient disturbances, such as interplanetary shocks and magnetic clouds. The same pattern as for recurrent solar wind disturbances emerges: the geomagnetic activity increases with increasing change in solar wind flux and is stronger if the interplanetary magnetic field or the field at the leading edge of the magnetic cloud has a southward component [196, 521], that is the magnetosphere has an open configuration. As a rule of thumb, an intense geomagnetic storm requires a southward component of the interplanetary magnetic field of more than 10 nT for at least 3 h.

8.5.5 Geomagnetic Activity on Longer Time Scales

Geomagnetic disturbances are not distributed uniformly in time. Instead, characteristic dependences can be observed which directly point to the physical mechanisms responsible for them. In the late 1930s Chapman and Bartels [94] showed that the number of geomagnetic disturbances is related directly to the number of sunspots and thus to solar activity. Figure 8.36 shows this close relation for the last 100 years using yearly averages of the sunspot number (lower curve) and the AA index (upper curve). Note that geomagnetic activity does not vanish during solar minima: while it is strongest during solar maximum due to the large number of transient disturbances, the geomagnetic activity during solar minima mainly is caused by corotating fast solar wind streams. This recurrent geomagnetic activity therefore shows a periodicity of 27 days.

In addition, there is an annual variation with enhanced geomagnetic activity during the equinoxes. Figure 8.37 shows the D_{st} index plotted versus time for 16 solar rotations during the 1974 solar minimum. Strong deviations of the D_{st} index to lower values are indicative of geomagnetic activity. For a better identification, these periods are blackened. Two systematic variations can be identified in this figure. Most obvious is the recurrence of the geomagnetic disturbances during each solar rotation. These disturbances are

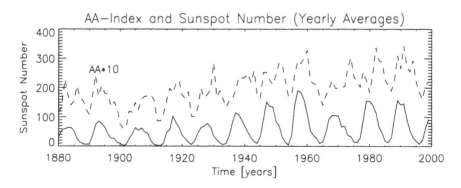

Fig. 8.36. Variation of sunspots and geomagnetic activity with the solar cycle

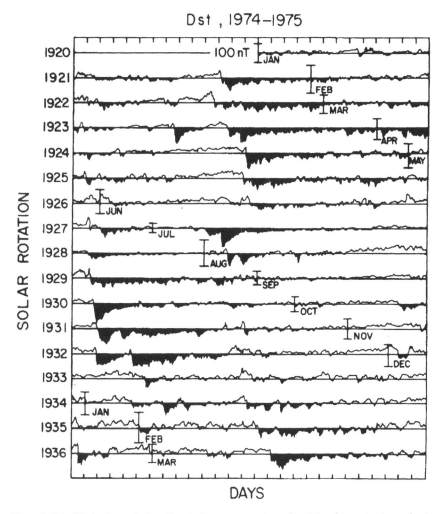

Fig. 8.37. Variation of the D_{st} index versus time for 16 solar rotations during the 1974 solar minimum. Reprinted from N.U. Crooker and G.L. Siscoe [115], in *Physics of the Sun, vol. III* (eds. P.A. Sturrock, T.E. Holzer, D.M. Mihalas, and K. Ulrich), Copyright 1986, with kind permission from Kluwer Academic Publishers

related to fast solar winds, which can be observed best during the solar minimum. But this recurrence is not observed for all 16 rotations: in the spring, the strongest geomagnetic disturbances are observed in the middle of each rotation, while in the autumn they are at the beginning of the rotation. In addition, during summer and winter the disturbances are weak, while in spring and autumn they are more pronounced.

To understand this change in pattern between the two equinoxes let us first look at one of these disturbances, the one beginning in the middle of

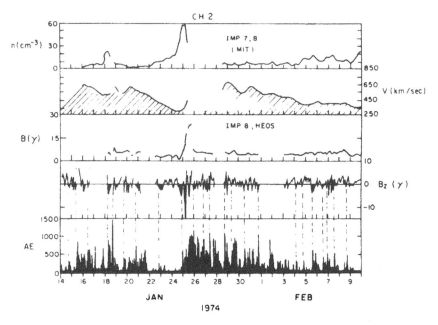

Fig. 8.38. Variation of solar wind density, speed, magnetic flux density, the north–south component of the interplanetary magnetic field with negative values indicating a southward interplanetary magnetic field, and geomagnetic activity over solar rotation 1921. The dashed vertical lines link the times of a southward excursions by the interplanetary magnetic field and enhanced geomagnetic activity. Reprinted from L. Burlaga and R.P. Lepping [71], *Planet. Space Sci* **25**, Copyright 1977, American Geophysical Union

rotation 1921 in Fig. 8.37. Figure 8.38 shows the density and speed of the solar wind, the flux density and north–south component of the interplanetary magnetic field, and the AE (auroral electrojet) index as a measure of geomagnetic activity related to aurorae and substorms. For solar rotation 1921 two fast streams can be identified, one starting in the data gap on 15 January, the other starting on 25 January. Immediately before the beginning of this stream the increase in plasma and magnetic flux density indicates the compression region in front of the fast stream. The envelope on the AE index traces the solar wind speed, thus changes in the solar wind are related to geomagnetic disturbances. But even at times of rather constant solar wind speed there can be intense although short geomagnetic disturbances (for instance the one on the evening of 18 January). These disturbances are strongest when the interplanetary magnetic field has a southward component.

With this information we can go back to the interpretation of Fig. 8.37. The fact that geomagnetic disturbances are recurrent is related to the existence of two recurrent fast solar wind streams, one at the beginning and the other in the middle of each solar rotation. The geomagnetic effectiveness, on the other hand, depends on whether the interplanetary magnetic field has a southward component or not, i.e. whether the magnetosphere is open or closed. As discussed earlier, an open magnetosphere also means an easier exchange of energy and matter between the interplanetary medium and the terrestrial environment, which is essential to drive geomagnetic activity. But the annual variation of geomagnetic effectiveness does not indicate that the fast streams change polarity during the course of the year. Since they are related to the coronal structure and magnetic field, their polarity stays constant. The change occurs in the position of the Earth and the terrestrial magnetic field relative to the interplanetary magnetic field. The solar wind and thus the interplanetary magnetic field is fixed with respect to the Sun's equatorial plane. The plane of ecliptic, which also defines the orbit of the Earth, is inclined by 7.2° with respect to this plane, with both planes intersecting in the equinoxes while the Earth is above (below) the solar equator at the summer (winter) solstice. In addition, the axis of the Earth's rotation is tipped at 23° to the ecliptic plane. The combined effects regulate the southward component of the interplanetary magnetic field with respect to the terrestrial field. If the magnetic field is directed away from the Sun, in spring it has a strong southward component in geomagnetic coordinates while in autumn it has a very weak southward or even a northward component. If the interplanetary magnetic field is directed inwards, the pattern is just the opposite. Thus a fast solar wind stream which has a high geomagnetic effectiveness in spring is less efficient in autumn, while a stream with an opposite direction of the interplanetary magnetic field shows the reverse pattern. At the solstice, in both streams the southward component is diminished, leading to a reduced geomagnetic effectiveness.

8.6 Aurorae

Aurorae, or polar lights, are a prime example of solar–terrestrial relationships. They also provide an interesting example of the development in geophysical research. And they are simply beautiful and, depending on his habitat, have fascinated or frightened mankind since historical times. Recent reviews about the plasma physical aspects of aurora are given in [330, 400]; the more popular aspects of aurora together with numerous examples are discussed in [54, 60, 142]. Internet resources on aurora are e.g. www.geo.mtu.edu/weather/aurora/, www.northern-lights.no/, www.polarimage.fi or www.pfrr.alaska.edu/aurora/INDEX.HTM, and on aurora prediction from satellite data sec.noaa.gov/pmap/.

    The aurora, also called the northern light, is a typical phenomenon of the high latitudes, where under suitable conditions (clear sky, no full moon), it can be observed almost constantly. Under normal conditions, the aurora is a colored arc extending roughly from east to west, changing its appearance in a

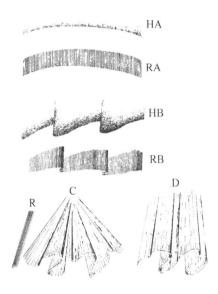

Fig. 8.39. Typical shapes of the aurora: homogeneous arc HA, rayed arc RA, homogeneous bands HB, rayed band RB, rays R, coronae C, and draperies D. Reprinted from A. Brekke and A. Egeland [60], *Northern lights*, Copyright 1983, Springer-Verlag

typical pattern during the night. Under geomagnetically disturbed conditions, the aurora brightens, becomes highly structured, moves equatorwards across the sky, and changes its appearance fast. Typical auroral structures are shown in Fig. 8.39, with the arcs and bands more typical of geomagnetically quiet conditions and the draperies, rays, and corona more often observed during geomagnetic activity. The aurora is less bright than the full moon, and thus although even in mid-Europe some aurorae occur each year, often they are difficult to detect because of a city's counterglow in the sky.

8.6.1 Historical Excursion

Records of the aurora can be traced back for at least 2500 years. The ancient Chinese described dragons winding in the sky, calling the aurora "flying dragons". More detailed records date back to the Romans. Although made at low latitudes (Mediterranean), these observations describe the many shapes and colors normally only observed in the auroral oval. But the most frequent aurora in low latitudes is a reddish glow at the horizon, often misinterpreted as a burning farmstead or village. For instance, Seneca writes that the emperor Tiberius sent troops to the village of Ostia because an aurora evoked the impression of the village being in flames. Such misinterpretation can be traced throughout history. Even in the twentieth century reddish glows have caused false alarms of distant forest fires. The ancient Greeks, too, saw the aurora and named it "chasmata", which means frightening apparition.

At middle and low latitudes the reddish color, combined with the small number of sightings, has led to interpretations of the aurora in terms of bad omens of war, famine, fire, and pestilence, or clerics interpreted it as a battle

Fig. 8.40. Fantastic illustration of an aurora observed from Bamberg, Germany, in December 1560. The flashing lights in the northern sky are interpreted as sparks from clashing swords in a heavenly battle

in the heavens between good and evil, with rays appearing as swords, spears, or faculae. A typical example of such an interpretation is shown in Fig. 8.40 for an aurora observed in 1560 from the German town of Bamberg.

Since aurorae were observed only occasionally, such a superstitious interpretation in an end-of-the world mood was typical of mid-Europe. Cultures, which from the very same scholars were regarded as primitive (note that at the time of the woodcut shown in Fig. 8.40 the discovery of America was already history), often had a healthier attitude towards these lights. One of the reasons obviously is the higher frequency of occurrence for people living at higher geomagnetic latitudes, taking away much of the horror and sometimes even encouraging the search for natural explanations. In the *King's Mirror*, a Norse chronicle of 1259 as many as three alternative explanations are offered, which in the framework of the then world picture are quite reasonable:

The men who have thought about and discussed these lights have guessed at three sources, one of which, it seems, ought to be true. Some hold that fire circles about the ocean and all the bodies of water that stream about on the outer side of the globe; and since Greenland lies on the outermost edge of the Earth to the north, they think it is possible that these fires shine forth from the fires that encircle the outer ocean. Others have suggested that during the hours of night, when the Sun's course is beneath the Earth, an occasional gleam of light may shoot up into the sky, for they insist that Greenland lies so far out on the Earth's edge that the curved surface which shuts out the sunlight must be less prominent there. But there are still others who believe (and it seems to me not unlikely) that the frost and the glaciers have become so powerful there that they are able to radiate forth these flames.

With the image of the Earth as a flat disk and the daily experience of glittering water and gleaming glaciers, all three explanations are quite reasonable.

Nonetheless, although aurorae were part of the daily experience, other cultures close to the northern Arctic Circle have offered less scientific and more mythological interpretations. Many Inuit tribes interpreted aurorae as torches in the hand of Gods leading the souls of their deceased or as the souls of the departed playing ball on the heavenly meadows. Other Eskimo tribes interpreted them as the dances of their Gods. A similar interpretation was offered by the Scotts about 2000 years ago (they called the aurorae 'merry dancers') or by the Aboriginals. Their neighbors, the Maori, interpreted the aurorae as fires set ablaze by their ancestors who, in their canoes, had drifted too far to the south and now had lightened fires to keep themselves warm.

But even cultures that rather often saw aurorae were not immune against superstition. The North American native Indians, for instance, also interpreted an aurora as a bad omen or fighting tribes. And even the Laps, who should be acquainted with the aurora rather well, interpreted it as a sign of violent death, as the souls of the victims.

8.6.2 Beginning of the Scientific Analysis

At the beginning of the scientific analysis, the aurora had been interpreted as the counterglow or reflection of a natural light source, for instance the reflection of the Sun, already below the horizon, from clouds or a flat surface of water, ice, or snow. Alternative interpretations included natural terrestrial sources such as volcanos, streams of lava, or bog fires. These explanations are particularly attractive for an aurora observed as a reddish glow just above the horizon, such as observed most often from mid-Europe.

This line of thought survived for rather a long time, although in the middle of the eighteenth century Cavendish, using triangulation, found the heights of the aurora to be at about 100 km. Thus the aurora cannot be explained by reflection from clouds at much lower altitudes. A very systematic analysis of the heights of the lower edges of the aurora was performed by Størmer. Figure 8.41 shows his results: aurorae are observed in a wide band of altitudes ranging from just above 80 km to more than 500 km. Nonetheless, most aurorae had their lower edge at an altitude between 90 km and 150 km.

In the 1860s, a spectroscopic analysis of the aurora by Ångström brought an end to all reflection theories. Since pressure in the high atmosphere is very low, the auroral emission consists of many forbidden lines, at that time unknown from laboratory experiments. But although these lines were identified only in the 1920s, the limitation of auroral emissions to a few lines gave evidence against the reflection of a continuous solar spectrum. Thus the aurora had been identified as an independent phenomenon, originating in the discharge of excited gases.

Hints at its origin had developed only slowly; in particular, the relationship between solar and auroral activity had long been debated. In 1730,

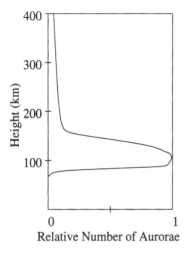

0   1   **Fig. 8.41.** Height distribution of aurorae,
Relative Number of Aurorae   based on [505]

Lewis proposed that sunspots were responsible for auroral activity because the number of aurorae and their southward extent, which also leads to higher detectability and therefore higher frequency at lower latitudes, increases with sunspot number. Since it was difficult to establish a causal chain between these phenomena, Lewis's hypothesis was disregarded. More than a century later it was revived by two different observations. In May 1859, Carrington observed a flare in white light, which not only gave rise to the first record of a flare but two days later also to a violent geomagnetic storm and strong auroral activity even at mid-latitudes. Carrington himself speculated on a relationship between the flare and the aurora; however, he also cautioned against hasty conclusions since one swallow does not make a summer. A more systematic analysis was published in 1873 by H. Fritz, see Fig. 8.42, confirming the close correlation between sunspot number and aurora proposed about 150 years earlier by Lewis. Fritz described the results of his analysis, which considered data from the past 100 years, as follows:

The aurora is a periodical phenomenon and closely related to the formation of dark spots on the Sun. The times of the richest exhibition of spots on the central body of our planetary system are characterized by rich and magnificent light phenomena around the poles of our Earth, while times of sunspot minima correspond to rareness, weak development, and small spatial extent of aurora. During these times, in the mid-latitudes of our planet the aurora vanishes while during maximum times it extends downward from both poles, occasionally even close to the equator.

Because of this close relationship between aurora and solar activity, the frequency of aurorae at mid-latitudes can be used as a proxy for solar activity, in particular when historical times are concerned where no or only scattered records of sunspots are available.

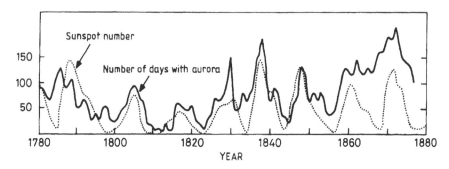

Fig. 8.42. Correlation between sunspot number and frequency of aurora. Reprinted from A. Brekke and A. Egeland [60], *Northern lights*, Copyright 1983, Springer-Verlag

Fritz not only studied the frequency of aurorae but also their spatial distribution, as did Muncke, Franklin, and Loomis before him. The distribution of auroral activity can be described in a map of isochasms (see Fig. 8.43); an isochasm is a line of equal frequency of an unusual or frightening apparition. The isochasms were found to be ovals around the geomagnetic pole with the maximum at geomagnetic latitudes around 70°. For the auroral oval around the North Pole, the southernmost extension of about 60° latitude is at about 90° western longitude. Then the isochasm deviates northward from the parallel and goes through Baffin Bay, around the southern tip of Greenland, crosses Iceland and the northern part of Spitzbergen, reaching the highest northern latitude at about 40° eastern longitude. Then it deviates southward from the parallel, turning back to its starting point on a route across the Siberian Ice Sea and just north of the Bering Strait. Farther north or south of this line, the aurora is less frequent. Nonetheless, the southernmost northern light recorded so far was observed on 15 September 1909, from Singapore, just a few degrees north of the equator.

Thus the spatial distribution of aurorae somehow is governed by the terrestrial magnetic field. A connection between aurora and magnetic field variations had been observed even earlier. For instance, in 1749, Celsius and Hiorter published reports on the relationship between magnetic field disturbances and the aurora, wondering "who would have thought that there is a relation between the aurora and the compass needle, and that the northern light, if moving southwards across the zenith, can cause a deviation of the magnetic needle by some degrees?".

8.6.3 Modern Interpretation

A simple, though incomplete, analogy of our understanding of the aurora is the cathode-ray tube in a television set: particles accelerated at the cathode (during a substorm in the plasma sheet in the magnetotail) are deflected by a

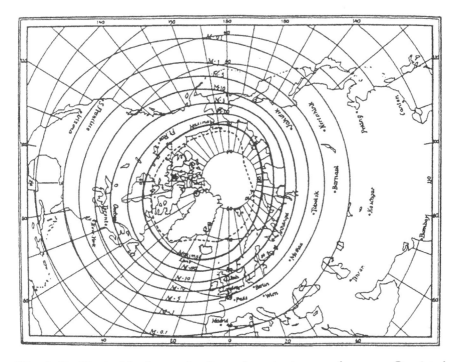

Fig. 8.43. Chart of isochasms, i.e. lines of constant aurora frequency. Reprinted from H. Fritz [173], *Das Polarlicht*, Copyright 1881, with kind permission from Spektrum Akademischer Verlag

field (the geomagnetic field) and emit light on hitting a fluorescent screen (the upper atmosphere). Early last century, long before the development of the TV, Birkeland [50] built the terrella, a model that allowed simulation of the auroral distribution in the laboratory. It completely illustrates this analogy: a cathode-ray tube in a vacuum chamber emits electrons towards a sphere, the terrella, covered with a fluorescent material. If an electromagnet inside the terrella is switched on, the terrella emits light only from two circles around its poles, while with the electromagnet turned off, light is emitted from all over the hemisphere viewing the cathode-ray tube.

8.6.4 Electron Acceleration

Although this analogy is vivid, it is incomplete. The basic difference between the aurora and a TV set lies in the motion of the particles just before the screen. As the particles are accelerated in the plasma sheet, they propagate towards the Earth, are deflected by the magnetic field from equatorial regions towards higher latitudes, and then have to penetrate into the atmosphere to excite the atoms and molecules which in turn emit the light seen as aurora. The fundamental problem in this chain of events is the propagation of the particles deep into the atmosphere. Moving closer towards the Earth, the particles find themselves in a converging magnetic field. The constancy of the magnetic moment then leads to a reflection back towards the magnetosphere. This reflection occurs at heights of about 1000 km, where densities are too small for recognizable light emission. Particles could penetrate deeper into the atmosphere if they were accelerated during their motion. Although the details of the acceleration are still under debate, the spectra of electrons measured by rockets at heights of some hundreds of kilometers clearly give evidence for it: outside of auroral arcs, the spectrum is roughly a power law up to energies of at least 10 keV. Over an auroral arc, however, there is a pronounced peak superimposed on this power law, exceeding the power law intensities by up to two orders of magnitude at energies of a few kiloelectronvolts. Thus the additional electron component is nearly monoenergetic which strongly suggests acceleration in an electric field.

Today it is assumed that at heights of some thousands of kilometers a potential structure develops along the field line with a higher positive potential at lower altitudes. Although such a structure would explain the observed electron spectra, an explanation of the structure itself is difficult. Since the plasma is collisionless, according to (8.25) the conductivity $\sigma_{\parallel}$ parallel to the magnetic field is infinite and the magnetic field lines are equipotential lines. Thus parallel electric fields $E_{\parallel}$ are canceled immediately.

Different processes for the development of the potential structure are discussed [52, 190, 209]. All mechanisms agree that the fundamental driving mechanism is an increase in the magnetospheric convection due to changes in solar wind properties. One of the mechanisms under discussion is the development of double layers. A double layer forms when currents are flowing between plasmas with different properties. In geospace, these are the cold and dense ionospheric plasma and the hot and rarefied magnetospheric plasma. For a double layer to form and to be stable, a field-parallel current must already flow. In the high-latitude ionosphere this would be the Birkeland current, connecting the ionosphere with the magnetosphere. While it is difficult to describe the formation of a double layer, we can at least explain how it can stay stable. If the double layer has formed, a potential drop exists. The particle populations encountering this drop can be divided into electrons and ions, ionospheric and magnetospheric plasma, and thermal and suprathermal particles. Let us start with the ionospheric population. This is a thermal population, consisting of ions and electrons. Ions moving upward from the ionosphere towards the double layer see a decreasing potential and thus are accelerated, leaving the ionosphere through the double layer. Ionospheric electrons, on the other hand, are reflected back into the ionosphere. For the magnetospheric plasma, the situation is just the opposite: flowing towards the Earth it encounters a positive potential drop, leading to an acceleration of electrons towards the ionosphere while the ions are deflected. In addition, electrons are created inside the double layer by the interaction between the accelerated magnetospheric ions and neutrals. These electrons are accelerated downwards, too. The net effect is an acceleration of the electrons without destroying the potential drop. Note that this works only if the flow speed of the magnetospheric electrons is larger than their thermal speed, that is a current already flows. This is the upward Birkeland current shown in Fig. 8.23. This is in agreement with the observation of aurorae being limited to regions where the Birkelands current flow upwards, i.e. where electrons move from the magnetosphere to the ionosphere.

A modification of the double layer concept is the electrostatic shock. This is a double layer which is not stationary like the one discussed above but moves with the average ion speed along the magnetic field line.

Observations suggest that a different mechanism also is important in the electron acceleration: the pitch angle distributions of electrons and ions above auroral arcs are different. Thus both populations are reflected at different positions in the converging magnetic field. Since the particles stay at the reflection point for rather a long time (their velocity parallel to the field vanishes), a charge separation and thus an electric field results which in turn accelerates the lighter species, the electrons.

8.6.5 Excitation of the Atmosphere

On hitting the denser atmosphere, the electrons cause electromagnetic emissions due to excitation $(M + e^- \rightarrow M^* + e^-)$ or excitation and ionization $(M + e \rightarrow M^{+*} + 2e^-)$ of the neutrals. Here M denotes an atmospheric constituent, such as N, N_2, O, and O_2. Auroral lines are emitted in the entire range from UV to IR. The most intense lines in the visible are the green oxygen line at 557.7 nm and the red double line of oxygen at 630 nm and 636.4 nm. Both are forbidden lines, thus early observers were not able to identify them. Another intense line is emitted by the nitrogen molecule at 427 nm, a weaker one at 470 nm. Both lines result from transitions between different vibrational states and can be observed only at the lower edge of the aurora since nitrogen molecules are rare above 120 km. Table 8.1 summarizes the most important auroral lines and the heights where they are emitted.

Not only electrons but also protons can excite the neutral atmosphere. Then the excitation is due to charge exchange: the proton is decelerated and becomes an excited hydrogen atom. This, in turn, emits either the $L\alpha$ line in UV or the $H\alpha$ line in the red. Thus proton aurorae are always red. In addition, they are less structured than electron aurorae, cover larger areas, and are observed only at quite high altitudes, i.e. between 300 km and 500 km. Proton and electron aurorae can occur simultaneously. A special example of a proton aurora is the polar glow that forms when solar protons penetrate into the dayside magnetosphere along the cusps.

Table 8.1. Frequent lines in the auroral emission. The Hα line is observed in proton aurorae only

Wavelength (nm)	Emitting species	Altitude (km)	Visual color
391.4	N$^+$	1000	violet-purple
427.8	N$^+$	1000	violet-purple
557.7	O	90–150	green
630.0	O	>150	red
636.4	O	>150	red
656.3	Hα	200–600	red
661.1	N$_2$	65–90	red
669.6	N$_2$	65–90	red
676.8	N$_2$	65–90	red
686.1	N$_2$	65–90	red

8.6.6 Shape and Local Time

But the aurora is not only limited to the cusps and the nightside. Satellite observations indicate that often a closed auroral oval can be observed and observations from the ground show that auroral activity is not only limited to a few hours around local midnight but can be observed during the entire night (and during the polar night even at day-time). In this case a variation of the aurora with local time can be observed, as shown in Fig. 8.44. In the shaded area between noon and midnight, diffuse aurora form or quiet and stable arcs. After about 20 LT, these arcs become more wavy, forming

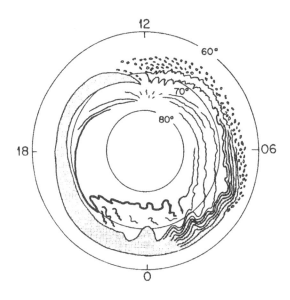

Fig. 8.44. Shape of the aurora in dependence on local time. Reprinted from K. Schlegel [460], in *Plasmaphysik im Sonnensystem* (eds. K.-H. Glassmeier and M. Scholer), Copyright 1992, with kind permission from Spektrum Akademischer Verlag

complex patterns of bands which particularly during a substorm move across the sky in a westward travelling surge. During the decay phase of a substorm, these bands resolve into isolated patches travelling eastward. These patches frequently are observed during the morning hours. The small arcs on the dayside at latitudes of about 75° are formed when solar particles or the solar wind penetrate into the polar cusps.

A special case is the theta aurora. It occasionally can be observed from high-flying satellites as a closed auroral oval supplemented by an arc extending across the polar cap from the dayside to the nightside, giving the aurora the shape of the greek letter Θ. It is observed only at times of a northward interplanetary magnetic field, i.e. at times of a closed magnetosphere. The existence of this arc is difficult to understand since the field lines from the polar cap connect back to the lobes where the plasma density is very low. Current interpretations involve boundaries along the Sun–Earth line with Pederson currents converging over the caps at this boundary and then flowing upwards.

8.7 Energetic Particles in the Magnetosphere

Particle populations in the magnetosphere have different sources and properties. A simple distinction is based on rigidity. Particles with high rigidity are able to traverse the magnetosphere, and thus no long-lived trapped particle components with high rigidities exists. High-rigidity particles coming from the outside, such as galactic cosmic rays, depending on their direction of incidence, penetrate deep into the magnetosphere and interact with the upper atmosphere or are deflected back into space. For low-rigidity particles, three cases can be distinguished, see the right-hand side in Fig. 8.45. Particles hitting the low-latitude magnetosphere from the outside perform half a gyro-orbit inside the magnetosphere and then are reflected back into space. Only at the polar cusps can these particles penetrate into the magnetosphere, and on interaction with the atmosphere produce polar cap absorption (PCA) events. The third low-rigidity particle component is different from all particle populations discussed above in so far as it is not a transient but a long-lived

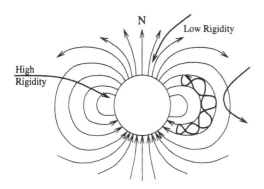

Fig. 8.45. Orbits of particles with low (*right*) and high (*left*) magnetic rigidity

component: particles are trapped inside the radiation belts. Their motion is regulated by the adiabatic invariants; nonetheless, radiation belts are not a static phenomenon but a dynamic one with sources and losses depending on the other particle populations and on geomagnetic activity.

8.7.1 The Radiation Belts

The discovery of the radiation belts was the first significant scientific result of space research with satellites. The first observations were made with a Geiger counter on board Explorer 1, launched on 31 January 1958. The discovery of the radiation belt was accidental. The Geiger counter had been designed by a group around J. van Allen from the University of Iowa to measure cosmic rays in the high atmosphere and had been adjusted to accommodate the expected fluxes. During short parts of the satellite orbit, the observations met the expectations; for long times, however, the observed fluxes either were much too large or way too small, raising doubts about the performance of the instrument. But from the pattern of expected and unexpected counting rates, it became evident that the unexpected signals always indicated particle fluxes much higher than expected, with the low counting rates being a saturation effect. Subsequent measurements with Explorer 3 and Sputnik 3, both in 1958, confirmed the existence of the radiation belts.

First Observations. The first measurements did not identify the particle species or energy. Electrons with energies between 50 keV and 150 keV were expected to be responsible for most of the counts. A strong dependence of counting rate on height was observed. Observations with Pioneer 3 in a highly elliptical orbit with an apogeum at 107 400 km suggested a double structure with an inner radiation belt starting at about 400 km with a maximum at about 1.5 r_E. The counting rates then decreased, but a second radiation belt was found between 3 r_E and 4 r_E with a maximum at about 3.5 r_E. The depleted region between these two radiation belts was called the slot.

Subsequent measurements changed this rather simple picture. First, it was discovered that the trapped particles not only are electrons in the tens and hundreds of kiloelectronvolt range but also protons with energies up to more than 30 MeV. And second, the two distinct radiation belts are fictitious.

Properties and Orbits of Radiation Belt Particles. Figure 8.46 gives a more detailed description of the radiation belts. The upper panel is concerned with protons, the lower ones with electrons. In the upper panel, solid lines give the distribution of protons with energies greater than 30 MeV, and the dotted lines represent protons with energies between 0.1 MeV and 5 MeV. The high energetic protons dominate the inner part of the radiation belt: its lower edge is at about 1.15 r_E. At lower altitudes the losses of particles due to interaction with the atmosphere are too large to allow for a stable trapped

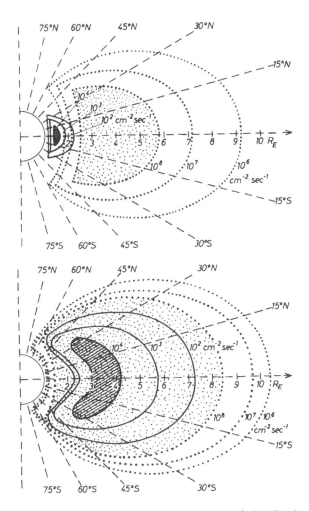

Fig. 8.46. Distribution of particles inside the radiation belts. In the upper panel, the solid lines give the distribution of protons with energies above 30 MeV, and the dotted lines represent protons with energies between 0.1 MeV and 5 MeV. In the lower panel, the solid lines give electrons with energies above 1.6 MeV, while the dotted lines correspond to energies between 0.04 MeV and 1 MeV. Reprinted from W. Kertz [284], *Einführung in die Geophysik*, Copyright 1971, with kind permission from Spektrum Akademischer Verlag

population. The maximum of the high energetic protons is at about 1.5 r_E; with increasing height the density decreases. The picture is different for the low energetic proton component. Here the fluxes are much higher and a broad maximum can be found around 3.5 r_E.

If we consider the electrons as shown in the lower panel in Fig. 8.46, the pattern is slightly different. Low-energy electrons (dotted line, 0.04–1 MeV) and high-energy electrons (solid lines, energy above 1.6 MeV) have their maxima at similar positions, i.e. about 3.5 r_E. Nonetheless, the high-energy component is confined to a smaller spatial region.

Note that in Fig. 8.46 the radiation belts are given in geomagnetic coordinates under the assumption of axisymmetry around the dipole axis. Since the dipole axis is offset with respect to the center of Earth, in spatial coordinates the radiation belts are asymmetric. In particular at the SAA off the coast of Brazil, the radiation belts can be found in rather low altitudes.

Although the upper panel of Fig. 8.46 still suggests a description in terms of two separate radiation belts, consideration of the intermediate energies not shown in the figure suggests a different picture. Today, we understand the radiation belt as a zone of trapped particles with the properties of the particles changing continuously with distance. The higher energies can be found predominately close to the Earth, thus in the inner part of the radiation belt the energy spectrum is rather hard while it steepens with increasing distance.

More recent observations, in particular by SAMPEX (Solar, Anomalous, and Magnetospheric Particle EXplorer, see e.g. `sunland.gsfc.nasa.gov/smex/sampex/` or `surya.umd.edu/www/sampex.html`), indicate the existence of a distinct trapped particle component inside the inner radiation belt. The particles of this new radiation belt differ from the other radiation belt particles insofar as their composition and charge states closely resemble those of the anomalous component instead of the galactic cosmic rays. The flux densities are about two orders of magnitude larger than those of the anomalous component outside the magnetosphere. The dynamics of the new radiation belt are similar to those of the van Allen belt.

Nonetheless, the radiation belt often is divided into two zones, an inner one with $L < 2$ and an outer one with $L > 2$. This distinction is not so much concerned with the properties of the particles as with stability: in the inner zone, the particle populations are very stable and long-lived, while the particle populations in the outer zone vary with solar and geomagnetic activity.

The basics of the motion of the radiation belt particles under undisturbed conditions is described by the adiabatic invariants (see Sect. 2.4). Equation (2.83) is the mirror condition: it gives the smallest pitch angle which will be reflected in a certain mirror configuration. In the magnetosphere (or more generally in a dipole field), the field is weakest at the magnetic equator and increases towards the poles. Thus, we are interested in the smallest pitch angle of a particle to keep it confined in the radiation belt and prevent it from being lost due to interaction with the atmosphere. With (8.11) and (2.83) we obtain

$$\sin^2 \alpha_{eq} = \frac{B_{eq}}{B_m} = \frac{\cos^6 \Lambda_m}{\sqrt{1 + 3\sin^2 \Lambda_m}} \tag{8.33}$$

where B_m is the magnetic field at the mirror point and Λ_m is the geomagnetic latitude of the mirror point. The bounce period between the two mirror points then can be determined as the integral of the full motion

$$\tau_b = 4 \int_0^{\lambda_m} \frac{ds}{v_\parallel} \ . \tag{8.34}$$

This equation can be solved numerically after inserting $v_\parallel$ and yields for the bounce period

$$\tau_b \approx \frac{LR_E}{\sqrt{W_{kin}/m}} \ (3.7 - 1.6 \sin \alpha_{eq}) \ . \tag{8.35}$$

Note that the bounce period only weakly depends on the equatorial pitch angle: the value in the parentheses is between 2.1 and 3.7, i.e. a variation of less than a factor of 2 in bounce period for particles almost standing at the equator (pitch angle close to $90°$) and particles travelling almost field-parallel. The time a particle stays in a stretch ds of the field line is longest for large pitch angles while it is small for small pitch angles. Thus the particle spends most of the bounce period close to the mirror points – and this time is the same for particles with large and small equatorial pitch angles.

The angular drift velocity can be obtained similarly. Again, the integral can be solved only numerically:

$$\langle v_D \rangle \approx \frac{6L^2 W_{kin}}{qB_E R_E} \ (0.35 + 0.15 \sin \alpha_{eq}) \ . \tag{8.36}$$

Equation (8.36) gives the drift velocity averaged over one bounce motion. For $\alpha_{eq} = 0$ we obtain for the equatorial ring current

$$j_d = \frac{3L^2 n W_{kin}}{B_E R_E} \tag{8.37}$$

where n is the particle density. From (8.36) we obtain the drift period

$$\langle \tau_D \rangle = \frac{2\pi L R_E}{\langle v_D \rangle} \approx \frac{\pi q B_E R_E^2}{3LW_{kin}} \ (0.35 + 0.15 \sin \alpha_{eq}) \ . \tag{8.38}$$

The solar-wind generated $\boldsymbol{E}$-field (dawn-to-dusk field or equatorial traverse field) causes an $\boldsymbol{E} \times \boldsymbol{B}$ drift with a drift speed

$$v_{eq} = \frac{E_{eq}}{B_{eq}} = \frac{E_{eq} L^3}{B_e} \ . \tag{8.39}$$

This drift is in the sunward direction. The grad$B$-drift, on the other hand, gives a westward drift for positive ions and an eastward drift for electrons, thus both species drift into opposite directions in the dawn side. Close to the Earth, the magnetic drift prevails and a symmetrical ring current arises. At larger distances, the $\boldsymbol{E} \times \boldsymbol{B}$ drift dominates and only a partial ring current forms.

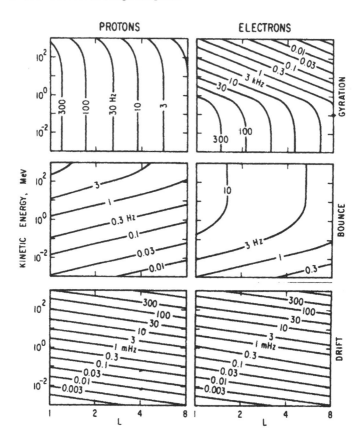

Fig. 8.47. Scales of particle motion in the Earth's magnetosphere. Reprinted from M. Schulz and L.J. Lanzerotti [468], *Particle diffusion in radiation belts*, Copyright 1974, Springer-Verlag

The typical time scales can be found in Fig. 8.47: the particles gyrate around their guiding field line with periods of the order of 10^{-5} s. They bounce back and forth along their guiding center field line with periods of about 1 s (north–south oscillation), and they drift around the Earth with a period of about 500 s, forming a ring current. Detailed discussions of particles in the radiation belts can be found in [320, 445, 468, 544].

The particles trapped inside the magnetosphere show a typical pitch angle distribution, the loss cone distribution: particles with small pitch angles are absent because during the north–south oscillation these particles are not mirrored back at high altitudes but penetrate deep enough into the atmosphere to be lost by interaction with atmospheric constituents.

Gains and Losses. The radiation belts would be stable were it not for two processes: first, magnetic field fluctuations scatter particles during their north–south oscillation into the loss cone, and second, during geomagnetic activity, the entire structure of the magnetosphere is distorted. Thus the adiabatic invariants can be violated and particles are fed into the loss cone. On the other hand, the radiation belts do exist, and thus there are sources replenishing them. Despite their structure and temporal variability, the average properties of the radiation belts are remarkably constant, thus an equilibrium between sources and sinks appears to exist.

The first indications for the lifetime of radiation belt particles came from atmospheric A-bomb tests. In the Starfish experiment in 1962, an artificial electron population had been injected into the radiation belt. After only about 10 years, this population had vanished into the background. However, the lifetime of trapped particles is not a universal constant, but depends on altitude and particle properties. The lifetime increases fast from the lower edge of the radiation belts to some years at an altitude of about 8000 km ($1.25 \, r_{\mathrm{E}}$) and decreases to minutes at the outer edge of the radiation belts [533]. The exact variation depends on the external circumstances, in particular the influence of solar and geomagnetic activity: under undisturbed conditions, the lifetime is larger than during strong solar or geomagnetic activity.

Particle losses are always due to the interaction of radiation belt particles with the atmospheric gas. Significant losses occur only when the atmosphere is sufficiently dense to support interactions, that is below an altitude of about 100 km. Thus, the losses happen at the mirror points where the particles come closest to the atmosphere. The chance of getting lost therefore is largest for particles with small pitch angles, and overall losses increase when pitch angle scattering increases. Under the simplifying assumption that interaction happens only at the surface, the equatorial loss cone can be described by (8.33). Using the L-shell parameter, the loss cone also can be written as

$$\sin \alpha_{\mathrm{eq}}^2 = \sqrt{4L^6 - 3L^5} \, . \tag{8.40}$$

Thus, the loss cone becomes rather small for $r > 3R_{\mathrm{E}}$ which validates the above distinction into an inner and outer radiation belt. The physical mechanisms for losses are:

- Charge exchange with a particle from the neutral atmosphere: a fast radiation belt proton captures an electron from the atmospheric hydrogen, leaving behind a slow proton and continuing itself as a fast hydrogen atom through the atmosphere. This mechanism is of particular importance in the inner radiation belt for protons with energies below about 100 keV. With increasing energy, the interaction time between radiation belt protons and atmospheric hydrogen decreases, making the interaction less likely.
- Nuclear collisions between protons and atmospheric atoms and molecules: these are important loss mechanism for protons with energies above 75 MeV in the inner radiation belts.

- Scattering into the loss cone is the main loss mechanism for electrons and protons in the outer radiation belt. Scattering can occur either by violation of the second adiabatic invariant due to changes in the field at time scales shorter than the north–south oscillation time, or due to pitch angle scattering at electrostatic or electromagnetic waves; see Sect. 7.3.5.

Sources. The sources of radiation belt particles can be divided into the "creation" of particles inside the radiation belts or the motion of particles into the radiation belts under violation of the second adiabatic invariant.

The high energetic particles in the inner radiation belts in general are created there. The main mechanism is CRAND (Cosmic Ray Albedo Neutron Decay). Nuclei from the galactic cosmic radiation penetrate deep into the atmosphere and interact with the atmospheric gas. A 5 GeV proton, for instance, on average creates about seven neutrons during these interactions. Some of the neutrons are slowed down, creating the cosmogenic nuclides such as radiocarbon (capture of thermal neutrons by atmospheric nitrogen: $^{14}N(n,p)^{14}C$) or ^{10}Be (spallation of nitrogen or oxygen due to the capture of fast protons or neutrons). Other neutrons simply escape without interaction. Since neutrons are neutrals, their motion is not influenced by the geomagnetic field, thus some of them might propagate into the radiation belts. But neutrons are not stable and decay into a proton, an electron, and an antineutrino. If this decay happens inside the radiation belt, electrons and protons are trapped, thus replenishing the radiation belt population. This process is called CRAND because the primaries are Cosmic Rays, creating secondaries which are partly reflected (Albedo = reflectivity) into the radiation belts, where the Neutrons Decay.

Another source, also based on in situ creation, is the influx of particles from the outer magnetosphere. In the outer magnetosphere, particles with low energies dominate. Occasionally, these particles can recombine, forming neutrals. These neutrals are no longer guided by the magnetic field and eventually propagate into the radiation belts. On the dayside, the neutrals are not stable but immediately become ionized by the Sun's hard electromagnetic radiation. If this process happens while the neutral is inside the radiation belt, an additional electron and ion are fed into the belt population.

In the outer radiation belt ($L > 3$), the sources and sinks become more complex and are much more intimately related to solar and geomagnetic activity. For instance, the fluxes of energetic electrons are increased during higher geomagnetic activity and relax only slowly after a strong geomagnetic storm. While the maximum of the radiation belt is depleted during the storm, at larger distances the flux densities are higher, slowly propagating towards smaller L-shells (see Fig. 8.48). Thus the radiation belt is refilled with particles from the outer magnetosphere.

For a fixed distance in space, changes in particle flux can be observed as a consequence of changes in the solar wind speed and the related geomagnetic activity. In Fig. 8.49 the dashed line gives the variation of the solar wind

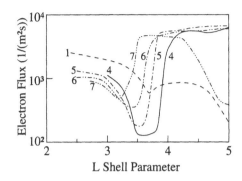

Fig. 8.48. Fluxes of trapped electrons with energies above 1.6 MeV for different times after a geomagnetic storm. The dashed curve (1) gives the undisturbed conditions on 7 Dec. 1962. The other curves are taken after the storm on 20 Dec. (4), 23 Dec. (5), 29 Dec. (6), and 8 Jan. (7). The motion of the electrons towards the inner magnetosphere is clearly visible. Based on [168]

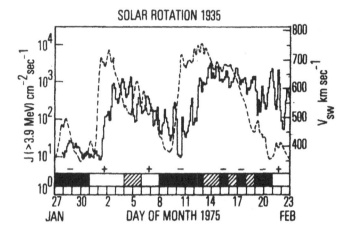

Fig. 8.49. Flux of energetic electrons above 3.9 MeV in a geostationary orbit (*solid line*) compared with the solar wind speed (*dashed line*). The polarity pattern of the interplanetary magnetic field is shown at the bottom. Reprinted from G.A. Paulikas and J.B. Blake [402], in *Quantitative modeling of magnetospheric processes* (ed. W.P. Olson), Copyright 1979, American Geophysical Union

speed, and the solid line the changes in electron fluxes in a geostationary orbit at a height of 36 000 km, well outside the maximum of the radiation belts. The electron flux increases in response to an increase in solar wind speed with an energy-dependent delay between about 1 and 2 days due to the inward propagation of the particles. Thus compression of the magnetosphere is connected with an increase in particle fluxes, although the latter is delayed depending on the energy of the particles and the position of the observer.

L-Shell Diffusion. Particles refilling the radiation belts have entered the magnetosphere at the polar cusps, either from the outside as solar wind plasma or solar energetic particles, or from the ionosphere. Although some of these particles might have rather high energies on entering the magnetosphere, most of them will have energies much closer to the plasma's thermal energy than to the energies typically observed in the radiation belts. Thus not only do we need a mechanism to transport these particles into the radiation belts but we also need to accelerate them.

Both transport as well as particle acceleration is induced by magnetic fluctuations, as is evident from the observations described in connection with Figs. 8.48 and 8.49. The transport mechanism basically is diffusion, although convection and drift also contribute to the transport [478]. The acceleration is related to the transport because during the inward transport the magnetic flux density increases. The acceleration mechanisms are the betatron effect and Fermi acceleration.

In the betatron effect a charged particle gyrates in a magnetic field with slowly increasing magnetic flux density. As the flux increases, the gyro-frequency increases, too (see (2.28)). Since the angular momentum $mvr_{\mathrm{L}} = mv^2/\omega_{\mathrm{c}}$ stays constant, the particle's kinetic energy increases as

$$W_{\mathrm{kin}} = mv^2/2 \sim \omega_{\mathrm{c}} \sim B \ . \tag{8.41}$$

Thus a particle moving inwards in the magnetosphere gains energy as the magnetic field increases.

Whereas the betatron effect accelerates particles moving perpendicular to the field towards regions of higher flux density, the Fermi effect accelerates the motion of a particle parallel to the field: according to the second adiabatic invariant (2.85), a particle gains energy as its travel path along the magnetic field line shortens. As a particle moves inwards in the magnetosphere, the field line along which it bounces shortens and the particle gains energy.

The inward motion of the particles is a diffusive process based on the violation of the third adiabatic invariant (flux invariant) when marked variations of the magnetosphere occur on time scales smaller than the drift period of the particles. Figure 8.50 gives a very simple model: assume a narrow particle distribution in the equatorial plane (a). A sudden variation in the solar wind leads to a fast compression (expansion) of the magnetosphere (b). Since this

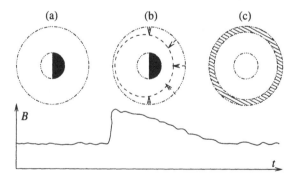

Fig. 8.50. Effect of an asymmetric sudden compression and slow relaxation of the geomagnetic field on a narrow band of equatorially trapped particles. After the recovery period, the particles fill the shaded band

is a fast change, the magnetic flux inside the particle's drift orbit is suddenly increased (decreased). The field now relaxes slowly (adiabatically) back to its original state. As the flux is now conserved, the drift orbit expands (shrinks) and the particle orbit has moved in radius. Since the magnetosphere is asymmetric, a change in solar wind leads to a stronger variation in the field on the dayside than on the nightside. Thus the effect on the particle not only depends on the properties of the disturbance but also on the particle's position in its drift orbit. Therefore the initial distribution is spread in radius (c). This process is diffusive, and it is called L-shell diffusion or radial diffusion.

In sum, the adiabatic heating creates 10–100 keV ring current ions from 1–10 keV plasma sheet ions. The adiabatic heating is divided into a transverse part

$$\frac{W_\perp}{W_{\perp,0}} = \left(\frac{L_0}{L}\right)^3 \tag{8.42}$$

and a longitudinal one

$$\frac{W_\parallel}{W_{\parallel 0}} = \left(\frac{L_0}{L}\right)^\kappa \tag{8.43}$$

where $\kappa = 2$ for $\alpha_{eq} = 0$ and $\kappa = 2.5$ for $\alpha_{eq} \to 90°$: the adiabatic heating due to the Fermi effect decreases with increasing path between the mirror points.

The process shown in Fig. 8.50 leads to a spread of the particle distribution but not necessarily to an inward motion. Since particle streaming in diffusion is driven by a gradient (see (7.10)), for undisturbed magnetospheric conditions, such as in the particle distributions in Fig. 8.46, we would not expect an inward streaming but rather one directed outwards. After a strong geomagnetic storm, however, the particle density increases in the outer magnetosphere due to compression as well as to an injection of plasma from the tail, leading to a density gradient and therefore a particle streaming towards the inner magnetosphere as is evident in the observations shown in Fig. 8.48.

8.7.2 Galactic Cosmic Rays – Størmer Orbits

Galactic cosmic rays have high rigidities and thus their gyro-radii are comparable to the size of the magnetosphere. Thus concepts such as the adiabatic invariants cannot be applied to the motion of these particles: instead, their equation of motion has to be integrated along the particle orbit. This was first done by C. Størmer; the orbits thus are called Størmer orbits.

In interplanetary space, a galactic proton has a gyro-radius of the order of 1/100 AU. If such a proton hits the magnetosphere, it senses a strong change in the magnetic field on a spatial scale much smaller than its Larmor radius. Thus the simplifying assumption of a uniform magnetic field made for the radiation belt particles cannot be applied here. The fate of the particle then depends on its pitch angle and the direction and location of incidence:

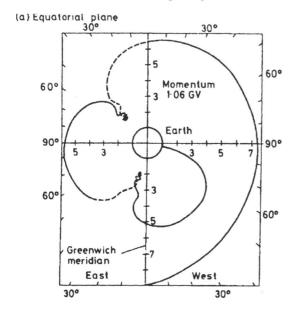

(a) Equatorial plane

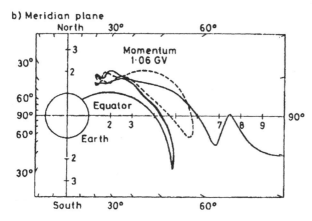

b) Meridian plane

Fig. 8.51. Orbit of a high energetic particle in a magnetic dipole field in (a) the equatorial plane and (b) in a meridional plane. Reprinted from J.K. Hargreaves [212] (based on [345], *The solar-terrestrial environment*, Copyright 1992, with kind permission from Cambridge University Press

particles propagating nearly field-parallel into the polar cusps are less influenced by the geomagnetic field than particles hitting the magnetosphere in equatorial regions perpendicular to the magnetic field.

The motion of such a high energetic particle is still determined by the Lorentz force; however, since the magnetic field varies strongly on a scale smaller than the gyro-radius, the equation of motion has to be integrated. An example is sketched in Fig. 8.51 where the orbit of a 1 GV particle coming from the east is shown in the equatorial plane (a) and the meridional plane (b) of a magnetic dipole field. The particle trajectory is irregular with wide excursions in latitude and longitude until it finally hits the Earth.

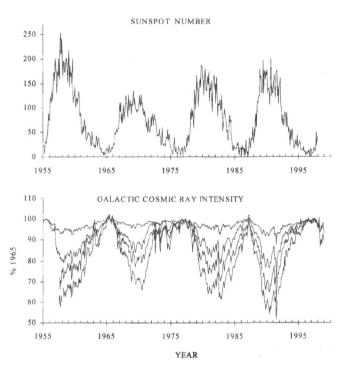

Fig. 8.52. Intensity of the galactic cosmic radiation observed with neutron monitors at different geomagnetic latitudes (from *top* to *bottom*): Huancayo ($R_c = 13$ GV), Climax (3 GV), Moscow (2.4 GV), and Murmansk (0.6 GV). Reprinted from G.A. Bazilevskaya, *Observations of variability in cosmic rays*, Copyright 2000, with kind permission of Kluwer Academic Publishers

The particle's orbit depends on the location and direction of incidence. The most important result of Størmer's work has been the definition of allowed and forbidden regions on the ground which can be reached by a charged particle travelling towards the Earth. For instance, to hit the Earth at a certain magnetic latitude ϕ_c, the particle's rigidity must be at least $P_c = 14.9 \cos^4 \phi_c$, with P_c in units of 10^9 V or 1 GV. This rigidity is also called the cutoff rigidity: particles with a rigidity R_c can reach all latitudes of ϕ_c and above, but particles with smaller rigidities can hit the ground only at higher latitudes. This latitude dependence can be seen, for instance, in the neutron-monitor counting rates measuring the galactic cosmic radiation. Figure 8.52 shows in its top panel sunspot numbers, and in its bottom panel counting rates for neutron monitors with different geomagnetic cutoff rigidities R_c: from top to bottom, the sites of these monitors are Huancayo/Haleakala ($R_c = 13$ GV), Climax ($R_c = 3$ GV), Moscow ($R_c = 2.4$ GV), and Murmansk ($R_c = 0.6$ GV). The modulation of galactic cosmic rays with the solar cycle affects only the lower energies; see Sect. 7.7 and Fig. 7.29. As a consequence, the modulation with the solar cycle almost vanishes at low-latitude stations such as Huancayo/Haleakala, while it increases with increasing geomagnetic latitude and decreasing geomagnetic cutoff.

The extent of the forbidden region is roughly the Størmer unit $C_{St} = \sqrt{M_E/(Br_L)} = \sqrt{M_E/P}$. For a 1 keV solar wind proton, the forbidden region extends for about $100r_E$ and it certainly will not reach the ground. For a solar wind electron, the forbidden region is even wider.

If a proton of the galactic cosmic radiation has been detected on the ground, its origin can be determined by the reverse procedure: we let a negatively charged particle with the same rigidity propagate out through the geomagnetic field towards infinity. Thus the direction of incidence of the galactic cosmic radiation can be determined. On Earth, the galactic cosmic radiation is isotropic, while during ground-level events solar protons with gigaelectronvolt energies can be detected coming predominantly along the interplanetary magnetic field line connecting the Earth to the Sun.

8.7.3 Solar Energetic Particles – Polar Cap Absorption

Except for ground-level events, the energies of solar energetic particles are too small to allow them to penetrate down to the ground. This is true for most of the magnetosphere. At the polar cusps, however, lower energetic solar particles also can reach the denser atmosphere. Their interaction with the atmosphere leads to increased ionization, which in turn leads to the absorption of radio waves. Since this absorption is limited to the polar regions, this phenomenon is called polar cap absorption (PCA). Owing to the propagation time between the Sun and the Earth, a PCA starts a few hours after the flare. A similar effect on the dayside ionosphere, the sudden ionospheric disturbance, on the other hand, starts immediately after flare onset because here the increased ionization is due to the flare's hard electromagnetic radiation.

The protons causing PCAs typically have energies between 1 MeV and 100 MeV and can penetrate down to heights between about 30 km and 90 km. Thus the ion population created during a PCA lies below the ionosphere in an atmospheric layer which normally is neutral, allowing for ion–chemical interactions that normally do not occur at these heights (see Sect. 10.4.2).

Figure 8.53 shows the areas typically affected by PCA. PCAs are limited to a small latitudinal ring around the geomagnetic pole. Outside this ring, PCAs only occur during very strong geomagnetic disturbances when the geomagnetic cut offs are reduced, allowing particles to precipitate into the denser atmosphere at lower geomagnetic latitudes.

Part of the energy transferred from the solar particles to the atmosphere leads to electromagnetic radiation in the visible range, called the polar glow aurora (PGA). In contrast to the normal aurora, which is highly variable and often well structured, the polar glow aurora is a diffuse red glow of the entire sky in high latitudes. Most often it is red since the excited atoms mainly are hydrogen, emitting in the visible the $H\alpha$ line and in the UV range the $L\alpha$ line. Even if its luminosity is well above the visibility threshold, it often is

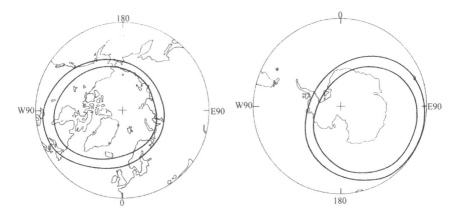

Fig. 8.53. Areas affected by PCA: the regions inside the inner curves give the polar plateaus while the regions outside the outer curves usually are unaffected, except during severe geomagnetic disturbances. Reprinted from G.C. Reid [439], in *Physics of the Sun, vol. III* (eds. P.A. Sturrock, T.E. Holzer, D.M. Mihalas, and K. Ulrich), Copyright 1986, with kind permission from Kluwer Academic Publisher

difficult to detect because of its total lack of any contrasts. Note that the polar glow aurora is the only visible effect caused by solar energetic particles; the much more impressive aurora on the other hand is caused by plasma from the solar wind and the plasma sheet in the geomagnetic tail.

8.8 What I Did Not Tell You

So far we have discussed some features of the magnetosphere in rather simple terms which allow the application of concepts introduced in earlier sections. A complete description of the magnetospheric fields and plasmas, however, requires a self-consistent approach. Let us illustrate some neglected effects with the example of the magnetopause. For instance, we have described the magnetopause as a discontinuity in the magnetic field without adjusting the current systems accordingly. In addition, we have ignored the polarization field inside the current sheet, which results from the different Larmor radii of solar wind electrons and ions but also influences the motion of the very same particles, as is sketched in Fig. 8.9. A self-consistent approach, however, would have to consider quasi-neutrality and the screening of charges in the field of other charges. The characteristic length of the particle motion in the magnetopause would then be the inertial length c/ω_{ce} of an electron instead of the gyro-radius. In the calculation of the magnetopause we have also ignored the magnetosheath, where the field and plasma properties are quite different from those in the solar wind, although both the field and the plasma are of solar wind origin. Putting all these effects together, the thickness of the magnetopause would be some hundreds to about 1000 km, much more than the ion gyro-radius of about 100 km but much closer to the observations.

All of these effects can be described analytically; the resulting set of equations, however, can be solved only by means of MHD simulations. In particular, the dynamic aspects, such as reconnection of planetary and interplanetary fields and the accompanying plasma exchange, can be incorporated into such models. Overviews of magnetospheric modeling can be found in [260, 543], and some articles can be found in [385]. The modeling of the magnetospheric plasma is discussed in [364], and models of the radiation belts can be found in [320]. Resources on the Web for modeling of the magnetosphere and/or the interaction between the solar wind and the magnetosphere include those of the Space Plasma Simulations Group at the University of California, Los Angeles, at www-spc.igpp.ucla.edu/, and the Space Plasma Physics Modeling Group at the University of Washington at www.geophys.washington.edu/Space/SpaceModel/modelling.html.

8.9 Summary

The magnetosphere is shaped by the interaction between the solar wind and the terrestrial magnetic field. Since both are variable the magnetosphere shows a slow evolution as well as fast fluctuations. The basic features in the magnetospheric field topology are the magnetopause, the polar cusps, and the tail. The magnetopause marks the separation between the planetary and the interplanetary magnetic fields. In front of the magnetopause a bow shock develops where the solar wind is slowed down to subsonic speeds. Thus the region between bow shock and magnetopause, the magnetosheath, is filled by a hot but slow solar wind flow.

Although the magnetopause separates the magnetic fields, the solar wind plasma can still penetrate through it, filling part of the outer magnetosphere. In the inner magnetosphere, the cold and dense ionospheric plasma fills the plasma sheet, as is evident from the high O^+ content. The region between this inner plasmasphere and the outer magnetosphere contains a rarefied plasma of mixed origin: under undisturbed conditions, the geosphere is dominated by the solar wind plasma while during geomagnetically active periods an outflow of ionospheric plasma is added.

The particle populations in the magnetosphere have different sources and can be divided into stable and transient populations. The only stable component are the radiation belt particles, forming the ring current. Its variations lead to geomagnetic activity, detected on the ground as geomagnetic storms. The radiation belts are only a quasi-stable population since particles are lost due to interaction with the denser atmosphere and are created due to cosmic rays (CRAND) or the transport and acceleration of lower energetic particles or solar wind plasma from the outer magnetosphere towards the Earth. Galactic cosmic rays have much higher rigidities and are not trapped inside the geomagnetic field. Instead, depending on their direction of incidence, they can either penetrate down to the denser atmosphere or are deflected back into space. Particles with lower rigidities, such as solar energetic particles or the solar wind, can penetrate into the magnetosphere only at the polar cusps and, on interaction with the atmosphere, lead to polar cap absorption.

The dynamic magnetosphere manifests itself in geomagnetic disturbances and aurorae. The energy source for these phenomena is the solar wind; the coupling between the solar wind and the magnetosphere is due to magneto-spheric convection. The level of geomagnetic activity is determined by the solar wind fluctuations and the direction of the interplanetary magnetic field. If the latter has a southward component, the magnetosphere has an open configuration, more energy and plasma are fed into it, and its responses to solar wind variations are much stronger. The variations in the solar wind responsible for geomagnetic disturbances are fast solar wind streams, often leading to recurrent geomagnetic disturbances, as well as shocks and the ejecta driving them. While the recurrent disturbances can best be observed during the solar minimum when stable and recurrent fast and slow solar wind streams have developed, geomagnetic disturbances in response to transient phenomena are much more frequent during the solar maximum, since flares, CMEs, and interplanetary shocks are more frequent at these times.

Exercises and Problems

8.1. Determine the magnetic flux density and the direction of the field for your home town (assume a simple dipole field). Compare with Fig. 8.3. To which L-shell is your home town connected magnetically? Determine also the cut off rigidity and the Størmer unit for this rigidity.

8.2. A 10 keV proton with an equatorial pitch angle of $40°$ moves from $L = 6$ to $L = 1.5$. Assume a dipole field and calculate the energy gained by Fermi acceleration (second adiabatic invariant) during this motion.

8.3. Assume a sinusoidal variation with period $T = 1$ h in the solar wind speed with an amplitude of ±40 km/s around an average of 400 km/s. Determine the speed of the magnetopause and the maximum and minimum stand-off distances.

8.4. Størmer orbits are calculated for a terrestrial dipole field. Give a qualitative statement about the errors made in neglecting the actual shape of the magnetosphere. Try to consider the influence of the shifted dipole as well as the different topologies in the noon and midnight directions.

8.5. Fig. 8.48 shows the inward L-shell diffusion of energetic electrons. Try to estimate the diffusion coefficient for this process (assume that the energy of the particles does not change during the inward motion).

8.6. The magnetopause is determined by the equilibrium between the gas-dynamic pressure of the solar wind and the magnetic pressure of the geomagnetic field. The magnetic pressure of the interplanetary magnetic field is neglected, as is the gas-dynamic pressure of the plasmasphere. Determine the error due to this approximation.

8.7. Give an order-of-magnitude estimate of the Chapman–Ferraro current.

9 Planetary Magnetospheres

Empty space is like a kingdom, and heaven and earth
are no more than a single individual person in that
kingdom ... How unreasonable it would be to suppose
that besides the heaven and earth which we can see
there are no other heavens and no other earths?

Tang Mu, 13th century

A magnetosphere is not a typical terrestrial phenomenon. Instead, magnetospheres can be found around all magnetized bodies embedded in a plasma
flow. Even around unmagnetized bodies (comets, planets without a magnetic
field) a cavity is formed by the interplanetary magnetic field frozen into the
deflected solar wind flow. In the solar system, all planets except Mars and
Venus have a magnetosphere. Although they are different in size, for most
of them the shape and the properties of the magnetosphere are similar to
those in the terrestrial one. Special features are the large size and the flat
inner structure of the Jovian magnetosphere and the oscillation of Neptune's
magnetosphere between pole-on and Earth-like.

9.1 The Planets

Table 9.1 lists the planets with their distance from the Sun, their sizes, and
basic orbital parameters. The planets are divided into two groups, the inner,
Earth-like planets Mercury, Venus, Earth, and Mars, and the outer, gaseous
giants Jupiter, Saturn, Uranus, and Neptune. The Earth-like planets all have
comparable radii with Mercury being the smallest, its radius is only about
one-third of the Earth's radius. Since they all have a solid crust and a heavy,
iron-rich core, their densities are relatively high, ranging between 4 g/cm^3
and 6 g/cm^3. All Earth-like planets have atmospheres, although these are
vastly different in density, composition, and temperature: Mercury has a very
thin atmosphere which can hardly be recognized as an atmosphere, while
the carbon-dioxide atmosphere on Venus is very thick, leading to a strong
greenhouse effect with surface temperatures of about 750 K. The inner planets
have few or no moons and no rings.

Table 9.1. Properties of the planets in the solar system

	Solar distance (AU)	Sidereal period	Spin period (days)	Average density (gm/cm^3)	Surface gravity (N/kg)
Mercury	0.39	88 d	56.8	5.4	3.6
Venus	0.72	225 d	243	5.1	8.7
Earth	1.00	365 d	1	5.5	9.8
Mars	1.52	1 yr 322 d	1.03	4.0	3.7
Jupiter	5.20	11 yr 315 d	0.41	1.3	26.0
Saturn	9.55	29 yr 167 d	0.44	0.7	11.2
Uranus	19.22	84 yr 7 d	0.72	1.18	9.4
Neptune	30.11	164 yr 280 d	0.67	1.56	15.0

The outer planets, Jupiter, Saturn, Uranus, and Neptune, again are similar. They are large, gaseous giants, mainly consisting of hydrogen and helium. Thus their average densities are low with values close to 1 g/cm^3. From the outer atmosphere towards the center of the planet, the pressure increases up to more than a million times the surface pressure on Earth. Here the hydrogen first becomes liquid and later metallic. It is still under debate whether the outer planets have solid cores. They have many moons and ring systems affecting the plasma and particle populations in the magnetospheres.

The energetics of the outer planets are surprising: they radiate more energy back into space than they receive from the Sun. Jupiter radiates back twice the energy received from the sun; on Saturn this ratio is about 3.5. The additional energy either is gravitational, released during the contraction of the planet, or a remnant from the creation of the solar system. Thus the large gaseous planets are more similar to the Sun than to the inner planets.

The outermost planet, Pluto, cannot be fitted into this scheme: it appears to be a solid piece of rock and shows no similarities with the gaseous giants. Pluto's orbit is unusual, too. It is highly eccentric and inclined with respect to the plane of the ecliptic. It is likely that Pluto was not formed together with the solar system but is a captured asteroid.

A review of planetary magnetospheres can be found in [65], and the magnetospheres of the outer planets are discussed in [452].

9.2 Planets with a Magnetic Field

Aside from the outer gaseous giants, only Earth and Mercury have a magnetic moment strong enough to support a magnetosphere. We will now discuss the structures of the magnetospheres together with the plasma populations and their relationships to moons and rings. A comparison between the different magnetospheres follows in the subsequent section.

9.2.1 Mercury

Mercury's magnetosphere is rather simple because both the planet's rotation axis as well as the magnetic field axis are roughly perpendicular to the plane of ecliptic. Since Mercury's magnetosphere is very small with a stand-off distance of only 1.1 r_p, it neither accommodates radiation belts nor a plasma population.

Despite this simplicity, Mercury's magnetosphere is very dynamic. Compared with the other planets, Mercury's orbit is highly eccentric ($\varepsilon = 0.206$) with a perihelion[1] at 0.308 AU. In the perihelion, fluctuations in the solar wind can create such a strong kinetic pressure that the typical magnetopause distance is smaller than the planet's radius. Thus a magnetosphere in the sense of a planetary magnetic field enclosing the planet no longer exists, and the magnetic field is swept into the magnetotail. This variability of the magnetosphere does not support stable radiation belts.

Nonetheless, Mercury's magnetosphere contains a thermal plasma consisting mainly of H, He, O, Na, and K ions as well as electrons. The lighter species, H and He, are swept up out of the solar wind. The heavier ions are created on the dayside of the planet, as indicated in Fig. 9.1: solar UV radiation, and to a lesser extent also the solar wind, knock Na, K, and O out of the planet's surface. These neutrals immediately become ionized by the solar UV radiation and then are kept in the planet's magnetic field.

The tail of Mercury's magnetosphere even shows dynamic phenomena which closely resemble the substorms known from the terrestrial magnetosphere. Because Mercury has neither an atmosphere nor an ionosphere, this observation indicates that the ionosphere is not a necessary condition for the occurrence of substorms. For the planetary consequences of a substorm, nonetheless, the existence of an ionosphere makes a difference.

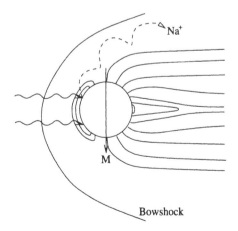

Fig. 9.1. Structure of Mercury's magnetosphere. The creation of a thermal plasma population due to incident solar UV radiation is indicated

---

[1] In an elliptical orbit the perihelion is the position closest to the Sun. Its counterpart is the aphelion.

9.2.2 Jupiter

While Mercury's magnetosphere is the smallest and simplest inside the solar system, the Jovian one is the largest and most complex magnetosphere. A very attentive and detailed description based on the Pioneer and Voyager fly-bys can be found in [131]; results obtained by Galileo are summarized in *Science* **272**. In addition, the Jovian magnetosphere appears to be highly variable, as can be inferred from a comparison of the magnetic field data obtained by Voyager in 1979, Ulysses in 1992, and Galileo in 1995.

The Jovian magnetic field can be approximated as a dipole with its axis inclined by 11° with respect to the planet's rotation axis and slightly offset with respect to it. The magnetic moment of Jupiter exceeds that of Earth by four orders of magnitude; the magnetic field at the planet's surface is about one order of magnitude larger. Since at Jupiter the solar wind intensity has decreased markedly, the magnetosphere extends far above the planet. The fast rotation of the planet with a period of slightly less than 10 h adds another feature to the Jovian magnetosphere: the centrifugal forces in the equatorial plane are very strong, resulting in an outward plasma flow. The planetary magnetic field frozen into this plasma is carried out, too, leading to the magnetic field configuration shown in Fig. 9.2. Therefore the inner Jovian magnetosphere is much flatter than the dipole-like inner magnetospheres of the other planets.

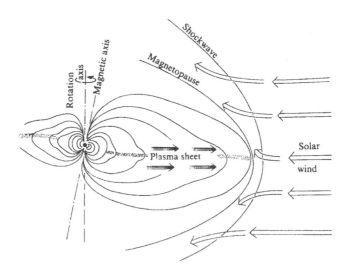

Fig. 9.2. Jovian magnetosphere: the planet's fast rotation leads to a concentration of plasma in the equatorial plane which carries out the magnetic field. Reprinted from J.A. Simpson and B.R. McKibben [486], in *Jupiter* (ed. T. Gehrels), Copyright 1976, with kind permission from The University of Arizona Press

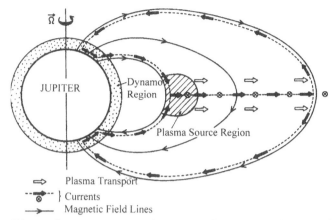

Fig. 9.3. Plasma flows, currents, and magnetic field lines in the inner Jovian magnetosphere. Based on F.M. Neubauer [374], in *Plasmaphysik im Sonnensystem* (eds. K.-H. Glassmeier and M. Scholer), Copyright 1991, reprinted with kind permission from Spektrum Akademischer Verlag

This flattening of the magnetosphere is associated with a current system (see Fig. 9.3): the outward plasma flow is accompanied by an electric current directed radially outward. Without this current the plasma would not corotate with the planet but stay behind. The current system is closed by field-parallel currents through the ionospheric dynamo region. An additional current, separating the fields of opposite polarity, flows around Jupiter in the equatorial plane. Currents inside the Jovian magnetosphere can exceed 10^9 A.

Within about 15 planetary radii, the dipole field is dominant. In this region, the radiation belts are observed; see Figs. 9.12 and 9.13. The particles gyrating inside the radiation belts emit synchrotron radiation with frequencies up to some megahertz, making Jupiter a strong radio source.

Some of Jupiter's moons are inside the magnetosphere. Since the relative speed between the moons and the magnetospheric plasma is smaller than the Alfvén speed, no bow shocks develop in front of the moons. They are unmagnetized bodies, and thus no magnetospheres develop around them. Therefore Jupiter's magnetospheric plasma can interact freely with the moon's surface. Since absorption exceeds the sputtering, the particle intensities in the radiation belts are reduced at the moon's orbit; see Figs. 9.12 and 9.13.

One moon, Io, is also a plasma and particle source: its volcanism injects particles, in particular S (see Table 9.2). This matter immediately becomes ionized and is accelerated to the speed of the corotating magnetosphere. Small differences in the initial conditions and in the interaction between charged particles and neutrals lead to the formation of a torus along the moon's orbit instead of just a cloud of ionized gas around the moon (see Fig. 9.4). Charged particles flowing field-parallel out of this torus create Whistler waves and, on

Table 9.2. Sources of thermal plasma in the magnetospheres of the outer planets. The contributions of the planetary ionospheres and moons can be identified by their unusual composition, in particular the large amount of ions heavier than He

	Jupiter	Saturn	Uranus	Neptune
Sources	Io ionosphere	ionosphere, icy moons, solar wind, Titan	ionosphere, H-corona	ionosphere, solar wind, Triton
Composition	H, Na, O, K, S, O, H_2O, N	H, OH, H_2, O,	H	H, N
Source strength	$\sim 3 \times 10^{28}$ i/s	10^{26} i/s	10^{25} i/s	3×10^{25} i/s
Life time	months, years	months	days	weeks

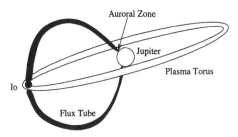

Fig. 9.4. Plasma torus around Io's orbit and flux tubes connecting to the auroral regions. Because the planet's magnetic axis is tilted with respect to the axis of rotation, Io weaves up and down within the plasma torus

interaction with Jupiter's ionosphere, a glowing aurora in the polar regions. Io's plasma torus has a sharp inner boundary, while its outer boundary is less well-defined and extends outwards into the magnetosphere.

9.2.3 Saturn

Saturn's magnetosphere is a rather simple one. It is axial symmetric because both the magnetic field axis and the axis of rotation are perpendicular to the plane of ecliptic. Since Saturn's magnetic field is much weaker than Jupiter's, the magnetosphere is smaller, more closely resembling the one at Earth. Plasma tori with different dominant species exist, as indicated in Fig. 9.5 with particle densities well below the ones in the Jovian magnetosphere.

Saturn's magnetosphere can be divided into four distinct parts. The outer part is filled with a plasma corotating with the planet. The highest plasma density is observed at about 6 planetary radii, which is still deep inside the magnetosphere. The plasma mainly consists of N, O, and OH. Part of this plasma most likely originates in the ring system, and trapped particles in the sense of a radiation belt are not observed.

Moving inwards, the radiation belts start with a slot region, extending from about 6.5 r_p inwards to about 4 r_p. Here the particle fluxes are reduced because of absorption by the moons Dione, Thetis, and Enceladus;

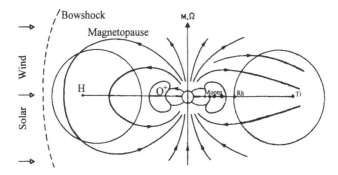

Fig. 9.5. Saturn's magnetosphere with the hydrogen torus created by Titan, the oxygen torus of Thetis, and the positions of the inner moons and the ring system. Reprinted from D.A. Bryant [65], in *Plasma Physics* (ed. R. Dendy), Copyright 1993, with kind permission from Cambridge University Press

see Figs. 9.12 and 9.13. In the region between 4 and 3.1 planetary radii, i.e. between the orbit's of Enceladus and Minas, the fluxes of radiation belt particles increase again, as does their energy. Farther in, in the region occupied by the ring system, particle fluxes drop sharply.

Titan, one of Saturn's moons, has a relatively thick atmosphere, mainly consisting of methane and nitrogen. As Titan loses particles out of this atmosphere, it creates a torus around its orbit with a width of about 2 Saturn radii inside the orbital plane and 7 to 23 planetary radii perpendicular to it. The interaction between Titan and Saturn's magnetosphere is not as continuous as the one between the Jovian magnetosphere and Io because part of Titan's orbit lies outside the magnetopause in the magnetosheath or even in interplanetary space.

Saturn's magnetosphere contains another plasma component which we have not encountered in the other magnetospheres: strongly ionized dust particles with sizes between a few microns and the size of a small moon. These dust particles modify the plasma properties. Surrounded by an electron cloud, they behave like the nuclei of large quasi-atoms.

9.2.4 Uranus

Uranus is a special planet in the solar system in so far as its axis of rotation nearly parallels the plane of ecliptic. Thus one pole of the planet always is directed towards the Sun. For the other planets, the axis of rotation is more or less perpendicular to the ecliptical plane, corresponding to our concepts about the formation of the solar system. It is speculated that a collision between Uranus and a relatively large and massive body might have tipped over the planet's axis of rotation.

On the basis of our observations at Mercury, Earth, Jupiter, and Saturn, as well as our understanding of the magnetohydrodynamic dynamo, we

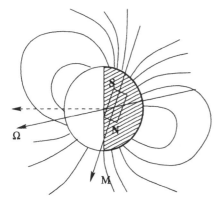

Fig. 9.6. Magnetic dipole of Uranus

would expect the magnetic field axis to be approximately parallel to the axis of rotation. If this were true, a pole-on magnetosphere should result: the magnetic field axis lies in the plane of ecliptic and the solar wind blows directly in the direction of the polar cusp. As a consequence, the magnetotail would be quite different: instead of one current sheet separating the northern and southern lobes, within each lobe a current sheet would result. A sketch of such a pole-on magnetosphere is shown in the right panel of Fig. 9.8.

The Voyager encounter with Uranus, however, revealed a magnetosphere similar to the one at Earth (see Fig. 9.10). The magnetometer measurements on Voyager indicated an inclination of the dipole axis of 58.6° with respect to the rotation axis. In addition, the magnetic dipole is offset with respect to the center of the planet (see Fig. 9.6). Thus the magnetosphere is not axial-symmetric as expected for a pole-on magnetosphere but rotates around the planet's rotation axis.

Two interpretations for the large angle between the dipole axis and the axis of rotation are suggested. First, Uranus just might undergo a polarity reversal. Second, the tilt between the two axes might originate in an unusual inner structure of Uranus, in particular in locations where the planetary dynamo operates. This second interpretation is plausible in so far as on Neptune, which has a similar inner structure, a similar tilt between the dipole and the rotation axes is observed, although Neptune's axis of rotation is inclined with respect to the ecliptic plane by only 28.8°.

Inside the magnetosphere, radiation belts exist; however, the particle fluxes and energies are rather small; see Figs. 9.12 and 9.13. In particular, in the electron fluxes the influence of the moon on the radiation belt populations is obvious: along the moon's orbit, the particle fluxes are reduced.

Compared with all other magnetospheres in the solar system, the magnetosphere of Uranus seems to be the only closed one: because the magnetospheric plasma does not contain any α-particles, the solar wind seems to be completely frozen-out. Plasma sources most likely are the ionosphere, ionization of the vast hydrogen corona of the planet, and, to a lesser extent, sputtering from the moon's surfaces.

9.2.5 Neptune

Neptune's magnetic field and the resulting magnetosphere proved to be even more complex. Neptune's dipole axis is inclined by $47°$ with respect to the axis of rotation, which in turn is inclined by $28.2°$ with respect to the plane of ecliptic (see Fig. 9.7). The dipole not only is eccentric but also shifted along its axis towards its south pole. During one rotation of the planet, the inclination of the dipole axis with respect to the plane of ecliptic varies between $90° - 28.8° - 47° = 14.2°$ and $90° + 28.8° - 47° = 71.8°$. Thus within 8 h, Neptune's magnetosphere oscillates between nearly pole-on and Earth-like (see Fig. 9.8). These oscillation were observed during the Voyager fly-by.

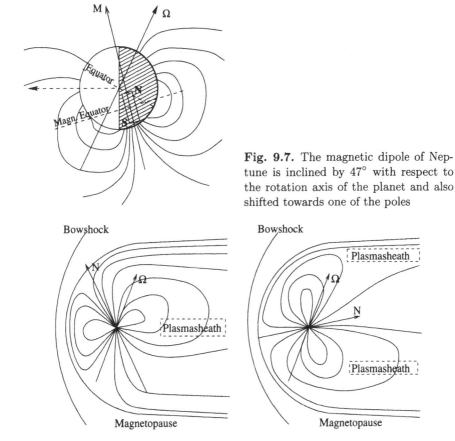

Fig. 9.7. The magnetic dipole of Neptune is inclined by $47°$ with respect to the rotation axis of the planet and also shifted towards one of the poles

Fig. 9.8. Neptune's magnetosphere can be similar to the terrestrial one (*left panel*) while half a rotation later it is a pole-on magnetosphere

Within the orbit of Triton, a radiation belt exists in Neptune's magnetosphere with fluxes much lower than those observed at Uranus. Plasma sources are the solar wind, the atmosphere, sputtering from the moons, and heavier ions from Triton. The plasma density in the outer magnetosphere is too small to distort the field significantly. Neptune's magnetosphere appears to be the least active in the solar system, being only weakly influenced by the variable solar wind.

9.3 Planets Without a Magnetic Field

The interaction of the solar wind with unmagnetized planets is studied best for Venus; our knowledge about the Martian magnetosphere at best is poor. Venus interacts with the solar wind in the same way comets do.

The dayside of Venus's "magnetosphere" is shown in Fig. 9.9. As the solar wind is slowed down to subsonic speed, a bow shock develops. In the magnetosheath behind the bow shock, the plasma and the interplanetary field are compressed. The magnetosheath is separated from Venus's ionosphere by the ionopause. Analogous to the magnetopause, the ionopause is defined as an equilibrium between the kinetic plasma pressure and the magnetic field pressure, only this time it is the planetary plasma and the combined pressure of the solar wind and the interplanetary magnetic field. The latter is frozen out of the ionospheric plasma. Although Venus has a magnetic field, it is too weak to support a magnetosphere in the classical sense: the subsolar magnetopause would lie inside the planet.

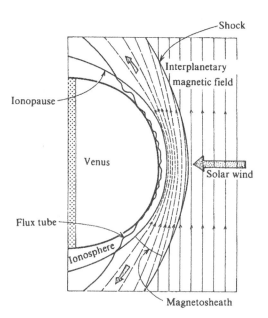

Fig. 9.9. The interaction between the solar wind and the dayside of Venus leads to an ionopause and a bow shock. Reprinted from T. Encrenaz and J.J. Bibring [146], *The solar system*, Copyright 1990, Springer Verlag

The solar wind deflected around the planet forms a cavity which is nearly void of any plasma. Since the magnetic field lines forming the magnetotail are of opposite polarity, a current system similar to the one observed in the geomagnetic tail evolves. Depending on the solar wind conditions, strong magnetic fields can be observed at a height of a few hundred kilometers above the planet's surface. These fields, however, are not of planetary origin but stem from the compressed and deflected interplanetary field.

9.4 Comparison of Planetary Magnetospheres

Although the physical mechanisms shaping a magnetosphere are the same for all planets, the properties of these magnetospheres can be different, depending on the magnetic moment of the planet, its rotation period, plasma sources and sinks, and the local solar wind properties. In this section we shall give a comparison of the magnetospheres discussed above.

9.4.1 Structures of Planetary Magnetospheres

In front of all magnetospheres a bow shock develops where the supersonic solar wind is slowed down to subsonic speeds. Behind the bow shock, structures are different. The two neighbors of Earth, Mars and Venus, both have only weak magnetic fields (see Table 9.3). Thus no well-structured magnetosphere develops, instead an ionopause/magnetopause separates the interplanetary and planetary plasmas. This does not result from an interaction between the solar wind and the planetary magnetic field but between the solar wind and

Table 9.3. Parameters of planetary magnetic fields and magnetospheres. The column 'angle between axes' gives the angle between the planet's magnetic field axis and its rotation axis. In the column 'plasma sources' the 'W' indicates solar wind, 'A' indicates planetary atmospheres and ionospheres, and 'S' satellites

	Equatorial field (gauss)	Dipole moment (gauss cm^3)	Angle between axes	Typical stand-off distance (r_p)	Calculated stand-off distance	Plasma sources
Mercury	0.002	3×10^{22}	$< 10°$	1.45	1.74	W
Venus	<0.0003	$< 10^{21}$	?	1.1	?	W,A
Earth	0.305	7.9×10^{25}	$11.5°$	10.7	10.7	W,A
Mars	0.0004?	1.4×10^{22}	?	?	?	?
Jupiter	4.2	1.5×10^{30}	$9.5°$	47–97	45	W,A,S
Saturn	0.2	4.3×10^{28}	$< 1°$	17–24	20	W,A,S
Uranus	0.23	3.8×10^{27}	$58.6°$	18–25	26	A
Neptune	0.06–1.2	2×10^{27}	$46.8°$	23–26.5	25	W,A,S

the planetary ionosphere. The interplanetary magnetic field then is frozen-out of the planet's ionospheric plasma.

The most common type of magnetosphere is Earth-like. It also can be found at Mercury, Saturn, Uranus, and, at least for part of its rotation, around Neptune. The complexity of these magnetospheres depends on the angles of the planetary dipole and the axis of rotation with respect to each other and to the solar wind flow. The sizes of the Earth-like magnetospheres are different, depending on the planet's magnetic moment.

The most complex magnetosphere can be found on Jupiter: although the axis of rotation and the magnetic field axis are parallel to each other and nearly perpendicular to the plane of the ecliptic, the magnetosphere is far more complex than the Earth-like ones because centrifugal forces from Jupiter's fast rotation stretch the magnetosphere outward in the equatorial plane.

The parameters of the planetary magnetospheres, such as the magnetic moment, the angle between the magnetic field axis and the axis of rotation, the typical distance of the magnetopause in planetary radii r_p as observed and as calculated from the pressure balance, as well as plasma sources are listed in Table 9.3 with 'W' for the solar wind, 'A' for planetary atmospheres and ionospheres, and 'S' for the planetary satellites. Note the relation between the rotation period and the dipole moment: with decreasing rotation period, the dipole moment increases because the MHD dynamo works more efficiently. Mars is an exception to this rule: most likely, the planet has cooled down too much to allow for a liquid core. Thus the dynamo process does not work. Nonetheless, the visit of Mars Global Surveyor to Mars revealed a magnetic field close to the planet about an order of magnitude larger than expected, and in size comparable to Mercury's field. But this field does not result from a planetary dynamo, but rather appears to originate in ferromagnetic rocks in the planet's crust [1].

9.4.2 Sizes

Jupiter not only has the most complex magnetosphere but also has the largest one. This is true not only if the subsolar distance of the magnetopause is expressed in units of the planet's radius but, since Jupiter is the largest planet, also in absolute numbers. The Jovian magnetosphere is also the largest structure in the solar system, even dwarfing the Sun in size; see Fig. 9.10.

One parameter that defines the size of a magnetosphere is the stand-off distance expressed in units of the planetary radius r_p. This "typical magneto-pause distance" is given twice in Table 9.3: first as the observed value and second as calculated from the balance between the magnetic pressure of the planetary field and the kinetic pressure of the solar wind. Both values agree quite well, and the variability of the observed magnetopause distances reflects the high variability in solar wind pressure.

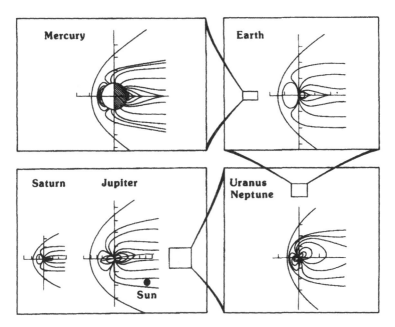

Fig. 9.10. Relative sizes of the planetary magnetospheres. Reprinted from C.T. Russell and R.J. Walker [452], *Introduction to Space Physics*, (eds. M.G. Kivelson and C.T. Russell), Copyright 1995, with kind permission of Cambridge University Press

The typical magnetopause distances order with the magnetic moment of the planet: the inner planets all have very small magnetospheres, extending less than one planetary radius above the planet's surface. The only exception is Earth with its relatively high magnetic moment. The outer planets have even higher magnetic moments, corresponding to larger stand-off distances. Although at the outer planets the stand-off distances expressed in planetary radii are not much different from the Earth's, these magnetospheres are much larger because the planets themselves are larger; see Fig. 9.10.

9.4.3 Plasma Sources

A magnetosphere often is called a magnetic cavity: the solar wind plasma cannot penetrate into this cavity. Nonetheless, as we have already seen in Chap. 8, the cavity is not void of any plasmas or particle populations. Plasma sources in the Earth's magnetosphere are the solar wind and the terrestrial atmospheres and ionospheres, see the last column in Table 9.3. The same sources are available in Venus's magnetosphere. Mercury, on the other hand, which is too small and too close to the Sun to support its own atmosphere, picks up all the magnetospheric plasma from the solar wind.

Going farther out to the gaseous giants, we find another plasma source: for these planets at least some of the moons and the ring systems are inside the magnetosphere. These plasma sources easily can be identified by their unusual composition: in Chap. 8 we mentioned that the relatively high contribution of O^+ points to a partly atmospheric origin of the terrestrial magnetospheric plasma. Moons and rings can contribute to the magnetospheric plasma populations by even more complex mechanisms, occasionally leading also to more exotic species. (a) On hitting the solid surface of a moon, energetic ions can knock out neutral atoms. These neutrals become ionized by charge-exchange with the surrounding plasma. This process, called sputtering, appears to be relevant in the inner magnetosphere of Saturn. (b) Interactions between the dust of the planetary ring system and the surrounding magnetospheric plasma might create an additional plasma component. This process also is discussed as a likely explanation of the spokes in Saturn's rings. (c) Vulcanism on a planetary moon directly injects particles into the magnetosphere. A prominent example is Io, a small moon in the Jovian system which is kneaded so thoroughly by Jupiter's strong gravitational pull that its core is molten, allowing for volcanism. Io mainly injects S^+, S^{2+}, and O^+ into the Jovian magnetosphere. In time, a plasma torus has formed along Io's orbit, consisting of these ions. (d) The atmospheres of moons also directly inject particles into planetary magnetospheres. In the solar system, two moons with atmosphere are known: Titan, with an atmosphere basically consisting of nitrogen and methane with methane being available in all three phases, comparable with the water on Earth, orbits around Saturn, and Triton, with a thin atmosphere of nitrogen and methane, revolves around Neptune. Along their orbits, these moons have also created plasma tori; however, the intensities are much smaller than the ones observed along Io's orbit. (e) And finally, moons also provide a loss mechanism for the magnetospheric plasma: on hitting the moon's surface, a particle might be absorbed.

9.4.4 Upstream of the Bow Shock: The Foreshocks

Since all planets in the solar system have a bow shock, the shock enthusiast might wonder whether all these shocks accelerate particles. If they do, we would expect to observe a foreshock with plasma turbulence and energetic particles similar to the one in front of the terrestrial bow shock (see Sect. 7.6.4). And indeed, a foreshock can be found in front of all planets. But like the magnetospheres themselves, these foreshocks are variable. Their properties depend on the angle between the interplanetary magnetic field and the bow shock. Since the curvature of the Archimedian field line increases with increasing radial distance, the angle between the field line and the bow shock changes, too. With increasing radial distances other parameters also change, such as the strength of the interplanetary magnetic field and the temperature and kinetic pressure of the solar wind plasma.

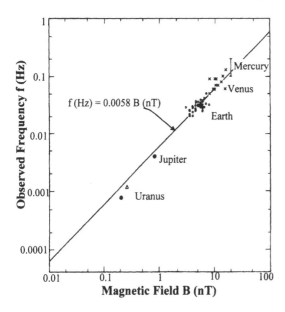

Fig. 9.11. Frequencies of waves upstream of planetary magnetosphere in dependence of the interplanetary magnetic field strength. Reprinted from C.T. Russell et al. [454], *J. Geophys. Res.* **95**, Copyright 1990, American Geophysical Union

Despite these vastly different local conditions, all planets have foreshocks. Although they differ from the terrestrial foreshock, they nonetheless show the same general features, such as a resonance between waves and particles and, depending on θ_{Bn}, regions with different particle distributions, reflecting the underlying acceleration mechanisms. Except for the Jovian electrons, the accelerated particles have rather low energies in the kiloelectronvolt to tens of kiloelectronvolt range. Depending on the planet's distance from the Sun and the magnetic field strength at this distance, the frequencies of the waves resonating with the particles change: the lowest frequencies are observed at the outermost planets while the foreshock of the innermost planet shows the highest frequencies (see Fig. 9.11).

9.4.5 Radiation Belts

Except for Jupiter, all the outer planets have radiation belts that closely resemble the terrestrial one. Radial diffusion transports particles perpendicular to the magnetic field and pitch-angle scattering scatters particles into the loss cone from where they can penetrate deep into the atmosphere. The Jovian magnetosphere is different: Io lies deep inside the magnetosphere and is a significant source of matter and energy. The resulting radiation belt is humangeous compared with the typical radiation belts of Earth or the other outer planets. It is not only the large spatial extent, but also the high energies and fluxes of the particles which make the Jovian radiation belts a hazard even for unmanned spacecraft.

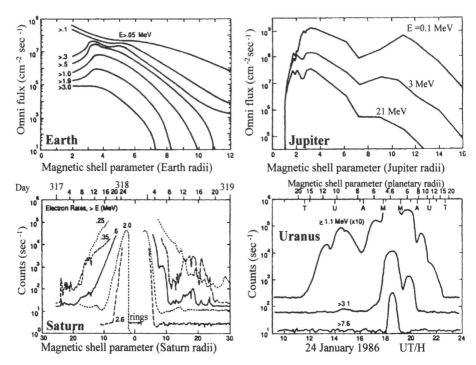

Fig. 9.12. Fluxes of energetic electrons in the magnetospheres of Earth, Jupiter, Saturn, and Uranus. Reprinted from C.T. Russell and R.J. Walker [452], in *Introduction to Space Physics* (eds. M.G. Kivelson and C.T. Russell), Copyright 1995, with kind permission from Cambridge University Press

Figures 9.12 and 9.13 show cuts through the radiation belts of electrons and protons for Earth, Jupiter, Saturn, and Uranus. Neptune's radiation is not shown; it is very similar to the one of Uranus, although the particle fluxes are significantly lower. The radiation belts of the different planets are quite similar: the highest fluxes always are observed just above the planetary atmosphere. Saturn is an exception to this rule in so far as the maximum of the radiation belts is just outside the ring system. In all magnetospheres particles with high energies are limited to the inner radiation belts while lower energy particles can also be observed farther out. Thus the energy spectrum is hardest close to the planetary atmosphere; with increasing radial distances the spectrum steepens. Particle fluxes and energies, on the other hand, are quite different. For instance, the flux of >3 MeV electrons in the Jovian magnetosphere exceeds the one in the Earth's atmosphere by 3 orders of magnitude, which in turn is still one order of magnitude higher than in the magnetosphere of Uranus. In addition, the electron energies in the Jovian magnetosphere are much higher than in the other magnetospheres. The differences in energies and fluxes of protons are even more pronounced.

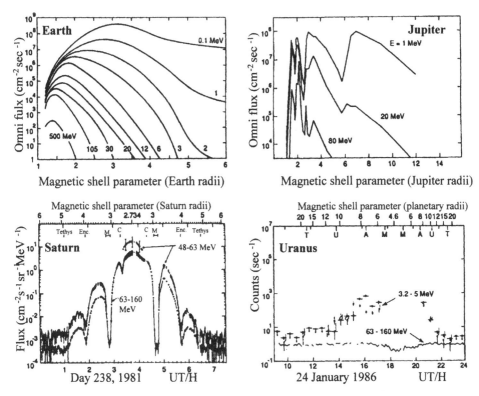

Fig. 9.13. Fluxes of energetic protons in the magnetospheres of Earth, Jupiter, Saturn, and Uranus. Reprinted from C.T. Russell and R.J. Walker [452], in *Introduction to space physics* (eds. M.G. Kivelson and C.T. Russell), Copyright 1995, with kind permission from Cambridge University Press

Magnetospheric size cannot be the parameter responsible for the different fluxes and energies because in the rather small terrestrial magnetosphere higher fluxes and more energetic particles can be observed than on Saturn, Uranus, and Neptune. The differences more likely result from the interaction between the magnetosphere and the solar wind as the ultimate source of energy. In particular, differences in the reconnection rates at the magnetopause might feed more energy into the Earth and Jupiter systems, which in turn leads to more efficient acceleration. Loss mechanisms for the accelerated particles, in particular pitch-angle scattering into the loss cone, might modify the particle distributions; however, they are not likely to explain the large differences in energy and fluxes.

9.5 Summary

Almost all planets in the solar system, except Mars and Venus, have a magnetosphere. The structures, sizes, and dynamics of these magnetospheres can be quite different. The most important parameters determining magnetospheric properties can be summarized as follows:

- The planet's magnetic moment combined with the local solar wind pressure determine the typical size of the magnetopause. If this is smaller than the planet's radius, no classical magnetosphere is observed.
- The structure of the magnetosphere is determined by the inclination of the dipole axis with respect to the axis of rotation and the plane of ecliptic, leading to Earth-like magnetospheres, pole-on magnetospheres, or oscillations between these two states, as observed in Neptune's magnetosphere.
- Jupiter's magnetosphere is the largest object in the solar system. Owing to strong centrifugal forces, the Jovian magnetosphere is rather flat with a plasma sheet in the equatorial plane.
- Sources of magnetospheric plasma include the solar wind (except for Uranus), the atmospheres and ionospheres (except for Mercury), and at the outer planets also the moons and rings.
- Radiation belts can be observed inside the Earth's magnetosphere and the magnetospheres of the outer planets. At the orbits of the moons particle fluxes are significantly lower because the moons absorb particles. Some moons create plasma tori along their orbits, such as Io in the Jupiter system and Titan orbiting Saturn.

Exercises and Problems

9.1. Determine the energies of electrons and protons in resonance with the waves in the foreshock of the different planets. Use the numbers given in Fig. 9.11.

9.2. Confirm the calculated stand-off distances given in Table 9.3 (assume that the solar wind density decreases as r^{-2}).

9.3. Discuss the relationship between the spin period and the magnetic moment of the planet in terms of the magnetohydrodynamic dynamo. Would this relationship be in agreement with the assumption of a similar process working in all planets?

9.4. Discuss the possibility of aurorae and their shapes and detectability on other planets. Use your knowledge about the aurorae on Earth.

9.5. Describe and discuss the plasma sources in the different magnetospheres.

10 Solar–Terrestrial Relationships

> It is a good morning exercise for a research scientist to discard a pet hypothesis every day before breakfast. It keeps him young.
>
> K. Lorenz, *The So-Called Evil*[1]

Solar–terrestrial relationships deal with the influence of the Sun and solar activity on our terrestrial environment. The driving force is the input of energy and matter into geospace. From the viewpoint of space plasma physics, the most important (and also the scientifically soundest) consequences have been discussed in Chap. 8. But there are also side-effects to these phenomena, such as the influence of solar energetic and auroral particles on the chemistry of the atmosphere, or the uproar caused in our technical environment due to severe geomagnetic disturbances. Other connections exist, too, relating solar cycle variations to weather and climate. These latter bear the seed for controversial discussions; nonetheless, some of the ideas will be reported here.

Recent reviews on solar–terrestrial relationships are given in [41, 67, 172, 498]. The Sun's role in climate change is discussed in [238] and in the very controversial popular account [74]. Internet resources include `www.estec.esa.nl/wmwww/wma/spweather/`, `www.windows.ucar.edu/spaceweather/basic_facts.html`, and `www.sel.noaa.gov/index.html`; additional links can be found at `space.rice.edu/ISTP/`.

10.1 Solar–Terrestrial Relationships: Overview

The Sun is a source of electromagnetic radiation, plasmas, fields, and energetic particles. These emissions have a constant component and a variable, solar-cycle-dependent one (see Fig. 10.1). Ordered by decreasing energy, they shape our terrestrial environment as follows:

(a) The continuous solar electromagnetic radiation, described by the solar constant, determines the structure of the atmosphere and the climate. Since the solar constant varies by less than 0.1% during the solar cycle, its variation

---

[1] Reprinted from K. Lorenz *On Aggression* (transl. M. Latzken), Routledge, London, Copyright 1983, with kind permission from Deutscher Taschenbuch Verlag.

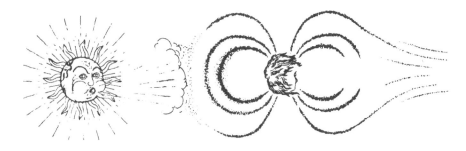

Continuous Emission:

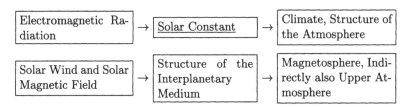

Solar-Cycle-Dependent Emission:

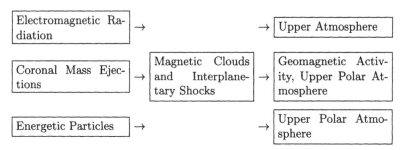

Fig. 10.1. Simplified overview of solar–terrestrial relationships. The top panel is from a poster for a 1978 workshop on solar–terrestrial influences on weather and climate at the Ohio State University

is not recognized in climate. Although the solar constant is roughly constant, the hard electromagnetic emission is enhanced by up to an order of magnitude during solar maximum [310], causing solar cycle variations in the atmospheric height, the thermospheric temperature, and the ionospheric layers.

(b) The solar wind and the frozen-in magnetic field shape interplanetary space and, on interaction with the terrestrial magnetic field, form the Earth's magnetosphere. The energy density in the solar wind is six orders of magnitude smaller than the one in solar electromagnetic radiation. Nonetheless, disturbances in the solar wind cause much stronger modifications in the terrestrial environment than does the flare's increased electromagnetic radiation. In addition, the structure of the heliosphere leads to modulation of the galactic cosmic radiation. These particles influence the chemistry and ionization of the atmosphere.

(c) Solar flares emit electromagnetic radiation from the gamma to the radio range. As a consequence, the temperature, circulation, and chemistry of the upper atmosphere, and even the upper stratosphere, can be modified, leading to an increased drag on satellites in low orbits or to sudden ionospheric disturbances. Since the travel time of light from the Sun to the Earth is only 8 min, these effects start immediately after the flare.

(d) Coronal mass ejections and the shocks driven by them cause geomagnetic disturbances. Their influence is delayed by 1 to 3 days with respect to the flare. Both transfer energy to the magnetosphere which, during auroral activity, partly is transferred to the upper polar atmosphere. Energetic particles accelerated at the shock can penetrate into the atmosphere at the polar cusps, causing additional ionization. This in turn influences the temperature, chemistry, and circulation of the thermosphere.

(e) Solar energetic particles arrive at Earth a few hours after the flare. Like the particles accelerated at the shock, they can interact with the upper atmosphere in the cusp regions.

In solar–terrestrial relationships, three "levels of knowledge" can be distinguished: (i) accepted relationships with unambiguous observations and plausible explanations (for instance, the variations in the upper atmosphere and the disruptions to our technical environment), (ii) relationships with sound observations but with insufficient explanation (such as the long-term relationship between solar activity and climate, as evidenced, for instance, in the Little Ice Age), and (iii) relationships with less sound observational evidence and lack of sufficient explanations (such as the influence of the Sun on the drought cycle in the western US). In the latter cases, a relationship often appears unreasonable since the energy fed into the terrestrial system is small compared with the energy contained in the phenomenon. However, it should be noted that energetics is a tricky argument because the climate system is highly non-linear. Thus small energy inputs can act as triggers or might be amplified to cause strong consequences.

In the remainder of this chapter, I shall give some examples of possible solar–terrestrial relationships. Since the energy argument is very strong, I shall start with a brief overview of the role of the solar electromagnetic radiation in our climate system, although strictly speaking this is beyond the scope of this book. But since all other effects have to be compared against variations in electromagnetic radiation, we cannot neglect such variations.

10.2 Solar Activity, Climate, and Culture

A controversial topic is the relationship between solar activity, weather, and climate. Recently, the discussion has become more aggressive since some authors claim that the temperature increase observed over the last hundred years can be explained solely by variations in the solar output while other claim that it is solely anthropogenic. Since the atmosphere is a complex system, strongly coupled to the oceans, it is not likely that there is only one cause; rather climate variability most likely is influenced by many factors which include internal oscillations of the climate system, extraterrestrial causes, as well as anthropogenic influences. To delve deeply into this debate would be the topic of another book; nonetheless, one observations relating climate and solar activity will be presented here as a starter, in particular, because it is often cited as reference.

Although evidence is only indirect, a widely accepted relationship between solar activity and terrestrial climate exists for the last 7000 years. This correlation is closely tied to the name of Eddy [144, 145]; however, attempts to correlate the Sun and climate can be traced back for at least 200 years, a review is given in [41].

Figure 10.2 correlates solar activity and climate parameters. In the upper panel, the deviation of the ^{14}C content from its long-term trend is shown on a reversed scale, that is an increase in ^{14}C, which indicates a decrease in solar activity, is pointing downwards. Since ^{14}C is a cosmogenic nuclide, it provides a measure of the intensity of the galactic cosmic radiation, which in turn is modulated by solar activity: high solar activity corresponds to a reduction in GCRs and therefore also in ^{14}C, while during the solar minimum both are enhanced. Thus the ^{14}C record can be used as a long-term record of solar activity, which is shown in the middle panel. The panel does not show individual cycles because it is difficult to extract them from the ^{14}C record since the isotope is produced in the stratosphere but stored in organic matter.[2] The combined transport and assimilation processes have time scales of some years, smearing out the individual solar cycles. The scale on ^{14}C is reversed because it should serve as a proxy for solar activity: a downward excursion from the average is meant as indicator for low solar activity. In the bottom panel, climate parameters are shown with T being the temperature, and W the winter index as an indirect measure of the strength of the winter, inferred from the time of freezing and thawing of certain lakes or rivers. The curve G gives the advance and retreat of glaciers in Europe.

The figure suggests that times of high solar activity are connected with a retreat in glaciers while lower solar activity corresponds to their advance. On shorter time scales of some hundreds to about a thousand years, the crude climate measure 'glacier' is supplemented by historical records and in the last 400 years also by direct measurements. Times of high solar activity not only are connected with higher temperatures but also with cultural growth. Examples are indicated by the vertical lines: (12) marks the Sumerian maximum, the first advanced civilization, between about 2700 BC and 2600 BC, (11) the Pyramid Maximum (2300 BC to 2050 BC), (10) the Stonehenge Max-

---

[2] For solar cycle analysis the cosmogenic nuclide ^{10}Be is more suitable because the time scales of its deposition are shorter. For a recent review see [37].

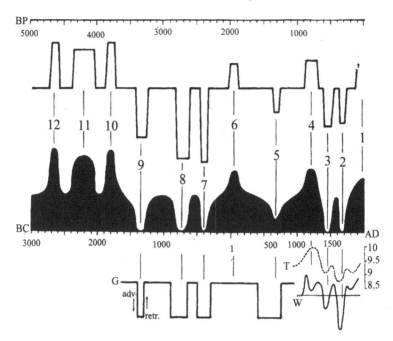

Fig. 10.2. Changes in solar activity as inferred from cosmogenic nuclides and related changes in climate. Reprinted from J.A. Eddy [144], in *Physics of solar planetary environments* (ed. D.J. William), Copyright 1976, American Geophysical Union

imum (1850 BC to 1750 BC), (6) the Roman Maximum, and (4) the Medieval Maximum, when the climate was warm enough to allow the Vikings to cross the Atlantic Ocean and to discover Greenland as a green land. Times of low solar activity, such as the Spörer Minimum (2) and the Maunder Minimum (3), both marking the Little Ice Age, or (5) the Medieval Minimum often are related to cultural depression and retreat or to migration of nations in the hope of finding a more suitable environment. Evidence for higher solar activity during the Roman Maximum is also given by the frequently observed and reported aurora which today is rare in the Mediterranean.

Although the observations are accepted, the mechanism(s) behind the correlation between weak solar activity and a colder climate is still not completely understood. When the first measurements of the solar constant over an entire solar cycle became available in the late 1980s, they showed only a small variation of 0.08% in the solar constant with a maximum during solar maximum. Thus despite the larger number of dark spots, the Sun emits more radiation at maximum conditions. This additional radiation comes from a bright chromospheric network related to the magnetic field structure. Nonetheless, for the climate system a variation in the solar constant by less than 0.1% is way too small to produce any recognizable signal, even if this reduction lasts for some tens of years.

Thus if we assume that the minima in solar activity indicated in Fig. 10.2 are nothing more than a prolonged ordinary solar minimum, changes in solar irradiance would not explain the variations in climate. But long-lasting minima might be different from the ordinary sunspot minima known today. A careful analysis of observations of the Sun during the Maunder Minimum in the late seventeenth century suggested a larger solar diameter and a slower but stronger differential rotation than observed during the more recent solar cycles. In addition, although no sunspots were observed, the cosmogenic nuclide ^{10}Be suggested the continuation of a solar cycle with a reduced period of about 9 years.[3] Thus the solar dynamo was still at work although at a faster pace and with weaker photospheric fields, as is evident from the lack of sunspots. From these changes in solar properties, reductions of the solar constant between 0.14% and 0.5% have been suggested. Presently it is still debated which of these numbers is more reliable and whether it alone would be sufficient to explain the Little Ice Age. However, there are also suggestions that, despite the lack of visible solar activity in the form of sunspots, the aurora as a consequence of solar activity still follows the 11-year solar cycle [467].

10.3 Solar Electromagnetic Radiation

The Sun's electromagnetic radiation is the ultimate energy input into the terrestrial system, shaping the climate system, the hydrological cycle, and also the biosphere. Thus variations in this energy input are expected to have consequences for the system of the Earth. Before we look at variations, first let us have a closer look at the role of the Sun in the climate system.

10.3.1 The Climate System

Climate is weather averaged over a long time period, or, as Pollack [415] puts it, "Climate is what you expect, weather is what you get." The climate system basically resides in the troposphere; the upper atmospheric layers are believed to have only a small influence on climate.

On time scales relevant for humans and societies, the Earth is believed to have a constant average surface temperature. Therefore, incoming and outgoing energy must be in an equilibrium. The incoming energy is described by the solar constant $S_\odot = 1380 \ \mathrm{W/m^2}$. This is intercepted over the area of a circle with a diameter equal to the radius of the Earth, which provides a power

---

[3] ^{10}Be records are stored in the Arctic and Antarctic ice sheets. In contrast to ^{14}C, the Be isotope is immediately washed out of the atmosphere by precipitation, allowing the detection of variations on time scales of about a few years.

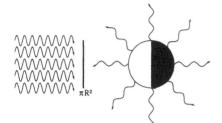

Fig. 10.3. The effective temperature of the Earth is determined by the equilibrium of incoming short-wave solar radiation and outgoing long-wave terrestrial radiation

equal to $S_\odot \pi r_E^2$. Part of the radiation is reflected directly back into space. Its amount is described by the albedo A, which, averaged over the entire Earth, is about 30%. Thus the incoming power is given by $P_{in} = (1 - A) S_\odot \pi r_E^2$. The outgoing radiation can be characterized by an effective temperature T_{eff} and a resulting radiation flux $q = \sigma T_{eff}^4$. This radiation is emitted over the entire Earth; see Fig. 10.3. Thus the outgoing power is $P_{out} = 4\pi r_E^2 \sigma T_{eff}^4$. Since this equals the incoming radiation, we obtain the following for the effective temperature of the Earth

$$T_{eff} = \sqrt[4]{\frac{(1 - A) S_\odot}{\sigma}} \approx 253 \text{ K} . \tag{10.1}$$

This is well below the observed annual mean temperature of 288 K. The difference results from the greenhouse effect: solar electromagnetic radiation is short-wave radiation which passes through the atmosphere nearly unhindered. The terrestrial radiation, on the other hand, is long-wave radiation well in the thermal infrared. This is absorbed mainly by water vapor and carbon dioxide in the atmosphere. Therefore, the atmosphere is heated and itself radiates. Part of this atmospheric radiation is again absorbed by the Earth. Thus the Earth receives the incoming solar short-wave radiation and recycled terrestrial long-wave radiation. Consequently, it has a higher temperature. Most of the above temperature difference of 35 K is due to absorption by water vapor; the remaining difference is due to absorption by carbon dioxide and other, partly human-made, greenhouse gases. These gases are receiving a lot of attention in the discussion of human influences on climate, in particular the anthropogenic greenhouse effect. The interested reader is referred to [229] or to the detailed reports by the Intergovernmental Panel on Climate Change (IPCC) [62, 230–233, 550].

10.3.2 Short-Term Variability of the Solar Constant

Obviously, the Earth's effective temperature and thus also the Earth's climate will change if any of the parameters in (10.1) changes, in particular, if the solar constant changes.

Direct evidence for variation of the solar constant can be inferred from satellite instruments such as the Earth Radiation Budget Experiment (ERBE) and the Active Cavity Radiometer Irradiance Monitor (ACRIM). Measurements are available since the late 1970s and reveal variations in the solar constant during the solar cycle by about 0.08%; for a recent review see [311]. The relative variation in the solar constant during the solar cycle is therefore only small. Using (10.1), we can relate changes in the solar constant to changes in temperature by

$$\frac{\delta T_{\text{eff}}}{T_{\text{eff}}} = \frac{1}{4} \frac{\delta S_{\odot}}{S_{\odot}} .$$
(10.2)

A 0.08% change in the solar constant then corresponds to a change in the effective temperature by 0.07 K over the solar cycle. This is much smaller than the year-to-year variability in temperature, and thus even if this influence were present it would probably be much too small to be detected; see the discussion in [238].

The picture changes if we look not at the solar constant, which is the integral over the solar emission spectrum, but at the spectrum itself. At the fringes of the black-body spectrum, in particular in the EUV and hard X-rays, the variability over the solar cycle is much larger, up to a factor of 10 to 1000 (see Fig. 10.4). These frequency ranges, however, are absorbed in the mesosphere and thermosphere and are responsible for the large solar-cycle-related atmospheric variations observed above a height of about 80 km, in particular, variations in the thermospheric temperature, electron content in the ionosphere, and ionospheric heights. Even in the UV, which is responsible for both the production and the destruction of ozone, the solar-cycle variability is about 15%. Nonetheless, again the argument is that the energy content in this part of the Sun's electromagnetic spectrum is way too small to produce any recognizable signal in the climate system. Before we start to speculate about effects related to these frequency ranges, let us first have a look at variations of the solar constant on longer time scales.

10.3.3 Long-Term Variability of the Solar Constant

Direct observations of the solar constant are limited to the last two solar cycles. The variability on longer time scales can be inferred by a number of different methods, for instance extrapolation from stellar activity; correlations between the Sun's energy output and the solar radius and/or rotation rate, both of which have been observed since the Maunder Minimum; and extrapolation from the CaII line, which is a measure of solar magnetic activity. Results from these methods are summarized in [41, 118, 496]. The various approaches give a reduction of the solar constant during the Maunder Minimum of the order of 0.5% or a few watts.

Although these variations are larger than the present solar-cycle variation of the solar constant, it appears doubtful whether they are sufficient to explain the Little Ice Age during the Maunder Minimum if the simple estimate of

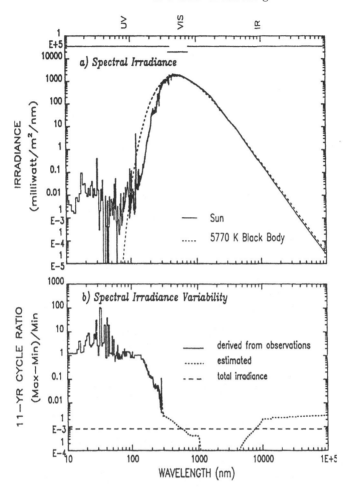

Fig. 10.4. Solar spectrum (*top*) and spectral variability of the solar constant (*bottom*) derived from satellite measurements during solar cycle 21. Reprinted from J. Lean, *Variations in the Sun's radiative output* [310], Copyright 1992, American Geophysical Union

(10.1) is used. However, if inferred reduced solar constants are used in a coupled ocean–atmosphere global climate model, they give a small global cooling and a stronger cooling in the north Atlantic region, leading to a marked variation of the North Atlantic Oscillation (NOA). The resulting patterns, in particular those of storm tracks, are close to the weather and climate reports from the Maunder Minimum [155, 442].

Nonetheless, this result is still a tentative one, because the chain of assumptions is long: the solar constant during the Maunder Minimum has to be guessed; the climate of the Little Ice Age has to be inferred from incomplete data (in particular, data limited mainly to Europe); the climate model, although extremely complex, is still incomplete and so on. But the result is important insofar as it suggests that variations in the Sun's electromagnetic output have consequences for the terrestrial climate and that other influences on climate, such as anthropogenic influences or space plasmas, have to be compared with this natural climate variability.

10.3.4 At the Fringes of the Electromagnetic Spectrum

The variability with the solar cycle is much larger at the fringes of the spectrum, although the total energy flux is small. For instance, the variability around 100 nm can amount to about 15%; the resulting change in irradiance, however, is only about 0.1 W/m^2. Nonetheless, the upper atmosphere reacts strongly to these variations: during the solar cycle, thermospheric temperatures vary by about 50–60 K at heights around 120 km and by some hundreds of K above 400 km, and the entire atmosphere expands during solar maximum. Even a lower part of the atmosphere, the stratosphere, reacts to the solar cycle: the total ozone column varies by a few percent, with a maximum around solar maximum [213,480] because the increased amount of hard electromagnetic radiation not only destroys but also creates ozone.

Solar Electromagnetic Radiation Summarized: The variation in total solar irradiance (solar constant) is too small to lead to a detectable solar cycle signal in weather and/or climate parameters. Nonetheless, the Maunder Minimum and Little Ice Age suggest that such a link might exist. The variations in solar irradiance are much stronger at the fringes of the spectrum and lead to detectable signals in the upper atmosphere, even down to the stratosphere. Since these harder wavelengths are absorbed well above the troposphere, they should not influence weather and climate. Nonetheless, with a suitable coupling mechanism between the upper atmosphere and the troposphere, the solar-cycle-induced variations at greater heights might have consequences for the troposphere, for instance because of wave coupling [481,482].

10.4 Energetic Particles and the Atmosphere

Solar cycle variations in the upper atmosphere as well as its responses to solar activity are supported by many observations and can be understood quite well. The solar-cycle-dependent input into the upper atmosphere comprises hard electromagnetic radiation (its variation with the solar cycle as well as enhancement during flares), auroral particles, solar energetic particles, and galactic cosmic radiation. The consequences of these inputs are heating and ionization. The energy input follows different spatial patterns: changes in the solar electromagnetic radiation directly affect the entire dayside atmosphere or, since the exchange between dayside and nightside is fast and the atmosphere rotates with the planet, often the entire atmosphere. The galactic cosmic radiation basically influences the entire atmosphere, too. The energy input of auroral and solar energetic particles, on the other hand, is limited to high latitudes. Consequences of variations in the electromagnetic radiation are, for instance, an increase in thermospheric temperature and density with increasing solar activity, changes in thermospheric composition, or an increase in ionospheric electron density with increasing solar activity.

10.4.1 Precipitating Energetic Particles: Primary Consequences

The primary consequence of the precipitation of energetic charged particles into the atmosphere is ionization. Some secondary effects resulting from this ionization are electron density increases in the ionosphere, modifications of the atmospheric chemistry, in particular the creation of NO_x and the destruction of ozone, and variations in the global electric circuit. All effects are well documented insofar as they can be observed directly in individual events and do not require correlative studies. In addition, the physical mechanisms are understood reasonably well.

The consequences of variations in the particle input into the atmosphere include ionization and heating, too. The particle energies range from a few kiloelectronvolts (auroral particles) to megaelectronvolts (solar energetic particles) and even into the gigaelectronvolt range (galactic cosmic rays). Let us consider the direct energy transfer only, i.e. we do not follow the path of the particles but assume the particles to be incident isotropically from the upward-looking hemisphere into the upper atmosphere. The primary energy transfer mechanism is ionization, as described by the Bethe–Bloch equation:

$$\frac{\mathrm{d}E}{\mathrm{d}x} = -\frac{e^4}{4\pi\varepsilon_0 m_e}\frac{Z^2}{v^2}\,n_e\left[\ln\frac{2m_e v^2}{\langle E_\mathrm{B}\rangle} - \ln(1-\beta^2) - \beta^2\right]. \tag{10.3}$$

The energy loss $\mathrm{d}E$ per distance travelled $\mathrm{d}x$ thus depends on some fundamental constants (the first fraction), the parameters of the incoming particle (the second fraction, where Z is the charge of the particle and v its velocity, and $\beta = v/c$), and the parameters of the material (the remaining part of the equation, where n_e is the electron density and $\langle E_\mathrm{B}\rangle$ the average ionization energy). The energy loss of an ionizing particle in a given material (that is, $n_e = \mathrm{const}$) therefore increases with the particle's charge and decreases as v^{-2}. The terms in the square brackets are corrections which become important at relativistic energies: the first term depends on the ratio of the particle energy to the average ionization energy, and the other terms are the relativistic corrections depending on the ratio β between the particle's speed and the speed of light. For $\beta \to 1$, the energy loss is constant and minimal: the particle is minimally ionizing, with the energy loss determined by the particle's charge (and the parameters of the absorbing material). For non-relativistic particles, the energy loss is

$$\frac{\mathrm{d}E}{\mathrm{d}x} \sim \frac{Z}{v^2}. \tag{10.4}$$

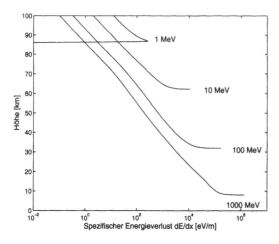

Fig. 10.5. Energy loss of a proton in the atmosphere for various energies between 1 MeV and 1000 MeV

Equation (10.3) gives the specific energy loss of the particle and therefore can be used to track the particle through the atmosphere. The equation has to be solved numerically. Figure 10.5 shows the energy loss for protons with various energies between 1 MeV and 1 GeV in the atmosphere. At a given height, dE/dx decreases with increasing particle energy because of the v^{-2} dependence of the energy loss. For the same reason, the energy loss increases towards the end of the particle range: most of the particle's energy is deposited close to its stopping height in the atmosphere. Since the energy is lost because of ionization of the atmosphere, the particle's energy loss is directly proportional to the number of ion pairs produced along its track, with the proportionality coefficient being the average energy required for one ionization.

During an SEP event, a large number of particles precipitates into the atmosphere. Each individual particle causes ionization as described by (10.3) and Fig. 10.5. For the atmospheric consequences, we are interested in either the total ionization (that is, the ionization of all particles is summed over the entire event) or the ion pair production rate, that is, the number of ion pairs produced per unit time interval. This latter quantity is required for chemistry models. It can be obtained from energetic-particle measurements, such as those shown for the Bastille Day event in Fig. 7.26, by folding the observed particle intensities with their respective energy losses [426]. Figure 10.6 shows the ion pair production rates for three days of the Bastille Day event. Curve 1 includes the onset of the event, where the highest particle energies are dominant. Consequently, ion pair production occurs at rather low altitudes. With increasing time, the intensities of high-energy protons decrease while the intensity at lower energies increases, shifting the ion pair production to greater heights and to a larger maximum. In particular, the bulk of ion

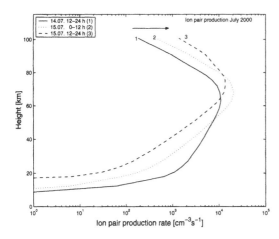

Fig. 10.6. Ion pair production rates (daily averages) during the Bastille Day event (see Fig. 7.26)

pair production is at heights between 30 km and 90 km, where normally the atmosphere is neutral: thus energetic charged particles provide a source of ionization at heights where normally no ions exist. Therefore they can have a rather strong influence on atmospheric chemistry.

Excursion 8. For a simple estimate, we can circumvent the numerical integration of the Bethe–Bloch equation with the following ansatz. As above, the ionization rate is the product of the flux F_e of precipitating protons and dE/dz, their energy loss, giving an ionization rate

$$q_e(z) = F_e \frac{dE}{dz} . \tag{10.5}$$

The energy loss can be approximated as

$$\frac{dE_0}{dz} = \kappa_e E_{ion} \sigma_n n . \tag{10.6}$$

This quantity depends on the collisional ionization rate κ_e, the energy E_{ion} of the precipitating ions, the interaction cross section σ_n and the number density n. Inserting (10.6) into (10.5), taking into consideration the barometric height formula (8.19), yields

$$q_e(z) = \kappa_e F_e E_{ion} \sigma_n n_0 \exp(-z/H) . \tag{10.7}$$

From (10.6) we also obtain

$$E_0 = \int_0^E dE = \int_\infty^{x_s} \kappa_e W_{ion} \sigma_n n_0 \exp(-z/H) . \tag{10.8}$$

Solving for z_s gives the stopping height:

$$z_s = H \ln \left(\kappa_e \sigma_n n_0 H \frac{E_{ion}}{E_0} \right) . \tag{10.9}$$

□

10.4.2 Precipitating Particles and Ozone

The consequence of ionization by precipitating particles is a modification of the atmospheric chemistry. The bulk of the ions produced by solar energetic protons or magnetospheric electrons in the middle atmosphere, i.e. at altitudes from a few tens of kilometers up to about 90 km, are N_2^+ and O_2^+, simply because N_2 and O_2 are the major species; smaller amounts of Ar^+, CO_2^+, O_3^+, O^+, and N^+ are produced too. The electrons released in these interactions either are lost by recombination with a positive ion or form negative ions with a previously neutral atmospheric constituent.

A Little Ozone Chemistry. The most important chemical reactions in the middle atmosphere involve the ozone layer. Some of the ions created by the incident protons are ozone-destroying radicals. The reduction of the ozone layer locally can be as strong as a factor of 2 to 4, as has been observed in the strong polar cap absorption (PCA) event following a flare in November 1969 [555]. Both O_2^+ and N_2^+ contribute to ozone destruction.

The destruction of ozone by O_2^+ is based on a chain of events. First, the ionized oxygen molecule has to attach itself to a water molecule. The latter then reacts with another water molecule:

$$O_2^+ \cdot H_2O + H_2O \rightarrow H_3O^+ \cdot OH + O_2 ,$$

forming a hydroxyl radical OH and an H^+. The HO_x species are predominately formed in the upper stratosphere and the mesosphere. They are a rather short-lived component which is destroyed by photochemical processes within less than a day.

The following pairs of reactions can result:

$$H + O_3 \quad \rightarrow OH + O_2 ,$$
$$OH + O \quad \rightarrow H + O_2 ,$$

$$OH + O_3 \rightarrow HO_2 + O_2 ,$$
$$HO_2 + O \rightarrow OH + O_2 ,$$

$$OH + O_3 \rightarrow HO_2 + O_2 ,$$
$$HO_2 + O_3 \rightarrow OH + 2O_2 .$$

In the upper two chains, the net effect is the recombination of an ozone molecule and an oxygen atom, yielding two oxygen molecules:

$$O + O_3 \rightarrow 2O_2 .$$

In the lower chain, the net effect is a recombination of two ozone molecules, yielding three oxygen molecules:

$$O_3 + O_3 \rightarrow 3O_2 .$$

In these reactions, H and OH act as catalysts. They are lost from the reaction chain only when two of them meet, forming either H_2 or H_2O. These reactions

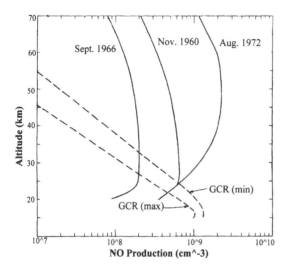

Fig. 10.7. Calculated NO-production in the atmosphere for three solar energetic-particle events and for the galactic cosmic radiation under solar minimum and maximum conditions. Reprinted from G.C. Reid [439], in *Physics of the Sun, vol. III* (eds. P.A. Sturrock, T.E. Holzer, D.M. Mihalas, and R.K. Ulrich), Copyright 1986, with kind permission from Kluwer Academic Publishers

have time constants of a few hours, and thus the reaction chain cannot go on indefinitely and the ozone will recover after several days. In addition, these reactions reduce the water content in the middle atmosphere, since during recombination more H_2 is formed than H_2O. Thus further ozone reduction, for instance during the next PCA, will be reduced. This is a kind of self-healing of the atmosphere.

Since they are long-lived, stable ions function as catalysts only; the main hazard for the ozone layer is the nitrogen oxides, in particular NO. These NO_x species are formed in both the stratosphere and the mesosphere. For instance, NO is produced by a two-step reaction from N_2^+. First, N_2^+ undergoes dissociative recombination, giving two neutral nitrogen atoms. If these interact with oxygen, then NO is formed:

$$N + O_2 \rightarrow NO + O .$$

Ozone destruction is then caused by the reactions

$$NO + O_3 \rightarrow NO_2 + O_2 ,$$

$$NO_2 + O \rightarrow NO + O_2 .$$

One or two large PCA events, lasting for 2 or 3 days, produce more nitrogen oxides than the galactic cosmic radiation produces during an entire year; see Fig. 10.7. Thus at times of high solar activity, the NO production in the mesosphere is determined basically by solar activity. The nitrogen oxides, however, are produced at higher altitudes than those produced because of ionization by galactic cosmic rays since the solar energetic particles have lower energies. But the removal of the ozone-destroying NO is slower at higher altitudes: the loss mechanism for NO above about 40 km is photodissociation, leading to the formation of N_2. NO produced by the galactic cosmic radiation close to the tropopause, on the other hand, can be washed out rather easily, forming nitric acids.

Observational Evidence: The Bastille Day Event. Evidence for a depletion of the ozone layer was first reported for the large SEP of August 1972 [117, 219]. A prominent recent example is the ozone depletion in the Bastille Day event of 14 July 2000. The proton intensity profiles have already been shown in Fig. 7.26, and the ion pair production rates are given in Fig. 10.6. Ozone observations by the HALOE instrument on board the UARS satellite show a marked reduction in ozone concentration inside the polar caps during the particle event; see Fig. 10.8. Ozone concentrations above 0.5 hPa (about 50 km) are reduced by 40% for about 2 days, mainly owing to the short-lived HO_x. Ozone depletion at lower heights is smaller but lasts longer because the long-lived NO_x destroys ozone.

Numerical models of atmospheric chemistry and circulation combined with ionization models have been developed well enough to model the basic observed features of particle-induced ozone depletion, namely the spatial and temporal patterns. The ionization can extend to geomagnetic latitudes below the polar cap: large SEP events are accompanied by travelling interplanetary shocks and CMEs. As these hit the Earth, geomagnetic disturbances lead to a rearrangement of the field and allow particles to have access to normally closed dipole field lines. This reduction in the geomagnetic cutoff latitude has been observed by SAMPEX [321].

Although this effect is well documented, we should be aware that ozone depletion due to energetic particles is limited to higher latitudes, namely inside the polar cap, and to heights above about 35 km. Thus SEP events affect the atmosphere but only in a limited spatial region.

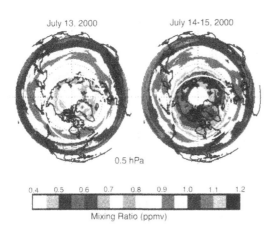

July 13, 2000   July 14-15, 2000

0.5 hPa

0.4   0.5   0.6   0.7   0.8   0.9   1.0   1.1   1.2

Mixing Ratio (ppmv)

Fig. 10.8. Ozone concentration above 0.5 hPa in the northern hemisphere prior to (*left*) and during (*right*) the Bastille Day event. The *white ring* marks the polar cap. Reprinted from C. Jackman et al. [252], *Geophys. Res. Lett.* **28**, Copyright 2001, American Geophysical Union

Precipitating Particles During the Solar Cycle. SEP events typically last for a few days and are more frequent during solar maximum than during solar minimum. Since the NO_x produced by the precipitating particles is long-lived [251, 536, 537], the NO_x production of subsequent events is superposed, leading to stronger atmospheric effects.

Figure 10.9 shows the changes in the total ozone column due to precipitating particles calculated from the observed proton fluxes at GOES using the SLIMCAT/TOMCAT chemistry model [99] for the time period 1988 to 2002. The solar cycle is obvious with maxima in ozone depletion around the solar maxima in 1990 and 2000. Although the particle precipitation is limited to latitudes well above 60°, ozone depletion is also visible at lower latitudes, in particular during times of high solar activity. This spatial spread of ozone depletion reflects the atmospheric transport: NO_x is produced in the polar caps at high latitudes but can be transported to lower latitudes. At mid-latitudes, the solar cycle variation in ozone due to precipitating energetic particles is comparable to that due to the variation in hard electromagnetic radiation, although the two effects have opposite signs. The differences between the northern and the southern hemisphere result from the concurrent photochemical processes. The first strong signal of ozone depletion in Fig. 10.9 is due to the October 1989 event; it is much more pronounced in the northern hemisphere because it occurs just at the beginning of the polar night and thus no photochemical reactions can destroy NO_x. In the southern hemisphere, on the other hand, ozone depletion is much smaller because here NO_x is destroyed by photochemical processes.

Since ozone also has radiative properties, precipitating energetic particles influence not only the atmospheric chemistry but also the radiative transport. As a consequence, temperatures in the stratosphere are modified [489]. Since circulation patterns are modified when the temperature gradients are

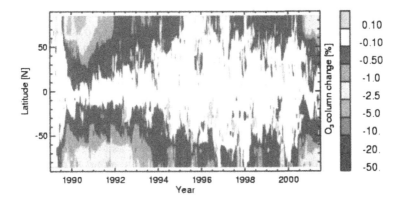

Fig. 10.9. Variation of total ozone column calculated with the SLIMCAT/TOMCAT model using proton intensities measured by GOES. Figure courtesy of M. Sinnhuber, University of Bremen; Copyright M. Sinnhuber

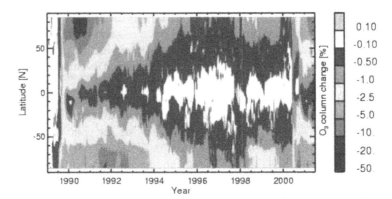

Fig. 10.10. As Fig. 10.9, but for a vanishing geomagnetic field. Figure courtesy of M. Sinnhuber, University of Bremen; Copyright M. Sinnhuber

changed, precipitating particles also have the potential to influence climate. An amplification of this process – or a damping – might also result from planetary waves (gravity waves) [481, 482].

This effect might become even more pronounced on longer time scales for two reasons. First, the present-day Sun appears to be unusually quiet with respect to particle event production, as can be inferred from thin nitrate layers in ice cores [347, 348]– the very same nitrate that we have discussed above in connection with ozone chemistry. In particular, Carrington's white light flare stands out in the record, suggesting a total particle fluence about eight times that in the largest event (October 1989) in spacecraft records. In addition, in the 1890s, a series of four events with particle fluences exceeding that of the October 1989 event is observed within a few years. Simulations suggest patterns of ozone destruction similar to those discussed so far; however, the magnitude of depletion is much larger. In particular, in the 1890s series of events, NO_x from the individual events is summed.

Secondly, although it is decreasing, the geomagnetic field is still strong and efficiently reduces the fluxes of particles precipitating into the atmosphere. If the field were to decrease to about 25% of its present-day value, as typically assumed for a field in reversal, particles would precipitate over a much larger area. Model simulations give a rather amazing result (see Fig. 10.10): although the particles now precipitate over the entire globe, ozone depletion at low latitudes is still rather weak, while it is strongly enhanced at high latitudes despite the fact that the particle fluxes have not changed there. This counter-intuitive result points to the problems in making simple estimates in a complex system and can be understood from the superposition of NO_x generation, transport, and depletion patterns.

Precipitating Magnetospheric Electrons. SEP events are not the only particle events that influence atmospheric chemistry and ozone. During strong geomagnetic storms, particles from the radiation belts are injected into the loss cone and precipitate into the atmosphere too. The electron component of the magnetospheric particles is energetic enough to precipitate down to altitudes of about 60 km. As in the case of the SEPs, these electrons ionize the atmosphere, which leads to NO_x production and ozone depletion [75–81]. It should be noted that the spatial precipitation pattern is different: while the solar energetic particles precipitate inside the polar cap, the magnetospheric particles precipitate inside the auroral oval.

10.4.3 Precipitating Particles and Thermospheric Circulation

We will now have a look at rather low energetic particles that stop in the thermosphere or upper mesosphere. The main consequence of these particles is an increase in ionospheric electron density and heating.

The latitudinal temperature gradient in the thermosphere drives a meridional circulation system. This gradient is caused by the latitudinal variation of both the solar electromagnetic radiation and energetic particles. As in the troposphere, more radiative energy is absorbed close to the equator than at high latitudes. In the thermosphere the energetic particles incident at high latitudes provide an additional heat source. Thus the thermospheric circulation system shows a seasonal dependence (latitude dependence of the electromagnetic radiation) as well as a dependence on the solar cycle.

Figure 10.11 shows model calculations for the thermospheric circulation for the equinoxes (left) and solstices (right) for different levels of solar activity: (a) an extremely quiet Sun, (b) average solar activity, and (c) a highly active Sun. The influence of the latitudinal dependence of the incident electromagnetic radiation can be seen best for very quiet solar conditions: the thermospheric circulation then consists of a thermally driven Hadley cell. During the equinoxes, heating is strongest close to the equator, leading to an updwelling of hot air which, at higher altitudes, is transported poleward where it cools, sinks, and moves as cold air equatorwards through the stratosphere (that would be at the bottom of the figure), closing the circulation cell. During solstices, only one Hadley cell evolves, spanning the entire globe. Heating is strongest at high latitudes in the summer hemisphere, thus air dwells up there and is transported across the equator towards the high latitudes of the winter hemisphere where it sinks.

With increasing solar activity, heat is supplied to the high-latitude upper thermosphere by energetic particles. Thus close to the poles, secondary circulation cells evolve providing heat transport towards lower latitudes. During equinoxes, two such cells are observed, one over each pole. With increasing solar activity, these cells expand, nearly suppressing the thermally driven circulation. During the solstices only one such cell develops in the winter hemisphere. Wind speeds in the thermospheric circulation can be up to 2500 km/h.

Figure 10.11 is a grossly simplified meridional cross-section. Since the Earth rotates, the real thermospheric circulation consists of a thermally

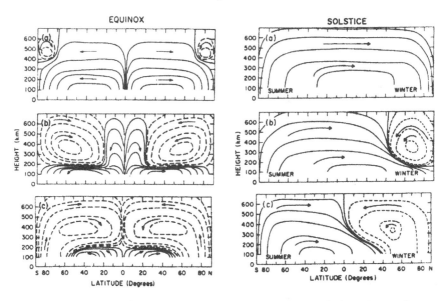

Fig. 10.11. Thermospheric circulation at the equinoxes and the solstices for extremely quiet solar conditions (*a*), average solar activity (*b*), and a very active Sun (*c*). Reprinted from R.G. Roble [444], *The upper atmosphere and magnetosphere*, Copyright 1977, with kind permission from National Academy Press

driven meridional component and a zonal deflection due to the Coriolis force. Whereas in the troposphere the Coriolis force breaks up the Headley cell into three cells in each hemisphere, the thermospheric circulation pattern basically consists of the one cell shown in Fig. 10.11. Only around the geomagnetic poles can closed zonal circulation patterns exist, depending on solar activity. In the southern hemisphere, closed vortices around the pole appear to be a dominant feature which is observed in the oceans, in the stratosphere (where it plays a prominent role in the seasonal appearance of the ozone hole), as well as in the thermosphere.

10.4.4 Precipitating Particles and the Global Electric Circuit

Precipitating energetic particles ionize the atmosphere. If the ionization is strong enough, even the electric properties of the atmosphere might be modified. However, we normally think of the atmosphere in terms of a neutral medium and atmospheric electricity as a phenomenon limited to thunderstorms. But the galactic cosmic rays can penetrate well down into the troposphere; in fact, they are the main source of ionization in the troposphere. Even directly above the ground, radioactive decay contributes only about half of the ionization. It has been suggested that energetic electrons can account for fair-weather lightning because with a sustained electron avalanche, they can provide the necessary charge transfer even at electric-field magnitudes much smaller than needed for the initial breakdown [206, 207]. In a well-developed thunderstorm, cosmic rays may introduce erratic variations in flash rate and type [341].

Thunderstorms and ionization by energetic particles have to be viewed in the broader context of the global electric circuit between the Earth's surface and the ionosphere. Between the ionosphere and the ground, there is a voltage drop of about 250 kV. Discharge through the atmosphere would destroy this potential drop within minutes. The most important generator in this circuit is the totality of global thunderstorms; smaller generators are tidal winds within the ionospheric dynamo region and the interaction of the solar wind with the magnetosphere.

Some kind of continuation of thunderstorms from the troposphere up to the ionosphere is observed in the form of sprites and elves, which are brief flashes of light above large thunderstorms [562]. The sprite has been identified as an electrical discharge; they tend to have a vertical extension and are some kind of electrical (not optical) mirror image of cloud-to-ground lightning. Elves are horizontally extended with a doughnut shape: the vertical lightning return stroke creates an electric field azimuthally symmetric around the axis of the lightning channel.

The solar modifications to the global electric circuit are due to ionizing radiation, that is, energetic particles and hard electromagnetic radiation. Both vary with the solar cycle: the ionizing radiation from galactic cosmic rays, which affects mostly the troposphere and lower stratosphere, is continuous and enhanced during solar maximum, while the ionization from SEPs and magnetospheric particles is sporadic. Ionization by SEPs occurs predominately in the stratosphere and mesosphere and is more frequent during solar maximum, while ionization by magnetsopheric particles occurs in the mesosphere and above. Therefore, the global electric circuit is a system that depends on the solar cycle and couples different layers of the atmosphere, from the troposphere to the upper atmosphere [518, 540].

The coupling between atmospheric electricity and climate is most likely due to cloud formation, the redistribution of charges within clouds, and the loss of charges by rain. Such a mechanism is attractive from the viewpoint of energy. As mentioned in the introduction to this chapter, the energy density in the space plasmas is much lower than that contained in the electromagnetic radiation. However, with the aid of the global electric circuit, the influence occurs not via the input of additional energy but via a redistribution of the available energy. Observational support also exists (see below).

The variation of the ionization by space plasmas would influence both parts of the global electric circuit: the generator, thunderstorms, and the load, the atmosphere. This in turn modifies droplet formation and the development of clouds. In particular, an increased electric field leads to larger droplets and heavier precipitation in fair-weather areas. If heavier than usual precipitation has occurred, the condensation of water vapor must have been stronger and, consequently, more latent heat has been released during cloud formation. Thus the higher troposphere has become warmer, which in turn affects mesoscale and synoptic circulation patterns – a possible influence on weather and climate.

10.4.5 Solar Cycle Length, Galactic Cosmic Rays, and Cloud Cover

The ionization–cloud-formation–climate change chain also underlies a hotly debated scenario linking solar activity to climate variations.

For the time period 1870 to 1990, Friis-Christensen and Lassen [171] report a correlation (with a 95% confidence level) between the global temperature trend and the length of the solar cycle: with decreasing cycle length, the surface temperature of the Earth increases. The two curves track each other quite well; in particular, the decrease in temperature between 1940 and 1970 is correlated with an increase in the solar cycle length.

A suggestion for the physical link has originated from an entirely different correlation, between the global cloud cover and the intensity of galactic cosmic rays [334, 509, 510]: with decreasing intensity of the galactic cosmic radiation, the cloudiness decreases too. Thus, it appears that the atmosphere acts as a giant cloud chamber. Subsequently, it has been shown that the galactic cosmic ray intensity also shows a long-term trend which is closely correlated with the solar cycle length [510]. Thus, it appears that the global temperature is influenced by the intensity of the galactic cosmic radiation, which in turn is determined by the state of the heliosphere. The causal link between galactic cosmic rays and temperature might be directly due to the cloud cover, as suggested in [509], or, in a more complex process, it might be related to global atmospheric electricity [517].

However, the picture might not be as simple as sketched here. For instance, some doubt has been raised about the original correlation between solar cycle length and temperature, which is basically concerned with the determination of the solar cycle length [372]. In addition, ground-based observations of cloud cover [301] do not seem to confirm the correlation reported in [334].

A more serious problem, however, exists not in the data analysis but in the causal chain: the correlation with clouds is best with low clouds (below 3 km) at low latitudes. At low latitudes, however, there is almost no modulation of the galactic cosmic radiation with the solar cycle; see Fig. 8.52. This does not disprove a possible causal relation between clouds and cosmic rays; however, if it were to exist, it would not be a simple, straightforward correlation. Instead, it must invoke the complexity of the climate system and its coupling with the global electric circuit – with the latter not being fully understood.

10.5 Sector Boundaries, Droughts, and Thunderstorms: Sun and Weather

Attempts at correlations between weather or climate and the solar cycle or the geomagnetic activity, such as the example discussed above, are many; a comprehensive, though somewhat older overview can be found in [344]. Some of these correlations, such as the droughts in the western US, can be traced back for centuries, while others cover only shorter time spans of 3 to 10 solar cycles because older records are not available.

In contrast to the particle influences discussed in Sect. 10.4.2 the data base is limited to correlations while the direct observation of the relevant details in the causal change and an understanding of this chain are missing.

10.5.1 Droughts in the Western US

A prominent example of the correlation of weather phenomena and the solar cycle are droughts in the western US, in the dust bowl. Droughts severely affect plant growth, and thus a drought index can be derived from tree rings: a wide tree ring indicates plenty of water available while a smaller one suggests a reduction in the water supply. To reduce the influence of temperature on tree ring size, plant species with different dependences on water supply and temperature can be compared. With these data in hand, a drought index for the western US can be traced back to about 1600, the time when the first systematic records of sunspots and solar activity started. There is a correlation between droughts and solar activity at the 99% confidence level; however, this correlation is not with the 11-year sunspot cycle but with the 22-year magnetic cycle. The largest extension of the area affected by drought is about 2 to 3 years after the minimum at which the Hale cycle starts. The severity of the droughts is also modulated with a 90-year cycle, which also modulates the sunspot cycle. This is the Gleißberg cycle with is also evident in the nitrate layers in ice cores,

The dependence on the 22-year cycle is amazing in so far as it suggests a relationship between climate and the magnetic properties of the Sun rather than with its radiation. Although the underlying processes are not completely understood, it appears likely that the energy that triggers the atmospheric change, i.e. the drought, is fed into the terrestrial system by geomagnetic activity and/or particles. During times of low solar activity, recurrent geomagnetic disturbances can be observed (see Fig. 8.37). A comparison of field variations in subsequent solar rotations allows the definition of a geomagnetic recurrence index. The latter one is large only if the geomagnetic disturbance can be observed again during the next rotation. Thus the recurrence index is anti-correlated with the solar cycle since at times of high solar activity transient disturbances modify the structure of interplanetary space and recurrent fast solar wind streams are observed less frequently. The recurrence index is larger and stays high for a longer time period in the minimum between an even and an odd solar cycle, i.e. at the beginning of a new magnetic cycle when the droughts in the western US are also observed.

10.5.2 Sector Boundaries and Weather Forecast

Evidence for a possible involvement of the interplanetary magnetic field also stems from a correlation between the vorticity area index (VAI) and geomagnetic activity. The VAI describes the size of the vortex field on the 500 mb surface and can be used as a measure of the strength of the high- and low-pressure regions. Thus the VAI is directly related to the large-scale atmospheric phenomena that determine our weather. On the basis of a data set of some hundred sector boundary crossings collected over 20 years, a pattern is established with a reduction of the VAI by up to 10% on the day after the sector boundary crossing. The reduction is strongest if the interplanetary magnetic field has a southward component (open magnetosphere); it is weaker if the interplanetary field is directed northwards. The strongest reductions in VAI are observed if the interplanetary magnetic field has a southward component and the sector boundary is accompanied by a stream of fast protons. Thus somehow energy fed from the interplanetary medium through the magnetosphere into the upper atmosphere modifies the flow pattern in the troposphere.

Modifications to the VAI directly influence the weather. This can be shown indirectly: the reliability of the 12 h and 24 h weather forecast in the middle latitudes of the northern hemisphere is reduced from 85% to 65% on the day following the sector boundary crossing. In addition, the thunderstorm area then is largest, too.

Both correlations give some clues to the possible mechanism. Since the tropospheric response is strongest for a southward interplanetary magnetic field accompanied by a stream of energetic protons, particles fed into the upper polar atmosphere appear to trigger the tropospheric changes, probably by modifying the stratospheric heat balance and circulation which then modifies the tropospheric circulation pattern. The thunderstorm area is larger because ionization caused by the precipitating particles modifies the global electric circuit which is directly related to thunderstorm activity.

10.5.3 Solar Cycle Signals in Tropospheric Winds

There are also suggestions for changes in the tropospheric wind patterns with the solar cycle. In the northern hemisphere, winter storm tracks are 2.5° farther south during solar maximum than during solar minimum. This is related to an equatorward motion of the jet stream, a fast zonal wind in the upper troposphere which guides the motion of the pressure systems. For an observer at northern mid-latitudes the weather depends on whether she

Fig. 10.12. Upper tropospheric circulation pattern for very low (*left*), average (*middle*), and high geomagnetic activity (*right*). Reprinted from V. Bucha [66], *Ann. Geophys.* **6**, Copyright 1980, Springer-Verlag

is north or south of the jet stream: south of the jet stream, she is met by the low-pressure regions and accompanying storms, giving a mild and rainy winter, while north of the jet stream the arctic high with clear and chilly skies dominates the weather.

Not only the average latitudes of the jet stream and the storm tracks shift but also their northward and southward excursions. Figure 10.12 shows the circulation pattern over the northern Atlantic ocean for extremely low (left), average (middle), and high (right) geomagnetic activity. The times of high geomagnetic activity are associated with strong westerlies over the Atlantic ocean, thus storms move from west to east, leading to mild and rainy weather in mid-Europe. With decreasing solar activity (middle), a blocking weather pattern develops over the Atlantic Ocean, leading to the penetration of Arctic air into mid-Europe, as indicated by the arrow in the middle panel. If the geomagnetic activity is very low, the blocking pattern becomes even stronger with westerlies only at relatively low latitudes and the intrusion of Arctic air well into mid-latitudes.

The changes in the tropospheric wind pattern are most likely a super-imposition of two effects: the increased ionization and heating of the upper atmosphere by hard electromagnetic radiation during solar maximum as well as the increased precipitation of particles into the atmosphere at high lati-tudes. The latter ones obviously change the thermospheric/stratospheric cir-culation but do not have enough energy to change the tropospheric pattern. Nonetheless, it appears possible that atmospheric waves might provide a suf-ficient coupling between the different atmospheric layers to trigger such a straightening of the tropospheric flow pattern.

There is another interesting aspect related to this interpretation. The change in the flow pattern might require only a relatively small amount of energy. In fact, changes between westerly and blocking patterns are frequent and natural, it is only the persistence and higher frequency of the blocking pattern at times of low solar activity that cause recognizable tropospheric effects. The consequences of this changed pattern, however, might evoke the impression of much larger energy inputs. In mid-Europe, a larger number of blocking weather patterns, which were observed, for instance, also during the Little Ice Age, always mean cold and relatively dry winters. Thus the average

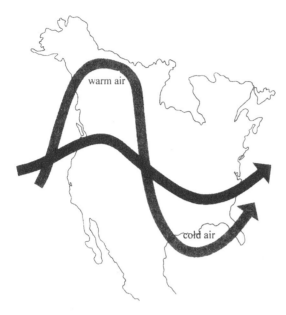

Fig. 10.13. Westerly flow and blocking weather pattern over North America. Reprinted from Roberts and Lansford [443], *The climate mandate*, Copyright 1979, with kind permission from Freeman Co.

temperature in mid-Europe would be reduced. But this does not necessarily imply a global temperature reduction. Instead, it is more a redistribution of heat; thus while in one place the temperatures are lower in another they might be higher. Figure 10.13 illustrates this. It shows the main direction of the upper tropospheric motion for a westerly flow and a blocking weather pattern over North America. The westerly flow brings mild and wet winters all over the US while the blocking pattern leads to a far northward excursion of warm air, leading to unusually high temperatures in Alaska. This air cools over Alaska and moves south towards Florida, leading to unusually cold eastern winters with a lot of snow. As a consequence, during one such blocking weather pattern in winter 1977, on some days in January temperatures were higher in parts of Alaska than in Florida. The western US, on the other hand, experiences a period of drought since the moist air coming from the Pacific Ocean has moved northwards. Thus even the dependence of the western US droughts on the solar cycle can be fitted, at least, qualitatively into this picture.

10.6 The Technical Environment and Solar Activity: Geomagnetic Storms

Solar-cycle-related disturbances in our technical environment mainly stem from magnetic storms caused by the variable solar wind. The changing magnetic flux at the Earth's surface causes induction currents in long leads, in particular in power lines or pipelines. Since these currents can amount to 50–100 A, they can trigger overload protectors or cause damage to transformers, leading to a disruption of the electrical power supply. In recent years, this problem has become more important since power supplies are connected over increasingly larger distances. Since a large contribution to the disturbed geomagnetic field is due to the enhanced auroral electrojet, high latitudes, such as Scandinavia, Canada, or the northern US, are more frequently affected by these events. For instance, the large geomagnetic storm on 13 March 1989, one of the largest ever recorded, caused a disruption of the entire 9000 MW power supply network of the province of Quebec for more than 9 hours. Computer networks also are affected or damaged by induction currents.

Pipelines for natural gas or oil are vulnerable to large geomagnetic storms because the induction currents cause corrosion. This effect has been observed best at the Alaska oil pipeline, which stretches for about 1300 km from north to south through the auroral oval. Where the pipeline is connected to the ground, the induction currents lead to strong corrosion.

The effects of geomagnetic disturbances are not only limited to the Earth's surface. Satellites also can be damaged by induction currents in their electronics or on sensitive surfaces, leading to the failure of the satellite or at least of some instruments. Occasionally, a satellite might lose its spatial orientation in a large geomagnetic storm: some satellites are oriented by reference to the geomagnetic field. If the latter is strongly disturbed, in particular if fluctuations are fast or if the magnetopause has moved so far inwards that the satellite suddenly finds itself in interplanetary space where the orientation of the magnetic field is entirely different, communication with the satellite might be lost because the satellite is not able to keep the proper orientation of its antenna. Although important at the time of the geomagnetic storm, the disorientation does not cause permanent damage since the satellite is able to orientate itself properly as soon as average magnetic conditions are restored.

Higher magnetospheric electron fluxes during magnetic storms also might cause satellite failure if the spacecraft becomes electrically charged. This, by itself, does not cause harm, but becomes harmful if discharge occurs between different components. In addition, satellites are also vulnerable to the direct action of energetic particles: high energetic particles, although only few in number, might cause radiation damage to the electronics. Obviously, increased fluxes of energetic particles also are a risk to manned spaceflight.

The disturbance of communication systems is not related to geomagnetic activity but is a direct consequence of the increased ionization in the ionosphere due to the hard electromagnetic radiation (sudden ionospheric disturbance SID) and to a lesser and spatially more limited extent also to energetic particles (polar cap absorptions PCA). Ground–ground communication with short and long waves is disturbed because these waves are no longer reflected by the ionosphere but are absorbed. Satellite–ground communication also can be affected due to enhanced absorption. Even if the signals pass through the ionosphere they might be distorted or their travel time might be elongated, making navaids such as the GPS less reliable at times of high solar activity.

The increased hard electromagnetic radiation during solar maximum conditions as well as in flares not only enhances ionization but also causes the atmosphere to expand outwards. Or, viewed from a fixed altitude, the density and temperature increase with increasing solar activity. A satellite flying at low altitudes, typically a few hundred kilometers, an altitude used for instance by astrophysical and solar research satellites, then experiences an increased drag, reducing its speed and therefore also its altitude. In time, the satellite will move closer to the denser atmosphere and finally plunge down. Prominent victims of increased solar activity have been Spacelab and the Solar Maximum Mission, SolarMax.

And the Biosphere? While the influence of solar activity on our technical environment is undoubted, the influences of solar activity is discussed controversly. If there are any links, most likely the magnetosphere and geomagnetic activity should provide the connection because solar-activity-related enhancements in hard electromagnetic radiation are absorbed high in the atmosphere. Some influences are obvious or can be tested easily in the laboratory. Migrating animals, such as pigeons and whales, use the geomagnetic field for orientation and navigation. They become disturbed during geomagnetic active periods. For instance, carrier pigeons launched in a magnetic storm need a much longer time to find their home (days compared with some hours). Some micro-organisms seem to have a magnetic field sensor, too, using the field for their orientation. Laboratory experiments show that changes in the magnetic field can affect the vital functions of these organisms.

The influence of geomagnetic disturbances on humans is under debate, too. Although we still might have a magnetic field sensor, as is suggested by some tests, our means of orientation and navigation rely much more on visual information. Thus most people probably would not even be able to deliberately access the information provided by such a field sensor. Nonetheless, since currents are the carriers of information in the human body, the influences of geomagnetic activity should not be ruled out completely, see, for example, [261]. An often cited example is the higher incidence of cardiovascular diseases, in particular cardiac infarction, during geomagnetic storms, which is suggested by different statistical studies at different places over the last 30 years. These correlations often have been doubted since the changes in the surface magnetic field are small compared with the magnetic fields created by our technical environment; however, recent research suggests that the crucial factor probably is not the magnitude but the frequency on the fluctuations: it appears that geomagnetic disturbances have frequencies in the range of slightly less than and up to a few hertz (geomagnetic pulsations) which in principle can lock with the cardiac rhythm. If such locking occurs during the vulnerable phase, cardiac arrest may result.

Caveat. I want to caution the reader that the above discussion represents only a tiny glimpse of a small facet of the complex climate system and its connection with solar activity. The correlations and interpretations given here are examples of investigations and lines of thought, pointing to mechanisms that might contribute to a connection between solar activity and climate. The real phenomenon "climate system", however, cannot be reduced to one parameter and a sole cause. And the tampering of humankind with the system Earth should not be ignored, too.

Part III

The Methods

11 Instrumentation

Measurement began our might.
W.B. Yeats, *Last Poems*

Observations in the solar–terrestrial environment are made from the ground, such as magnetic field measurements, in situ from rockets and satellites, and by means of remote sensing, for instance radio sounding of the ionosphere. Measured parameters cover fields and their fluctuations, plasmas, energetic particles, and electromagnetic radiation. This chapter gives a brief overview of the basic principles of field, plasma, and particle measurements. It does not show examples of instruments because each instrument is designed for a very special purpose and therefore has unique specifications. Collected papers on measurement techniques in space plasmas can be found in [407, 408].

To study complex phenomena, not a single parameter but a set of parameters has to be determined. To study the propagation of energetic particles, for instance, in addition to the particles the magnetic field and its fluctuations must be measured, too. Thus not only does each instrument have to be designed cleverly to fulfill the specifications for space-flight (such as small mass, low power consumption) but also the combination of instruments on the spacecraft or rocket has to be chosen to yield the maximum of information for the topic under study. Some special publications contain summaries of mission goals as well as detailed technical descriptions of the instruments, e.g. SOHO [161], Ulysses [531], Wind [565], Polar [414], and Galileo [180]; internet resources are listed in the Appendix under missions.

11.1 Field Instruments

The measurement of the electromagnetic field can divided into the measurement of the magnetic and the electric fields. Since the spacecraft itself generates electric and magnetic fields, field sensors are always mounted on booms that extend up to some tens of meters from the spacecraft. These booms consist of thin strong wires which, during the launch of the spacecraft, were tightly wound on a reel. They are extended by the centrifugal force caused be the spacecraft's rotation. Thus the sensors can be at very long beams in a plane perpendicular to the spacecraft's axis of rotation, while parallel to the spin axis only rather short, folded metal rods can be used as booms. This latter constraint holds for all axis if the spacecraft is not spinning. A modern electric field instrument, such as the one on board the Freja satellite, uses a set of six probes extended from booms with lengths between 5 and 15 m; the probes on the Viking satellite are mounted on 40 m long wire booms.

11.1.1 The Magnetic Field

The magnetic field is described by the magnetic field vector, i.e. the field direction and the flux density. The measurements can be done either by observing changes in the magnetic flux, for instance in the pulsation or fluxgate magnetometer, or by utilizing effects at the atomic level, such as proton precession or the Zeeman effect. These latter effects are used for absolute ground-based magnetic field measurements only.

Pulsation Magnetometer. A pulsation magnetometer is a tri-axial arrangement of three coils. According to Faraday's induction law (2.6), changes in the magnetic flux through a surface cause an electromotoric force in its circumference. Thus the fluctuating field causes an induction current in the coil which is proportional to the change in the magnetic flux. The pulsation magnetometer is used to measure magnetic field fluctuations in three perpendicular directions with each coil giving one fluctuating magnetic field component. Since it cannot be used to determine the absolute flux density, it is only used in ground-based observatories, preferentially at high geomagnetic latitudes, to give detailed records of magnetic field fluctuations.

Fluxgate Magnetometer. Similar to the pulsation magnetometer, a fluxgate magnetometer is a tri-axial arrangement of three sensors. Fluxgate magnetometers not only measure the magnetic field fluctuations but also the components of a constant field. However, a fluxgate magnetometer is not an absolute instrument but has to be calibrated.

The sensor is a small transformer, wound around a high-permeability core. Its primary winding is excited by a high frequency-current, typically some kilohertz. Current and permeability are chosen such as to drive the core to saturation during each half-cycle of excitation. The secondary winding then detects a time-varying voltage which is related to the input signal by the hysteresis curve of the core material. For a high-permeability core, these curves are very non-linear and the output signal is highly distorted, containing all higher harmonics of the input signal. If there was no external magnetic field along the axis of the transformer, the hysteresis loop would be traversed in a symmetric manner. In this case, only the odd harmonics of the input frequency would show up at the output. In the presence of a magnetic field, however, saturation is acquired earlier in one half-cycle of excitation than in the other. This asymmetry adds even harmonics to the output signal. The amplitudes and phases of the even harmonics are then proportional to the field component parallel to the axis of the transformer.

Since the odd harmonics are much stronger than the even ones, their signals must be eliminated. This can be done with the ring-core or race-track transformer: two cores, each with its own primary coil, are operated in parallel with a common secondary winding. The primary windings are excited by equal currents of opposing direction. Thus if the external magnetic field was zero, the output at the secondary coil would be zero, too, because the odd harmonics of both primaries cancel. If an external field is present, the odd harmonics still cancel, reducing the output signal to the desired quantities, the even harmonics.

Alkali-Vapor Magnetometers. The alkali-vapour magnetometer utilizes the Zeeman effect, i.e. the splitting of atomic lines in the presence of a magnetic field. The distance between these splitted lines is proportional to the magnetic field strength.

The sensor consists of two cells filled with atomic vapor, for instance Rb, aligned with a photocell. As the first cell is heated, it emits light with frequencies corresponding to the transition between the different sublevels of the two lowest atomic levels. This light passes through a circularly polarizing filter into the second cell. Here it is absorbed, raising the electrons to specific sublevels of the next highest level. Falling back to the lower level, these electrons emit light. Since this light is circularly polarized, only certain transitions between different sublevels are allowed, and after a brief time all electrons will fill the highest sublevel of the lowest level. Since transitions from the highest sublevel are forbidden, no further absorption occurs. The photocell thus detects the following chain of events: as the first cell is heated, light is emitted and detected by the photocell. As absorption sets in inside the second cell, this light is reduced until finally all electrons occupy the highest sublevel and the full stream of light hits the photocell again.

The second cell is surrounded by a coil which allows the transmission of a radio signal into the cell, redistributing the electrons to different sublevels. Thus the pumping process can be repeated. For redistribution to occur, the frequency of the radio signal must correspond to the frequency difference between the higher states. It is generated by using the output of the photocell to control the frequency of an oscillator. Thus the entire system oscillates between absorption during pumping and radiation when all electrons occupy the highest sublevel. The frequency of the radio signal required to maintain the oscillation depends on the frequency of the splitted lines and therefore is a measure for the magnetic field strength.

The alkali–vapor magnetometer is an extremely fast and accurate instrument; however, it only gives the total field but not its components.

Proton-Precession Magnetometers. Proton-precession magnetometers, too, are absolute instruments measuring field strength only. They utilize the fact that the magnetic moment of the proton makes it a small bar magnet. Thus protons can be aligned field-parallel, for instance by putting a container with a liquid rich in protons into a coil through which a strong current passes. The protons than align along this field. If the current is switched off, they try to re-align to the external magnetic field, the one to be measured. Since the electrons rotate, they behave like small gyroscopes, precessing around the field line. This precession frequency is inversely proportional to the gyromagnetic ratio g_p and proportional to the magnetic field. It can be determined from the magnetic effects of the precessing protons on the coil.

11.1.2 Electric Field Measurements

The basic idea in electric field measurement is simple: use two (or more) probes extending on booms from the spacecraft and measure the potential between them. But this method alone is not sufficient to determine the electric field because two effects modify this potential difference. First, the motion of the satellite through a magnetic field, which always is present in space plasmas, causes an electric $\boldsymbol{v} \times \boldsymbol{B}$ induction field. In the inner magnetosphere, this might be as large as 0.5 V/m. Since this is a well-defined effect, it can easily be corrected for, using either simultaneously performed magnetic field measurements or the field strength at the satellite's position predicted from a magnetic field model. The second effect is more serious and has to be taken into account in instrument design: the probes (and the satellite) interact with the ambient plasma, and thus the potential measured at the probe does not reflect the electric field alone. Electrons and ions accumulate at the probe, forming a plasma sheet around it. If the probe, or part of it, is lit by sunlight, electrons also can be removed by photo-ionization. The sum of these currents gives the probe's floating potential (see Fig. 11.1). This floating potential can be balanced by applying a negative bias current to the probe.

An entirely different method for the electric field measurement utilizes the drift of electrons in a crossed electric and magnetic field (see Sect. 2.3.2). Such an instrument is far more complex; however, its measurements are not

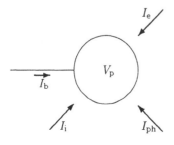

Fig. 11.1. Currents influencing the potential V_p measured by an electric field probe. Currents are due to photo-ionization of the probe (I_{ph}) and flows of ions (I_i) and electrons (I_e) towards the probe. These latter currents create the floating potential, which has to be balanced by a bias current I_b

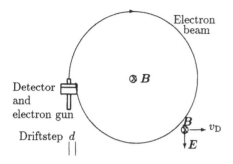

Fig. 11.2. Basic measurement principle to determine the electric field from the drift of electrons

influenced by the interaction between the spacecraft and the plasma. Such an electric field sensor consists of an electron gun producing electrons with a well-defined energy, a system of coils to generate a homogeneous magnetic field perpendicular to the electron beam, and a detector with high spatial resolution. The latter is mounted perpendicular to the direction of electron injection close to the position of the electron gun (see Fig. 11.2). For a vanishing electric field the electron performs one gyration and is intercepted by the detector. In the presence of an electric field, the gyro-orbit is not a closed cycle: the $E \times B$-drift (2.51) modifies the gyro-orbit and the electron is intercepted at some distance from the starting point. From this offset, the drift speed and the direction of drift can be determined. Combined with the prescribed electron speed and magnetic field strength, this gives the electric field.

11.1.3 Wave Measurements

To identify waves in space we have to measure the electric and magnetic fields, as well as variations in the plasma parameters, such as density and temperature. The maximum frequency which can be measured is determined by the temporal resolution of the field and plasma instruments. It should be sufficiently high to allow the identification of all waves of interest in the physical question.

The measurement of waves in space plasmas is a difficult task. In a textbook, a wave is sinusoidal of indefinite length. Space plasmas consist of a mixture of different waves and fluctuations; the waves often can be bursty, localized wave packets, occasionally even solitary waves. In addition, the observer is not fixed with respect to the wave field but moves relative to it, either because of his own motion, which would be the case in the quiet magnetosphere, or because the plasma and fields are swept over him, such as in interplanetary space or in the active magnetosphere. Thus the observed field and plasma variations consist of both temporal and spatial fluctuations.

The detection of waves in space plasmas therefore requires plasma and field measurements with various sensors and high sampling rates. Multipoint measurements are advantageous. For small wavelengths, these can be realized with antennas of various orientation and length. But even if very accurate, all these measurements give the fluctuations only. To identify the waves, a sound knowledge of wave physics and good guesses on the expected waves are required, too. Occasionally, particle measurements can be used to check the interpretation of the field fluctuations in terms of waves. One example was discussed in Sect. 7.4.

Multipoint measurements to study linger waves require a large separation between the individual instruments and therefore different satellites. This can be either done with mini-satellites carrying only filed instruments. These mini-satellites are launched with one carrier and deployed at different positions. While such mini-satellite projects still are under development, multipoint measurements with conventional satellites also exist, for instance the Helios measurements (see Fig. 7.25 for an example concerning particles and shock waves). Tomographic measurements of the magnetosphere, which also include the study of waves, are performed by the four CLUSTER-II satellites (sci.esa.int/science-e/www/area/index.cfm?fareaid=8)

11.2 Plasma Instruments

Plasma instruments are designed to measure the density, temperature, speed, and composition. They can be divided into two groups: in dense plasmas particle energies are low and the instruments measure the collective behavior of the plasma, while in rarefied plasmas individual particles are collected.

11.2.1 Instruments for Dense Plasmas

The plasma densities in the ionosphere are high, and thus plasma properties can be determined from the plasma's collective behavior, such as currents. Rather simple probes and traps can be used to measure densities and temperatures. The measurement principle is based on a sensor extending into the plasma, and drawing electrons or ions from it, depending on the potential of the sensor. A trap utilizes the motion of the spacecraft relative to the plasma, like a snowplow piling up electrons and ions in front of it.

Langmuir Probes. The basic plasma sensor is a Langmuir probe. It can even detect particles with very low energies by the current they carry. Figure 11.3 gives the current–voltage characteristic of a Langmuir probe. The space potential V_s is used as a universal reference potential. Its meaning can be visualized as follows: imagine a large open-wire grid placed inside an ionized medium. Its potential relative to a large but remote body, such as Earth, can be varied. When the flow of electrons and ions through the holes in the grid no longer is modified by its presence, the grid has the space potential.

Let us now substitute the grid by a plate at the same potential. Then electrons and ions are collected at its surface. The electron current is $n_e e v_e$,

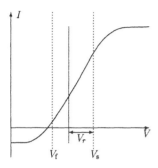

Fig. 11.3. Voltage–current characteristic of a Langmuir probe

the ion current $n_i e v_i$. Assuming quasi-neutrality, we have $n = n_e = n_i$. Thus the net current towards the plate is $ne(v_i - v_e)$. Since the electron speed by far exceeds the proton speed, a negative current towards the plate results, representing the preferred electron current. If we now disconnect the plate from its original potential, it soon will be charged negatively up to the floating potential V_f. At this point the net flow vanishes and $e(n_i v_i - n_e v_e)$ becomes zero. Note that now the densities of electrons and ions are different. Since $v_e > v_i$ still holds, close to the plate the density of the electrons must be smaller than the one of the ions. Thus at the floating potential a negatively charged plate is surrounded by a cloud of positively charged ions, controlling the direction of the current towards the plate, comparable to the space charge in a diode. The thickness of the sheath is the Debye length.

The physical principle of the Langmuir probe is thus based on the different masses and speeds of electrons and ions and the resulting dependence of the net current on the probe's potential. The characteristic in Fig. 11.3 has three well-defined regions which can be utilized for measurement:

(a) If the probe's potential V exceeds the space potential V_s, electrons are drawn towards the probe while ions are repulsed. Thus the current is determined by the electron density n_e only.

(b) If the probe's potential is between the space potential and the floating potential, the current still is determined by the electrons. But only the faster electrons are collected at the probe since the retarding potential $V_r = V_s - V$ prevents the slower ones from reaching it. Thus the retarding potential determines the minimum kinetic energy an electron must have to be captured by the probe. Variation of V_r allows us to determine the change in electron density with energy, yielding the electron distribution. Since the electron distribution is a Maxwellian, the temperature can be determined.

(c) If the probe's potential V is smaller than the floating potential V_f, electrons are repulsed while ions are collected and the ion density can be determined. Since the speed of the ions in general is smaller than the spacecraft's speed v_{sc}, the theoretical treatment is more difficult than for the electrons. The collection condition than becomes $eV_r \leq m_i v_{sc}^2 / 2$, allowing us to determine ion masses and abundances.

Retarding Potential Analyzer. The retarding potential analyzer basically is a Langmuir probe supplemented by some grids. Since with this configuration a more detailed analysis of the ions becomes possible, it is also called an ion trap. At the upper grid, all ions arrive with approximately the same speed, the speed of the satellite. The main purposes of the grids are to deflect the electrons and to screen the detector from potential changes on other grids. But one grid serves a special purpose. Its potential can be varied between 0 and some tens of volts, deflecting also the lighter ions. Thus the ion composition can be determined. The analysis of the current–voltage characteristics gives the ion temperatures as well.

As well as retarding potential analyzers Langmuir probes can be built in different shapes, either planar, as discussed above, or spherical or cylindrical. The operating principle, however, is the same.

Impedance and Resonance Probes. An entirely different approach is chosen in the impedance and resonance probes. These probes do not measure the plasma itself but the electrical properties of the medium, which then allow us to derive the plasma density from theory.

In the impedance probe, the dependence of the dielectric constant ϵ on the electron density and radio frequency is used. Thus the probe has to measure the dielectric constant in space. In the laboratory, the dielectric constant of a material is determined by placing it inside a capacitor and then measure the change in its capacity. The same principle is used in space plasmas. Here the capacitor is designed to be part of an oscillator. Since the frequency depends on the capacity, the electron density can be inferred from it. Impedance probes often are flown on rockets. Since they are fast instruments, the allow for a very detailed analysis of the height profile of atmospheric ionization.

Resonance probes are based on a tuneable transmitter and a receiver. There are characteristic frequencies at which the medium between the transmitter and the receiver starts to resonate. Plasma parameters than can be determined from these resonances.

11.2.2 Instruments for Rarefied Plasmas

We will now discuss plasma instruments that do not utilize collective effects of the plasma but detect individual particles. These instruments therefore also can be used in the rarefied plasmas in interplanetary space.

Mass Spectrometer. A standard instrument for the analysis of a gaseous medium is the mass spectrum meter. Mass spectrometers in space obey the same principles as earth-bound ones, but have two advantages: first, in space a spectrometer does not need a vacuum system since above an altitude of about 100 km collisions have become negligible. Second, there is no need to ionize the sample because all matter is ionized.

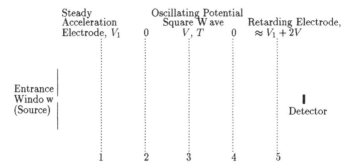

Fig. 11.4. Basic principle of a time-of-flight spectrometer

The basic principle of a mass spectrometer is the deflection of particles in a magnetic field by the Lorentz force. Using an acceleration voltage and a system of shutters, the particle speed can be prescribed. The deflection then depends on the particle's mass-to-charge ratio m/q. In the upper atmosphere, ions normally are singly ionized, and thus their mass can be determined.

Mass spectrometers are only rarely used on spacecraft since the main part of the spectrometer is a large magnet, adding to the weight of the instrument. In addition, the high magnetic fluxes might disturb other instruments.

Time-of-Flight Spectrometer. The time-of-flight spectrometer avoids the problems just outlined. The principle of a time-of-flight spectrometer is shown in Fig. 11.4. To reach the detector, the ions must pass through a series of grids, some of them biased by a steady potential, the central one having an oscillating potential. The potential drop V_1 between grids 1 and 2 would accelerate the ion to a speed $\sqrt{2q_iV_1/m_i}$ which depends on the mass-to-charge ratio m/q. However, depending on their initial speed, the ions still have different speeds. The oscillating potential at grid 3 only accelerates ions with speeds close to $2d/T$, with d being the distance between grid 2 and 3 and T being the oscillation period. Grid 5 now is used as a retarding electrode, preventing all ions that have not acquired the maximum possible speed from reaching the detector. Thus the ion beam passing grid 2 is velocity filtered and only ions with a certain mass-to-charge ratio reach the detector.

Time-of-flight spectrometers often are combined with electrostatic deflection systems. The flight path then more closely resembles the one in a mass spectrometer. One example of an extremely complex, but very advanced, combination of different sensors is CELIAS, the Charge Element and Isotope Analysis System on board SOHO [161].

Modern time-of-flight spectrometers are used to measure ions with energies up to some MeV/nucl. Thus the instruments not only measure the plasma composition but also energetic particles. However, only the lower end of the energetic particle population can be analyzed in a time-of-flight spectrometer since the higher energies would require longer travel paths and therefore larger instruments.

11.2.3 Energetic Particle Instruments

The measurement principle for energetic particles is based on their interaction with matter. Energy losses are due to ionization, bremsstrahlung, Cerenkov radiation, or nuclear interactions [514]. The energy loss inside the detector can be measured, giving us clues to the particle's properties. The most commonly used particle detector is a semiconductor.

If an energetic charged particle traverses matter, electrons will be shifted to higher energy levels (excitation) or will be removed from the atom (ionization). The energy loss of the energetic particle is described by the Bethe–Bloch equation (10.3) and depends on the particle's parameters, i.e. charge and speed, but not on the particle's mass. Thus with only one detector, particles cannot be identified. Instead, a particle telescope is used, consisting of a stack of detectors surrounded by an anticoincidence detector (see Fig. 11.5). The anticoincidence detector serves two purposes: together with the upper two detectors, it defines an entrance aperture, i.e. the solid angle out of which particles can be detected. Particles coming from outside this angle might also hit the telescope and lose energy, for instance in detectors D2 and D3. But these particles can be singled out since they also have produced a signal in the anticoincidence detector. Since the anticoincidence detector only needs to detect the particle without determining its energy loss, often scintillation counters are used.

The information obtained with such a particle telescope is the energy loss dE/dx for each detector and the residual energy dE in the detector where the particle comes to rest. Combined, this gives the particle's energy, which of course could also have been determined with a single sufficiently thick detector. But the use of a stack of detectors also allows the identification of the particle. An initial separation between different particle species can be done with a suitable choice of the first detector D1. The thickness of this

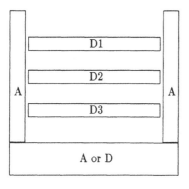

Fig. 11.5. Basic principle of a particle telescope: a stack of detectors D is surrounded by an anticoincidence detector A

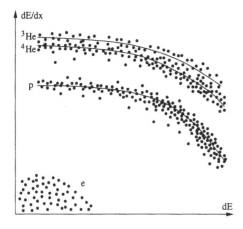

Fig. 11.6. Pulse height matrix for a particle instrument

detector can be chosen such that even a slow electron loses less energy than a fast proton and even a slow proton loses less energy than a fast α particle. Thus, by defining certain thresholds for the energy loss, electrons, protons, and heavier ions can be distinguished. This is a crude but simple method and can be used in onboard data analysis to assign the particles to energy channels. A more careful analysis can be based on a combination of the energy losses in individual detectors in a pulse height matrix. Here, even different heavier ions and isotopes such as ^{3}He and ^{4}He can be distinguished.

A sketch of a pulse height matrix is shown in Fig. 11.6. All particles that stop in detector D3 are considered, and the specific energy loss dE/dx in D2 is plotted versus the residual energy in D3. The dots represent entries from individual particles; the lines give the loci expected for different particle species. Let us start with the locus for protons. If the proton has a rather low energy and just manages to make a signal in D3, its energy loss in D3 is small. In addition, it was already rather slow in D2 and thus has experienced a rather large specific energy loss there. Such a proton therefore sits at the left end of the locus. A faster proton which is stopped just before leaving D3, on the other hand, must have been rather fast in D2. It therefore has experienced a smaller specific energy loss in D2 but has deposited a rather large residual energy in D3. It is therefore located at the right end of the locus. Other ions show the same pattern; however, depending on Z, the energy losses are higher and the curves are therefore shifted upwards. Two examples, ^{3}He and ^{4}He, are indicated. The electrons cluster in the lower left corner: owing to the lower energy required to reach D3, their energy losses and residual energies are small.

The dots representing individual particles show a scatter around the locus. The reasons are twofold. (a) Ionization is a stochastic process, and not a continuous energy loss as suggested by the Bethe–Bloch equation. The Bethe–Bloch equation therefore gives only an average energy loss, around which the real losses scatter. (b) The locus is calculated for a particle moving along the detector axis, that is, an angle of incidence of 0°. Particles in fact hit the detector at different angles. If they cross the detector at a rather large angle, their travel path in the detector material is longer and, consequently, their energy loss is larger. This gives an additional deviation from the idealized locus.

As a consequence, in the example in Fig. 11.6, a clear separation of ^{3}He and ^{4}He is not possible. Of the two effects leading to the scatter, we cannot influence the statistical nature of the energy loss process. But we could correct for different angles of incidence by collecting particles with different angles of incidence in different pulse height matrices. To do this, we must know the angle of incidence, that is, we have to know the places in which the particle hits D1 and D2. These positions can be obtained by using sectorized detectors.

But even the separation of electrons and protons is not always unambiguous. The ionization of an atom by a proton requires the interaction of a heavy particle (the incident proton) and a particle of much smaller mass (an electron in an atomic shell). During this interaction the travel direction of the incident particle is not changed significantly; the proton flies straight ahead. If the incident particle is an electron, two particles of equal mass interact, and the electron is deflected from its original travel path. The resulting zigzag path leads to larger energy losses than does a straight travel path. Since the deflection angles are distributed stochastically, a broad scatter in energy losses results: the cloud in the lower left corner in Fig. 11.6.

If the particles are identified from the pulse height matrices, this multiple scattering does not pose a problem in the discrimination of electrons and ions. However, since often not all pulse height words are stored and processed on board, in many instruments the energy loss in D1 is used to discriminate electrons and protons: electron energy losses are below a certain threshold, proton losses above that value. However, the threshold is determined for a straight travel path: owing to multiple scattering, an electron might lose more energy than a minimally ionizing proton following a straight path and therefore might be identified as a proton. This can be circumvented by inserting a detector element into the telescope that uniquely identifies the electrons. For a similar range in matter (that is, the same stopping detector), electrons must be relativistic, while ions are non-relativistic. Thus any detector that can identify relativistic particles, such as a Cerenkov detector, can be inserted into the stack in our sample detector in Fig. 11.5 and will provide a clear separation between electrons (which produce Cerenkov radiation) and protons (which do not).

Since particle telescopes have a well-defined aperture, they can be used to determine the angular distributions of energetic particles too. On an axis-stabilized spacecraft, this can be done only by combining instruments looking in different directions. On a spinning spacecraft, the situation is simpler: if mounted perpendicular to the spin axis, the telescope scans all directions in a plane perpendicular to the rotation axis during one rotation. Thus sectorized data can be obtained, which in turn allow us to determine the anisotropy.

The illustration of a telescope in Fig. 11.5 is only a very simple sketch. The thickness and number of detectors determine the maximum particle energy to be measured. For higher energies, either more detectors have to be added or absorbers can be placed between the detectors. If these are passive absorbers, there will be a gap in the energy spectrum obtained by the instrument. Active absorbers, on the other hand, do not necessarily give the energy loss exactly, but at least some information about whether the particle has stopped in the last detector before the absorber or whether it has managed to pass into the absorber and has stopped there. Thus a continuous energy spectrum extending over a larger energy range can be obtained.

11.3 Supplementary Ground-Based Observations

Historically, ground-based observations are the oldest means for observing geospace. Although today rockets and satellites are attractive means to study the terrestrial environment in situ, ground-based observations nonetheless still are important, basically for two reasons. First, in situ observations are one-point measurements, giving extremely accurate data for one isolated spot without considering its surroundings. Ground-based observatories, on the other hand, often give a more regional overview or combined even the global picture. For some questions, such as the extent and strength of geomagnetic disturbances, this global information is more useful then the in situ observation at one spot only. Thus ground-based observations often can be used to complement in situ measurements. In addition, ground-based measurements in many cases can provide a continuous baseline which allows the study of long-term variations. Second, ground-based measurements can provide access to regions which cannot be covered by in situ measurements, in particular the lower ionosphere up to a few hundred kilometer altitude where the atmospheric density is still to high to allow for long-living satellites.

Ground-based measurements comprise the magnetic field measurements, which in turn give information about the ionospheric currents, optical observations of the aurora, and measurements of ionospheric parameters by radio sounding and radar. Instruments for magnetic field measurements have already been described at the beginning of this chapter; the principle of radio sounding was discussed in Sect. 4.5. Auroral observations from the ground are performed with different cameras, in particular all-sky cameras showing the evolution of the aurora during a geomagnetic storm.

The most comprehensive information on ionospheric parameters can be obtained by radar. The basic principle in ionospheric radar is the same as used in radar on ships and airports: a radio wave is scattered back by the object of interest. During this scattering process the properties of the wave change, depending on the properties of the scattering object/medium. Radar can be operated in two modes: in a monostatic radar, the transmitter and the receiver are located at the same position; in bistatic mode, the transmitter and the receiver are located at different positions. In the latter case the two antenna beams define the scattering volume while in the former case the scattering volume is defined by the transmitter modulation in the range direction.

In the first instance, the reflected signal is used to derive quantities describing the spectrum, such as width, power, or Doppler shift. From knowledge of how the properties of the medium influence the spectral shape, the parameters of the medium, such as densities, temperatures, or velocities, can be inferred. With the EISCAT radar system (www.eiscat.com/about.html) located in northern Scandinavia, the following plasma parameters can be measured: electron density, temperature, and velocity, ion temperature and velocity, ion mass, the relative number of negative ions, the gyro-frequency of ions, and the ion neutral collision frequency. From these parameters, other parameters can be deduced, for instance the electric field, the velocities of the neutral component, the ion composition, the current density, the energy deposition $E \cdot j$, and the particle's energy. Although this list is impressive, we should be aware of the fact that data are noisy and imperfect and thus assumptions have to be made about some of the parameters, allowing us to infer the others. For instance, temperature and collision frequency cannot be fitted independently. Instead, for heights below about 105 km, assumptions about the temperature are made and the collision frequency can be inferred, while at higher altitudes the modeled collision frequencies are used to infer the temperature.

12 Science in a Complex Environment

> To see a World in a Grain of Sand,
> and a Heaven in a Wild Flower,
> hold Infinity in the palm of your hand
> and Eternity in an hour.
> W. Blake, *Auguries of Innocence*

Space physics is the application of physical concepts to a complex and highly variable system of environments: the Sun, the interplanetary medium, the planets, and the interactions between them. Space physics is therefore different from laboratory physics – it asks different questions, uses other methods, and builds models on a different basis. In other words, space physics has a different view of a physical system from that which laboratory physics has. This last chapter is not meant as a comprehensive overview of scientific methodology or philosophy. Instead, it just gives hints about some of the methods of space physics which might be worth to be evaluated a little bit further. Thus this chapter is meant to help one think about the facts of space physics as presented in the other chapters, as well as to think about thinking.

12.1 Physics in a Complex Environment

At school, and often even at university level, physics is taught as a set of laws that can be applied to more or less complex situations – mostly in laboratory experiments, sometimes also to real-world phenomena. Thus physics often evokes the impression of certainty: ask a question and either there is (exactly) one correct answer to it or there is a measurement method to answer it. But that is not what physics is concerned with: physics aims to understand basic concepts and principles – not details or numbers.

A complex environment is different from laboratory physics: there is neither a simple measurement nor a simple answer. Instead, a complex environment is determined by a multitude of different parameters with non-linear interactions between them. Thus a simple input–output model, with a correspondingly straightforward laboratory experiment, cannot be developed.

Instead, in a complex environment we must retreat further to the original methods of science. For instance, as Popper suggests, the scientific method relies on three ideas:

- a hypothesis can in principle be falsified,
- a hypothesis must be based on objective tests, and
- the results must be repeatable.

The development of scientific ideas and concepts thus is an interplay between model, observation or experiment, and predictions. Although model and experiment together can greatly contribute to the development of science, on their own they are at risk of being caught in an infinite circle: only the application of model predictions to independent observations can provide a test of the model, because the model–experiment combination is at risk of selective perception: "we see only what we want to see". On the other hand, controlled experimental settings ensure the repeatability of results.

The natural environment influences all aspects of the scientific method:

- The measurement problem: in a complex natural system we are observers not experimenters. In a laboratory experiment, we modify one parameter and study the reaction of the experiment. In space, as in any complex natural system, nature continuously changes a lot of parameters simultaneously (and in most cases without telling us explicitly) and the system adjusts accordingly: each time we measure, we measure in a different system.
- The data interpretation problem: we observe a phenomenon and some related parameters while nature performs the experiment. But we do not always know which parameters are relevant and which not and we normally observe only a few parameters, not all.
- The modeling problem: a complex system is often formally described by a set of coupled partial differential equations that can be solved only numerically. In addition, many of the parameters in these equations are not constants but can vary over a broad range, often orders of magnitude.

12.2 The Measurement Problem

Regarding measurement and experimental methods, space physics can be likened to other fields of geophysics, such as earthquake and volcano research: it is observational science, not experimental science. We can "order" neither a magnitude 7 quake at a certain fault nor a gradual γ-ray flare with a large CME at the western limb of the Sun. Instead, we observe the quake or the flare. And although we observe the phenomenon, nature does not easily supply all the additional information required, such as the exact location, magnitude, and kind of the energy release or other supplementary parameters.

And we cannot even repeat the experiment. Although San Francisco will in time experience another strong earthquake, it will never be an exact repetition of the 1906 quake even if the magnitude and the location of the seismic focus are almost identical: the 1906 quake has already changed the configuration of plates that led to its occurrence, and that configuration will not return. The same is true for flares and the magnetic field configuration in the photosphere and corona.

Some of the measurement problems in space physics include:

- Noisy data: even a simple instrument, such as a thermometer in atmospheric research or an electric-field instrument in space research, is influenced by its surroundings.
- Incomplete data:
  - incomplete time series,
  - one-point observations,
  - unknown/unobserved parameters.
- Shifts in instrument response with time.

Note that in other fields of geophysical research, including atmospheric and oceanographic science, the measurement problems are quite similar.

The sources of noise in the data are manifold. There are instrumental sources (heating due to solar radiation, and drifts due to aging of components) and errors in the transmission (the data are transmitted through the ionosphere, and thus, in particular during a flare and the accompanying sudden ionospheric disturbance, transmission errors might occur). However, the data are also noisy in a different sense: parameters that in laboratory physics could be regarded as constants are anything but constant, as will be discussed in the next section.

There are also a number of problems in data acquisition. Some of them already have been addressed in the "What I Did Not Tell You" sections. The most important constraint is the limited number of measurements, with the limitation showing up in three respects. Most quantities in the natural environment are time-dependent fields, while most observations are one-point measurements only. This field character applies not only for the quantities are which classically regarded as fields, such as the electromagnetic field, but also for quantities which are normally treated more like material parameters such the diffusion coefficient, which is a transport parameter. Thus our observation might be as representative as that of an alien space probe that in an attempt to study the Earth's atmosphere was unfortunately dropped into the sulphurous ash cloud emitted by an erupting volcano.

To avoid such bad luck, repeated measurements are helpful because they at least allow one to determine average parameters. In space science, this repeated measurement is basically the time series obtained by an instrument during its lifetime of many years, often even many ten years. In that case, we face the long-term evolution of both the observables of and the performance of the instrument: we cannot exclude drifts and degeneration in the detectors and electronics which might lead to systematic or stochastic errors in the aging instrument. For long-term analysis, a time series is often constructed from a sequence of observations from different spacecraft of one family, such as the

IMP satellites or the GOES satellites. Although the instrument specification is often identical, the actual instrument performance might be different. Thus the time series is not necessarily homogeneous. It should be noted that most terrestrial long-term time series used for correlation, such as those for surface temperature, suffer from the same problem; see for example [41].

In addition, owing to malfunction of instruments or problems in data transmission, time series are often incomplete. Thus if we want to work with them, besides the data we need the metadata which contain the information about the data, the measurement principle, and the measurement practice.

The above incompleteness of data is basically a measurement problem: with more instruments at more positions in space, our measurements could take account of the field character of the observed quantities and would help to distinguish between spatial and temporal variations of a given quantity. But there is another kind of incompleteness in the data: although a satellite is designed to measure many different parameters related to a problem under study, not all relevant parameters are measured, simply because the problem is too complex and our material constants are not exactly constants.

12.3 The Data Interpretation Problem

Data analysis in laboratory experiments often involves a comparison between two, occasionally more, measured quantities. Often a correlation analysis is used: we measure a distance s travelled by a falling ball and the time t required for the fall, plot one against the other, try to find a formal relation between them, namely $s = gt^2/2$, and determine the relevant constant g.

In principle, we do the same with measurements in a natural environment. However, the situation is different in that normally each measured parameter depends on many other parameters. Thus we never get a clean correlation such as that in the above laboratory experiment. The situation is more comparable to the same experimental setting of a falling body, but using a feather or leaf outdoors on a day when a gusty wind is blowing: we shall probably still find some indication that with increasing travelled distance the travel time increases, but neither will there be a simple formal relation (merely because the wind is not constant, the travel time depends on the angle at which the body starts to fall) nor will we be able to determine a constant g^* which can be applied universally in the relation between s and t. At best, we might be able to give some average g^*.

Thus, even if data are not noisy, we shall not be able to find simple, clear correlations. Nonetheless, as in the above example, we shall probably find that there is some relation between the parameters and try to prove this formally by using some kind of correlation analysis; for methods see for example [419]. As a measure of the quality of the correlation, we normally state some level of significance, for instance "more than 95%", which means that the chance that the correlation is accidental is smaller than 5%. This latter number is also called the error probability.

In space physics, such a correlation analysis can be applied to many sets of data. For instance, we can correlate the intensities in ESP events with properties of the accompanying shock or correlate the intensities of SEPs with the γ-emission of the parent flares. In both cases, correlation analysis makes sense, because ESPs are accelerated at the shock and particles accelerated in flares partly interact with the Sun, producing γ-emission, and partly escape, being detected as SEPs. Thus we expect some physical relation between the relevant variables: since other parameters, for instance interplanetary transport, might also influence particle measurements in space, the correlation most likely will look more like that for the falling feather than that obtained in a well-defined laboratory experiment. But on physical grounds, we can nevertheless expect these parameters to be correlated. And if the correlation is at a 95% level (which, with a large data set, can be obtained even if the scatter of the data is strong), we feel comfortable with the correlation because we can understand it in physical terms and it supports our model.

The situation becomes more complicated if we lack a good model, as is the case in many attempts to study solar–terrestrial relationships. There is a tendency to correlate one indicator of solar activity, for instance the sunspot number, with almost any parameter in the terrestrial environment, such as sea surface temperature, vorticity index, drought index, wheat prices, alcohol consumption, the Dow-Jones index, and many more. Let us assume that a scientist has access to many different databases and manages to do 100 correlations. The scientist is happy to find five correlations with confidence level 95% (or error probability less than 5%) and, although he/she has absolutely no idea why these parameters should correlate with the sunspot number, he/she writes a paper on these five correlations and submits it. The first referee is impressed by five correlations, each at a 95% confidence level, and recommends publication. The second referee is more sceptical, asks the author for the number of parameters tested, and, on learning about the 100 attempts to find a correlation, rejects the paper. The second referee obviously has a better understanding of probabilities and significance: if I do one correlation and get an error probability below 5%, this is fine. If I do five correlations, each with an error probability below 5%, this is also fine. But if I do 100 correlations, out of which five have an error probability below 5%, I have probably just found the accidental ones: the error probability of 5% also means that, on average, 5% of my correlations are positive by chance. Therefore reports about correlations without any idea about a reasonable relation between the parameters should be treated with caution.

In solar–terrestrial relationships, two time series are often compared, e.g. Figs. 7.30 and 8.42. If the data stretch is short, for instance 4 years in the declining phase of the solar cycle, the sunspot number will correlate (or anticorrelate) with almost any parameter which shows a steady temporal evolution.

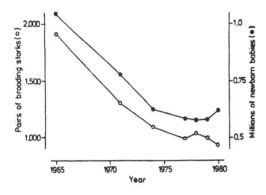

Fig. 12.1. Correlation between pairs of breeding storks (*open symbols*) and millions of newborn babies (*closed symbols*). Reprinted from H. Sies [483], *Nature* **332**, Copyright 1998, with kind permission Nature Publishing Group

Although the chance of an accidental correlation decreases with increasing length of the data stretch, we still have a remaining risk and, if we make enough correlations, will shall find one by chance.

However, again we feel more comfortable with such a correlation between time series if we expect a correlation from our understanding of the system. A good example of such a correlation has been published in *Nature* and is reprinted in Fig. 12.1. The author, Helmut Sies [483], notes: "There is concern in West Germany over the falling birth rate. The accompanying graph might suggest a solution that every child knows makes sense." But the author has not made full use of all the information contained in that correlation. Beck-Bornholdt and Dubben [38] have noticed that the time lag between the curves also shows that storks start their work only at the age of two, and not earlier.

The above examples are extreme. They should caution the reader: science in a complex environment is often at risk of being fooled or of fooling itself by misunderstood applications of basic statistical concepts. Part of this reflects the desire of mankind to describe its environment, often by use of grouping and categorizing. But we should limit correlations, grouping, and categorizing to parameters and systems where they can be applied reasonably. Benestad [41] refers to a comment by D.B. Stephenson (University of Reading): "We have grouped stars that we know are randomly distributed into stellar constellations which we have named." Again this should serve to caution the reader.

Not only does data analysis face challenges not normally encountered in laboratory experiments, but the results often also have a different meaning. Again, take the falling-body experiment. In the laboratory experiment, we do the experiment to determine a constant, the gravitational acceleration g. With the falling feather in windy weather, there is no constant that describes the motion, but some kind of acceleration parameter that consists of the combined effects of gravitational acceleration and drag, and varies depending on the variations in the wind and therefore in the drag. The situation is similar in many respects to space physics. For instance, if we study interplanetary transport, formally the diffusion coefficient in the transport equation can be regarded as some kind of material constant, similar to the index of refraction. But if we remind ourselves of the nature of the transport process, the pitch-angle scattering in wave fields, we realize that the diffusion coefficient depends on the ever-changing properties of the wave field and also on the particle properties – it is more similar to the g^* of the falling feather than to the g of the laboratory experiment.

For the scientist interested in interplanetary propagation, this makes science interesting. For the researcher interested in modulation studies or shock acceleration, however, this makes science difficult because here it would be much simpler if the diffusion coefficient could be treated as a constant material parameter and not as a quantity that varies in time and space. Thus estimates such as that of the characteristic acceleration time in diffusive shock acceleration in example 23 have to be treated with caution: although for average scattering conditions, the estimate suggests that the acceleration of 10 MeV protons is rather slow, this does not exclude the possibility that there are a few shocks in which this process works nicely and extremely fast because scattering is much stronger.

12.4 The Modeling Problem

In physics, a model is formulated in terms of causes and consequences: the electric field exerts a force on an electron and accelerates it. The ideas of cause and consequence also hold in the physics of natural environments, the only difference being that one cause can have many consequences and one consequence is a result of many causes: we cannot reduce the system to just one cause–consequence chain as we would attempt to do in a laboratory experiment. Consequently, models are more complex and, formally, often consist of sets of coupled partial differential equations. Although analytical approximations are valuable for understanding the characteristics of the solutions and for obtaining an idea of the dependence of the solution on various parameters, most models rely on numerical solutions.

Thus a physical model in space science is developed from physical ideas and observations, often in the form of correlations. Its formal treatment and its test require numerical simulations. As mentioned above, these simulations often require assumptions about certain parameters, such as the diffusion coefficient. Therefore, these numerical models are often used for parameter studies to understand the influence of the variation of each parameter on the solution. This approach is similar to the one normally chosen in laboratory physics: keep all parameters constant and well defined and study the reaction of the system to just one parameter. Therefore, in the case of a natural system, laboratory experiments are often replaced by numerical experiments. An example is shown in Fig. 7.12, which studies the energy dependence of the effects of the solar wind in interplanetary transport for a fixed solar wind, a fixed observer's position, fixed particle injection, fixed interplanetary propagation conditions, etc.

As with laboratory experiments, we must be careful that our numerical instruments do not lose touch with the natural system. Therefore, we have to test our simulations. This, basically, can be done in two ways. The interplanetary-transport equation, for instance, has been applied to a large number of particle events; one example is shown in Fig. 7.13. The fact that the model is able to fit the observations does not prove the model, but at least it increases our confidence in it. Thus the confidence gained from multiple successful application of a model to data is comparable to the reproducibility of laboratory instruments.

In addition, the validity of a model can be assessed by its ability to make correct predictions. For instance, it was recognized early that with average parameters, diffusive shock acceleration is too slow for a large energy gain. It was also recognized rather early that particles streaming away from the shock might generate waves which would increase the acceleration efficiency. Thus the prediction was: if you see particles upstream of the shock, you should find waves too. A closer look at the data confirmed this.

Another example of prediction and, in this case, the disproof of a model is the following. Originally, the heliosphere was assumed to be rather irregular in the lower latitudes above the streamer belts, because here different solar wind streams interact and transient disturbances (shocks and CMEs) constantly modify the structure. Over the poles, on the other hand, there is no activity, and the solar wind out of the coronal hole is fast and contains almost no turbulence. With only a small amount of turbulence, there is not much scattering of particles and, in particular, galactic cosmic rays should have easy access over the poles of the Sun, similarly to the solar energetic particles precipitating at the cusps of a planetary magnetosphere. Therefore, a mission over the poles of the Sun should allow one to observe galactic cosmic rays without too much modulation by the heliosphere, and thus allow one to determine the local interstellar spectrum more accurately. As a consequence of this prediction, Ulysses was designed. Once it was over the poles, the observations disproved our previous concepts of the structure of the heliosphere and the physics of modulation.

12.5 Is This Still Physics?

I am working in a physics department with colleagues all of whom do laboratory physics or theoretical physics. During seminars and colloquia, sometimes the question arises: is this still physics? All these uncertainties, the unknown parameters, the incompleteness of observations. Can this be physics?

Yes, of course it is physics. It is the application of physical concepts to a natural system. Space physics works according to the same principles of science as laboratory physics does: models, observations, repeatability, and prediction. The only difference is the broader scope, the larger number of parameters, and the broader scatter in them. It is sometimes difficult for a laboratory physicist to grasp that an order-of-magnitude estimate in space physics may be quite good while he/she fights to determine some other constant with an accuracy of less than one per thousand.

All these "problems" and uncertainties in space physics, however, also make it an attractive topic to study or to do research in it: like any work in a complex environment, it is an intellectual challenge. To understand a simple phenomenon such as a flare, a large number of scenarios may be developed, and with each new observation another piece of a jigsaw puzzle is provided – which might fit into its own picture or into an entirely different one.

For whatever reason you have read this book, it may be study, research, or just interest, I wish you a lot of fun and success in this intellectually challenging field of space physics.

Appendix

A.1 List of Symbols

α	pitch angle
β	plasma-β
γ	spectral index
γ	Lorentz factor $\gamma = \sqrt{1 - v^2/c^2}$
γ_a	adiabatic exponent, c_p/c_V
η	(dynamic) viscosity
φ	electric potential
Φ	magnetic flux
$\kappa(\mu)$	pitch angle diffusion coefficient
λ	mean free path
λ	wave length
λ_c	Coulomb logarithm $\lambda_c = \ln \Lambda$
λ_D	Debye length
λ_{De}	electron Debye length
μ	magnetic moment
ν	kinematic viscosity
ν	frequency
ν_c	collision frequency
ν_{ei}	electron–ion collision frequency
ω	angular frequency
Ω	angular speed
Ω	solid angle
ω_c	cyclotron frequency
ω_{ce}	electron cyclotron frequency
ω_{cp}	proton cyclotron frequency
ω_p	plasma frequency
ω_{pe}	electron plasma frequency
ω_{pp}	proton plasma frequency
ψ	azimuthal angle in a circular motion
Ψ	stream function
ρ	mass density
ρ_c	charge density

ρ_e	electron density
σ	conductivity
σ	scattering cross-section
τ	collision time
θ_{Bn}	angle between shock normal and magnetic field direction
χ	thermal conductivity
χ	degree of ionization
A	area
$\boldsymbol{A}$	vector potential
$\boldsymbol{B}$	magnetic field
B_{MP}	magnetic field at the mirror point
c	speed of light
c_p	specific heat at constant pressure
c_V	specific heat at constant volume
d_{min}	minimum distance of charges in a plasma
D	spatial diffusion coefficient
D	diffusion tensor
$\boldsymbol{E}$	electric field
E_{BO}	energy of the ground state (Bohr model)
E_{el}	electrostatic energy
E_F	Fermi energy
E_{rel}	relativistic electron energy
E_{th}	thermal energy
$\boldsymbol{e}$	unit vector
$\boldsymbol{F}$	force
$\boldsymbol{f}$	force density
$f(\boldsymbol{r}, \boldsymbol{v}, t)$	distribution function (phase space density)
$\boldsymbol{g}$	gravitation
G	universal constant of gravitation
h	Planck's constant
$\boldsymbol{j}$	current density
$\boldsymbol{j}_D$	drift current density
J	differential flux
J_i	action integral
$\boldsymbol{k}$	wave vector
k	wave number
k_B	Boltzmann's constant
L	characteristic length scale
$\boldsymbol{l}$	vector along a path
m	mass
m_e	electron mass
m_p	proton mass
n	number density
n_e	number density of electrons

n_p	number density of protons
$\boldsymbol{p}$	momentum
p_i	generalized momentum
p	pressure
p_M	magnetic pressure
P	stress tensor
P	rigidity
q	charge
q_i	generalized spatial coordinate
$\boldsymbol{r}$	radius vector
r_c	radius of curvature
r_L	Larmor radius
r_Le	electron Larmor radius
R_M	magnetic Reynolds number
R_MP	mirror ratio
r_0	impact parameter
$\boldsymbol{S}$	vector defining a surface
s	path length (along the Archimedean spiral)
$\boldsymbol{S}_\mathrm{P}$	Poynting vector
T	temperature
$\boldsymbol{t}_\mathrm{B}$	tangent vector to the magnetic field
T_e	electron temperature
T_c	gyration time (cyclotron period)
$\boldsymbol{u}$	bulk velocity
U	particle flux
v_sowi	solar wind speed
$\boldsymbol{v}$	velocity of a particle
$v_\perp$	speed perpendicular to the field
$v_\parallel$	speed parallel to the field
v_A	Alfvén speed
$\boldsymbol{v}_\mathrm{D}$	drift velocity
$\boldsymbol{v}_\mathrm{e}$	electron speed
v_g	group speed
$\boldsymbol{v}_\mathrm{gc}$	velocity of the guiding center
$\boldsymbol{v}_\mathrm{p}$	proton speed
v_ph	phase speed
v_s	sound speed
v_th	thermal speed
$v_\mathrm{th,e}$	thermal speed of electrons
W_kin	kinetic energy
$W_{\mathrm{kin},\parallel}$	kinetic energy parallel to the field
$W_{\mathrm{kin},\perp}$	kinetic energy perpendicular to the field
W_th	thermal energy
Z	atomic number

A.2 Equations in the SI and cgs System

Maxwell's equations:

$$\nabla \cdot \boldsymbol{E} = \varrho_{\mathrm{c}}/\epsilon_0 \qquad\qquad \nabla \cdot \boldsymbol{E} = 4\pi \varrho_{\mathrm{c}} \qquad (\mathrm{A.1})$$

$$\nabla \cdot \boldsymbol{B} = 0 \qquad\qquad \nabla \cdot \boldsymbol{B} = 0 \qquad (\mathrm{A.2})$$

$$\nabla \times \boldsymbol{E} = -\frac{\partial \boldsymbol{B}}{\partial t} \qquad\qquad \nabla \times \boldsymbol{E} = -\frac{1}{c}\frac{\partial \boldsymbol{B}}{\partial t} \qquad (\mathrm{A.3})$$

$$\nabla \times \boldsymbol{B} = \mu_0 \boldsymbol{j} + \frac{1}{c^2}\frac{\partial \boldsymbol{E}}{\partial t} \qquad\qquad \nabla \times \boldsymbol{B} = \frac{4\pi}{c}\boldsymbol{j} + \frac{1}{c}\frac{\partial \boldsymbol{E}}{\partial t} \qquad (\mathrm{A.4})$$

Field transformations:

$$\boldsymbol{E}' = \boldsymbol{E} + \boldsymbol{v} \times \boldsymbol{B} \qquad\qquad \boldsymbol{E}' = \boldsymbol{E} + \frac{1}{c}\boldsymbol{v} \times \boldsymbol{B} \qquad (\mathrm{A.5})$$

$$\boldsymbol{B}' = \boldsymbol{B} - \frac{1}{c^2}\boldsymbol{v} \times \boldsymbol{E} \qquad\qquad \boldsymbol{B}' = \boldsymbol{B} - \frac{1}{c}\boldsymbol{v} \times \boldsymbol{E} \qquad (\mathrm{A.6})$$

Ohm's law:

$$\boldsymbol{j} = \sigma\left(\boldsymbol{E} + \boldsymbol{v} \times \boldsymbol{B}\right) \qquad\qquad \boldsymbol{j} = \sigma\left(\boldsymbol{E} + \frac{1}{c}\boldsymbol{v} \times \boldsymbol{B}\right) \qquad (\mathrm{A.7})$$

Energy densities and fluxes:

$$\epsilon_{\mathrm{B}} = \frac{B^2}{2\mu_0} \qquad\qquad \epsilon_{\mathrm{B}} = \frac{B^2}{8\pi} \qquad (\mathrm{A.8})$$

$$\epsilon_{\mathrm{E}} = \frac{\epsilon_0 E^2}{2} \qquad\qquad \epsilon_{\mathrm{E}} = \frac{E^2}{8\pi} \qquad (\mathrm{A.9})$$

$$\boldsymbol{S} = \frac{\boldsymbol{E} \times \boldsymbol{B}}{\mu_0} \qquad\qquad \boldsymbol{S} = \frac{\boldsymbol{E} \times \boldsymbol{B}}{4\pi} \qquad (\mathrm{A.10})$$

Single particle motion:

$$\boldsymbol{F} = q\left(\boldsymbol{E} + \boldsymbol{v} \times \boldsymbol{B}\right) \qquad\qquad \boldsymbol{F} = q\left(\boldsymbol{E} + \frac{\boldsymbol{v} \times \boldsymbol{B}}{c}\right) \qquad (\mathrm{A.11})$$

$$\omega_{\mathrm{c}} = \frac{qB}{m} \qquad\qquad \omega_{\mathrm{c}} = \frac{qB}{mc} \qquad (\mathrm{A.12})$$

$$r_{\mathrm{L}} = \frac{mv_\perp}{|q|B} \qquad\qquad r_{\mathrm{L}} = \frac{mv_\perp c}{|q|B} \qquad (\mathrm{A.13})$$

$$P = \frac{p_\perp}{q} \qquad\qquad P = \frac{p_\perp c}{q} \qquad (\mathrm{A.14})$$

$$\boldsymbol{v}_{\mathrm{Drift}} = \frac{\boldsymbol{F} \times \boldsymbol{B}}{qB^2} \qquad\qquad \boldsymbol{v}_{\mathrm{Drift}} = \frac{c}{q}\frac{\boldsymbol{F} \times \boldsymbol{B}}{B^2} \qquad (\mathrm{A.15})$$

Magnetic pressure

$$p_{\text{M}} = \frac{B^2}{2\mu_0} \qquad\qquad p_{\text{M}} = \frac{B^2}{8\pi} \qquad (\text{A.16})$$

Alfvén speed

$$v_{\text{A}} = \frac{B_0}{\sqrt{\mu_0 \varrho_0}} \qquad\qquad v_{\text{A}} = \frac{B_0}{\sqrt{4\pi\varrho_0}} \qquad (\text{A.17})$$

Electron plasma frequency

$$\omega_{\text{pe}} = \sqrt{\frac{n_e e^2}{m_e \varepsilon_0}} \qquad\qquad \omega_{\text{pe}} = \sqrt{\frac{4\pi n_e e^2}{m_e}} \qquad (\text{A.18})$$

Ion plasma frequency

$$\omega_{\text{pi}} = \sqrt{\frac{n_0 e^2}{m_i \epsilon_0}} \qquad\qquad \omega_{\text{pi}} = \sqrt{\frac{n_0 e^2}{m_i}} \qquad (\text{A.19})$$

A.3 Useful Relations

$$\boldsymbol{A} \times (\boldsymbol{B} \times \boldsymbol{C}) = \boldsymbol{B}(\boldsymbol{A} \cdot \boldsymbol{C}) - \boldsymbol{C}(\boldsymbol{A} \cdot \boldsymbol{B}) \qquad (\text{A.20})$$
$$(\boldsymbol{A} \times \boldsymbol{B}) \cdot (\boldsymbol{C} \times \boldsymbol{D}) = (\boldsymbol{A} \cdot \boldsymbol{C})(\boldsymbol{B} \cdot \boldsymbol{D}) - (\boldsymbol{A} \cdot \boldsymbol{D})(\boldsymbol{B} \cdot \boldsymbol{C}) \qquad (\text{A.21})$$

A.3.1 Vector Calculus

$$\nabla = \boldsymbol{e}_x \frac{\partial}{\partial x} + \boldsymbol{e}_y \frac{\partial}{\partial y} + \boldsymbol{e}_z \frac{\partial}{\partial z} \qquad (\text{A.22})$$

$$\nabla \boldsymbol{A} = \frac{\partial A_x}{\partial x} + \frac{\partial A_y}{\partial y} + \frac{\partial A_z}{\partial z} \qquad (\text{A.23})$$

$$\nabla \times \boldsymbol{A} = \begin{pmatrix} \boldsymbol{e}_x & \boldsymbol{e}_y & \boldsymbol{e}_z \\ \frac{\partial}{\partial x} & \frac{\partial}{\partial y} & \frac{\partial}{\partial z} \\ A_x & A_y & A_z \end{pmatrix} \qquad (\text{A.24})$$

$$\nabla \cdot (\nabla \times \boldsymbol{A}) = 0 \qquad (\text{A.25})$$
$$\nabla \times (\nabla \times \boldsymbol{A}) = \nabla(\nabla \cdot \boldsymbol{A}) - \nabla^2 \boldsymbol{A} \qquad (\text{A.26})$$
$$\nabla \times \nabla \phi = 0 \qquad (\text{A.27})$$
$$\nabla \cdot \nabla \phi = \nabla^2 \phi \qquad (\text{A.28})$$
$$\nabla \cdot (\boldsymbol{A} \times \boldsymbol{B}) = \boldsymbol{B} \cdot (\nabla \times \boldsymbol{B}) - \boldsymbol{A} \cdot (\nabla \times \boldsymbol{B}) \qquad (\text{A.29})$$
$$\nabla \times (\boldsymbol{A} \times \boldsymbol{B}) = \boldsymbol{A}(\nabla \cdot \boldsymbol{B}) - \boldsymbol{B}(\nabla \cdot \boldsymbol{A}) + (\boldsymbol{B} \cdot \nabla)\boldsymbol{A} - (\boldsymbol{A} \cdot \nabla)\boldsymbol{B} \qquad (\text{A.30})$$

$$\nabla(\boldsymbol{A} \cdot \boldsymbol{B}) = \boldsymbol{A} \times (\nabla \times \boldsymbol{B}) + \boldsymbol{B} \times (\nabla \times \boldsymbol{A}) + (\boldsymbol{A} \cdot)\boldsymbol{B} + (\boldsymbol{B} \cdot \nabla)\boldsymbol{A} \quad (\text{A.31})$$

V is a volume enclosed by a surfac $\boldsymbol{S}$. $\mathrm{d}\boldsymbol{S} = \boldsymbol{e}_n \mathrm{d}S$ is a vector normal to this surface. Then we have

$$\int_V = \nabla\phi \, \mathrm{d}V = \int_S \phi \, \mathrm{d}\boldsymbol{S} \quad (\text{A.32})$$

Divergence theorem (Gauss's theorem):

$$\int_V \nabla \cdot \boldsymbol{A}\mathrm{d}^3 x = \int_S \boldsymbol{A} \cdot \mathrm{d}\boldsymbol{S} \quad (\text{A.33})$$

$$\int_V \nabla \cdot \mathsf{T} \, \mathrm{d}V = \int_S \mathsf{T} \cdot \mathrm{d}\boldsymbol{S} \quad (\text{A.34})$$

$$\int_V \nabla \times \boldsymbol{A} \, \mathrm{d}V = \int_S \mathrm{d}\boldsymbol{S} \times \boldsymbol{A} \quad (\text{A.35})$$

$$\int_V (\phi\nabla^2\rho - \rho\nabla^2\phi) \, \mathrm{d}V = \int_S (\phi\nabla\rho - \rho\nabla\phi) \cdot \mathrm{d}\boldsymbol{S} \quad (\text{A.36})$$

$$\int_V (\boldsymbol{A} \cdot \nabla \times \nabla \times \boldsymbol{B} - \boldsymbol{B} \cdot \nabla \times \nabla \times \boldsymbol{A}) = \int_S (\boldsymbol{B} \times \nabla \times \boldsymbol{A} - \boldsymbol{A} \times \nabla \times \boldsymbol{B}) \cdot \mathrm{d}\boldsymbol{S} \quad (\text{A.37})$$

$\boldsymbol{S}$ is an open surface enclosed by a curve C with length element $\mathrm{d}\boldsymbol{l}$:

$$\int_S \mathrm{d}\boldsymbol{S} \times \nabla\phi = \oint_d \mathrm{d}\boldsymbol{l}\phi \quad (\text{A.38})$$

Stokes' law:

$$\int_S (\nabla \times \boldsymbol{A}) \cdot \mathrm{d}\boldsymbol{S} = \oint_C \boldsymbol{A} \cdot \mathrm{d}\boldsymbol{l} \quad (\text{A.39})$$

$$\int_S (\mathrm{d}\boldsymbol{S} \times \nabla) \times \boldsymbol{A} = \oint_C \mathrm{d}\boldsymbol{l} \times \boldsymbol{A} \quad (\text{A.40})$$

$$\int_S (\mathrm{d}\boldsymbol{S} \cdot (\nabla\phi)) \times \nabla\rho = \oint_c \phi \, \mathrm{d}\rho = -\oint_C \rho \, \mathrm{d}\phi \quad (\text{A.41})$$

A.3.2 Cylindrical Coordinates

$$\nabla \cdot \boldsymbol{A} = \frac{1}{r}\frac{\partial}{\partial r}(rA_r) + \frac{1}{r}\frac{\partial A_\theta}{\partial \theta} + \frac{\partial A_z}{\partial z} \quad (\text{A.42})$$

$$\nabla\phi = \left(\frac{\partial\phi}{\partial r}, \frac{1}{r}\frac{\partial\phi}{\partial \theta}, \frac{\partial\phi}{\partial z}\right) \quad (\text{A.43})$$

$$\nabla \times \boldsymbol{A} = \left(\frac{1}{r}\frac{\partial A_z}{\partial \theta} - \frac{\partial A_\theta}{\partial z}, \frac{\partial A_r}{\partial z} - \frac{\partial A_z}{\partial r}, \frac{1}{r}\frac{\partial(rA_\theta)}{\partial r} - \frac{1}{r}\frac{\partial A_r}{\partial \theta}\right) \quad (\text{A.44})$$

$$\nabla^2\phi = \frac{1}{r}\left(\frac{\partial\phi}{r}\right) + \frac{1}{r^2}\frac{\partial^2\phi}{\partial\theta^2} + \frac{\partial^2\phi}{\partial z^2} \quad (\text{A.45})$$

$$\nabla^2 \boldsymbol{A} = \left(\nabla^2 A_r - \frac{2}{r^2}\frac{\partial A_\theta}{\partial \theta} - \frac{A_r}{r^2}, \nabla^2 A_\theta + \frac{2}{r^2}\frac{\partial A_\theta}{\partial \theta} - \frac{A_\theta}{\partial r^2}, \nabla^2 A_z\right) \quad (\text{A.46})$$

A.3.3 Spherical Coordinates

$$\nabla \cdot \boldsymbol{A} = \frac{1}{r^2}\frac{\partial}{\partial r}(r^2 A_r) + \frac{1}{r\sin\theta}\frac{\partial}{\partial \theta}(A_\theta \sin\theta) + \frac{1}{r\sin\theta}\frac{\partial A_\phi}{\partial \phi} \tag{A.47}$$

$$\nabla \rho = \left(\frac{\partial \rho}{\partial r}, \frac{1}{r}\frac{\partial \rho}{\partial \theta}, \frac{1}{r\sin\theta}\frac{\partial \rho}{\partial \phi}\right) \tag{A.48}$$

$$\nabla \times \boldsymbol{A} = \left(\frac{1}{r\sin\theta}\frac{\partial(A_\phi \sin\theta)}{\partial \theta} - \frac{1}{r\sin\theta}\frac{\partial A_\theta}{\partial \phi},\right.$$
$$\left.\left(\frac{1}{r\sin\theta}\frac{\partial A_r}{\partial \varphi} - \frac{1}{r}\frac{\partial(rA_\phi)}{\partial r}, \frac{1}{r}\frac{\partial(rA_\theta)}{\partial r} - \frac{1}{r}\frac{\partial A_r}{\partial \theta}\right)\right. \tag{A.49}$$

$$\nabla^2 \rho = \frac{1}{r^2}\frac{\partial}{\partial r}\left(r^2\frac{\partial \rho}{\partial r}\right) + \frac{1}{r^2\sin\theta}\frac{\partial}{\partial \theta}\left(\sin\theta\frac{\partial \rho}{\partial \theta}\right) + \frac{1}{r^2\sin^2\theta}\frac{\partial^2 \rho}{\partial \phi^2} \tag{A.50}$$

A.4 Useful Numbers

A.4.1 Fundamental Constants

c	speed of light in a vacuum	2.998×10^8 m/s
ϵ_0	permittivity in a vaccum	8.854×10^{-12} F/m
μ_0	permeabilty in a vaccum	$4\pi \times 10^{-7}$ Vs/Am
e	elementary charge	1.602×10^{-19} C
m_e	electron mass	9.109×10^{-31} kg
	electron rest energy	511 keV
m_p	proton mass	1.673×10^{-27} kg
	pront rest energy	939 MeV
m_p/m_e	proton-to-electron mass ratio	1.836×10^3
σ	Stefan–Boltzmann number	5.6708×10^{-8} Jm^{-2}s^{-1}K^{-1}
k_B	Boltzmann number	1.381×10^{-23} J/K
eV	electronvolt	1.602×10^{-19} J
eV/k_B	temperature equivalent for 1 eV	1.160×10^4 K
G	gravitational constant	6.673×10^{-11} Nm^{2}kg^{-2}
h	Planck's constant	6.623×10^{-34} Js

A.4.2 Numbers in Plasmas

In the following numerical expressions the plasma parameters have to be inserted in SI units [as long as no other units are indicated in squre brackets]: the electron density n_e has to be inserted in m^{-3}, B in Tesla, $v_\perp$ in m/s and the electron temperature T_e in K. The fundamental frequencies in plasmas then are given as:

ω_{pe}	electron plasma frequency	$0.564 \times \sqrt{n_e}$ rad/s
ω_{pp}	proton plasma frequency	$1.32 \times \sqrt{n_p}$ rad/s
ω_{ce}	electron cyclotron frequency	$1.76 \times 10^{11} B$ rad/s $= 28B[\text{nT}]$ Hz
ω_{cp}	proton cyclotron frequency	$9.58 \times 10^7 B$ rad/s $= 0.01525B[\text{nT}]\,\text{Hz}$

These are angular frequencies. Frequencies can be obtained by division by 2π.

Important length scales and speeds in a plasma can be calculated according to the following rules:

$v_{Th,e}$	electron thermal speed, $\sqrt{2k_B T_e/m}$	$5.51 \times 10^3 \sqrt{T_e}$ m/s
$v_{Th,i}$	ion thermal speed, $\sqrt{2k_B T_i/M_i}$	$129 \times \sqrt{T_i \cdot m_p/M_i}$ m/s
λ_{De}	electron Debye length, $\sqrt{k_B T_e/m}/\omega_{pe}$	$69 \times \sqrt{T_e/n_e}$ m
r_{Le}	electron Larmor radius, $v_\perp/\omega_{ce}$	$5.68 \times 10^{-12} v_\perp/B$ m
r_{Lp}	proton Larmor radius, $v_\perp/\omega_{cp}$	$1.04 \times 10^{-8} v_\perp/B$ m
v_s	sound speed	$0.117\sqrt{T_e}$ km/s

A.4.3 Conversion of Units

Occasionally, electrical units have to be converted into mechanical ones or vice versa. The following table gives some physical quantities in different units, which should help in unit conversion. It is neither complete nor systematic. Note that factors are not included, just different ways to express a unit!

B	magnetic field	T	$\text{kg s}^{-2}\text{A}^{-1}$	$\text{cm}^{-1/2}\text{g}^{1/2}\text{s}^{-1}$
E	electric field	V/m	$\text{m kg s}^{-3}\text{A}^{-1}$	$\text{cm}^{-1/2}\text{g}^{1/2}\text{s}^{-1}$
I	current	A		$\text{cm}^{1/2}\text{g}^{1/2}\text{s}^{-1}$
Q	electrical charge	C	A s	$\text{cm}^{3/2}\text{g}^{1/2}\text{s}^{-1}$
S_P	Poynting vector	W/m^2	kg s^{-3}	g s^{-3}

A.5 Useful Internet Resources

All information contained in the following list can be found via "clickable" links at **www.physik.uni-osnabrueck.de/sotere/spacebook/intro.html**.

A.5.1 Space Physics – General

Some useful general resources for space physics are:

- The Naval Research Laboratory has a general plasma physics page giving all relevant equations:
  wwwppd.nrl.navy.mil/nrlformulary/nrlformulary.html
- Some collections of links to other space physics resources are provided by:

- Geospace Environment Data Analysis System:
  `gedas22.stelab.nagoya-u.ac.jp/website/website.html`
- the space science web page of the University of Iowa:
  `www-pw.physics.uiowa.edu/links/space.html`
- links to space physics educational sites are provided by the Space Environment Center of NOAA at:
  `www.sec.noaa.gov/Education/ed_sites.html`

A.5.2 Textbooks and other Educational Material

There are some textbooks on space physics available on the Internet:

- The University of Oulu provides a space physics textbook at:
  `www.oulu.fi/~spaceweb/textbook/`
- Bo Thiede from the University of Uppsala provides a textbook on classical electrodynamics at: `www.plasma.uu.se/CED/Book/`

Additional educational tools on the net include:

- "What is Geospace" is an educational page by NASA:
  `http://www-istp.gsfc.nasa.gov/istp/outreach/geospace.html`
- NOAA's primer on space weather can be found at:
  `http://www.sel.noaa.gov/primer/primer.html`
- an exploration of the Earth's Magnetosphere is possible with D.P. Stern and M. Peredo at: `www-spof.gsfc.nasa.gov/Education/Intro.html`
- an encyclopedia of the atmospheric environment can be found at:
  `www.doc.mmu.ac.uk/aric/eae/english.html`
- the Global Climate Change Student Guide is at
  `www.doc.mmu.ac.uk/aric/gccsg/`
- a space physics tutorial from the Space Science Group of the University of California, Los Angeles, is provided at:
  `www-ssc.igpp.ucla.edu/ssc/tutorial.html`

A.5.3 Missions

Information about space missions can be obtained from the homepages of NASA and ESA. Homepages for some of the larger missions are listed below:

- Solar and Heliospheric Observatory (SOHO) homepage at NASA:
  `sohowww.nascom.nasa.gov`
- SOHO homepage at ESA: `sohowww.estec.esa.nl`
- Homepage of the Extreme ultraviolet Imaging Telescope (EIT) on SOHO:
  `umbra.gsfc.nasa.gov.eit`
- LASCO homepage: lasco-www.nrl.navy.mil/lasco.html
- The TRACE homepages are `lmsal.com/solarsites.html` and `vestige.lmsal.com/TRACE/POD/TRACEpodoverview.html`

- Yohkoh homepage: `www.lmsal.com/SXT`
- Ramaty High Energy Solar Spectroscopic Imager (RHESSI) homepage at: `hesperia.gsfc.nasa.gov/hessi/`
- STEREO `stp.gsfc.nasa.gov/missions/stereo/stereo.htm`
- Advanced Composition Explorer (ACE): `www.srl.caltech.edu/ACE/`
- Wind: `www-spof.gsfc.nasa.gov/istp/wind/`
- SAMPEX (Solar, Anomalous, and Magnetospheric Particle EXplorer): `sunland.gsfc.nasa.gov/smex/sampex/` and `surya.umd.edu/www/sampex.html`
- Ulysses: `ulysses.jpl.nasa.gov/`
- Interplanetary Monitoring Platform (IMP-8): `nssdc.gsfc.nasa.gov/space/imp-8.html`
- IMAGER (Imager for Magnetopause to Aurora Global Explorations): `image.gsfc.nasa.gov/`
- POLAR: `http://www-spof.gsfc.nasa.gov/istp/polar/`
- CLUSTER-II: `sci.esa.int/science-e/www/area/index.cfm?fareaid=8`
- Oersted/CHAMP: `web.dmi.dk/projects/oersted/` and `op.gfz-potsdam.de/champ/index_CHAMP.html`
- A compilation of links to solar and space physics missions is given at: `members.aol.com/gca7sky/mission3.htm`

A.5.4 The Sun

- General information concerning the Sun and a large number of useful links to other sites are given by Bill Arnett on: `seds.lpl.arizona.edu/nineplanets/nineplanets/sol.html`
- "The Sun: a pictorial introduction" is provided by the High Altitude Observatory (HAO) at `www.hao.ucar.edu/public/slides/slides.html`
- Big Bear Solar Observatory: `www.bbso.njit.edu`

A.5.5 Solar–Terrestrial Relationships

- Space Environment Center: `www.sel.noaa.gov/index.html`
- NOAA's space weather: `www.spaceweather.noaa.gov/`
- ESA's space weather page: `www.estec.esa.nl/wmwww/wma/spweather/`
- NASA's Sun Earth Connections page: `sec.gsfc.nasa.gov/`
- Windows to the Universe: Basic Facts about Space Weather: `www.windows.ucar.edu/spaceweather/basic\_facts.html`
- Lund Space Weather Center: `www.lund.irf.se/`
- Center for Integrated Space Weather Modeling (CISM): `www.bu.edu/cism/`
- Space weather resources compiled by Rice University: `space.rice.edu/ISTP/`

- Space weather resources at NSSDC:
  `spdf.gsfc.nasa.gov/space_weather/Space_Weather_at_SSD00.html`

A.5.6 Aurorae

Some useful links to pictures and general aspects of aurorae are

- Poker Flat Research Range, Fairbanks, Alaska:
  `www.pfrr.alaska.edu/aurora/INDEX.HTM`
- University of Michigan aurora page: `www.geo.mtu.edu/weather/aurora/`
- Norwegian Northern Light page: `www.northern-lights.no/`
- Jan Curtis aurora page, probably the best photographs on the Web:
  `www.geo.mtu.edu/weather/aurora/images/aurora/jan.curtis/` and the more recent version: `climate.gi.alaska.edu/Curtis/curtis.html`
- Polar image page, also covers noctilucent clouds: `www.polarimage.fi`

Predictions of aurorae from particle measurements in space can be found at `sec.noaa.gov/pmap/`.

A.5.7 Data

- Space Physics Interactive Data Resource (SPIDR) at the National Geophysical Data Center: `spidr.ngdc.noaa.gov/spidr/index.html`
- Solar Data Analysis Center at Goddard Space Flight Center:
  `umbra.nascom.nasa.goc.sdac.html`
- Solar and Upper Atmospheric Data Services:
  `www.ngdc.noaa.gov/SOLAR/solar.html`
- National Oceanographic and Atmospheric Administration (NOAA):
  `www.SpaceWeather.com`

References

1. Acuña, M.H. and 19 others (1998): Magnetic field and plasma observations at Mars: initial results from the Mars Global Surveyor Mission, *Science* **279**, 1676

2. Ahmad, Q.R., R. Allen, T.C. Andersen et al. (2001): Measurement of the rate of nu(e)+d -> p+p+e(-) interactions produced by B-8 solar neutrinos at the Sudbury neutrino observatory, *Phys. Rev. Lett.* **87**, 071301

3. Ahmad, Q.R., R. Allen, T.C. Andersen et al. (2001): Measurement of day and night neutrino energy spectra at SNO and constraints on neutrino mixing parameters, *Phys. Rev. Lett.* **89**, 011301

4. Akasofu, S.-I. (1989): Substorms, *EOS* **70**, 529

5. Alfvén, H. (1981): *Cosmic plasmas*, Reidel, Dordrecht

6. Anderson, K.A. and R.B. Lin (1969): Observation of interplanetary field lines in the magnetotail, *J. Geophys. Res.* **74**, 3953

7. Antia, H.M. (2003): Solar interior and seismology, in [9], p. 80

8. Antia, A.M., and S. Basu (2001): Temporal variations of the solar rotation rate at high latitudes, *Astrophys. J.* **559**, L67

9. Antia, H.M., A. Bhatnagar, and P. Ulmschneider (eds.) (2003): *Lectures on solar physics*, Lecture Notes in Physics 619, Springer, Berlin, Heidelberg

10. Antia, H.M., and S.M. Chitre (1998): Determination of temperature and chemical composition profiles in the solar interior from seismic models, *Astron. Astrophys.* **339**, 239

11. Armstrong, T.P., M.E. Pesses, and R. B. Decker (1985): Shock drift acceleration, in [524], p. 271

12. Aschwanden, M.J., and C.E. Parnell (2002): Nanoflare statistics from first principles: fractal geometry and temperature synthesis, *Astrophys. J.* **572**, 1048

13. Axford, W.I. (1982): Acceleration of cosmic rays by shock waves, in *Plasma Astrophysics* (eds. T.D. Tiyenen and T. Levy), ESA-SP 151

14. Axford, W.I. (1984): Magnetic field reconnection, in *Magnetic reconnection in space and laboratory plasmas*, Geophys. Res. Monogr. 30, American Geophysical Union, Washington, DC

15. Axford, W.I., and C.O. Hines (1961): A unifying theory of high-latitude geophysical phenomena and geomagnetic storms, *Can. J. Phys.* **39**, 1433

16. Axford, W.I., E. Leer, and G. Skadron (1977): The acceleration of cosmic rays by shock waves, *Proc. 15th Int. Cosmic Ray Conf.* **11**, 132

17. Babcock, H. (1961): The topology of the Sun's magnetic field and the 22-year cycle, *Astrophys. J.* **133**, 572

18. Bagenal, F. (1992): Giant planet magnetospheres, *Annu. Rev. Earth Planet Sci.* **20**, 289

19. Bahcall, J.N. (1979): Solar neutrinos: theory versus observation, *Space Sci. Rev.* **24**, 227

20. Bahcall, J.N., M.P. Pinsonneault, and S. Basu (2001): Solar models: Current epoch and time dependences, neutrinos, and helioseismological properties, *Astrophys. J.* **555**, 990

21. Bahcall, J., and R. Ulrich (1988): Solar models, neutrino experiments, and helioseismolog, *Rev. Mod. Phys.* **60**, 297

22. Bai, T. (1986): Two classes of γ-ray/proton flares: impulsive and gradual, *Astrophys. J.* **311**, 437

23. Bai, T. (1986a): Classification of solar flares and the relationship between first and second phases, *Adv. Space Res.* **6**, 203

24. Baker, D.N., R.C. Anderson, R.D. Zwickl, and J.A. Slavin (1987): Average plasma and magnetic field variations in the distant magnetotail associated with near-Earth substorm effects, *J. Geophs. Res.* **92**, 71

25. Baliunsa, S., and R. Jastrow (1990): Evidence for long-term brightness changes of solar-type stars, *Nature* **348**, 520

26. Balogh, A., A.J. Smith, B.T. Tsurutani, D.J. Southwood, R.J. Forsyth, and T.S. Horbury (1995): The heliospheric magnetic field over the south polar region of the sun, *Science* **268**, 1007

27. Balogh, A., J.T. Gosling, J.R. Jokipii, R. Kallenbach, and H. Kunow (eds.) (2000): *Corotating Interaction Regions*, Kluwer, Dordrecht

28. Barnden, L.R. (1973): The large-scale magnetic field configuration associated with Forbush decreases, *Proc. 13th Int. Cosmic Ray Conf.* **2**, 1277

29. Barnes, A. (1970): Theory of bow-shock associated hydromagnetic waves in the upstream interplanetary medium, *Cosmic Electrodyn.* **1**, 90

30. Barnes, A. (1992): Acceleration of the solar wind, *Rev. Geophys.* **30**, 43

31. Barnes, A., R.E. Hardtle, and J.H. Bredekamp (1971): On the energy transport in stellar winds, *Astrophys. J. Lett.* **166**, L53

32. Barnes, C.W., and J.A. Simpson (1976): Evidence for interplanetary acceleration of nucleons in corotating interaction regions, *Astrophys. J.* **207**, 977

33. Basu, S. (1997): Seismology of the base of the solar convection zone, *Mon. Not. R. Astron. Soc.* **288**, 572

34. Batchelor, D.A., C.J. Crannell, H.J. Wiehl, and A. Magun (1985): Evidence for collisionless conduction fronts in impulsive flares, *Astrophys. J.* **295**, 258

35. Bath, M. (1974): *Spectral analysis in geophysics*, Elsevier, Amsterdam

36. Baumjohann, W., and R.A. Treumann (1996): *Basic space plasma physics*, Imperial College Press, London

37. Beer, J. (2000): Long-term indirect indices of solar variability, in [172], p. 53, also *Space Sci. Rev.* **94**, 53

38. Beck-Bornholdt, H.-P., and Dubben, H.-H. (2002): *Der Hund, der Eier legt*, rororo, Rowohlt Taschenbuch Verlag, Reinbek

39. Bell, A.R.I. (1978): The acceleration of cosmic rays in shock fronts, *Mon. Not. R. Astron. Soc.* **182**, 147

40. Bendat, J.S. and A.G. Piersol (1971): *Random data: analysis and measurement procedures*, Wiley–Interscience, New York

41. Benestad, R.E. (2003): *Solar activity and earth's climate*, Springer-Praxis, London

42. Benz, A.O. (1993): *Plasma astrophysics*, Kluwer, Dordrecht

43. Benz, A.O. (2003): Radio diagnostcs of flare energy release, in [291], p. 80

44. Benz, A.O., and S. Krucker (2002): Energy distribution of microevents in the quiet solar corona, *Astrophys. J.* **568**, 413
45. Bequerel, H. (1879): *Aurora, their characters and spectra*, Spon, London
46. Bieber, J.W., W.H. Matthaeus, C.W. Smith, W. Wanner, M.-B. Kallenrode, and G. Wibberenz (1994): Proton and electron mean free paths: the Palmer consensus range revisited, *Astrophys. J.* **420**, 294
47. Bieber, J.W., E. Eroshenko, P. Evenson, E.O. Flckiger, and R. Kallenbach (2000): *Cosmic rays and Earth*, Kluwer, Dordrecht
48. Biermann, L. (1951): Kometenschweife und solare Korpuskularstrahlung, *Z. Astrophys.* **29**, 274
49. Bingham, R. (1993): Space plasma physics: microprocesses, in *Plasma physics* (ed. R. Dendy), Cambridge University Press, Cambridge, p. 233
50. Birkeland, K. (1908): *The Norwegian aurora polaris expedition 1902 - 1903, vol. 1: On the cause of magnetic storms and the origin of terrestrial magnetism*, H. Ashebourg, Christiania
51. Blandford, R.D., and J.P. Ostriker (1978): Particle acceleration by astrophysical shocks, *Astrophys. J. Lett.* **221**, L29
52. Block, L.P. and C.G. Fälthammar (1990): The role of magnetic field aligned electric fields in auroral acceleration, *J. Geophys. Res.* **95**, 5877
53. Bohm, D. and E.P. Gross (1949): Theory of plasma oscillations, A. Origin of medium-like behavior, *Phys. Rev.* **75**, 1851
54. Bone, N. (1991): *The aurora – Sun–Earth interactions*, Ellis Horwood, New York
55. Borrini, G., J.T. Gosling, S.J. Bame, W.C. Feldman, and J.M. Wilcox (1981): Solar wind helium and hydrogen structure near the heliospheric current sheet: a signal of coronal streamers at 1 AU, *J. Geophys. Res.*, **86**, 4565
56. Bothmer, V., and R. Schwenn (1995): Eruptive prominences as sources of magnetic clouds in the solar wind, *Space Sci. Rev.* **70**, 215
57. Bougeret, J.-L. (1985): Observations of shock formation and evolution in the solar atmosphere, in *Collisionless shocks in the heliosphere: reviews of current research* (eds. B.T. Tsurutani and R.G. Stone), Geophys. Monogr. 35, American Geophysical Union, Washington, DC
58. Boyd, T.J.M. and J.J. Sanderson (1969): *Plasma dynamics*, Barnes and Noble, New York
59. Brekke, A. (1997): *Physics of the upper polar atmosphere*, Wiley, Chichester
60. Brekke, A. and A. Egeland (1983): *Northern lights*, Springer, Berlin
61. Brown, M.R. (1999): Experimental studies of magnetic reconnection, *Phys. Plasmas* **6**, 1717
62. Bruce, J., H. Lee, and E. Haites (eds.) (1996): *Climate change 1995: economics and social dimensions of climate change*, Cambridge University Press, Cambridge
63. Brückner, G.E., and 14 others (1995): The large angle spectroscopic coronograph (LASECO), in [161], p. 357
64. Bruno, R., B. Bavassano, and U, Villante (1985): Evidence for long-period Alfvén waves in the inner solar system, *J. Geophys. Res.* **90**, 4373
65. Bryant, D.A. (1993): Space plasma physics: basic processes in the solar system, in *Plasma physics: an introductory course* (ed. R. Dendy), Oxford University Press, Oxford, p. 209
66. Bucha, V. (1980): The influence of solar activity on atmospheric circulation types, *Ann. Geophys.* **6**, (5)513

67. Burch, J.L., R.L. Carovillano, and S.K. Antiochos (eds.) (1999): *Sun–Earth Plasma Connections*, Geophysical Monograph Series, 109, American Geophysical Union, Washington, DC

68. Burgess, D. (1995): Collisionless shocks, in [290], p. 129

69. Burkepile, J. and O.C. St. Cyr (1993): *A revised and expanded catalogue of coronal mass ejections observed by the Solar Maximum Mission coronograph*, High Altitude Observatory, NCAR Technical Note, NCAR/TN-369+STR, Boulder, CO

70. Burlaga, L.F. (1991): Magnetic clouds, in [475], p. 1

71. Burlaga, L.F. and R.P.Lepping (1977): The causes of recurrent geomagnetic storms, *Planet. Space Sci.* **25**, 1151

72. Burlaga, L.F., F.B. McDonald, M.L. Goldstein, and A.J. Lazarus (1985): Cosmic ray modulation and turbulent interaction regions near 11 AU, *J. Geophys. Res.* **90**, 12 127

73. Burlaga, L.F., F.B. McDonald, and N.F. Ness (1993): Cosmic ray modulation and the distanc heliospheric magnetic field: Voyager 1 and 2 observations from 1986 to 1989, *J. Geophys. Res.* **98**, 1

74. Calder, N. (1997): *The manic sun – weather theories confounded*, Pilkington, Yelvertoft

75. Callis, L.B., and M. Natarajan (1986): Ozone and nitrogen dioxide changes in the stratosphere during 1979–1984, *Nature* **323**, 772

76. Callis, L.B, R.E. Boughner, D.N. Baker, R.A. Mewaldt, J.B. Blake, R.S. Selesnick, J.R. Cummings, M. Natarajan, G.M. Mason, and J.E. Mazur (1996): Precipitating electrons: evidence for effects on mesospheric odd nitrogen, *Geophys. Res. Lett.* **23**, 1901

77. Callis, L.B., D.N. Baker, M. Natarajan, J.B. Blake, R.A. Mewaldt, R.S. Selesnick, and J.R Cummings (1996): A 2-D model simulation of downward transport of NO_y into the stratosphere: effects on the austral spring O_3 and NO_y, *Geophys. Res. Lett.* **23**, 1905

78. Callis, L.B., M. Natarajan, J.D. Lambeth, and R.E. Boughner (1997): On the origin of midlatitude ozone changes: data analysis and simulations for 1979–1993, *J. Geophys. Res.* **102**, 1215

79. Callis, L.B., M. Natarajan, D.S. Evans, and J.D. Lambeth (1998): Solar atmospheric coupling by electrons (SOLACE), 1. Effects of the May 12, 1997 solar event on the middle atmosphere, *J. Geophys. Res.* **103**, 28405

80. Callis, L.B., M. Natarajan, J.D. Lambeth, and D.N. Baker (1998): Solar atmospheric coupling by electrons (SOLACE), 2. Calculated stratospheric effects of precipitating electrons, 1979–1988, *J. Geophys. Res.* **103**, 28241

81. Callis, L.B., M. Natarajan, and J.D. Lambeth (2001): Solar atmospheric coupling by electrons (SOLACE), 3. Comparison of simulations and observations, 1979–1997, issues and implications, *J. Geophys. Res.* **106,** 7523

82. Campbell, W.H. (1997): *Introduction to geomagnetic fields*, Cambridge University Press, Cambridge

83. Cane, H.V. (1995): The structure and evolution of interplanetary shocks and the relevance for particle acceleration, *Nucl. Phys. B (Proc. Suppl.)* **39A**, 35

84. Cane, H.V., R.E. McGuire, and T.T. von Rosenvinge (1986): Two classes of solar energetic particle events associated with impulsive and long-duration soft X-ray flares, *Astrophys. J.* **301**, 448

85. Cane, H.V., D.V. Reames, and T.T. von Rosenvinge (1988): The role of inter-planetary shocks in the longitude distribution of energetic particles, *J. Geophys. Res.* **93**, 9555

86. Cane, H.V., I.G. Richardsen, and T.T. von Rosenvinge (1993): Cosmic ray decreases and particle acceleration in 1978 - 1982 and the associated solar wind structures, *J. Geophys. Res.* **98**, 13295

87. Carpenter, D.L. (1966): Whistler studies of the plasmapause in the magneto-sphere, *J. Geophys. Res.* **71**, 693

88. Celsius, A. (1847): Bemerkungen über der Magnetnadel stündliche Veränder-ungen in ihrer Abweichung, *Svenska Ventensk. Handl.* **8**, 296

89. Chao, J.K. and Y.H. Chen (1985): On the distribution of θ_{Bn} for shocks in the solar wind, *J. Geophys. Res.* **90**, 149

90. Chao, J.K. and B. Goldstein (1972): Observations of slow shocks in interplan-etary space, *J. Geophys. Res.* **75**, 6394

91. Chapman, S. (1919): An outline of a theory of magnetic storms, *Proc. R. Soc. Ser. A* **95**, 61

92. Chapman, S. (1957): Note on the solar corona and the terrestrial ionosphere, *Smithsonian Centr. Astrophys.* **2**, 1

93. Chapman, S. (1959): Interplanetary space and the earth's outermost atmo-sphere, *Proc. R. Soc. Ser. A* **253**, 462

94. Chapman, S. and J. Bartels (1940): *Geomagnetism*, Oxford University Press, Oxford

95. Chapman, S. and V.c.A. Ferraro (1931): A new theory of magnetic storms, *Terr. Magn.* **36**, 77

96. Chappell, R., K.K. Harris, and G.W. Sharp (1970): A study of the influence of magnetic activity on the location of the plasmapause as measured by OGO 5, *J. Geophys. Res.* **75**, 50

97. Chen, F.F. (1984): *Plasma physics and controlled fusion*, vol. 1, Plenum Press, New York

98. Chenette, D.L., T.F. Conlon, K.R. Pyle, and J.A. Simpson (1977): Observa-tions of jovian electrons at 1 AU throughout the 13 month jovian synodic year, *Astrophys. J.* **215**, L95

99. Chipperfield, M., 1996: The TOMCAT offline chemical transport model, Uni-versity of Leeds, UK

100. Chitre, S.M. (2003): Overview of solar physics, in [9], p. 3

101. Choudhuri, A.R. (1998): *The physics of fluids and plasmas*, Cambridge Uni-versity Press, Cambridge

102. Choudhuri, A.R., M. Schüssler, and M. Dikpati (1995): The solar dynamo with meridional circulation, *Astron. Astrophys.* **303**, L29

103. Cleveland, B.T., T. Daily, R. Davis Jr., J.R. Distel, K. Lande, C.K. Lee, P.S. Wildenhain, and J. Ullman (1998): Measurement of the solar electron neutrino flux with the Homestake chlorine detector, *Astrophys. J.* **496**, 505

104. Cliver, E.W., D.J. Forrest, H.V. Cane, D.V. Reames, R.E. McGuire, T.T. von Rosenvinge, S.R. Kane, and R.J. MacDowell (1989): Solar flare nuclear gamma rays and interplanetary proton events, *Astrophys. J.* **343**, 953

105. Cohen, C.M.S., A.C. Cummings, R.A. Leske, R.A. Mewaldt, E.C. Stone, B.L. Dougherty, M.E. Wiedenbecke, E.R. Christian, and T.T. von Rosenvinge (1999): Inferred charge states of high energy solar particles from the Solar Isotope Spectrometer on ACE, *Geophys. Res. Lett.* **26**, 149

448   References

106. Conlon, T.F. (1978): The interplanetary modulation and transport of jovian electrons, *J. Geophys. Res.* **83**, 541

107. Cornwall, J.M. (1986): Magnetospheric ion acceleration processes, in *Ion acceleration in the magnetosphere and ionosphere* (ed. T. Chang), Geophys. Monogr. 38, American Geophys. Union, Washington, DC

108. Courant, R. and K.O. Friedrichs (1991): *Supersonic flow and shock waves*, Springer, Berlin, reprint of the second Interscience edition, 1948

109. Covas, E., R. Tavakol, D. Moss, and A. Tworkowski (2000): Torsional oscillations in the solar convection zone, *Astron. Astrophys.* **360**, L21

110. Cowley, S.W.H. (1985): *Magnetic reconnection*, in [421], p. 121

111. Cowling, T.G. (1957): *Magnetohydrodynamics*, Interscience, New York

112. Cox, A., W.C. Livingston, and M.S. Matthews (eds.) (1991): *Solar interior and atmosphere*, University of Arizona Press, Tucson, AZ

113. Cravens, T.E. (1997): *Physics of solar system plasmas*, Cambridge University Press, Cambridge

114. Crooker, N.U. (1977): The magnetospheric boundary layers: a geometrically explicit model, *J. Geophys. Res.* **82**, 3629

115. Crooker, N.U., and G.L. Siscoe (1986): The effect of the solar wind on the terrestrial environment, in *Physics of the Sun*, vol. III (eds. P.A. Sturrock, T.E. Holzer, D.M. Mihalas, and R.K. Ulrich), Reidel, Dordrecht

116. Crooker, N., J. A. Joselyn, and J. Feynman (eds.) (1997): *Coronal Mass Ejections*, Geophysical Monograph 99, Amer. Geophys. Union, AGU, Washington, DC

117. Crutzen, P.J., I.S.A. Isaksen, and G.C. Reid (1975): Solar proton events: stratospheric sources of nitric oxide, *Science* **189**, 457.

118. Cubasch, U., and R. Voss (2000): The influence of total solar irradiance on climate, in [172], p. 185, also *Space Sci. Rev.* **94**, 185

119. Daiborg, E.I., V.G. Kurt, Yu.I. Logachev, V.G. Stolpovskii, V.F. Melnikov, and I.S. Podstrigach (1987): Solar cosmic ray events with low and high e/p ratios: comparison with X-ray and radio emission data, *Proc. 20th Int. Cosmic Ray Conf.* **3**, 45–48

120. Davis, R. (1964): Solar neutrinos. 2. Experimental, *Phys. Rev. Lett.* **12**, 302

121. Decker, R.B. (1981): The modulation of low-energy proton distributions by propagating interplanetary shock waves: a numerical simulation, *J. Geophys. Res.* **86**, 4537

122. Decker, R.B. (1983): Formation of shock spike events at quasi-perpendicular shocks, *J. Geophys. Res.* **88**, 9959

123. Decker, R.B. (1988): Computer modelling of test particle acceleration at oblique shocks, *Space Sci. Rev.* **48**, 195

124. Decker, R.B. and L. Vlahos (1985): Shock drift acceleration in the presence of waves, *J. Geophys. Res.* **90**, 47

125. Decker, R.B. and L. Vlahos (1986): Modeling of ion acceleration through drift and diffusion at interplanetary shocks, *J. Geophys. Res.* **91**, 13349

126. Decker, R.B. and L. Vlahos (1986): Numerical studies of particle acceleration at turbulent oblique shocks with an application to prompt ion acceleration during solar flares, *Astrophys. J.* **306**, 710

127. de Hoffman, F. and E. Teller (1950): Magnetohydrodynamic shocks, *Space Sci. Rev.* **80**, 692

128. Dendy, R.O. (1990): *Plasma dynamics*, Oxford Science Publications, Oxford

129. Denskat, K.U., H.J. Beinroth, and F.M. Neubauer (1983): Interplanetary magnetic field power spectra with frequencies from $2.4 \cdot 10^{-5}$Hz to 470 Hz from Helios-observations during solar minimum conditions, *J. Geophys. Res.* **87**, 2215

130. Desai M.I., G.M. Mason, J.R. Dwyer, J.E. Mazur, C.W. Smith, and R.M. Skoug, (2001): Acceleration of He-3 nuclei at interplanetary shocks, *Astrophys. J.* **553**, L89

131. Dessler, J.A. (eds.) (1983): *Physics of the Jovian magnetosphere*, Cambridge University Press, Cambridge

132. Dessler, A.J., and J.A. Fejer (1963): Interpretation of K_p index and M-region geomagnetic storms, *Planet. Space Sci.* **11**, 505

133. Diehl, R., E. Parizot, R. Kallenbach, and R. von Steiger (eds.) (2002): *The astrophysics of galactic cosmic rays*, Kluwer, Dordrecht

134. Dorman, L.I. and G.I. Freidman (1959): *Problems of magnetohydrodynamics and plasma dynamics*, Zinätne, Riga

135. Drake, R.P. (2000): The design of laboratory experiments to produce collisionless shocks of cosmic relevance, *Phys. Plasmas* **7**, 4690

136. Droege, W., and 11 others (1986): Effects of corotating interaction regions and Ulysses high energy particles, *Solar Wind Eight*, (eds. D. Winterhalter et al.), p. 515, AIP Press, Woodbury, NY, p. 515

137. Drury, L.O'C. (1983): An introduction to the theory of diffusive shock acceleration of energetic particles in tenuous plasmas, *Rep. Prog. Phys.* **46**, 973

138. Dryer, M., S.T. Wu, G. Gislason, S.F. Han, Z.K. Smith, D.F. Smart, and M.A. Shea (1984): Magnetohydrodynamic modeling of interplanetary disturbances between sun and earth, *Astrophys. Space Sci* **195**, 187

139. Dungey, J.W. (1961): Interplanetary magnetic field and the auroral zones, *Phys. Rev. Lett.* **6**, 47

140. Dungey, J.W. (1967): The theory of the quiet magnetosphere, in *Solar-terrestrial physics* (ed. J.W. King and W.S. Newman), Academic Press, London, p. 91

141. Dwivedi, B.N. (2003): The solar corona, in [9], p. 281

142. Eather, R.H. (1983): *Majestic Lights*, American Geophysical Union AGU, Washington

143. Eddington, A.S. (1926): *The internal constitution of the stars*, Cambridge University Press, Cambridge

144. Eddy, J.A. (1976): The Sun since the Bronze Age, in *Physics of solar planetary environments, vol. II* (ed. D.J. Williams), p. 958, American Geophysical Union, Washington, DC

145. Eddy, J.A. (1976): The Maunder-Minimum, *Science* **192**, 1189

146. Encrenaz, T. and J.J. Bibring, with M. Blanc (1990): *The solar system*, Springer, Berlin

147. Eguchi, K., S. Enomoto, K. Furuno et al. (2003): First results from KamLAND: Evidence for reactor antineutrino disappearance, *Phys. Rev. Lett.* **90**, 021802

148. Evenson, P., D. Huber, E, Tuska Patterson, J. Esposito, D. Clements, and J. Clem (1995): Cosmic electron spectra 1987–1994, *J. Geophys. Res.* **100**, 7873

149. Faber, T.E. (1995): *Fluid dynamics for physicists*, Cambridge University Press, Cambridge

150. Fermi, E. (1949): On the origin of cosmic radiation, *Phys. Rev.* **5**, 1169

450   References

151. Fermi, E. (1954): Galactic magnetic fields and the origin of cosmic radiation, *Astrophys. J.* **119**, 1

152. Feynman, J. and A.J. Hundhausen (1994): Coronal mass ejections and major solar flares: the great active center of March 1989, *J. Geophys. Res.* **99**, 8451

153. Fichtner, H. (2000): Anomalous cosmic rays, in [459], p. 191

154. Fichtner, H. (2001): Anomalous cosmic rays: messengers from the outer heliosphere, *Space Sci. Rev.* **95**, 639

155. Fischer-Bruns, I., U. Cubasch, H. von Storch, E. Zorita, J. Fidel Gonzales-Rouco, and J. Luterbacher (2002): Modelling the Late Maunder Minimum with a 3-dimensional OAGCM, *Exchanges* **25**, Clivar – Selected Research Papers http://w3g.gkss.de/G/Mitarbeiter/storch/pdf/clivar.pdf

156. Fisk, L.A. (1979): The interaction of energetic charged particles with the solar wind, in *Solar system plasma physics, vol. I* (eds. C.F. Kennel, L.J. Lanzerotti, and E.N. Parker), North-Holland, Amsterdam, p. 177

157. Fisk, L.A. (1986): The anomalous component, its variation with latitude and related aspects of modulation, in *The sun and the heliosphere in three dimensions* (ed. R.G. Marsden), Reidel, Dordrecht, p. 401

158. Fisk, L.A., and W.I. Axford (1969): Anisotropies of solar cosmic rays, *Solar Phys.* **7**, 486

159. Fisk, L.A. and M.A. Lee (1980): Shock acceleration of energetic particles in corotating interaction regions in the solar wind, *Astrophys. J.* **237**, 260

160. Fisk, L.A., J.R. Jokipii, G.M. Simnett, R. von Steiger, and K.-P. Wenzel (eds., 1998): *Cosmic Rays in the Heliosphere*, Kluwer

161. Fleck, B., V. Domingo, and A.I. Poland (eds.) (1995): *The SOHO mission*, Kluwer, Dordrecht

162. Flückiger, E. (1985): Forbush decreases, geomagnetic and atmospheric effects, cosmogenic nuclides, *Proc. 19th Int. Cosmic Ray Conf.* **9**, 301

163. Flückiger, E. (1991): Solar and terrestrial modulation, *Proc. 22nd Int. Cosmic Ray Conf.* **5**, 273

164. Forbush, S.E. (1946): Three unusual cosmic ray increases possibly due to charged particles from the Sun, *Phys. Rev.* **70**, 771

165. Forman, M.A. and G.M. Webb (1985): Acceleration of energetic particles, in [506], p. 91

166. Forman, M.A., J.R. Jokipii, and A.J. Owens (1974): Cosmic ray streaming perpendicular to the mean magnetic field, *Astrophys. J.* **192**, 535

167. Forman, M.A., R. Ramaty, and E.G. Zweibel (1986): The acceleration and propagation of solar energetic flare particles, in *Physics of the Sun, vol. II* (ed. P.A. Sturrock), Reidel, Dordrecht

168. Frank, L.A., et al. (1964): A study of charged particles in the Earth's outer radiation belts with Explorer 14, *J. Geophys. Res.* **69**, 2171

169. Franklin, J. (1975): *A narrative of a journey to the shores of the polar sea in the years 1819, 1820, 1821 and 1822*, Hurtig, Edmonton

170. Fraser-Smith, A.C. (1987): Centered and eccentric magnetic dipoles and their poles, 1600–1985, *Rev. Geophys.* **25**, 1

171. Friis-Christensen, E., and K. Lassen (1991): Length of the solar cycle: an indicator of solar activity closely associated with climate, *Science* **254**, 698

172. Friis-Christensen, E., C. Fröhlich, J.D. Haigh, M. Schüssler, and R. von Steiger (eds.) (2000): *Solar variability and climate*, Kluwer, Dordrecht

173. Fritz, H. (1881): *Das Polarlicht*, Brockhaus, Leipzig

174. Fujii, Z., and F.B. McDonald (1997): Radial intensity gradients of galactic cosmic rays (1972–1995) in the heliosphere, *J. Geophys. Res.* **102**, 24 201

175. Fukuda, Y., T. Hayakawa, K. Inoue et al. (1996): Solar neutrino data covering solar cycle 22, *Phys. Rev. Lett.* **77**, 1683

176. Fukuda, Y., T. Hayakawa, E. Ichihara et al. (1999): Constraints on neutrino oscillation parameters from the measurement of day–night solar neutrino fluxes at Super-Kamiokande, *Phys. Rev. Lett.* **82**, 1810

177. Fuselier, S.A., S.M. Petrinec, K.J. Trattner, and W.K. Peterseon (2001): Origins of energetic ions in the cusp, *J. Geophys. Res.* **106**, 5967

178. Gadsden, M., and W. Schröder (1989): *Noctilucent clouds*, Springer, Berlin

179. Gaizouskas, V. (1983): The relation of solar flares to the evolution and proper motion of magnetic fields, *Adv. Space Res.* **2**, (11)11

180. Galileo Instruments (1992): *Space Sci. Rev.* **60**, 79

181. Garcia-Munoz, M., et al. (1991): The dependence of solar modulation on the sign of cosmic ray particle charge during the 22 year solar magnetic cycle, *Proc. 22nd Int. Cosmic Ray Conf.* **3**, 497

182. Gauß, C.F. (1866-1933): *Werke*, ed. Kgl. Gesell. d. Wissenschaften zu Göttingen, 8 Bde, Leipzig

183. Gazis, P.R. (1996): Solar cycle variation in the heliosphere, *Rev. Geophys.* **34**, 379

184. Geiss, J., G. Gloeckler, R. von Steiger, H. Balsiger, L.A. Fisk, A.B. Galvin, F.M. Ipavich, S. Livi, J.F. McKenzie, K.W. Ogilvie, and B. Wilken (1995): The southern high-speed stream: results from the SWICS instrument on Ulysses, *Science* **268**, 1033

185. Gilbert, W. (1600): *De magnete, magneticisque corporibus, et de magno magnete tellure; Physiologia nova, plurimis et argumentis, et experimentis demonstrata*, London (transl. by Fleury Mottelay, Dover, New York, 1958)

186. Glaßmeier, K.-H. (1991): ULF pulsations, in *Handbook of atmospheric electrodynamics II* (ed. H. Volland), 463, CRC Press, Boca Raton

187. Gleeson, L.J., and W.I. Axford (1968): Solar modulation of galactic cosmic rays, *Astrophys. J.* **154**, 1011

188. Gloeckler, G. (1984): Characteristics of solar and heliospheric ion populations observed near Earth, *Adv. Space Res.* **4** (2-3), 127

189. Gloeckler, G., J. Geiss, E.C. Roelof, L.A. Fisk, F.M. Ipavich, K.W. Ogilvie, L.J. Lanzerotti, R. Von Steiger, and B. Wilken (1994): Acceleration of interstellar pickup ions in the disturbed solar wind observed with Ulysses, *J. Geophys. Res.* **99**, 17637

190. Goertz, C.K. (1979): Double layers and electrostatic shocks in space, *Rev. Geophys. Space Phys.* **17**, 418

191. Goertz, C.K. and R.J. Strangeway (1995): Plasma waves, in [290],

192. Goldston, R.J., and P.H. Rutherford (1995): *Plasma physics*, IOP, Bristol

193. Golub, L., and J.M. Pasachoff (1997): *The solar corona*, Cambridge University Press, Cambridge

194. Golub, L., J. Bookbinder, E. DeLuca, M. Karovska, H. Warren, C.D. Schrijver, R. Shine, T. Tarbell, A. Title, J. Wolfson, B. Handy, and C. Kankelborg (1999): A new view of the solar corona from the transition region and coronal explorer (TRACE), *Phys. Plasmas* **6**, 2205

195. Golub, L., A.S. Krieger, J.K. Silk, A.F. Timothy, and G.S. Vaiana (1974): Solar X-ray bright points, *Astrophys. J. Lett.* **189**, L93

196. Gonzalez, W.D. and B.T. Tsurutani (1992): Terrestrial response to eruptive solar flares: geomagnetic storms, in *Eruptive solar flares* (eds. Z. Svestka, B.V. Jackson, and M.E. Machado), Springer, Berlin, p. 277.

197. Gopalswamy, N., S. Yashiro, M.L. Kaiser, R.A. Howard and J.L. Bougeret (2001): Radio signatures of coronal mass ejection interaction: Coronal mass ejection cannibalism?, *Astrophys. J.* **548**, L91

198. Gopalswamy, N., S. Yashiro, G. Michalek, M.L. Kaiser, R.A. Howard, D.V. Reames, R. Leske, and T.T. von Rosenvinge (2002): Interacting coronal mass ejections and solar energetic particles, *Astrophys. J.* **572**, L103

199. Gosling, J.T. (1992): In situ observations of coronal mass ejections in interplanetary space, in [511], p. 258

200. Gosling, J. (1993): The solar flare myth, *J. Geophys. Res.* **98**, 18 937

201. Gosling, J., J.R. Asbridge, S.J. Bame, W.C. Feldman, R.D. Zwickl, G. Paschmann, N. Sckopke, and R.J. Hynds (1981): Interplanetary ions during an energetic storm particle event – The distribution function from solar–wind thermal energies to 1.6 MeV, *J. Geophys. Res.* **86**, 547

202. Gosling, J., (1993): Three-dimensional reconnection, *J. Geophys. Res.* **99**, 5334

203. Gough, D.O., A.G. Kosovichev, J. Toomre et al. (1996): The seismic structure of the sun, *Science* **272**, 1296

204. Gringauz, K.I., V.V. Bezrukikh, V.D. Ozerov, and R.E. Rybchinskii (1960): A study of the interplanetary gas, high-energy electrons, and corpuscular radiation from the sun by means of a three-electrode trap for charged particles on the Soviet cosmic rocket, *Sov. Phys. Dokl.*, Engl. transl., **5**, 361

205. Gubbins, D. (1994): Geomagnetic polarity reversals: a connection with secular variation and core–mantle interaction, *Rev. Geophys.* **32**, 61

206. Gurevich, A.V., G.M. Milikh, and R.A. Roussel-Dupre (1992): Runaway electron mechanism for air breakdown and preconditioning during a thunderstorm, *Phys. Lett. A* **165**, 463

207. Gurevich, A.V., G.M. Milikh, and R.A. Roussel-Dupre (1994): Nonuniform runaway air breakdown, *Phys. Lett. A* **167**, 197

208. Habbal, E.R., R. Esser, J.V. Hollweg, and P.A. Isenberg (1999): *Solar Wind 9*, Amer. Inst. of Physics, Conf. Proc. 47, Woodbury, NY

209. Harendel, G. (2001): Auroral acceleration in astrophysical plasmas, *Phys. Plasma* **8**, 2365

210. Haerendel, G., and G. Paschmann (1982): Interaction of the solar wind with the dayside magnetosphere, in *Magnetospheric plasma physics* (eds. A. Nishida), Reidel, Dordrecht, p. 117

211. Hampel, W., J. Handt, G. Heusser et al. (1999): GALLEX solar neutrino observations: results for GALLEX IV, *Phys. Lett. B* **447**, 127

212. Hargreaves, J.K. (1992): *The solar–terrestrial environment*, Cambridge Atmospheric and Space Science Series 5, Cambridge University Press, Cambridge

213. Harrison, R.G., and K.P. Sine (1999): *A review of recent studies of the influence of solar changes on Earth's climate*, Tech. Rep. HCTN6, Hadley Centre for Climate Prediction and Research, UK Meteorological Office

214. Harvey, K.L. (1985): The relationship between coronal bright points as seen in the He-I-lambda-10830 and the evolution of the photospheric network magnetic field, *Aust. J. Phys.* **38**, 875

215. Hasselmann, K. and G. Wibberenz (1968): Scattering of charged particles by random electromagnetic fields, *Z. Geophys.* **34**, 353

216. Hasselmann, K., and G. Wibberenz (1970): A note on the parallel diffusion coefficient, *Astrophys. J.* **162**, 1049
217. Hathaway, D.H., P. Gilman, J.W. Harvey et al. (1996): GONG observations of solar surface flows, *Science* **272**, 1306
218. Haxton, W.C. (1995): The solar neutrino problem, *Annu. Rev. Astron. Astrophys.* **33**, 459
219. Heath, D.F., A.J. Krueger, and P.J. Crutzen (1977): Solar proton event: influence in stratospheric ozone, *Science* **197**, 886
220. Heber, B., and 9 others (1996): Spatial variation of > 40 MeV/n nuclei fluxes observed during the Ulysses rapid latitude scan, *Astron. Astrophys.* **316**, 538
221. Heber, B., et al. (1998): Latitudinal distribution of > 106 MeV protons and its relation to the ambient solar wind in the inner southern and northern heliosphere: Ulysses cosmic and solar particle investigatopn Kiel Electron Telescope results, *J. Geophys. Res.* **103**, 4809
222. Heras, A.M., B. Sanahuja, Z.K. Smith, T. Detman, and M. Dryer (1992): The influence of the large-scale interplanetary shock structure on a low-energy particle event, *Astrophys. J.* **391**, 351
223. Hiorter, O.P. (1847): Von der Magnetnadel verschiedener Bewegungen, *Svenska Vetensk. Handl.* **8**, 27
224. Hoeksema, J.T. (1991): Large-scale structure of the heliospheric magnetic field, *Adv. Space Res.* **11**, 15
225. Hoeksema, J.T., and P.H. Scherrer (1986): *The solar magnetic field – 1976 through 1985*, Report UAG-94, World Data Center A (NOAA), Boulder, CO
226. Hoffmeister, C. (1943): Physikalische Untersuchungen auf Kometen I, Die Beziehungen des primären Schweifstrahl zum Radiusvektor, *Z. Astrophys.* **22**, 265
227. Hoffmeister, C. (1944): Physikalische Untersuchungen auf Kometen II, Die Bewegung der Schweifmaterie und die Repulsivkraft der Sonne beim Kometen, *Z. Astrophys.* **23**, 1
228. Hones, E.W. (ed.) (1984): *Magnetic reconnection*, Geophys. Monogr. 30, American Geophys. Union, Washington, DC
229. Houghton, J., 1997: *Global warming*, Cambridge University Press, Cambridge
230. Houghton, J.T., G.J. Jenkins, and J.J. Ephraums (eds.) (1992a): *Climate change – the IPCC scientific assessment*, Cambridge University Press, Cambridge
231. Houghton, J.T., B.A. Callander and S.K. Varney (eds.) (1992b): *Climate change 1992 – the supplementary report to the IPCC scientific assessment*, Cambridge University Press, Cambridge
232. Houghton, J.T., L.G. Meira Filho, B.A. Callender, N. Harris, A. Kattenberg, and K. Maskell (eds.) (1996): *Climate change 1995: the science of climate change*, Cambridge University Press, Cambridge
233. Houghton, J.T., Y. Ding, D.J. Griegs, M. Noguer, P.J. van der Linden, X. Dai, K. Maskell, and C.A. Johnson (2001): *Climate change 2001: the scientific basis*, Intergovernmental Panel on Climate Change, available at www.ipcc.ch
234. Horwitz, J.L., D.L. Gallagher, and W.K. Peterson (eds.) (1998): *Geospace Mass and Energy Flow: Results From the International Solar-Terrestrial Physics Program*, Geophysical Monograph Series, volume 104, Amer. Geophys. Union, AGU, Washington, DC
235. Howard, R., and B.J. LaBonte (1980): The Sun is observed to be a torsional oscillator with a period of 11 years, *Astrophys. J.* **239**, L33

236. Howard, R.A., N.R. Sheeley, Jr., M.J. Koomen, and D.J. Michels (1985): Coronal mass ejections 1979–1981, *J. Geophys. Res.* **90**, 8173
237. Howe, R., R. Komm, and F. Hill (1999): Solar cycle changes in GONG p-mode frequencies, 1995–1998, *Astrophys. J.* **524**, 1084
238. Hoyt, D.V., and K.H. Schatten (1997): *The role of the sun in climate change*, Oxford University Press, Oxford
239. Hsieh, K.C., and J.A. Simpson (1970): *Astrophys. J.* **162**, L191
240. Hughes, W.J., (1995): The magnetopause, magnetotail, and magnetic reconnection, in [290], p. 227
241. Hultqvist, B., and M. Øieroset (eds.) (1997): *Transport across the boundaries of the magnetosphere*, Kluwer, Dordrecht
242. Hultqvist, B., M. Øieroset, G. Paschmann, and R. Treumann (eds.) (1999): *Magnetospheric plasma sources and losses*, Kluwer, Dordrecht
243. Humboldt, A. (1808): Die vollständigste aller bisherigen Beobachtungen über den Einfluß des Nordlichts auf die Magnetnadel, *Gilb. Ann.* **29**
244. Hundhausen, A.J. (1972): *Coronal expansion and the solar wind*, Springer, Berlin
245. Hundhausen, A.J. (1985): Some macroscopic properties of shock waves in the heliosphere, in [506]
246. Hundhausen, A.J. (1988): The origin and propagation of coronal mass ejections, in *Proc. 6th Int. Solar Wind Conf., vol. I* (eds. V.J. Pizzo, T.E. Holzer, and D.G. Sime), NCAR, Boulder, CO, p. 181
247. Hundhausen, A.J., and J.T. Gosling (1976): Solar wind structure at large heliocentric distances: an interpretation of Pioneer 10 observations, *J. Geophys. Res.* **81**, 1845
248. Hundhausen, A.J., T.E. Holzer, and B.C. Low, 1987: Do slow shocks precede some coronal mass ejections?, *J. Geophys. Res.* **92**, 11173
249. Ijiima, T. and T.A. Potemra (1976): Field-aligned currents in the dayside cusps observed by Triad, *J. Geophys. Res.* **81**, 5971
250. International Association of Geomagnetism and Aeronomy (1985): International geomagnetic reference field revision 1985, *EOS Trans. AGU* **67**, 523
251. Jackman, C.H., E.L. Fleming, and F.M. Vitt (2000): Influence of extremely large solar proton events in a changing stratosphere, *J. Geophys. Res.* **105**, 11659
252. Jackman, C.H., R.D. McPeters, G.J. Labow, C.J. Praderas, and E.L. Fleming (2001): Measurements and model predictions of the atmospheric effects due to the July 2000 solar proton event, *Geophys. Res. Lett.* **28**, 2883
253. Jacobs (1984): *Reversals of the earth's magnetic field*, Hilger, Bristol
254. Jaekel, U., W. Wanner, R. Schlickeiser, and G. Wibberenz (1994): Magnetic field fluctuation geometry as a possible solution of the proton mean free path discrepancy problem, *Astron. Astrophys.*, 291 (3): L35-L38
255. Johnson, C.Y. (1969): Ion and neutral composition of the ionosphere, *Ann. IQSY* **5**, MIT Press, Cambridge, MA
256. Jokipii, J.R. (1966): Cosmic ray propagation. I: Charged particles in random electromagnetic fields, *Astrophys. J.* **146**, 480
257. Jokipii, J.R. (1990): The anomalous component of cosmic rays, in *Physics of the outer heliosphere* (eds. E. Grzedzielsky and E. Page), COSPAR Colloq. Series vol. 1, Pergamon Press, Oxford, p. 169

258. Jokipii, J.R., H. levy, and W.B. Hubbard (1977): Effects of particle drifts on cosmic ray transport, I, general properties, application to solar modulation, *Astrophys. J.* **213**, 861

259. Jones, F.C. and D.C. Ellison (1991): The plasma physics of shock acceleration, *Space Sci. Rev.* **58**, 259

260. Jordan, C.E. (1994): Empirical models of the magnetospheric magnetic field, *Rev. Geophys.* **32**, 139

261. Joselyn, J.A. (1992): The impact of solar flares and magnetic storms on humans, *EOS Trans. American Geophysical Union*, **73/7**, 81

262. Kadomtsev, B.B. (1992): *Tokamak plasma: a complex physical system*, IOP, Bristol

263. Kahler, S.W. and A.J. Hundhausen (1992): The magnetic topology of solar coronal structures following mass ejections, *J. Geophys. Res.* **97**, 1619

264. Kai, K. (1987): Microwave source of solar flares, *Solar Phys.* **111**, 81

265. Kallenrode, M.-B. (1993): Particle propagation in the inner heliosphere, *J. Geophys. Res.* **98**, 19037

266. Kallenrode, M.-B. (1995): Particle acceleration at interplanetary shocks – observations at a few tens of keV vs. some tens of MeV, *Adv. Space Res.* **15**, (8/9)375

267. Kallenrode, M.-B. (1996): A statistical survey of 5 MeV proton events at transient interplanetary shocks, *J. Geophys. Res.* **101**, 24393

268. Kallenrode, M.-B. (2000): Galactic cosmic rays, in [459], p. 165

269. Kallenrode, M.-B (2001): Shock as a black box II: Eeffects of adiabatic deceleration and convection included, *J. Geophys. Res.* **106**, 24 989

270. Kallenrode, M.-B. (2001): Charged particles, neutrals, and neutrons, in *Solar encounter: the first solar orbiter workshop*, ESA SP-493, Nordwijk

271. Kallenrode, M.-B. (2003): Current views on impulsive and gradual solar energetic particle events, *J. Phys. G* **29**, 965

272. Kallenrode, M.-B., E.W. Cliver, and G. Wibberenz (1992): Composition and azimuthal spread of solar energetic particles from impulsive and gradual flares, *Astrophys. J.* **391**, 370

273. Kallenrode, M.-B., and E.W. Cliver (2001): Rogue SEP events: observational aspects, *Proc. 27th Int. Cosmic Ray Conf.*, 3314

274. Kallenrode, M.-B., and E.W. Cliver (2001): Rogue SEP events: modeling, *Proc. 27th Int. Cosmic Ray Conf.*, 3318

275. Kallenrode, M.-B. and G. Wibberenz (1997): Propagation of particles injected from interplanetary shocks: a black-box model and its consequences for acceleration theory and data interpretation, *J. Geophys. Res.* **102**,

276. Kallenrode, M.-B., G. Wibberenz, and S. Hucke (1992): Propagation conditions for relativistic electrons in the inner heliosphere, *Astrophys. J.* **394**, 351

277. Kan, J.R. (1991): Synthesizing a global model of substorms, Geophys. Monogr. 64, American Geophysical Union, Washington DC, p. 73

278. Kane, S.R. (1974): Impulsive (flash) phase of solar flares: hard X-ray, microwave, EUV, and optical observations, in *Coronal disturbances* (ed. G. Newkirk), Proc. 57th IAU Symp., Reidel, Dordrecht, p. 105

279. Kane, S.R., C.J. Crannell, D. Datlowe, U. Feldman, A. Gabrie, H.S. Hudson, M.R. Kundu, C. Mähler, D. Neidig, V. Petrosian, and N.R. Sheeley, Jr. (1980): Impulsive phase of solar flares, in *Solar flares* (ed. P. Sturrock), p. 187

280. Karpen, J.T. (1999): Formation of the slow solar wind in streamers, in [208], p. 47–52

281. Kaufmann, M. (2003): *Plasmaphysik und Fusionsforschung*, Teubner, Braunschweig

282. Kayser, S.E., J.L. Bougeret, J. Fainberg, and R.G. Stone (1987): Comparison of interplanetary type-III-storm footpoints with solar features, *Sol. Phys.* **109**, 107

283. Kennel, C.F., F.V. Coroniti, F.L. Scarf, W.A. Livesay, C.T. Russell, E.J. Smith, K.-P. Wenzel, and M. Scholer (1986): A test of Lee's quasi-linear theory of ion acceleration by traveling interplanetary shocks, *J. Geophys. Res.* **91**, 11917

284. Kertz, W. (1971): *Einführung in die Geophysik*, BI, Mannheim

285. Kippenhahn, R. and C. Möllenhoff (1975): *Elementare Plasmaphysik*, BI-Wissenschaftsverlag, Mannheim

286. Kippenhahn, R. and A. Schlüter (1957): *Z. Astrophys.* **43**, 36

287. Kirkwood S., V. Barabash, B.U.E. Brandstrom, A. Mostrom, K. Stebel, N. Mitchell, and W. Hocking (2002): Noctilucent clouds, PMSE and 5-day planetary waves: a case study, *Geophys. Res. Lett.* **29**, (10)art1411

288. Kivelson, M.G. (1995): Physics of space plasmas, in [290], p. 27

289. Kivelson, M.G. (1995): Pulsations and magnetohydrodynamic waves, in [290], p. 330

290. Kivelson, M.G. and C.T. Russell (eds.) (1995): *Introduction to space physics*, Cambridge University Press, Cambridge

291. Klein, L. (ed.) (2003): *Energy conversion and particle acceleration in the solar corona*, Lecture Notes in Physics 612, Springer, Berlin, Heidelberg

292. Klein, L., K. Anderson, M. Pick, G. Trottet, N. Vilmer, and S. Kane (1983): Association between gradual X-ray emission and metric continua during large solar flares, *Solar Phys.* **84**, 195

293. Koshio, Y., 1999: Solar neutrino results and oscillation analysis from Super-Kamiokande, *Proc. 26th Int. Cosmic Ray Conf.* **2**, 217

294. Kosovichev, A.G., J. Schou, P.H. Scherrer et al. (1997): Structure and rotation of the solar interior: initial results from the MDI medium-l program, *Solar Phys.* **170**, 43

295. Kosovichev, A.G., and V.V. Zharkova (1998): X-ray flare sparks quake inside Sun, *Nature* **393**, 317

296. Kosugi, T., B.R. Dennis, and K. Kai (1988): Energetic electrons in impulsive and extended solar flares as deduced from flux correlations between hard X-rays and microwaves, *Astrophys. J.* **324**, 1118

297. Koutchmy, S., J.B. Zirker, R.S. Steinolfson, and J.D. Zhugzda (1991): Coronal activity, in [112], p. 1044

298. Krall, N.A. and A.W. Trivelpiece (1986): *Principles in plasma physics*, San Francisco Press, San Francisco, CA.

299. Krause, F. and K.-H. Rädler (1980): *Mean-field magnetohydrodynamics and dynamo theory*, Akademie-Verlag, Berlin

300. Krimigis, S.M., E.C. Roelof, T.P. Armstrong, and J.A. van Allen (1971): Low energy ($\geq$ 30 MeV) solar particle observations at widely separated points ($>$ 0.1 AU) during 1967, *J. Geophys. Res.* **76**, 5921

301. Kristjansson, J.E., and J. Kristansen (2000): Is there a cosmic ray signal in recent variations in global cloudiness and cloud radiative forcing, *J. Geophys. Res.* **105**, 11851

302. Kuiper, G.P., ed. (1953): *The Sun*, University of Chicago Press, Chicago

303. Kuiper, T. (1992): *Theoretische Mechanik*, VCH, Weinheim
304. Kundu, M.R., and L. Vlahos (1982): Solar microwave bursts – a review, *Space Sci. Rev.* **32**, 405
305. Kunow, H., G. Wibberenz, G. Green, R. Müller-Mellin, and M.-B. Kallenrode (1991): in [475], p. 235
306. Kunow, H., and 11 others (1995): High energy cosmic ray results on Ulysses: 2. Effects of a recurrent high-speed stream from the southern polar coronal hole, *Space Sci. Rev.* **72**, 397
307. Landau, L.D. and E.M. Lifschitz (1982): *Mechanics*, Butterworth–Heinemann, Oxford
308. Lang, K. (1995): *Sun, earth and sky*, Springer, Berlin, Heidelberg
309. Lang, K.R. and C.A. Whitney (1991): *Wanderers in space*, Cambridge University Press, Cambridge
310. Lean, J. (1992): Variations in the sun's radiative output, *Rev. Geophys.* **29**, 505
311. Lean, J. (2000): Short term, direct indices of solar variability, in [172], p. 39, also *Space Sci. Rev.* **94**, 39
312. Lee, M.A. (1982): Coupled hydromagnetic wave excitation and ion acceleration upstream of the earth's bow shock, *J. Geophys. Res.* **87**, 5063
313. Lee, M.A. (1983): Coupled hydromagnetic wave excitation and ion acceleration at interplanetary traveling shocks, *J. Geophys. Res.* **88**, 6109
314. Lee, M.A. (1983): The association of energetic particles and shocks in the heliosphere, *Rev. Geophys. Space Phys.* **21**, 324
315. Lee, M.A. (1986): Acceleration of energetic particles at solar wind shocks, in *The sun and the heliosphere in three dimensions* (ed. R.G. Marsden), Reidel, Dordrecht, p. 305
316. Lee, M.A. (1995): Ulysses race to the pole: symposium summary, *Space Sci. Rev.* **72**, 485
317. Leibacher, J.W., and R.F. Stein (1971): New description of solar 5-minute oscillation, *Astrophys. Lett.* **7**, 191
318. Leighton, R.B., R. Noyes, and G.W. Simon (1962): Velocity fields in the solar atmosphere. 1. Preliminary report, *Astrophys. J.* **135**, 474
319. Lemaire, J.F., and K.I. Gringauz (1998): *The Earth's plasmasphere*, Cambridge University Press, Cambridge
320. Lemaire, J.F., D. Heyndrerickx, and D.N. Baker (eds.) (1997): *Radiation belts – models and standards*, Geophys. Monogr. 97, American Geophysical Union, Washington, DC
321. Leske, R.A., R.A. Mewaldt, E.C. Stone, and T.T. von Rosenvinge, (2001): Observations of geomagnetic cutoff variations during solar energetic particle events and implications for the radiation environment at the Space Station, *J. Geophys. Res.* **106**, 30011
322. Levy, E.H., S.P. Duggal, and M.A. Pomeranz (1976): Adiabatic Fermi acceleration of energetic particles between converging interplanetary shock-waves, *J. Geophys. Res.* **81**, 51
323. Lieberman, M.A., and A.J. Lichtenberg (1994): *Principles of plasma discharge and material processing*, Wiley and Sons, New York
324. Lites, B. (2000): Remote sensing of solar magnetic fields, *Rev. Geophys.* **38**, 1
325. Litvinenko, Y.E. (2003): Particle acceleration by magnetic reconnection, in [291], p. 213

458   References

326. Lockwood, G.W., B.A. Skiff, A.L. Baliunas, and R. Radick (1992): Long-term solar brightness changes estimated from a survey of Sun-like stars, *Nature* **360**, 653

327. Lockwood, J.A., and W.R. Webber (1997): A comparison of cosmic ray intensities near the Earth at the sunspot minima in 1976 and 1987 and during 1995 and 1996, *J. Geophys. Res.* **102**, 24 221

328. Loomis, E. (1860): On the geographical distribution of auroras in the northern hemisphere, *Amer. J. Sci. Arts* **30**, 89

329. Luhn A., B. Klecker, D. Hovestadt, and E. Möbius (1987): The mean ionic charge of silicon in He-3-rich solar flares, *Astrophys. J* **317**, 951

330. Lysak, R. (ed.) (1993): *Auroral Plasma Dynamics*, Geophys. Monogr. 80, American Geophysical Union, Washington, DC

331. Mandzhavidze N., R. Ramaty, and B. Kozlovsky (1999): Determination of the abundances of subcoronal He-4 and of solar flare-accelerated He-3 and He-4 from gamma-ray spectroscopy, *Astrophys. J.* **518** 918

332. Marsch, E. (1991): MHD turbulence in the solar wind, in [475], p. 159

333. Marsch, E. (1994): Theoretical models for the solar wind, *Adv. Space Phys.* **14** (4), 103

334. March, N., and H. Svensmark (2000): Cosmic rays, clouds, and climate, in [172], p. 215, also *Space Sci. Rev.* **94**, 215

335. Mason, G.M., R. von Steiger, R.B. Decker, M.I. Desai, J.R. Dwyer, L.A. Fisk, G. Gloeckler, J.T. Gosling, M. Hilchenbach, R. Kallenbach, E. Keppler, B. Klecker, H. Kunow, G. Mann, I.G. Richardson, T.R. Sanderson, G.M. Simnett, Y.M. Wang, R.F. Wimmer-Schweingruber, M. Franz, and J.E. Mazur (1999): Origin, injection, and acceleration of CIR particles: observations – report of Working Group 6, *Space Sci. Rev.* **89**, 327

336. Mason, G., J.E. Mazur, and J.R. Dwyer (1999): He-3 enhancements in large solar energetic particle events, *Astrophys. J.* **525**, L133

337. Matthaeus, W.H., M.L. Goldstein, and D.A. Roberts (1990): Evidence for the presence of quasi-two-dimensional nearly incompressible fluctuations in the solar wind, *J. Geophys. Res.* **95**, 20673

338. Maxwell, J.C. (1873): A treatise on electricity and magnetism, London (current print: Dover Publ., 1991)

339. Mazur, J.E., G.M. Mason, M.D. Looper, R.A. Leske, and R.A. Mewaldt (1999): Charge states of solar energetic particles using the geomagnetic cutoff technique: SAMPEX measurements in the 6 November 1997 solar particle event, *Geophys. Res. Lett.* **26**, 173

340. Mazur, J.E., G.M. Mason, and R.A. Mewaldt (2002): Charge states of energetic particles from corotating interaction regions as constraints on their source *Astrophys. J.* **566**, 555

341. McCarthy, M.P., and G.K. Parks (1992): On the modulation of X-ray fluxes in thunderstorms, *J. Geophys. Res.* **97**, 5857

342. McComas, D.J., P. Riley, J.T. Gosling, A. Balogh, and R. Forsyth: Ulysses' rapid crossing of the polar coronal hole boundary, *J. Geophys. Res.* **103**, 1955

343. McComas, D.J., J.L. Phillips, A.J. Hundhausen, and J.T. Burkepile (1991): Observation of disconnection of pen coronal magnetic structures, *Geophys. Res. Lett.* **18**, 73

344. McCormac, B.M., and T.A. Seliga (eds.) (1979): *Solar–terrestrial Influences on weather and climate*, Reidel, Dordrecht

345. McCracken, K.G. (1962): The cosmic ray flare effect: 3. Deductions regarding the interplanetary magnetic field, *J. Geophys. Res.* **67**, 447

346. McCracken, K.G., U.R. Rao, and R.P. Bukata (1967): Cosmic ray propagation processes, 1, a study of the cosmic ray flare effects, *J. Geophys. Res.* **72**, 4293

347. McCracken, K.G., G.A.M. Dreschhoff, E.J. Zeller, D.F. Smart, and M.A. Shea (2001): Solar cosmic ray events for the period 1561–1994. (1) Identification in polar ice, *J. Geophys. Res.*, 21 585

348. McCracken, K.G., G.A.M. Dreschhoff, D.F. Smart, and M.A. Shea (2001): Solar cosmic ray events for the period 1561–1994. (2) The Gleissberg periodicity, *J. Geophys. Res.*, 21 599

349. McKenzie, J.F., and H.J. Völk (1982): Non-linear theory of cosmic ray shocks including self-generated Alfvén waves, *Astron. Astrophys.* **116**, 191

350. McKibben, R.B. (1987): Galactic cosmic rays and anomalous components in the heliosphere, *Rev. Geophys.* **25** (3), 711

351. McLean, D.J. and N.R. Labrum (eds.) (1985): *Solar radiophysics*, Cambridge University Press, Cambridge

352. McPherron, R.L. (1995): Magnetospheric dynamics, in [290], p. 400

353. Menvielle, M., and A. Berthelier (1991): The k-derived planetary indices: description and availability, *Rev. Geophys.* **29**, 413

354. Merrill, R.T., and M.W. McElhinny (1983): *The earth's magnetic field*, Academic Press, London

355. Merill, R.T., M.W. McElhinny, and P.L. McFadden (1998): *The magnetic field of the Earth*, Academic Press, San Diego

356. Meyer, P., E.N. Parker, and J.A. Simpson (1956): Solar cosmic rays of February 1956 and their propagation through interplanetary space, *Phys. Rev.* **104**, 768

357. Meyer, P., R. Ramaty, and W.R. Webber (1974): Cosmic rays – astronomy with energetic particles, *Phys. Today* **27** (10),

358. Mihalov, J.D. (1987): Heliospheric shocks (excluding planetary bow shocks), *Rev. Geophys.* **25**, 697

359. Miroshnichenko, L.I. (2001): *Solar cosmic rays*, Kluwer, Dordrecht

360. Moebius E., M. Popecki, B. Klecker, L.M. Kistler, A. Bogdanov, A.B. Galvin, D. Heirtzler, D. Hovestadt, E.J. Lund, D. Morris, and W.K.H. Schmidt (1999): Energy dependence of the ionic charge state distribution during the November 1997 solar energetic particle event, *Geophys. Res. Lett.* **26**, 145

361. Moon, Y.-J., G.S. Choe, Y.D. Park, H. Wang, P.T. Gallagher, J. Chae, H.S. Yun, and P.R. Goode (2002): Statistical evidence for sympathetic flares, *Astrophys. J.* **574**, 434

362. Moon, Y.-J., G.S. Choe, H. Wang, and Y.D. Park (2003): Sympathetic coronal mass ejections, *Astrophys. J.* **581**, 117

363. Moore, T.E., and D.C. Delcourt (1995): The geopause, *Rev. Geophys.* **33**, 175

364. Moore, T.E., and J.H. Waite, Jr. (eds.) (1988): *Modeling magnetospheric plasma*, Geophys. Monogr. 44, American Geophysical Union, Washington, DC

365. Montgomery, D.C. (1983): Theory of hydrodynamic turbulence, in *Solar Wind 5* (ed. M. Neugebauer), NASAE-CP-2280, p. 107

366. Montmerle, T. and M. Spiro (eds.) (1986): *Neutrinos and the present day universe*, Commisariat à l'Energie Atomique, Saclay, France

367. Morfill, G., and M. Scholer (1977): Influence of interplanetary shocks on solar energetic particle events, *Astrophys. Space Sci.* **46**, 73

368. Morrison, D.G.O. (1992): Updated review of solar models and solar neutrino experiments, *Proc. Europ. Cosmic Ray Symp.*, Geneva

369. Munro, R.H., J.T. Gosling, E. Hildner, R.M. MacQueen, A.I. Poland, and C.L. Ross (1979): The association of coronal mass ejection transients with other forms of solar activity, *Solar Phys.* **61**, 201

370. Murphy, R.J. (1985): *Gamma-rays and neutrons from solar flares*, PhD Thesis, NASA Goddard Space Flight Center, Greenbelt, Md.

371. Murphy, R.J., and R. Ramaty (1984): Solar flare neutrons and gamma rays, *Adv. Space Res.* **4**, 127

372. Mursula, K., and T. Ulich (1998): A new method to determine the solar cycle length, *Geophys. Res. Lett.* **25**, 1837

373. Nandy, D., and A.R. Choudhuri (2002): Explaining the latitudinal distribution of sunspots with deep meridional flow, *Science* **296**, 1671

374. Neubauer, F.M. (1991): Die Magnetosphären der Planeten, in *Plasmaphysik im Sonnensystem* (eds. K.-H. Glassmeier and M. Scholer), Spektrum, Heidelberg, p. 184

375. Ness, N.F., M.H. Acuna, L.F. Burlaga, J.E.P. Connerney, R.P. Lepping, and F.M Neubauer (1989): Magnetic fields at Neptune, *Science* **246**, 1473

376. Ness, N.F., C.S. Scearce, and J.B. Seek (1964): Initial results of the IMP 1 magnetic field experiment, *J. Geophys. Res.* **69**, 3531

377. Newell, P.T., and T Onsager (eds.) (2003): *Earth's Low-Latitude Boundary Layer*, Geophys. Monogr.133, American Geophysical Union, Washington, DC

378. Nishida, A., D.N. Baker, and S.W.H. Cowley (eds.) (1998): *New Perspectives on the Earth's Magnetotail*, Geophys. Monogr. 105, American Geophysical Union, Washington, DC

379. Nitta, N., and T. Kosugi (1986): Energy of microwave emitting electrons and hard X-ray/microwave source models in solar flares, *Solar Phys.* **107**, 73

380. Ng, C.K., and L.J. Gleeson (1971): The propagation of solar cosmic ray bursts, *Solar Phys.* **20**, 116

381. Ng, C.K., and L.J. Gleeson (1975): Propagation of solar flare cosmic rays along corotating interplanetary flux tubes, *Solar Phys.* **43**, 475

382. Northrop, T.G. (1963): *The adiabatic motion of charged particles*, Interscience, New York

383. Ohki, K., T. Takakura, B. Tsumata, and N. Nitta (1983): General aspects of hard X-ray flares observed by Hinotori: gradual events and impulsive bursts, *Solar Phys.* **86**, 301

384. Ohtani, S.-I., R. Fujii, M. Hesse, and R.L. Lysak (eds.) (2000): *Magnetospheric Current Systems*, Geophys. Monogr. 118, American Geophysical Union, Washington, DC

385. Olson, W.P. (ed.) (1979): *Quantitative modeling of magnetospheric processes*, American Geophysical Union, Washington, DC

386. Orlowski, D.S., C.T. Russel, and R.P. Lepping (1992): Wave phenomena in the upstream region of Saturn, *J. Geophys. Res.* **97**, 19187

387. Pallavicini, R., S. Serio, and G.S. Vaiana (1977): A survey of soft X-ray limb flare images: the relation between their structure and other physical parameters, *Astrophys. J.* **216**, 108

388. Palmer, I.D. (1982): Transport coefficients of low-energy cosmic rays in interplanetary space, *Rev. Geophys. Space Phys.* **20**, 335

389. Praderie, F., and H.S. Hudson (eds.) (1995): Solar and stellar activity, *Advances in Space Research* **6**

390. Parker, E.N. (1955): Hydromagnetic dynamo models, *Astrophys. J.* **122**, 293
391. Parker, E.N. (1957): Newtonian development of the dynamical properties of ionized gases of low density, *Phys. Rev* **107**, 924
392. Parker, E.N. (1957): Sweet's mechanism for merging magnetic fields in conducting fluids, *J. Geophys. Res.* **62**, 509
393. Parker, E.N. (1958): Dynamics of the interplanetary magnetic field, *Astrophys. J.* **128**, 664
394. Parker, E.N. (1958): Cosmic ray modulation by the solar wind, *Phys. Rev.* **110**, 1445
395. Parker, E.N. (1963): *Interplanetary dynamical processes*, Wiley, New York
396. Parker, E.N. (1965): The passage of energetic charged particles through interplanetary space, *Planet. Space Sci.* **13**, 9
397. Parks, G.K. (1991): *Physics of space plasmas – an introduction*, Addison-Wesley, Redwood City, CA
398. Parnell, C.E., and P.E. Jupp (2000): Statistical analysis of the energy distribution of nanoflares in the quiet Sun, *Astrophys. J.* **529**, 554
399. Paschmann, G. (1979): Plasmastructure of the magnetopause and boundary layer, in *Magnetospheric boundary layers* (ed. B. Battrick), ESA-SP-148, p. 25
400. Paschmann, G., S. Haaland, and R. Treumann (eds.) (2003): *Auroral plasma physics*, Kluwer, Dordrecht
401. Paschmann, G., N. Sckopke, I. Papamastorakis, J.R. Asbridge, S.J. Bame, and J.T. Gosling (1981): Characteristics of reflected and diffuse ions upstream of the Earth's bow shock, *J. Geophys. Res.* **86**, 4355
402. Paulikas, G.A. and J.B. Blake (1979): Effects of the solar wind on magnetospheric dynamics: energetic electrons in the synchronous orbit, in [385], p. 180
403. Perrault, P., and S.-I- Akasofu (1978): A study of geomagnetic storms, *Geophys. J. R. Astron. Soc.* **54**, 547
404. Perko, J.S., and L.F. Burlaga (1992): Intensity variations in the interplanetary magnetic field by Voyager 2 and the 11-year solar-cycle modulation of cosmic rays, *J. Geophys. Res.* **97**, 4305
405. Petchek, H.E. (1964): Magnetic field anihilation, in *Symp. on Physics of Solar Flares*, 425, NASA SP-50, Goddard Space Flight Center
406. Petrovay, K. (2000): What makes the Sun tick? The origin of the solar cycle, in *The solar cycle and terrestrial climate*, ESA, Nordwijk, p. 3
407. Pfaff, R.F., J.E. Borovsky, and D.T. Young (eds.) (1998): *Measurement techniques in space plasmas: fields*, Geophys. Monogr. 103, American Geophysical Union, Washington, DC
408. Pfaff, R.F., J.E. Borovsky, and D.T. Young (eds.) (1998): *Measurement techniques in space plasmas: particles*, Geophys. Monogr. 102, American Geophysical Union, Washington, DC
409. Phillips, J.L., S.J. Bame, W.C. Feldman, B.E. Goldstein, J.T. Gosling, C.M. Hammond, D.J. McComas, M. Neugebauer, E.E. Scime, and S.T. Suess (1995): Ulysses solar wind plasma observations at high southerly latitudes, *Science* **268**, 1031
410. Pick, M. and M.E. Machado (eds.) (1993): *Fundamental problems in solar activity*, Adv. Space Res. 13, no. 9, Pergamon Press, Oxford
411. Pilipp, W.G. and G. Morfill (1978): The formation of the plasma sheet resulting from plasma mantle dynamics, *J. Geophys. Res.* **83**, 5670

462   References

412. Pizzo, V.J. (1985): Interplanetary shocks on the large scale: a retrospective on the last decades theoretical efforts, in [524], p. 51
413. Pizzo, V.J. (1991): The evolution of corotating stream fronts near the ecliptic plane in the inner solar system, 2. Three-dimensional tilted dipole fronts, *J. Geophys. Res.* **96**, 5405
414. Polar Instrumentation (1995): *Space Sci. Rev.* **71**
415. Pollack, M. (2003): *Uncertain science ... uncertain world*, Cambridge University Press, Cambridge
416. Potgieter, M. (1993): Modulation of cosmic rays in the heliosphere, *Proc. 23rd Int. Cosmic Ray Conf.*, Invited Paper, p. 213
417. Potgieter, M. (1998): The modulation of galactic cosmic rays in the heliosphere: theory and observations, in [160], 147
418. Potgieter, M.S., J.A. le Roux, L.F. Burlaga, and F.B. McDonald (1993): The role of merged interaction regions and drifts in the heliospheric modulation of cosmic rays beyond 20 AU: a computer simulation, *Astrophys. J.* **403**, 760
419. Press, W.H., S.A. Teukolsky, W.T. Vetterling, and B.P. Flamery (1992): *Numerical recipes*, Cambridge University Press, also at http://www.nr.com/
420. Priest, E.R. (1982): *Solar magnetohydrodynamics*, Reidel, Dordrecht
421. Priest, E.R., (ed.) (1985): *Solar system magnetic fields*, Reidel, Dordrecht
422. Priest, E.R., C.E. Parnell, and S.F. Martin (1994): A converging flux model of an X-ray bright point and an associated canceling magnetic feature, *Astrophys. J.* **427**, 459
423. Priest, E.R., and T.M. Forbes (2002): The magnetic nature of flares, *Astron. Astrophys. Rev.* **10**, 313
424. Priest, E.R., and A.W. Hood (eds.) (1991): *Advances in solar system magnetohydrodynamics*, Cambridge University Press, Cambridge
425. Proctor, M.R.E., and D.A. Gilbert, 1994: *Lectures on solar and planetary dynamos*, Cambridge University Press, Cambridge
426. Quack, M., M.-B. Kallenrode, M. von König, K. Künzi, J. Burrows, B. Heber, and E. Wolff (2001): Ground level events and consequences for stratospheric ozone, *Proc. 27th Int. Cosmic Ray Conf.*, 4023
427. Raadu, A.M., and A. Kuperus (1973): Thermal instability of coronal neutral sheets and the formation of quiescent prominences, *Astrophys. J.* **344**, 1010
428. Raadu, A.M., and A. Kuperus (1974): The support of prominences formed in neutral sheets, *Astron. Astrophys.* **31**, 189
429. Raddick, R.R., G.W. Lockwood, and S.L. Baliunas (1990): Stellar activity and brightness variations – a glimpse at the sun's history, *Science* **247**, 39
430. Ramaty, R. (1969): Gyrosynchrotron emission and absorption in magnetoactive plasma, *Astrophys. J.* **158**, 753
431. Ramaty, R., and R.E. Lingenfelter (1983): Gamma ray line astronomy, *Space Sci. Rev.* **36**, 305
432. Rao, U.R., K.G. McCracken, F.R. Allum, R.A.R. Palmeira, W.C. Bartley, and I. Palmer (1971): Anisotropy characteristics of low energy cosmic ray population of solar origin, *Solar Phys.* **19**, 209
433. Reames, D.V. (1990): Energetic particles from impulsive flares, *Astrophys. J. Suppl.* **73**, 253
434. Reames, D.V. (1990): Acceleration of energetic particles by shock waves from large solar flares, *Astrophys. J. Lett.* **358**, L63
435. Reames, D.V. (1999): Particle acceleration at the Sun and in the heliosphere, *Space Sci. Rev.* **90**, 413

436. Reames, D.V., H.V. Cane, and T.T. von Rosenvinge (1990): Energetic particle abundances in solar electron events, *Astrophys. J.* **357**, 259

437. Reames, D.V., B.R. Dennis, R.G. Stone, and R.P. Lin (1988): X-ray and radio properties of solar He-3-rich events, *Astrophys. J.* **327**, 998

438. Reames D.V., J.P. Meyer, and T.T. von Rosenvinge, (1994): Energetic particle abundances in impulsive solar flare events, *Astrophys. J. Suppl.* **90**, 649

439. Reid, G.C. (1986): Solar energetic particles and their effect on the terrestrial environment, in *Physics of the Sun*, vol. III (eds. P.A. Sturrock, T.E. Holzer, D.M. Mihalas, and R.K. Ulrich), Reidel, Dordrecht

440. Reinhard, R., P. van Nes, T.R. Sanderson, K.-P. Wenzel, E.J. Smith, and B.T. Tsurutani (1983): A statistical study of interplanetary shock associated proton intensity increases, *Proc. 18th Int. Cosmic Ray Conf.* **3**, 160

441. Richter, A.K. (1991): Interplanetary slow shocks, in [475], p. 235

442. Rind, D. (2002): The Sun's role in climate variations, *Science* **296**, 673

443. Roberts, W.O., and H. Lansford (1979): *The climate mandate*, Freeman, San Francisco

444. Roble, R.G. (1977): *The upper atmosphere and magnetosphere*, Nat. Acad. Sciences, Washington, DC

445. Roederer, J.G. (1970): *Dynamics of geomagnetically trapped radiation*, Springer, Berlin

446. Roelof, E.C. (1969): Propagation of solar cosmic rays in the interplanetary magnetic field, in *Lectures in high energy astrophysics* (eds. H. Ögelmann and J.R. Wayland), NASA SP-a99, p. 111

447. Rosenbauer, H., H. Grünwaldt, M.D. Montgomery, G. Paschmann, and N. Sckopke (1975): Heos 2 plasma observations in the distant polar magnetosphere: the plasma mantle, *J. Geophys. Res.* **80**, 2723

448. Ruffolo, D. (1995): Effect of adiabatic deceleration on the focused transport of solar cosmic rays, *Astrophys. J.* **442**, 861

449. Russell, C.T., (2001): Solar wind and interplanetary magnetic field: a tutorial, in [498], p. 71

450. Russell, C.T. and R.C. Elphic (1979): ISEE observations of flux transfer events at the dayside magnetopause, *Geophys. Res. Lett.* **6**, 33

451. Russell, C.T., and M.M. Hoppe (1983): Upstream waves and particles, *Space Sci. Rev.* **34**, 155

452. Russell, C.T. and R.J. Walker (1995): The magnetospheres of the outer planets, in [290], p. 503

453. Russell, C.T., E.J. Smith, B.T. Tsurutani, J.T. Gosling, and S.J. Bame (1993): Multiple spacecraft observations of interplanetary shocks: characteristics of the upstream ULF turbulence, in *Solar Wind Five*, (ed. M. Neugebauer), NASA CP 2280, Washington, DC, p. 385

454. Russell, C.T., R.P. Lepping, and C.W. Smith (1990): Upstream waves at Uranus, *J. Geophys. Res.* **95**, 2273

455. Saagdev, R.Z., and C.F. Kennel (1991): Collisionless shock waves, *Sci. Am.*, April 1991, 40

456. Sanderson, T.R. (1984): ISEE-3 observations of energetic protons associated with interplanetary shocks, *Adv. Space Res.* **4**, (2-3)305

457. Schatten, K.H., J.M. Wilcox, and N.F. Ness (1969): A model of interplanetary and coronal magnetic fields, *Solar Phys.* **6**, 442

458. Schatzmann, E. (1963): On the acceleration of particles in shock fronts, *Ann. Astrophy.* **26**, 234

459. Scherer, K., H. Fichtner, and E. Marsch (eds.) (2000): *The outer heliosphere: beyond the planets*, Copernicus Ges., Katlenburg-Lindau

460. Schlegel, C. (1991): Das Polarlicht, in *Plasmaphysik im Sonnensystem* (eds. K.-H. Glassmeier and M. Scholer), BI, Mannheim

461. Scholer, M. (1985): Diffusive acceleration, in [524], p. 287

462. Scholer, M. (1987): Observational overview of energetic particle populations associated with interplanetary shocks, in *Solar Wind VI* (eds. V. Pizzo, T.E. Holzer, and D.G. Sime), NCAR/TN306+Proc., Nat. Center for Atm. Research, Boulder, CO

463. Scholer, M. (2003): Reconnection on the Sun and in the magnetosphere, in [291], p. 9

464. Scholer, M. and G.E. Morfill (1977): Modulation of energetic solar particle fluxes by interplanetary shock waves, in *Contributed papers to the study of traveling interplanetary phenomena* (eds. M.A. Shea, D.F. Smart, and S.T. Wu), Air Force Geophys. Lab., Hanscom AFB, Spec. Rep. 209, p. 221

465. Schou, J., H.M. Antia, S. Basu et al. (1998): Helioseismic studies of differential rotation in the solar envelope by the solar oscillations investigation using the Michelson Doppler Imager, *Astrophys. J.* **505**, 390

466. Schrijver, C.J., and C. Zwaan (2000): *Solar and stellar magnetic activity*, Cambridge University Press, New York

467. Schröder, W. (1992): On the existence if the 11-year cycle in solar and auroral activity before and during the so-called Maunder Minimum, *J. Geomag. Geoelectr.* **44**, 119

468. Schulz, M. and L.J. Lanzerotti (1974): *Particle diffusion in radiation belts*, Springer, Berlin

469. Schulze, B.M., A.K. Richter, and G. Wibberenz (1977): Influence of finite injections and of interplanetary propagation on time-intensity and time-anisotropy profiles of solar cosmic rays, *Solar Phys.* **54**, 207

470. Schwabe, S.H. (1838): Über die Flecken der Sonne, *Astronom. Nachr.*

471. Schwabe, S.H. (1844): Sonnenbeobachtungen im Jahre 1843, *Astronom. Nachr.*

472. Schwenn, R. (1983): Direct correlation between coronal transients and interplanetary disturbances, *Space Sci. Rev.* **34**, 85

473. Schwenn, R. (1990): Large-scale structure of the interplanetary medium, in [474], p. 99

474. Schwenn, R. and E. Marsch (eds.) (1990): *Physics of the inner heliosphere, vol. I*, Springer, Berlin

475. Schwenn, R. and E. Marsch (eds.) (1991): *Physics of the inner heliosphere, vol. II*, Springer, Berlin

476. Sheeley, N.R., Jr., R.T. Stewart, R.D. Robinson, R.A. Howard, M.J. Koomen, and D.J. Michels (1984): Association between coronal mass ejections and metric type II bursts, *Astrophys. J.* **279**, 839

477. Sheeley, N.R., Jr., and 18 others (1997): Measurements of flow speeds in the corona between 2 and 30 solar radii, *Astrophys. J.* **484**, 472

478. Sheldon, R.B. and T.E. Eastman (1997): Particle transport in the magnetosphere: a new diffusion model, *Geophys. Res. Lett.* **24**, 811

479. Shibata, K., and S. Moriyasu (2003): manuscript in preparation

480. Shindell, D.T., D. Rind, R. Balachandran, J. Lean, and P. Lonergan (1999): Solar cycle variability, ozone and climate, *Science* **184**, 305

481. Shindell, D.T., G.A. Schmidt, R.L. Miller, and D. Rind (2001): Northern hemisphere winter climate response to greenhouse gas, ozone, solar and volcanic forcing, *J. Geophys. Res.* **106**, 7193

482. Shindell D.T., G.A. Schmidt, M.E. Mann, D. Rind, and A. Waple (2001): Solar forcing of regional climate change during the Maunder Minimum, *Science* **294**, 2149

483. Sies, H. (1998): A new parameter for sex education, *Nature*, **332**, 495

484. Simpson, J.A. (1989): Evolution of our knowledge of the heliosphere, *Adv. Space Res.* **9** (4), 5

485. Simpson, J.A. (1998): Recurrent solar modulation of the galactic cosmic rays and the anomalous nuclear component in three dimensions of the heliosphere, in [160], 7

486. Simpson, J.A. and B. McKibben (1976): Dynamics of the jovian magnetosphere and energetic particle radiation, in *Jupiter* (ed. T. Gehrels), University of Arizona Press, Tucson, AZ

487. Simpson, J.A., and 19 others (1995): Cosmic ray and solar particle investigations over the south polar region of the sun, *Science* **268**, 1019

488. Simpson, J.A., M. Zhang, and S. Bame (1996): A solar polar north-south asymmetry for cosmic ray propagation in the heliosphere: the Ulysses pole-to-pole rapid transit, *Astrophys. J.* **465**, L69

489. Sinnhuber, M., J.P. Burrows, M.P. Chipperfield, C.H. Jackman, M.-B. Kallenrode, K.F. Künzi, and M. Quack (2003): Particle events as possible sources of large ozone losses during magnetic polarity excursions, *Geophys. Res. Lett.*, **30**, doi:10.1029/2003/GL017265

490. Smith, E.J., and J.H. Wolfe (1976): Observation of interaction regions and corotating shocks between one and five AU: Pioneers 10 and 11, *Geophys. Res. Lett.* **3**, 137

491. Smith, E.J., B.T. Tsurutani, and R.L. Rosenberg (1978): Observations of the interplanetary sector structure up to heliographic latitudes of $16°$: Pioneer 11, *J. Geophys. Res.* **83**, 717

492. Smith, E.J., R.G. Marsden, and D.E. Page (1995): Ulysses above the sun's south pole: an introduction, *Science* **268**, 1005

493. Smith, E.J., J.R. Jokipii, and J. Kota (1997): Modelling of Ulysses pole-to-pole cosmic ray observations, *EOS Trans. AGU 78*, F547

494. Smith, Z.K., and M. Dryer (1990): MHD study of temporal and spatial evolution of simulated interplanetary shocks in the ecliptic plane within 1 AU, *Solar Phys.* **129**, 387

495. Snodgrass, H.B. (1991): A torsional oscillation in the rotation of the solar magnetic field, *Astrophys. J.* **383**, L85

496. Solanki, S., and M. Fligge (2000): Reconstruction of past solar irradiance, in [172], p. 127, also *Space Sci. Rev.* **94**, 127

497. Song, P., B.U.Ö. Sonnerup, and M.F. Thomsen (eds.) (1995): *Physics of the magnetopause*, Geophys. Monogr. 90, American Geophysical Union, Washington, DC

498. Song, P., H.J. Singer, and G.L. Siscoe (eds.) (2001): *Space weather*, Geophys. Monogr. 125, American Geophysical Union, Washington, DC

499. Spreiter, J.R., A.L. Summers, and A.Y. Alksne (1966): Hydromagnetic flow around the magnetosphere, *Planet. Space Sci.* **14**, 223

500. Srivastava, N., and R. Schwenn (2000): The origin of the solar wind: an overview, in [459], p. 13

501. St. Cyr, O.C., R.A. Howard, N.R. Sheeley, Jr., S.P. Plunkett, D.J. Michels, S.E. Paswaters, M.J. Koomen, G.M. Simnett, B.J. Thompson, J.B. Gurman, R. Schwenn, D.F. Webb, E. Hildner, and P.L. Lamy (2000): Properties of coronal mass ejections: SOHO LASCO observations from January 1996 to June 1998, *J. Geophys. Res.* **105**, 18 169

502. Steinolfson, R.S. and A.J. Hundhausen (1990): MHD intermediate shocks in a coronal mass ejection, *J. Geophys. Res.* **95**, 6389

503. Stix, M. (1987): in *Solar and stellar physics*, (eds. E.H. Schröter and M. Schüssler), Lect. Notes Phys., vol. 292, p. 15, Springer, Berlin

504. Stix, T.H. (1992): *Waves in plasmas*, American Institute of Physics, New York

505. Størmer, C. (1955): *The polar aurora*, Oxford University Press, London

506. Stone, R.G., and B.T. Tsurutani (eds.) (1985): *Collisionless shocks in the heliosphere: a tutorial review*, Geophys. Monogr. 34, American Geophysical Union, Washington, DC

507. Strohbach, K. (1991): *Unser Planet Erde*, Borntraeger, Berlin

508. Sturrock, P.A. (1994): *Plasma physics*, Cambridge University Press, Cambridge

509. Svensmark, H., and E. Friis-Christensen (1997): Variation of cosmic ray flux and global cloud coverage – a missing link in solar climate relationships, *J. Atm. Terr. Phys.* **59**, 1225

510. Svensmark, H. (1999): Influence of cosmic rays on Earth's climate, *Phys. Rev. Lett.* **81**, 5027

511. Svestka, Z., B.V. Jackson, and M.E. Machado (eds.) (1992): *Eruptive solar flare*, Springer, Berlin

512. Swanson, D.G. (1989): *Plasma waves*, Academic Press, Boston

513. Sweet, P.A. (1958): in *Electromagnetic phenomena in cosmical physics* (ed. B. Lehnert), Cambridge University Press, Cambridge MA

514. Tait, W.H. (1980): *Radiation detection*, Butterworth, London

515. Takakura, T. (1960): *Publ. Astron. Soc. Japan* **12**, 235

516. Thompson, M.J., J. Toomre, E. Anderson, H.M. Antia, G. Berthomieu, D. Burtonclay, S.M. Chitre, J. ChristensenDalsgaard, T. Corbard, M. DeRosa, C.R. Genovese, D.O. Gough, D.A. Haber, J.W. Harvey, F. Hill, R. Howe, S.G. Korzennik, A.G. Kosovichev, J.W. Leibacher, F.P. Pijpers, J. Provost, E.J. Rhodes, J. Schou, T. Sekii, P.B. Stark, and P.R. Wilson (1996): Differential rotation and dynamics of the solar interior, *Science* **272**, 1300

517. Tinsley, B.A. (1997): Do effects of global atmospheric electricity on clouds cause climate changes?, *EOS* **78**, 341

518. Tinsley, B.A. (2000): Influence of solar wind on the global electric circuit, and inferred effects on cloud microphysics, temperature, and dynamics of the troposphere, in [172], p. 231, also *Space Sci. Rev.* **94**, 231

519. Toptygin, I.N. (1983): *Cosmic rays in interplanetary magnetic fields*, Reidel, Dordrecht

520. Treumann, R.A., and W. Baumjohann (1996): *Advanced space plasma physics*, Imperial College Press, London

521. Tsurutani, B.T., and W.D. Gonzalez (1997): The interplanetary causes of magnetic storms: a review, in [522]

522. Tsurutani, B.T., W.D. Gonzalez, Y. Kamide, and J.K. Arballo (eds.) (1997): *Magnetic storms*, Geophys. Monogr. 98, American Geophys. Union, Washington, DC

523. Tsurutani, B.T., and R.P. Lin (1985): Acceleration of >47 keV ions and >2 keV electrons by interplanetary shocks at 1 AU, *J. Geophys. Res.* **90**, 1

524. Tsurutani, B.T., and R.G. Stone (eds.) (1985): *Collisionless shocks in the heliosphere: reviews of current research*, Geophys. Monogr. 35, American Geophysical Union, Washington, DC

525. Tu, C.-Y., and E. Marsch (1995): MHD structures, waves and turbulence in the solar wind: observations and theories, *Space Sci. Rev.* **73**, 1

526. Tu, C.-Y., and E. Marsch (2001): Wave dissipation by ion cyclotron resonance in the solar corona, *Astron. Astrophys.* **368**, 1071

527. Tverskoy, B.A. (1967): *Sov. Phys. JETP* **25**, 317

528. Tylka, A.J., W.F. Dietrich, and R. Boberg (1997): *Proc. 25th Int. Cosmic Ray Conf.* **1**, 101

529. Ulrich, R.K. (1970): 5-Minute oscillations on the solar surface, *Astrophys. J.* **162**, 993

530. Ulmschneider, P. (2003): The physics of chromospheres and coronae, in [9], p. 232

531. Ulysses Instruments (1992): *Astron. Astrophys.* **92** (2)

532. Usmanov, A.V., M.L. Goldstein, B.P. Besser, and J.M. Fritzer (2000): A global MHD solar wind model with WKB Alfvn waves: Comparison with Ulysses data, *J. Geophys. Res.*, **105**, 12 675

533. van Allen, J.A. (1991): Why radiation belts exist, *EOS* **72** (34), 361

534. van Kampen, N.G., and B.U. Felderhof (1967): *Theoretical methods in plasma physics*, North-Holland, Amsterdam

535. Vilmer, N. (1987): Hard X-ray emission processes in solar flares, *Solar Phys.* **111**, 207

536. Vitt, F.M., and C.H. Jackman (1996): A comparison of sources of odd nitrogen production from 1974 through 1993 in the earth's middle atmosphere as calculated using a two-dimensional model, *J. Geophys. Res.* **101**, 6729

537. Vitt, F.M., T.E. Cravens, and C.H. Jackman (2000): A two-dimensional model of the thermospheric nitric oxide sources and their contributions to the middle atmospheric chemical balance, *J. Atmo. Solar-Terr. Phys.* **62**, 653

538. Völk, H.J. (1987): Particle acceleration in astrophysical shock waves, *Proc. 20th Internat. Cosmic Ray Conf.* **7**, 157

539. Volland, H. (1984): *Atmospheric electrodynamics*, Physics and Chemistry in Space vol. 11, Springer, Berlin, Heidelberg

540. Volland, H. (1996): Electrodynamic coupling between neutral atmosphere and ionosphere, in *Modern ionospheric science* (eds. H. Kohl, R. Rüster, and K. Schlegel), Copernicus, Katlenburg-Lindau

541. von Steiger, R., J. Geiss, and G. Gloeckler (1995): Composition of the solar wind, in *Cosmic winds and the heliosphere* (eds. J. R. Jokipii, C. P. Sonett, and M. S. Giampapa), University of Arizona Press, Tucson

542. Vrsnak, B. (2003): Magnetic 3-D configurations of energy release in solar flares, in [291], p. 28

543. Walker, R.J. (1979): Quantitative modeling of planetary magnetospheric magnetic fields, in *Quantitative modeling of magnetospheric processes* (ed. W.P. Olson), American Geophysical Union, Washington, DC, p. 1

544. Walt, M. (1994): *Introduction to geomagnetically trapped radiation*, Cambridge University Press, Cambridge

545. Wang, Y.-M., and N.R. Sheeley Jr. (1990): Solar wind speed and coronal flux tube expansion, *Astrophys. J.* **355**, 726

546. Wang, Y.-M., N.R.Sheeley, Jr., J.H. Walters, G.E. Brückner, R.A. Howard, D.J. Michels, P.L. Lamy, R.Schwenn, and G.M. Simnett (1998): Origin of streamer material in the outer corona, *Astrophys. J. Lett.* **498**, L165

547. Wanner, W. and G. Wibberenz (1993): A study of the propagation of solar energetic protons in the inner heliosphere, *J. Geophys. Res.* **98**, 3513

548. Wanner, W., M.-B. Kallenrode, W. Dröge, and G. Wibberenz (1993): Solar eneergetic proton mean free paths, *Adv. Space Res.* **13(9)**, 359

549. Wanner, W., U. Jaekel, M.-B. Kallenrode, G. Wibberenz, and R. Schlickeiser (1994): Observational evidence for a spurious dependence of slab QLT mean free paths in the magnetic field angle, *Astron. Astrophys.* **290**, L5

550. Watson, R.T., M.C. Zinyowera, and R.H. Moss (1996): *Climate change 1995: impacts, adaptions and mitigation of climate change*, Cambridge University Press, Cambridge

551. Webb, D.F. (1991): The solar cycle variation of rates of CMEs and related activity, *Adv. Space Res.* **11**, 1(37)

552. Webber, W.R. (1987): Cosmic rays in the heliosphere, in *Essays in space science* (eds. R. Ramaty, T.L. Cline, and J.F. Ormes), NASA Goddard Space Flight Center, Greenbeld, MD, p. 125

553. Weber, W.E., and C.F. Gauß (1840): *Atlas des Erdmagnetismus*, Leipzig

554. Weber, W.E., and C.F. Gauß (1853): *Über die Anwendung der magnetischen Induktion auf die Messung der Inklination mit dem Magnetometer*, Göttingen

555. Weeks, L.H., R.S. Cuikay, and J.R. Orbin (1972): Ozone measurements in the mesosphere during the solar proton event of 2 November 1969, *J. Atmos. Sci.* **29**, 1138

556. Weiss, N.O. (1966): Expulsion of magnetic flux by eddies, *Proc. R. Soc. London*, **A293**, 310

557. Wenzel, K.-P., R. Reinhard, T.R. Sanderson, and E.T. Sarris (1985): Characteristics of energetic particle events associated with interplanetary shocks, *J. Geophys. Res.* **90**, 12

558. Wessen, J. (1989): *Tokamaks*, Clarendon Press, Oxford

559. White, R.B. (1989): *Theory of Tokamak plasmas*, North-Holland, Amsterdam

560. Wibberenz, G. (1973): Propagation of cosmic rays in the interplanetary space, in *Lectures on space physics, vol. 1* (eds. A. Bruzek and H. Pilkuhn), Bertelsmann, Gütersloh

561. Wibberenz, G. (1998): Transient effects and disturbed conditions: observations and simple models, in [160], 310

562. Williams, E.R. (2001): Sprites, elves, and glow discharge tubes, *Physics Today*, November, 41

563. Willis, D.M. (1971): Structure of the magnetopause, *Rev. Geophys. Space Phys.* **9**, 953

564. Wilson, P.R. (1994): *Solar and stellar activity cycles*, Cambridge University Press, Cambridge

565. Wind Instrumentation (1995): *Space Sci. Rev.* **71**

566. Woodgate, B.E., M.-J. Martres, J.B. Smith, K.T. Strong, M.K. McCabe, M.E. Machado, V. Gaisauskas, R.T. Stewart, and P.A. Sturrock (1984): Progress in the study of homologous flares on the Sun – part II, *Adv. Space Res.* **4**, (7)11

567. Wollin, G., D.B. Ericson, and W.B.F. Ryan (1971): Variations in magnetic intensity and climatic changes, *Nature* **232**, 549

568. Zank, G.P., W.K.M. Rice, J.A. le Roux, I.H. Cairns, and G.M. Webb (2001): The injection problem for quasi-parallel shocks, *Phys. Plasmas* **8**, 4560

569. Zare, R. (1964): *Chem. Phys.* **40**, 1934

570. Zhang, C., H. Wang, J. Wang, and Y. Yan (2000): Sympathetic flares in two adjacent active regions, *Solar Phys.* **195**, 135

571. Zhao, X. and J.T. Hoeksema (1993): Unique determination of model coronal fields using photospheric observations, *Solar Phys.* **143**, 41

572. Zirin, H. (1988): *Astrophysics of the sun*, Cambridge University Press, Cambridge

573. Zöllich, F., G. Wibberenz, H. Kunow, and G. Green (1981): Corotating events in the energy range 4–13 MeV as observed on board Helios 1 and 2 in 1975 and 1976, *Adv. Space Res.* **1**, 89

Index

Printed in the United States
By Bookmasters